T0190112

THE FRONTIERS COLLECTION

Series editors

THE FRONTIERS COLLECTION

The books in this collection are devoted to challenging and open problems at the forefront of modern science, including related philosophical debates. In contrast to typical research monographs, however, they strive to present their topics in a manner accessible also to scientifically literate non-specialists wishing to gain insight into the deeper implications and fascinating questions involved. Taken as a whole, the series reflects the need for a fundamental and interdisciplinary approach to modern science. Furthermore, it is intended to encourage active scientists in all areas to ponder over important and perhaps controversial issues beyond their own speciality. Extending from quantum physics and relativity to entropy, consciousness and complex systems—the Frontiers Collection will inspire readers to push back the frontiers of their own knowledge.

More information about this series at http://www.springer.com/series/5342

For a full list of published titles, please see back of book or springer.com/series/5342

George Ellis

HOW CAN PHYSICS UNDERLIE THE MIND?

Top-Down Causation in the Human Context

Springer

George Ellis
Department of Mathematics and Applied
Mathematics
University of Cape Town
Rondebosch
South Africa

ISSN 1612-3018 ISSN 2197-6619 (electronic)
THE FRONTIERS COLLECTION
ISBN 978-3-662-57036-4 ISBN 978-3-662-49809-5 (eBook)
DOI 10.1007/978-3-662-49809-5

Softcover reprint of the hardcover 1st edition 2016

Printed on acid-free paper

This Springer imprint is published by Springer Nature
The registered company is Springer-Verlag GmbH Berlin Heidelberg

Advance Praise for This Book

A stark lacuna lies at the heart of science: half the causal narrative has been omitted! Ellis makes a cogent and compelling case that the causal architecture of the universe is subtler and richer than the austere reductionist picture dictates. In this impressively scholarly volume, the author assembles evidence and argument from across the great sweep of intellectual inquiry, from pure mathematics and computation to neuroscience and engineering, and weaves them into a formal, systematic framework for understanding physical reality as we observe it, and for taking seriously human agency and moral choice. This book will set the agenda for the next leap forward in humanity's attempt to make sense of how the world actually works.

Paul Davies, Beyond Center, Arizona State University

Physics went through a major revolution in its conceptual foundations a century ago with the arrival of quantum mechanics and the theories of relativity. All this passed by biology with virtually no impact. Ellis's book makes it very clear why a major conceptual change is required also in biology, through the incorporation of top-down causation. The sweep of the book is enormous as it details the evidence and the impact in each area of science. It forms a major landmark, and it does so at an exciting time, when the purely gene-centric views of biology are being seriously challenged.

Denis Noble, CBE FRS, University of Oxford

The culmination of three decades of work, Ellis's *magnum opus* makes the most comprehensive case yet for top-down causation in the natural world. Encyclopedic in scope, yet guided by a single sustained argument, this defense of "strong emergence" sets a high, perhaps unreachable bar for scientific reductionists.

As Ellis rightly notes, our entire conception of ourselves and our world depends on the outcome of this debate.

Philip Clayton, author of *Mind and Emergence*

Reductionism has been an extremely successful strategy in science. But, as George Ellis demonstrates in this important and provocative book, reductionism can't be the whole story. Instead, other modes of explanation, including those based on emergence and top-down causation, are vital for a fully orbed account of the natural world.

Ard Louis, Rudolph Peierls Centre for Theoretical Physics
University of Oxford

An admirable, systematic approach to the issue of emergence from physics to sociology, of great originality, broad scope, and deep understanding. George Ellis argues with admirable charity of thought that much of the world we live in is governed not by the blind dance of atoms, but by high-level causes and purposes.

Giulio Tononi, University of Wisconsin-Madison

An essential antidote to the shallow forms of reductionism that dominate both popular and academic thinking about our world. A carefully crafted argument, steeped in the scholarly literature, yet accessible to the ordinary reader.

Alister E. McGrath, University of Oxford

Preface

As the title suggests, the subject of this book is the emergence of complexity and the mind, focusing on the role of top-down causation. The aim is to engage with the complexity of the emergence of life and the mind out of the underlying physics. What makes this possible?

The world of biology, where purpose and adaptation abound, is quite different from the natural world of rocks, oceans, atmospheres, planets, stars, and galaxies, where impersonal forces hold sway. Yet both are based on the same underlying physics. How can such different outcomes emerge from the same underlying particles and forces? Can we identify the key enabling principles?

Top-Down Causation

I will make a case that, in addition to bottom-up causation, top-down causation is a key element in what is going on, enabling genuine emergence to take place, with higher levels of structure having genuine causal powers in their own right. As well as in bottom-up causation, which is crucial, emergent entities exert downward influences on their components, and this is the basis for true complexity (Chap. 3). In some cases the less contentious phrase 'contextual effect' might be preferred, and that certainly often takes place. However, I will make a stronger claim that 'top-down causation' is appropriate in some cases, and specifically when the mind is involved.

Brain and Mind. The ultimate focus is the brain and the mind (Chap. 7). However, most of the book discusses a much wider range of issues, setting the scene for the discussion of the brain and the emergence of the mind. This broader context is crucial in order to look at the brain properly: the foundation stones for that study will have been properly laid by the time we get there.

Across all the Sciences. The aim is therefore an integrative view to show how this holds in all sciences including chemistry and physics, and is of particular significance in understanding digital computers, life and brain. Thus this book engages with sciences across the board.

Please note that I am not an expert in all the areas discussed here. However, what I am able to do is to comment on the larger patterns of causation that occurs in these various contexts, and how they relate to the theme of this book. This builds up to an integrative view of the whole.

Key Issues

Four key issues emerge:

- **Key question: Who does the work, who decides what will be done?** These are different kinds of causal effects; both occur in any complex system. The lower levels do the physical work, but the higher levels decide what work should be done. This theme will recur throughout the text, with Sect. 2.7 discussing the key example of digital computers.
- **Key issue: How is there a causal room at the bottom?** This will be dealt in depth in Chap. 4, exploiting the fact that top-down effects can change the nature of lower level elements, or even determine whether they exist or not.
- **Key concept: Multiple realisability.** A key concept in the whole schema is the multiple realisability of higher level structures and functions in terms of lower level components fulfilling higher level functions, leading to the real effective causal entities being equivalence classes of lower level entities. This is discussed in Sect. 3.5.
- **Key concept: Supervenience** Exactly identical lower level structures and excited states may lead to identical higher level effects in a bottom-up way, as is captured by the idea of supervenience. However, the relevant complex lower level states and excitations in the case of living systems can only come into being if top-down processes, and in particular adaptive effects, shape them according to their context. They cannot come into existence purely by bottom-up processes. This is discussed in Sect. 3.5.3.

Applicability. I strongly believe that science should be able to relate to the complexities of what happens in the everyday world, as well as in the laboratory; and propose that this is only possible if one takes into account the top-down strands of causation as well as the bottom-up ones. I will give many examples. Laboratory experiments seem to shield the system from top-down effects—until one realizes that the occurrence of the experiment is only possible because of the top-down effect of the human mind on the physical world, i.e., the human mind that created the laboratory and the experimental apparatus in the first place.

Effects. This is not just an academic topic. Views on how causation work affects our mental models of how things work and how we should interact with them, and in particular how the brain works and how to deal with its complexities. Consequently, there are implications in particular in health care, mental welfare, and education. These are complex topics. I will briefly consider the case of education and literacy in Sect. 8.6.

Mathematical Models. Mathematical models are needed to give depth and credibility to the discussion. They are commented on, but are largely segregated from the main text, so they can be skipped if one just wants to follow the main line of argument. There is a great deal of evidence for what is proposed here that is a valid support for the present proposal, independent of any mathematical models.

Respecting the Physics. My argument will not in any way deny the nature of the science that enables and underlies our existence: nothing I propose in any way suggests that science is overridden by the processes I describe. Rather what I put forward is a broadening of our understanding of how causal effects work in accordance with the underlying scientific laws, when higher levels of causation are taken into account as well as the level of particles and forces that is the focus of fundamental physics. Physiology is a science just as much as physics is: it is compatible with physics and operates within the constraints imposed by the underlying physics, but it is not determined by physics. It is shaped by the logic of physiological needs, which determine physiological structures and function.

Novelty. A great deal of the book is a survey of well-established results presented from a particular perspective involving the interaction of bottom-up and top-down effects. For example, I give a discussion of digital computing from this viewpoint in Chap. 2. However, the discussion also has various new aspects. What is novel is noted in Sect. 1.6.2.

Controversial Aspects. Some people, such as those working on integrative aspects of the brain or physiology, will take what I say as quite uncontroversial. Others, primarily working in physics or molecular biology, may find it either trivial or simply wrong. I believe the latter viewpoints are answered adequately in the main text. A brief note on which items in my discussion are controversial is given in Sect. 1.6.1.

Chapters and References

The chapters of this book have, at the request of the publisher, been written so that they can to a large degree be read independently. Therefore some important topics are covered several times, and references for each chapter have been given separately at the end of the chapter. This results in some duplication of references. The payoff is that you can refer to them easily by turning to the end of the chapter, if that is all you have downloaded.

Origins

The origin of this book was a series of Vatican Observatory–CTNS discussions organized by George Coyne, Bob Russell, Nancey Murphy, and Bill Stoeger. My interest in top-down causation arose through discussions there with Arthur Peacocke, Phil Clayton, and Nancey Murphy, and led to a number of papers on the topic referred to in the text, as well as a book with Nancey Murphy.[1] This interest developed further through a Wawona meeting convened by Mary Ann Meyers, leading to a book jointly edited with Nancey Murphy and Tim O'Connor,[2] a meeting with Gennaro Auletta and Luc Jaeger in Rome and Cape Town, and a London meeting also convened by Mary Ann Meyers, leading to proceedings jointly edited with Denis Noble and Tim O'Connor.[3] Interactions with Paul Davies and Sarah Walker at the Beyond Centre, Arizona State University, have taken it further, as has a Manchester Gödel Centenary meeting (which I attended thanks to Hyung-Choi) in the case of digital computers, and an FQXI essay competition in case of physics.

One should note that, although these are the origins of what is written here, it then developed its own logic over a period of some decades, a logic which is presented here. Of course, the arguments given must stand or fall on their own merits, irrespective of how they arose.

Thanks

I thank all the colleagues mentioned above for valuable discussions and insights. I particularly thank the Vatican Observatory–CTNS collaboration (George Coyne, Bill Stoeger, Bob Russell, Nancey Murphy) for their very enlightening conferences which were crucial in my thinking, Phil Clayton for important discussions on causal closure, Gennaro Auletta, and Luc Jaeger for key discussions on multiple realizability, Tim Maudlin for helpful comments, Hyung Choi and Mary-Ann Meyers for support of various of these events via the Templeton Foundation, Paul Davies and Sara Walker for their ASU meetings, and Angela Lahee for the suggestion to

[1]N Murphy and G F R Ellis (1996) *On the Moral Nature of the Universe: Cosmology, Theology, and Ethics* (Fortress Press, Minneapolis).

[2]N Murphy, G F R Ellis, and T O'Connor (Eds) (2009) *Downward Causation and the Neurobiology of Free Will* (New York: Springer).

[3]G F R Ellis, D Noble, and T O'Connor (Eds) (2012) "Top-down causation: An integrating theme within and across the sciences?" *Royal Society Interface Focus* Special issue **2**:1–140.

publish this as a Springer book. I thank Mark Solms for very helpful discussions as regards Chap. 7, and two referees who made comments that led me to improve the text.

And above all, I thank my wife Carole for her loving support. Discussions with her on the case of learning to read have been invaluable, and she co-authored Sect. 8.6 on this topic with me.

Cape Town
April 2016

George Ellis

Contents

List of Figures

List of Tables

Chapter 1
Complexity and Emergence

One of the most astonishing things in the physical world is the way that mind emerges from matter. Atoms obeying fundamental impersonal physical laws form stars, rocks, oceans, planets, amoeba, mice, whales, the human brain. Somehow the brain enables creation of Bach concertos, supercomputers, Jumbo jet aircraft, roast lamb, the Mona Lisa, the rules of chess, global warfare, Einstein's theories of relativity, the Eiffel Tower, and Shakespeare's sonnets. How on earth can this be possible?

From a physics viewpoint, physics underlies all [75]. The law-like behaviour of matter investigated by Galileo, Newton, and Laplace suggests the world is determinate and describable by mathematical equations. Newton's second law:

$$\mathbf{F}_i = m_i \mathbf{a}_i = m_i \frac{\mathrm{d}^2 \mathbf{x}_i}{\mathrm{d}t^2}, \tag{1.1}$$

which says that the force on a particle i equals its mass times its acceleration, implies that given the forces $\mathbf{F}_i$ on each particle i and full initial data, i.e., the initial positions $\mathbf{x}_i(t_0)$ and velocities $\mathrm{d}\mathbf{x}_i/\mathrm{d}t|_{t_0}$ of all relevant particles, you can calculate the acceleration of every particle and hence determine the outcome. All is determinate![1] The body is made of particles and so falls under this rubric. The brain is part of the body and if the mind is an outcome of the workings of the brain, as assumed by present day neuroscience, its operations are determined simply by physics (which determines the flow of electrons in neural dendrites and axons [53]). Free will is an illusion, consciousness is an epiphenomenon [82]. In the end, production by the brain of both a Bach concerto and a theory such as the standard model of particle physics is nothing but the outcome of complex interactions of electrons and protons. Physics reigns supreme [77].

This book will present arguments that counter that understanding. It certainly does not claim to solve the hard problem of consciousness: indeed at present science has no idea how to tackle that issue (despite some claims [26] that it has been solved).

---

[1] Poincâre and others discovered that, if the equations are chaotic, the outcome is unpredictable in practice, but that is not important for what follows. It does not relate to the emergence of complexity.

G. Ellis, *How Can Physics Underlie the Mind?*, The Frontiers Collection,
DOI 10.1007/978-3-662-49809-5_1

What I aim to do is support the view that, even though physical laws underlie all material entities, there exist higher level causal relations that allow the brain to act as a means of creating theories, searching for meaning, expressing tenderness, and doing all the other myriad things that make us human, without contradicting or overwriting those lower level physical laws. Consequently, physics does not control the mind, it enables the mind. The same is true for genetics and neurobiology: they both to some degree shape what the mind does, but neither by itself determines the outcome, because the mind has a logic of its own (for example, the understandings contained in 'folk psychology'[2]). We are genuinely fully human, even though we emerge through the interactions of fundamental particles.

The lower level physical interactions enable the propagation of signals encoded in action potentials in neurons in our brains, these signals being part of the causal nexus enabled by the myriad connections between neurons, which in turn enables consciousness, feelings, and thoughts to emerge from matter. That is the extraordinary outcome that needs explanation [46]. I will will give grounds to show that it is not impossible that these higher level factors do indeed shape what happens in our brains and hence our bodies, even though they arise through the agency of lower level interactions. Thus my view agrees with that of Fordor [7, pp. 395–409].

Francis Crick famously said [21]:

> You, your joys and your sorrows, your memories and your ambitions, your sense of personal identity and free will, are in fact no more than the behavior of a vast assembly of nerve cells and their associated molecules.

This is the classic reductionist view. I will revisit this quotation in Sect. 7.7.3 in the light of the discussion in the rest of this book, showing how this is an inadequate position because it represents an arbitrary partial reductionism. I will further argue that the reductionist claims of 'nothing but' are fallacious because they ignore important aspects of causation. In fact, we are much more than the sum of our parts.

In this introductory chapter, I deal briefly in turn with:

- Section 1.1. The issue considered.
- Section 1.2. A basic viewpoint.
- Section 1.3. Key points of the argument.
- Section 1.4. Is it real? Testing the proposal.
- Section 1.5. Significant implications.
- Section 1.6. An outline of the book.
- Section 1.7. The necessity of the conclusion.

---

[2]See the entry *Folk Psychology as a Theory*, Stanford Encyclopedia of Philosophy, Plato.stanford.edu.

1.1 The Issue Considered

There are at least two kinds of causation at work in the world: blind physical forces doing their thing in an algorithmic and meaningless way, as explored by physics and physical chemistry, and living beings doing their thing in a purposeful and meaningful way, as explored by biology, the humanities, psychology, and sociology. Both kinds of causation are clearly active and causally effective in the real world.

So how do they fit together? Does one kind in fact supplant the other when we examine it closely, making the other an illusion? Then there is really only one kind of causation at work (as the diehard reductionists claim), inter alia implying that we have no free will: we are just automata, mindlessly obeying underlying algorithms and deceiving ourselves that we have meaningful control over our lives [82]. Or do both kinds of causation co-exist somehow? Can the brain truly function as the engine of meaningful aspects of life [28], functioning according to the psychology of social interactions and allowing the logic of scientific investigation, while it also obeys the strict laws of physics and neurobiology? If so, how can this happen? How is there space for both?

The digital computers that dominate in the world around us are based, at the bottom level, on a binary coding system: that is, every programme, and all the data it uses, are nothing but a sequence of zeros and ones. A Bach sonata played on your digital system by Yo-Yo Ma will ultimately be just a sequence of digits: 001101010001110001010100001011 1.... A Rembrandt self-portrait displayed on your screen, or your holiday photos stored on your computer, will be other such sequences, and so will the data used in a company accounting system, the signals in the computer controlling the flight of an airliner, all the emails you get, as well as the digital TV programmes and films you watch.

And here we already see the problem with the reductionist stance as regards the nature of the world around us. In the end, they are all 'nothing but' a sequence of zeros and ones. But these sequences store in their precise details the most astonishing variety of things: images, books, films, economic data, signals used in automated factories, and so on. The components are the same, working strictly according to the laws of physics, but radically different outcomes emerge depending on context. And that is a model of how complexity works.

The 'nothing but' story is true in a certain way—at the bottom, all digital data are just comprised of zeros and ones—but misses the essential core of what is going on. It is the specific organisation of the zeros and ones that crucially matters: they encode the meaning of the signal stored in the computer, and this meaning depends on the context. The correct context in each case (an appropriate high level programme running as required) interrogates the data and produces its intended meaning. If you run the wrong programme with the data (read the Yo-Yo Ma music with Photoshop, for example) you will get nonsense. Context is everything, turning the details into higher level meaningful entities.

In the influential book *What Is Life*, written in 1945, Erwin Schrödinger wrote [80, p. 81]:

> From all we have learnt about the structure of living matter, we must be prepared to find it working in a manner that cannot be reduced to the ordinary laws of physics. And that not on the ground that there is any 'new force' or what not, directing the behaviour of the single atoms within a living organism, but because the construction is different from anything we have yet tested in a laboratory.

Paradoxically, while the higher-level properties emerge from the lower-level processes, they have a degree of causal independence from them: they operate according to their own higher-level logic. According to Philip Anderson in his famous paper *More Is Different* [4]:

> Large objects such as ourselves are the product of principles of organisation and of collective behaviour that cannot in any meaningful sense be reduced to the behaviour of our elementary constituents. Large objects are often more constrained by those principles than by what the principles act upon.

The view put here will be that in accordance with this quote, physics makes possible, but does not causally determine, the higher-order layers of structure and meaning. It cannot replace psychology, sociology, politics, and economics as autonomous subjects of study. Physics underlies emergent biological complexity, including the physicist's mind, but does not comprehend it, because it has its own organisational principles [47].

The key point is that the emergent higher levels of causation are indeed causally effective and underlie genuinely complex existence and action, even though these kinds of causation are not included within the usual physics picture of the world [31]. The essential proof that this is so is the fact that coherent, experimentally supported scientific theories, such as present-day theoretical physics, molecular biology, and neuroscience, exist. They have emerged from a primordial state of the universe characterized by random perturbations that cannot in themselves have embodied such higher-level meanings (Sect. 8.1). What enables this to occur is emergence of true complexity, with autonomous higher level laws of behaviour, such as rational argumentation, determining the outcomes. These laws are enabled by the combination of bottom-up and top-down causation in the hierarchy of complexity [33, 35]. How this takes place is the focus of this book.

1.2 A Basic Viewpoint

Modular hierarchical structures are the basis of true complexity [8, 13, 40, 84]: for example, the human brain is comprised of various cortical and other areas that contain networks of neurons, each of which is made of components such as axons and dendrites that in turn are made of biomolecules, made up of atoms, made up of protons and electrons, and so on. Here I shall deal in turn with:

- The main claim (Sect. 1.2.1).
- Hierarchies (Sect. 1.2.2).
- Emergence (Sect. 1.2.3).

1.2.1 The Main Claim

Complex structures, with their own intrinsic higher level laws of behaviour, emerge out of combinations of simple components with simpler behaviour [19]; living beings are a particular case [16, 69, 79]. The basic thesis of this book is as follows:

> **Thesis.** It is the combination of bottom-up and top-down causation that allows genuinely complex behaviour to emerge out of simple components combined together to form modular hierarchical structures. As well as bottom-up causation, top-down causation takes place in these structures [15, 36, 89] through the crucial role of context in determining the outcomes of lower level causation [12]. This occurs in the natural world of inanimate objects; in the biological world of plants, animals, and intelligent beings, and in the manufactured world of artefacts. It takes place in evolutionary, developmental, and functional contexts.

It is in this way that high level intentions and understandings have arisen in the cosmic context, and can be causally effective, given the underlying physics. This does not contradict the underlying physical causation:

> **Explication.** Top-down causation in the hierarchy of causation works by breaking symmetries and so setting constraints for lower level causation, thus channeling lower level interactions. This paradoxically creates new possibilities of complex behaviour, while respecting the lower level physics.

The claim will be that emergence based on bottom-up action by itself, while it can produce impressive outcomes such as flocks of birds and abstract patterns such as those that occur in Conway's game of Life, and even the results of interactions between swarms of intelligent agents, can only go so far: it cannot produce complexity such as that embodied in living cells or digital computers. That requires top-down coordination of the activity of the parts. The argument that this must at least sometimes be the case is given by three clear examples:

- Complex goal-oriented social organisation, such as is required to construct a Jumbo jet airliner. The actions of thousands of workers must be very carefully coordinated to produce the outcome. A swarm of interacting intelligent agents cannot possibly create such a complex object without such central planning and coordination (a beehive or ant's nest is not of comparable complexity).
- The functioning of the human body [8]. If the actions of the arms and legs were not centrally coordinated by the brain, in turn coordinating flows of electrons in muscular tissue, we could not walk or act.
- The functioning of a digital computer. The actions of the gates and memory registers is coordinated by the CPU in accordance with the applications program loaded [87]; top-down coordination enables it to play music or display a picture or process numbers or edit text, according to the logic of the program loaded.

The thesis of this book will be that top-down causation—contextual effects, if you prefer—is very widespread not only in those cases, but also throughout biology (e.g., in Darwinian evolution, epigenetics, the physiology of the heart, the functioning of the brain), in chemistry (e.g., in reagent purification), and even in physics (e.g., in state vector preparation and in the determination of the arrow of time). It is all around us when you look for it.

1.2.2 Hierarchies

The basis of complexity is modular hierarchical structures (see Chap. 3) [40, 84], leading to emergent higher levels of structure and function based on lower level networks. Each of these aspects (modularity, hierarchy, and structure) is crucial for the emergence of complexity out of interactions between simpler units [13, 33]. Hierarchies occur within complex networks of interactions, and modules may themselves contain sub-hierarchies: so extraordinarily complex networks of causation occur in social systems, microbiology, physiology, and the brain.

The hierarchy of structure and causation for both the natural sciences and the life sciences is shown in Table 1.1. This table gives a simplified representation of this hierarchy of levels of reality (as characterised by the corresponding academic subjects) in natural systems (left) and living beings (right); see [16, 32, 69] for a more detailed description of this hierarchical structure. There is a corresponding hierarchy for complex artificial systems [84], e.g., language [88] and hence writing, for computer systems [13, 87], and for large organizations [8]. The case of digital computers is considered in the next chapter. We should make three comments here.

Interlocking Hierarchies. First, there are in fact interlocking hierarchies of structure and causation. Many examples will be given below. In the many complex webs of interactions and interacting systems in the real world, such as gene interaction networks, computer systems, ecologies, and the human brain, one can find various different hierarchies that interact with each other. The comments that follow will be applicable to any of them. So there is not one linear hierarchy: there are many interlocking hierarchies when one looks at detailed structure and interactions. Nevertheless the broad overall hierarchy as indicated in Table 1.1 is fundamental, and essential for understanding natural systems [32] (left) and the origin and functioning of life [16, 24] and the mind [81] (right).

Table 1.1 The basic hierarchy of structure and causation for inanimate matter (*left*) and for life (*right*) as characterized by academic disciplines

	Inanimate matter	Living matter
Level 10	Cosmology	Sociology/Economics/Politics
Level 9	Astronomy	Psychology
Level 8	Space science	Physiology
Level 7	Geology, Earth science	Cell biology
Level 6	Materials science	Biochemistry
Level 5	Physical chemistry	Chemistry
Level 4	Atomic physics	Atomic physics
Level 3	Nuclear physics	Nuclear physics
Level 2	Particle physics	Particle physics
Level 1	Fundamental theory	Fundamental theory

Structure and Causation. Second, in order to encompass all that will be considered in this book, one must understand that the more fundamental of these interlocking hierarchies of structure and function is the hierarchy of causation, as will be illustrated in many examples that follow. The higher levels of causation need not be physical, indeed it is a major theme of this book that in many cases they will not be so (Sect. 1.3.5). This hierarchy is characterised by the way the higher levels determine what happens in the lower levels by setting the context for their operation. This is what determines which should be regarded as higher and which lower levels in any specific hierarchy. It is not essential to this understanding that one use the metaphor of higher and lower in discussing these interactions; some prefer to talk of whole–part constraints. But in practice it is helpful at least to set out the relations in terms of ordered levels as in the table above.

Top and Bottom. Third, there need not be any known topmost or bottommost level in a hierarchy. The relations discussed here hold between any neighbouring levels in a hierarchy, and hence imply causal relations between any two connected levels. They do not imply we have to know either the top or bottom levels, which may or may not exist (the brain is a case in point). If they do exist, we may not know what they are (fundamental physics refers here). Indeed the whole point of the argument is that, because intermediate levels have genuine existence in their own right, it is fallacious to think that we have to know the topmost or bottommost levels in order to understand the relations between intermediate levels. We have to investigate ultimate levels if we are interested in ultimate causation. That is an entirely different affair to the one we discuss here, raising deep philosophical issues which are not the topic of this book.

1.2.3 Emergence

Emergence occurs when phenomena arise from and depend on some more basic phenomena and yet are simultaneously autonomous from that base [7]. A phenomenon is emergent if it cannot be reduced to, explained, or predicted from its constituent parts [38]. The ultimate interest of this project is the emergence of mind, as the end result of Darwinian evolutionary processes on Earth, leading to initial life, and then plants, animals, and ultimately humans. One should note here the different contexts and timescales of emergence occurring through three different kinds of processes:

1. **Evolutionary processes** in the universe and on Earth (long timescale diachronic emergence). There was no life 13 billion years ago. Indeed, even the elements out of which living beings emerge did not exist then. Order has emerged from primordial chaos. The timescale is billions of years (preparing the context and starting life) to millions of years (emergence of intelligence).
2. **Developmental processes** for each class of living beings, including humans (short timescale diachronic emergence). Each of us started as a single cell which then divided many times to create the organized set of 10^{13} cells that constitutes a

human being. Huge alterations in body structure took place during this process, and this had to happen in such a way that biological functioning could be continuously maintained as development took place and systems transformed. The timescale involved is decades.
3. **Functional processes** keep us alive on a minute by minute basis. Here electrons, protons, and neutrons (that by themselves contain no trace of life) work together to form biomolecules, tissues, brains, and organisms, where the whole is more than the sum of its parts (synchronic emergence). The timescale involved is microseconds to hours.

These processes are interdependent: none could happen without the other. The issue of emergence arises in each case:

- High level functions do not exist initially in the first two cases, but they do in the end.
- In the third case, the component parts do not display high level functions, but their combination in the system does.

The viewpoint suggested here is that, taking bottom-up causation for granted, processes of top-down causation are crucial to the emergence of genuine complexity in each of these cases. In particular, a key aspect of biological emergence is that entities that initially are able to survive on their own become embodied or enmeshed in higher level organisational entities in such a way that *they are no longer able to function on their own: they can only survive in the higher level context.* Cells can only survive as part of the body that they comprise, and animals can only survive in the ecosystem of which they are a part. Thus contextual effects are not an optional add-on to individual functioning. They are essential to the existence of the lower level entities, which are adapted to their role in the overall system (Sect. 5.4).

1.3 Key Points of the Argument

The nature of causation is a core issue for science, which can be regarded as the move from a demon-centered world to a world based on reliable cause and effect, tested by experimental verification [44, 78]. Causes are separated from effects by searching for correlations between phenomena such that manipulation of one (the cause) can be shown, in a specific context, to reliably result in specific changes in the other (the effect) at a later time. One has to search for this correlation in the midst of internal and environmental noise [70]. Laboratory tests of isolated systems allow an understanding of the elements of causation, which are interactions between the particles that underlie all physical existence. In multiple combinations, these interactions underlie the emergence of complex phenomena such as life.

Thus physics is the basic science, characterized by mathematical descriptions [72] that allow predictions of physical behavior to astonishing accuracy. Moreover, it underpins the other sciences (see [39, Chap. 4] and [48]). The key question is whether other forms of causation such as those investigated in biology, psychology, and the social sciences are genuinely effective, or whether they are rather epiphenomena grounded in purely physical causation. The latter view is suggested by strong reductionist views drawing on the fact that all the physical entities we see around us, including ourselves, are based on the same chemical elements [37], composed of the same kinds of elementary particles, interacting with each other only through the four fundamental physical forces [67, 72]. How can there then be room for any other type of causation?

I deal in turn with:

- Multiple types of causation (Sect. 1.3.1).
- Hierarchy and causation (Sect. 1.3.2).
- Types of top-down causation (Sect. 1.3.3).
- The nature of variables (Sect. 1.3.4).
- The causal efficacy of non-physical entities (Sect. 1.3.5).
- Room at the bottom (Sect. 1.3.6).
- Supervenience (Sect. 1.3.7).

1.3.1 Multiple Types of Causation

I will claim here that there are indeed other types of causation at work in the real world, described quite well by Aristotle's four types of causes (discussed in Sect. 8.3.1 below). There are many contexts in which different kinds of causality are experienced: in physics and chemistry, where particles and forces interact in a way described by variational principles and symmetries; in biochemistry and cell biology, where information is important and adaptation takes place; in zoology, where purpose, planning, and anticipation are important; and in psychology and sociology, where analytic reflection, symbolic understanding, values and, meaning all are causally effective. These undeniably all occur in the real world, and must be recognized as such if we want a complete account of causation.

All the Factors. Thus an important part of the argument is that we must recognize and acknowledge all these forms of causation, rather than denying that they exist as effective causes. Not only do chance and necessity exist as causal factors, but so also does purpose.

The Cause. Perhaps it is useful here to realize that, in trying to understand systems of great complexity (or even simple ones), we take for granted a great many things that are also part of the causal web. In effect, we just assume that they are there without further comment. The reader understands this standpoint from the context: we are investigating neural effects in the brain, so right now we won't discuss how chemistry emerges from physics, how quantum theory leads to classical behaviour, how evolutionary processes led to the genetic code, how the Earth came into existence as a habitat for life, and so on. We focus on the item we want to understand and label it 'the cause'. And this is fine as long as we do not in fact regard it as the only causal influence at work. It is indeed 'the cause' as long as we accept and take for granted all the other causal influences at play, without which it would not have happened as it did. But in relating to broader issues, we need to remain aware that they are in fact only part of the causal nexus that led to a specific outcome.

1.3.2 Hierarchy and Causation

A simplified version of the hierarchies of complexity and causality for inanimate matter and animate matter is given in Table 1.1. Each of the different levels of the hierarchy function according to laws of behaviour appropriate to that level, and are describable only in terms of language suited to that level (the concepts that are basic to molecular biology, such as genes and proteins, cannot be described in the language of a particle physicist, such as quarks and gluons). Ideas applicable to lower level causation do not by themselves succeed in explaining the higher level behaviours, for the concepts employed are simply not appropriate to the higher level kinds of causation. Higher level entities, such as plans and intentions, have causal power in their own right, which determine what happens at lower levels in the hierarchy (billions of atoms move in accord with our intentions when we raise our arm).

Effective Levels. How does it all fit together? Coarse-graining and consequent loss of detailed information relates lower levels to next higher levels. This structuring leads to the emergence of effective (phenomenological) laws at each of the higher levels, with apparent autonomy from the lower levels [4]. It is this independence from the details of lower level causation that allows phenomenological laws to be good effective theories of higher level interactions. Thus for example motor mechanics and neurosurgeons do not have to understand particle physics or nuclear physics in order to ply their trade.

We do not need to know the details of the bottom level to investigate and understand the effective higher level emergent laws of behaviour. This is just as well, because no one knows what the bottom level is. If we were truly reductionist we would not be able to say anything about what is 'really' happening until the current debate about the nature of quantum gravity is concluded—which may take centuries. Happily the causal effectiveness of higher levels saves us from this depressing predicament.

Interlevel Causation. In terms of the usual bottom-up understanding of causation, each lower level underlies what happens at each higher level. Thus the atomic structure of matter leads to the behaviour of gases and solids; the bonding between atoms creates molecules and so underlies chemistry; the reading of the molecular information in genes underlies the existence of animals; it is the flow of action potentials in axons connecting neurons that leads to the functioning of the brain; the behaviour of individuals is what creates societies. Overall, each lower level underlies what happens at the next higher level.

Top-down causation, however, also takes place [36]. The emergent higher levels act down on the lower level to direct what happens at those levels, by setting the context for their action. The actual physical work is done at the lower levels. But what work is done is determined by the higher levels, which determine relations between lower level configurations and set crucial variables selecting what happens at lower levels. This intermingling of top-down causation and bottom-up causation allows interlevel feedback loops that characterize genuine complexity. It is taken for granted in physiology [65] and in studies of the mind [41], where social neuroscience shows how the social setting shapes synaptic connections [3, 14] and so enables culture to shape minds [11], and top-down influences play a key role in perception [54] and the planning of speech behavior [86]. However, it is far more widespread than that: it occurs, for example, also in microbiology [51], as is clear from epigenetic studies [43], and even in physics [34].

That is what I will discuss here. Top-down causation can be shown to be causally effective in the physical world in all these domains, *inter alia* allowing effective causation by non-physical entities such as the value of money, social conventions, and ethical standpoints, such as attitudes toward nuclear war and environmental issues.

1.3.3 Types of Top-Down Causation

Is there only one type of top-down causation, or does it have various manifestations? I suggest that there are five different types of top-down causation that can take place, depending on the context:

TD1 Deterministic top-down causation.
TD2 Non-adaptive information control.
TD3 Adaptive selection.
TD4 Adaptive information control.
TD5 Adaptive selection of selection criteria.

Each differs from the others in significant ways. They will all be discussed in detail in Chap. 4. There could possibly be others, but I claim that at least these can all be regarded as well-established and essentially different from each other. Intelligent top-down causation, namely top-down processes facilitated by symbolic reasoning, is crucial for TD5.

A specific point to note here is that dynamical systems with attractors (TD1) can appear to be similar to goal-driven feedback control systems (TD2), but they are essentially different from each other in terms of the mechanisms in operation: TD2 involves the causal efficacy of information, whereas TD1 does not. Cases TD3–TD5 are cases of *adaptive selection* [42, 49], with the key property that they allow new information to be gathered and brought into play as dynamic variables; this does not happen in TD1 and TD2, which proceed on the basis of pre-existing variables only.

1.3.4 The Nature of Variables

A key element in the analysis is to consider the relation between variables at different levels.

Coarse-Graining. The simplest relation is coarse-graining, whereby higher level variables are derived by some form of averaging over lower level variables. The mass of a body is the sum of the masses of its constituent particles, and its momentum and angular momentum are similarly derived. The electric current flowing in a wire is the sum of the currents carried by the electrons flowing in the wire. In the kinetic theory of gases, density, pressure, and temperature are derived by suitable integrals over the masses and velocities of the molecules comprising the gas. Naming the coarse-grained variable (density, pressure, temperature) identifies it as a dynamically significant higher level variable.

Specific higher level variables characterize the macroscopic state of the system at a specific level, and occur in effective laws of behaviour at that level. These are the handles by which we can influence the system. Varying them changes the state of myriad lower level variables in a coordinated way. An important part of physics is identifying what these higher level variables are and how they are related to lower level variables, e.g., identifying the forms of energy applicable at different levels.

Equivalence Classes. Information hiding plays a crucial role in hierarchical complex systems [13]. Coarse-graining loses a great deal of information about the lower level states. Indeed, higher level variables necessarily represent only very broad aspects of the lower level situation. A key point then is that many lower level states will correspond to the same higher level state. For example, billions of different lower level molecular states correspond to the same combination of pressure, density, and temperature of a gas. The number of lower level states that correspond to a specific higher level state characterizes the entropy of that state [71, 72]. We identify all the lower level states that correspond to a particular higher level state (as characterized by a specific set of meaningful variables) as an equivalence class of lower level states instantiating that higher level state. In a context where top-down causation takes place, they are the real dynamical variables at lower levels. The multiple realizability of higher level states characterized by the existence of these functional equivalence classes is a core conceptual aspect of top-down action [6] (see Sect. 3.5).

Filtering, Statistics, and Pattern Recognition. Higher level variables can be obtained in many more sophisticated ways from lower level variables than by simple coarse-graining. These include:

- **Filtering**. For example, selecting energy densities in specific frequency ranges in the incoming variables and neglecting the rest (any real sensing system does this).
- **Statistical Analysis**. Data is analyzed to find statistical patterns, e.g., using Bayesian analysis of incoming data (the human mind is adept at this).
- **Pattern Recognition**. For example, recognizing faces and attaching a name to them. The name is a higher level variable.

Each of these identifies significant higher level variables arising in some way out of lower level variables. They are emergent variables in the specific context considered.

Not Emergent. However there are also some effective higher level variables that are not coarse-grained or otherwise emergent from lower level variables: they represent irreducible high level properties and relations. The top-down influence of such variables is a key aspect of the argument of this book. Those variables include mental features like emotions and feelings, abstract entities like theories and plans, and socially determined effects like the value of money (Sect. 1.3.5). Those higher level features are demonstrably causally efficacious, but cannot be regarded either as coarse-grainings of lower level variables, or as inevitable outcomes of blindly working lower level forces. They are essentially higher level variables.

The inevitable conclusion from the existence of such variables is that there are other forms of causation than those encompassed by physics and physical chemistry. A full scientific view of the world must recognise this fact, or else it will ignore important aspects of causation in the real world, and so will give a causally incomplete view of things [31].

1.3.5 The Causal Efficacy of Non-physical Entities

Non-Physical Entities. The following kinds of non-physical entities can be demonstrated to exist and have real causal powers in the sense that they change the physical outcomes of networks of interactions:

- Mathematical entities such as the number π, trigonometric functions, and Pythagoras' theorem, which underlie aspects of engineering practice.
- Our mental understandings of the laws of physics underlying the behaviour of matter, as for instance expressed in Maxwell's equations for electromagnetism. These underlie the existence of radio, radar, TV, cellphones, and so on.
- Computer programs and data, underlying for example ATMs, internet banking, aircraft control systems, automated factories, and myriad other applications.
- Human plans and intentions for everyday objects such as the plans for a computer, a Jumbo Jet airliner, an airport, a teapot, or a pair of spectacles, which consequently result in manipulation of huge numbers of constituent particles.
- Human plans for experimental manipulation of microentities, such as molecular synthesis, nanotechnology, state vector preparation in quantum mechanics, and particle pair creation in colliders such as the LHC. All of these planned microevents are the result of human intentions.
- Expectations and predictions, or what we think is likely to happen, e.g., expectations about price changes on the stock market, which result in money being made or lost.
- Social constructions, such as the rules of chess, the value of money, and a legal system, without which society could not function.

- Roles in society, such as being a teacher, a judge, a student, or a policeman, which shape our expectations and actions, and role models, who guide us as to how to behave in those roles.
- Information, as evidenced by the existence of an IT industry.
- Beauty, as evidenced by estate agent brochures charging a great deal more for houses that have beautiful views than for those without.
- Language, without which we could not think and be intelligent beings.

What Exists. The claim that all these entities exist—that they are ontologically real—rests on a philosophical analysis of what kinds of things must be recognised as existing. The view taken here is as follows [30]:

> **Existence**. We must recognise the existence of any kind of entity that demonstrably has a causal influence on physical systems.

The reason is that if we do not include such entities as being real, we will have a causally incomplete view of the universe: some events or entities that actually occur will then be uncaused. We will have to believe in magic in order to explain some things that exist (such as a digital computer) or events that demonstrably happen (such as an aircraft flying past). They are both the result of intelligent top-down causation TD5 from abstract conceptions (the idea of an aircraft, developed into a detailed construction plan) to physical entities: without such abstract effective variables, they could not exist (Sect. 7.5).

1.3.6 Room at the Bottom

How can there be room at the bottom for top-down causation to take place? Isn't there over-determination because the lower level physics interactions already determine what will happen from the initial conditions?

There are various ways that top-down causation can be effective without violating lower level physical operations: the lower levels do the work, and the higher levels choose what work will be done by shaping not only lower level conditions, but also the entities that interact. This happens in the following ways:

- **By setting constraints on lower level interactions**. These constraints break symmetries and so create the possibility of channeled and structured interactions.
- **By changing the nature of the constituent entities**. The higher level context often changes the nature of the underlying entities, shaping them to fit higher level purposes.
- **By creating constituent entities**. In many cases the lower level entities would not exist without the higher level structure. Emergence of higher level entities has clearly occurred when lower level entities cannot exist outside their higher level context (again, a common effect in biology, where symbiosis is rife).

- **By deleting lower level entities**. There is no fixed, unchanging set of lower level entities when selection creates order out of disorder by deleting unwanted lower level entities or states: top-down action selects what the lower elements will be.
- **By statistical fluctuations and quantum uncertainty**. Lower level physics is not determinate: random fluctuations and quantum indeterminism result in an ensemble of lower level states from which a preferred outcome is selected according to higher level selection criteria. Thus top-down selection leading to increased complexity is enabled by the randomness of lower level processes.

Together these effects allow top-down causation to take place in a way that guides the underlying physical processes without in any way violating their nature. This will all be discussed in Chap. 5.

1.3.7 Supervenience

An argument against genuine emergence is based on the idea of supervenience [76]: the claim that the higher level states emerge uniquely from the lower level states, so all we need to do is set the lower level states appropriately and emergence will occur [75]. If both structure and excitations of two systems S, S' at a lower level L_1, L_1' are identical in every detail, then at a higher level H, the two system states must necessarily also be identical in every detail: the lower level states imply the higher level states without remainder. Thus if we prepare L_1, L_1' to be identical we can derive H_1 and H_1' in a bottom-up way to be identical. Hence bottom-up emergence is all we need to generate any higher level state at all.

However this is based on synchronic emergence, that is, the instant by instant emergence of the higher level from the lower. If the system is complex, for example H_1 is an entire living being or a functioning digital computer, it is not possible to prepare the state L_1 by self-assembly or by pure chance (as characterized by statistical physics). In order for the required state L_1 to occur, you require top-down mechanisms (either natural selection, as in the case of animals, or design, as in the case of digital computers). That is, diachronic emergence based on preparation of the required bottom level state cannot occur spontaneously: it is not possible to assemble L_1 in the way needed to lead to H_1 by any random or statistical process for a single living entity such as a cell (think of the difficulties facing origin of life theories), much less for billions of living beings that thrive on a day to day basis. A top-down process (either adaptive selection or design) must have led to the initial conditions that enable supervenience of L_2 to take place. This is discussed in Sect. 3.5.3.

Indeed, while snowflakes or sandpiles can form in a purely bottom-up way, discussing them does not encompass how life works. Biomolecules such as DNA [16] or proteins such as enzymes, hemoglobin, or kinesin [74], cannot form in a purely bottom-up way—for they have evolved to perform particular biological functions [47, 50, 74] and have been selected for this purpose [91]. Physics per se cannot predict their existence or function (although it can predict that their existence is allowed [91]).

Related to this is the *exclusion principle* [7], stating that if a higher-level property F supervenes on a physical property F* that is causally sufficient for a property G, then F cannot cause G. List and Menzies give a full counter to that claim in [59].

1.4 Is It Real? Testing the Proposal

The idea of top-down causation is intimately related to concepts of emergence. Indeed, it is a key factor in strong theories of emergence [7]. But some physicists believe strongly that everything is determined bottom-up: if we had full data of things at an early stage in the universe and if we just had sufficient computing power, we could predict everything that is happening today. In that case, 'top-down causation' is just an unnecessarily complex way of referring to bottom-up causation. Some philosophers have put essentially similar views: the higher levels dance to the tune of the lower levels, and the impression of high level causal powers is just an illusion. The fundamental issue that arises is as follows:

Fundamental issue. Does top-down causation actually occur? Is it real, or is it just an epiphenomenon?

There are differing views, both in science and in philosophy. The argument of this book is that top-down causation is indeed real. Evidence is of various kinds. I deal in turn with:

- Causal effects (Sect. 1.4.1).
- Experimental tests (Sect. 1.4.2).
- Kinds of data (Sect. 1.4.3).
- There is no other option (Sect. 1.4.4).

1.4.1 Causal Effects

The view here is based on an interaction picture:

Interaction picture. The basic point is that one demonstrates existence of top-down causation whenever manipulating a higher level variable can be shown to reliably alter lower level variables.

Manipulation of higher level variables generally alters lower level variables. However, it cannot generically determine which specific microstate will result as a consequence of manipulation of some macro variable. By such manipulation we can only access the underlying equivalence class. For example if we change the temperature of a system, this will change the microstate to any one of the huge class of microstates that correspond to the new temperature.

Such deterministic top-down causation (TD1) is commonplace in physics, chemistry, biology, and engineering, and its existence is for example supported by all the evidence that statistical physics and physical chemistry are correct. However, this only leads to interesting dynamics when lower level causation is channeled by carefully fashioned complex structures, as in computers and the brain. Then, for example, altering the program loaded in a computer alters the flow of electrons in gates at the microlevel.

What about the other types of top-down causation discussed in Chap. 4? Various lines of argumentation provide evidence they too involve real top-down dynamics:

- **Homeostasis or Feedback Control**. Whenever what happens is determined by preset goals rather than by initial data, the final state is dynamically determined in a top-down way by these goals. This is the case TD2. Altering the goal (say the temperature in a thermostat) alters the microstates of the systems (the motion of molecules in hot water).
- **Adaptive Selection**. It occurs whenever adaptive selection takes place [42, 49], because adaptive criteria are higher level features of the system guiding preferred outcomes. This includes cases TD3–TD5. Altering either the selection criteria or the context in which they operate alters the lower level outcomes. For example, altering the temperature on Earth by filling the atmosphere with carbon dioxide alters genes in animals as they adapt to this change.
- **Induced Entities and Symbiosis**. Top-down influences must have occurred when the very occurrence of lower entities is induced by higher level structures. They would not occur without that context. This is the case in particular when symbiosis occurs, where the individual components of a symbiotic relationship cannot exist without the others. This occurs throughout biology. For example, the cells in a human body live only if the body supplies them with oxygen and sugars.

Top-down effects are taken for granted in holistic sciences based on complex interactions. One cannot understand the brain or physiology or ecology or evolution or epigenetics without assuming top-down causation to be real [3, 14, 15, 41, 65]. Bottom-up explanation by itself cannot do the job. The very existence of these effects and outcomes is evidence for top-down causation.

1.4.2 Experimental Tests

These considerations are in my opinion conclusive, but are based on understanding and explaining what one already knows. What one likes in an experimental science is a prediction of something new that can then be verified by experiment or observation. So an important further question is this:

> What new experiments or observations can we propose that will substantiate or disprove the causal efficacy of higher level variables?

There seem to be four main streams of possibility here:

Convergent Evolution. When top-down causes drive what happens in evolutionary contexts, one often gets convergent evolution: different evolutionary pathways devise similar means of meeting the same higher level need [20, 63]. A famous case is the development of eyes by various evolutionary paths, driven by the need of animals to see, which clearly improves their survival capacity. One cannot explain such convergent evolution by bottom-up causation alone: it is driven by a combination of specific higher level needs in conjunction with restrictions on how they can be achieved physiologically [90]. Hence new evidence of convergent evolution is evidence for top-down causation.

Computer Simulations. Top-down causation can be demonstrated by computer simulations of complex systems where higher level variables are shown to determine the outcome, for example the simulations of heart physiology by Noble [66]. Changing the higher level variables demonstrably changes the lower level dynamics and hence the outcome.

Equivalence Classes. As has been emphasized above (Sect. 1.3.4), the concept of equivalence classes of lower level variables is crucial to the nature of and physical implementation of top-down causation [6]; and indeed their existence can be taken as convincing evidence that top-down causation is at work.

Now one might claim that this was already very well established in some cases, for example in statistical mechanics [2], where entropy is a measure of how many equivalence classes exist for a given macro state [73]. Nevertheless, it is better to have a new prediction of as yet undiscovered equivalence classes, that can then be verified by experiment. This is at least in principle possible in microbiology, where the existence of very interesting cases of equivalence is already established (see Jaeger and Calkins [51]), and one can hope to plan experiments that create new kinds of lower level members of an equivalence class satisfying some specific higher level need in cellular biology. This is a very promising area for future work. It is also possible in physics, in cases where the existence of equivalence classes is shown to be the key to a full understanding of dynamics, such as in Crutchfield's proposal for computational dynamics [22].

Direct Proof of the Power of Intrinsically Higher Level Variables. As has been mentioned above (Sect. 1.3.5), some variables are intrinsically higher level variables that cannot emerge by coarse graining of lower level variables. One can hope to show that some of these variables affect the structure of lower level entities. Robert Laughlin would claim that this kind of situation has already been shown to be the case in physical effects such as superconductivity and fractional quantization [57], and the next chapter will show that this occurs in digital computer systems (computer programs are not physical entities and are not emergent variables).

One might reasonably claim that the rise of epigenetics is convincingly demonstrating such causation in the case of developmental biology [23, 43], and it is becoming clear that one can show it to be the key to the historical origin of life [92, 93]. As for the brain itself, exciting work giving such proof is currently being done in social neuroscience [3, 14], where social variables can be shown to affect neuronal connections, and hence the flow of electrons in the dendrites and axons of neurons in the brain. This is top-down action from the social level to the level of neurons; so work in social neuroscience is evidence for top-down causation. Additionally, work on perception is clear evidence for top-down processing, as will be discussed later (Chap. 7), as is the evidence for placebos [9]. One can devise tests of these effects: Shea [83] gives a testable explication of the notion of top-down influence in the context of psychological processes, and there are many tests of the effects of placebos [44, 61].

1.4.3 Kinds of Data

One important point that arises in considering this issue concerns the kinds of data that will be taken into account in the quest to understand the types of causation that are at work in the real world.

1.4.3.1 Only Physics Data

Many writers on topics of causation are in effect claiming that one only needs to take into account evidence from the hard sciences (see, e.g., [5]). The implicit claim is that physics, biochemistry, microbiology, and neurology constitute the complete knowledge base we need to understand humans, and hence that the associated kind of data is all the data we need for such an investigation.

This is actually just another form of reductionist thinking, and is not adequate for the investigation we are undertaking here. We are ourselves creatures living in the universe, and our life experiences are data about the universe. They have to be interpreted with considerable caution of course, but for a complete view of things we need to take into account evidence from everyday life and the humanities as well as evidence from the hard sciences and human sciences. Examples of such careful observation can be found in the work of Kahneman [52] on economic behaviour and by Grandin [45] on animal behaviour, both with implications for the way the brain works.

1.4.3.2 The Existence of Artefacts

Teapots exists and so do spectacles and aircraft and buildings. These are artificial entities [84]. There is no causal process whereby they can have self-assembled with precisely the properties they have, for example, a pair of spectacles shaped to give clear vision for my individual eyes [31], for the eyes have a particular shape that is not captured in any universal laws. They are the result of the autonomous action of a mind, shaping physical entities according to our chosen goals and in response to particular circumstances (Sect. 4.7.6). This is top-down causation from the mind to the physical world.

1.4.4 *There Is No Other Option*

Finally, once we consider how things work in the cosmic context, there is no other option! There are two reasons for this.

The first is that the kinds of detailed initial data needed to make a functional digital computer or brain cannot plausibly self-assemble in a bottom-up way (Sect. 1.3.7). Self-assembly can create crystal structures or snow flakes, patterns such as those given by the reaction–diffusion equation, and so on, but it cannot create computers, or even biomolecules such as kinesin that perform biologically crucial functions in living cells [50]. These require the top-down effects of either adaptive selection [15] or purposeful design [84].

Secondly, there is no plausible way that the words in Einstein's 1915 paper on General Relativity can have been uniquely implied in a bottom-up way by the data on the so-called last scattering surface in the early universe, for these fluctuations are usually believed to be random Gaussian fluctuations. The argument is given in detail in Sect. 8.1. Like all the other theories developed by the human mind, those words must be the product of an emergent mind and brain working according to the precepts of logical argumentation—that is, genuine emergence must occur. And that requires top–down causation to occur [7].

1.5 Significant Implications

Finally, this investigation is not just an academic debate: there are significant implications of this dialogue for various practical areas. I will consider just three:

- Health care (Sect. 1.5.1),
- Mental health (Sect. 1.5.2),
- Education: learning to read (Sect. 1.5.3).

1.5.1 *Health Care*

The care of human physical health has long been the scene of contestation between bottom-up approaches, based on medicine and surgery alone, and top-down approaches, characterized as holistic medicine. This has important applications both in individual health care and in social health care.

I will just mention one such topic here: it seems well established that placebos can have important effects in improving the state of health of individuals [44, 61]. This is a top-down influence of the mind to affected organs. A scientific basis for aspects of such an effect can be found in the influence of the mind on the immune system, because some neurotransmitters are also immune system molecules [85]. That may not be the only causal channel, but it is sufficient to prove that it does indeed have a plausible physiological basis.

Health care measures need to take such top-down effects (positive and negative) into account, as well as the bottom-up (mechanistic) aspects, otherwise they will miss crucial aspects of what is going on. This topic is pursued in depth in [58].

1.5.2 Mental Health

Similar issues arise in mental health, which is influenced in both bottom-up ways, through genes and molecules to neurons and neural networks, and top-down ways, by social relations and interactions of many kinds; famously, one's relationship with one's parents. Correspondingly, there are a range of approaches in use, ranging from bottom-up methods (drugs of various kinds) to top-down methods ('the talking cure') [53]. The issue is discussed in depth in [10].

1.5.3 Education: Learning to Read

Education is an area of contestation between bottom-up approaches to learning (grasp the components of the topic first, worry about how it fits together later) and a top-down approach (try to get the big picture into view first, then worry about the parts later). The first approach concentrates on parts and mechanisms, the second on the whole and meaning. Both aspects are needed for effective education, but it makes a big difference which is the centre of an educational approach.

This is exemplified in science education. Here one can follow a traditional route of 'naming of the parts' (of cells, animals, etc.) with no proper concept of what their holistic function is; or an enquiry-based approach aimed at first discovering the purpose of things, clarifying the overall features of explanation, filling in the detail of how it works later on [64]. A similar crucial example is the teaching of reading and writing, where a holistic top-down approach is advocated by some [55], while a mainly bottom-up phonics-based approach is advocated by others. It is in particular claimed by some that neuroimaging studies support the bottom-up approach [25, 60]. In Sect. 8.6, I well make the case that the latter claims are basically flawed: they do not take seriously the core issue of what language is about, namely conveying meaning from one person to another, and they also ignore crucial data on how people actually read [29].

Education practice needs to take into account both top-down effects, particularly related to meaning and purpose, and bottom-up effects, inter alia related not just to technical issues but also to the emotional climate of the classroom, otherwise they will miss crucial aspects of what is going on. The topic is pursued in Sect. 8.6.

1.6 An Outline of the Book

The topics in the book are outlined here (Sect. 1.6.1), followed by brief comments on what is new (Sect. 1.6.2) and what is controversial (Sect. 1.6.3).

1.6.1 The Contents

The chapters in the book are as follows:

Chapter 1. **Complexity and Emergence**. The present chapter deals with basic topics in complexity and emergence, emphasizing the role of top-down causation in addition to bottom-up causation.

Chapter 2. **Digital Computer Systems**. This serves as a warm up to the main text by considering the case of digital computer systems. The outcome of the analysis is two-fold: a demonstration that top-down causation is taking place, as evidenced by the existence of lower level equivalence classes, and a proof of the causal effectiveness of non-physical entities (namely computer programs and data).

Chapter 3. **The Basis of Complexity**. This considers the nature of modular hierarchical structures, and then successively, bottom-up action, emergence, higher-level variables, and top-down action. The key concept focused on here is the existence of equivalence classes of lower level entities relative to higher level functions, which characterise the existence of top-down causal effects. I also deal with counterarguments based on the idea of supervenience.

Chapter 4. **Bottom-Up and Top-Down Effects**. Here I consider in detail the five classes of top-down action identified in this book. They are:

TD1 Deterministic top-down causation.
TD2 Non-adaptive information control.
TD3 Adaptive selection.
TD4 Adaptive information control.
TD5 Adaptive selection of selection criteria.

They can each be shown to exist by many examples and they are all different from each other. The last three are examples of adaptive selection.

Chapter 5. **Room at the Bottom**? This chapter considers how top-down causation can be possible if there already exists a complete set of causal processes at the lowest level. How is there room at the bottom then? The system seems over-determined. The argument will be that higher level structures exert a crucial top-down influence. Moreover, the very nature of the lower level elements is influenced by top-down effects, while micro-indeterminism allows space for adaptive selection to act on the lower levels in a top-down way.

Chapter 6. **The Foundations**: **Physics**. While top-down effects are obvious in subjects such as physiology, ecology, and psychology, can we also find them in the foundations: do they also occur in physics? The chapter will argue that they do, citing the cases of statistical mechanics, computational mechanics, cosmology, and quantum physics. In the latter case, for example, top-down effects occur in the process of state vector preparation and when topological effects occur.

Chapter 7. **The Mind**: **Intellect, Emotion, and Adaptation**. This chapter puts the preceding analysis to work by turning attention to the brain. After a brief look at the basics of the brain, adaptive selection and developmental processes are considered and then related to the evolutionary origins of the brain. The effect of these processes is the existence of basic patterns of understanding, with bottom-up and top-down processes intermingling as the mind in effect judges competing claims of intuition, emotion, reason, and values as to what should be done. Predictions and expectations play a key role in these processes.

Two key things emerge clearly: top-down processes occur from society to the brain, as evidenced by recent work in social neuroscience, and there is a causal effectiveness of non-physical entities through the workings of the mind. In particular, thoughts, plans, and social constructions such as language and money are causally effective.

Chapter 8. **The Broader View**. This final chapter summarizes the argument by considering causation in relation to genuine emergence. Following [62], causation in physics and biology is usually classified as being due to either chance or necessity. However, when biology is concerned, function or purpose is also involved [47]. The chapter looks at the relation between these kinds of causation and how they complement each other. In particular, it is the existence of random processes at lower levels that enables purposeful actions at higher levels to take place through selection of preferred outcomes according to higher level selection criteria. This enables processes of adaptation and learning in accordance with higher level logic. As regards necessity, I propose that the profound basis of necessity is the existence of Platonic possibility spaces and the associated causal efficacy of non-physical entities. Mathematical objects and relations provide an example of something that is transcendent rather than emergent, these being causally effective via the human mind [18].

Because the brain underlies social interactions, there are many implications for society, including learning to read and write. This chapter considers that issue.

1.6.2 *What Is New*

A great deal of this book is concerned with a survey of well established results presented from a particular perspective involving the hierarchy of structure and causation, and the interaction of bottom-up and top-down effects in that hierarchy. Apart from the overall integration of all the topics considered into a reasonably coherent whole from this viewpoint, the novel elements are as follows:

1. Discussion of digital computers from this viewpoint (Chap. 2).
2. Classification of top-down effects into five essentially different types (Chap. 4).
3. A comprehensive response to the supervenience and over-determination arguments against top-down causation (Sect. 3.4.6 and Chap. 5, respectively).
4. A large-scale overview of brain function presented in Sect. 7.2.3, with both purpose and meaning on the one hand, and primary emotions on the other, as key drivers of brain function (Sect. 7.4.1).
5. A classification of primary emotions in Sect. 7.2.4, extending the work of Panksepp [68].
6. A proposal, following Churchland's seminal work [18], regarding the causal power of Platonic entities in a top-down way over physical entities through the operation of the human brain (Sects. 2.7.5, 2.7.6, and 7.6).
7. An argument against deterministic causation of present day life on Earth from initial data in the expanding universe (Sect. 8.1).

Items 2, 3, and 7 are significant as regards the nature of causality in complex emergent systems in general. Items 1, 4, and 6 are significant in terms of applications to specific complex systems, viz., digital computers and the brain. Item 5 is significant in that it shores up a key part of 4.

1.6.3 *What Is Controversial*

A great deal of what is presented here is uncontroversial, representing rigorous statements based on established science. However, some is speculation, representing my own view on how various features fit together in a coherent way. As regards the latter, first there is the top-down thesis itself, and second there are various further aspects of what I discuss. I here just note these issues as a fair warning to the reader that these are controversial topics: I am ready to defend them all, but in each case substantial disagreement exists.

1.6.3.1 The Top-Down Thesis Itself

The top-down thesis is the subject of considerable controversy, particularly from reductionist physicists, but it is completely uncontroversial for holistic physiologists and neuroscientists, and even cell biologists move beyond that view—as do physicists concerned with superconductivity or topological effects.

The View from Physics. An example of the reductionist view is a statement by Sean Carroll in a recent book [17]. Having discussed bottom-up causation, he continues:

> And the converse, downward causation of human scale properties influencing the microscopic behaviour of particles, is simply misguided. A standard example given by proponents of top-down causation is the formation of snowflakes. Snowflakes are made of water molecules interacting with other water molecules to form a crystalline structure. But there are many possible structures, determined by the initial seed from which the snowflake grows. Therefore, it is claimed, the macroscopic shape of the snowflake is 'acting downwards' to determine the precise location of individual water molecules.
>
> We should all resist the temptation to talk that way. Water molecules interact with other water molecules, and other molecules in the air, in precise ways that are determined by the rules of atomic physics. Those rules are unambiguous: you tell me what other molecules an individual water molecule is interacting with, and the rules will say precisely what will happen next. The relevant molecules may indeed be a large part of a crystalline structure, but that knowledge is of precisely zero import when studying the behaviour of the water molecule under consideration. The environment in which the molecule is imbedded is of course relevant, but there is no obstacle to describing the environment in terms of its own molecular structure. The individual molecule has no idea it's part of a snowflake, and could not care less.[3]

This is a classic statement of the supervenience argument [7] from a physics viewpoint. Even as viewed from within physics, there are two problems with it. Firstly, it omits from consideration those cases where topological effects occur, such as the fractional quantum Hall state [56]. The lower level states are then crucially affected by non-local higher level structures. Secondly, it does not take into account those cases where key lower level elements such as phonons and Cooper pairs do not even exist unless the higher level state has a specific form that leads to their existence. This is discussed in Chaps. 5 and 6.

Modular Cell Biology. A further point is that Carroll regards it sufficient to dismiss top-down causation by discussing only snowflakes. They can easily self-assemble, albeit with some variation caused by environmental fluctuations, and so are not included here as examples. However, the real point as regards the supervenience argument is that one must account for how the lower level physical elements got to be what they are in the first place. In more interesting cases such as the components of a living cell, this cannot happen in a purely bottom-up way (which is why the emergence of life is so difficult to explain). Biological macromolecules such as RNA and DNA [16] and proteins [74] cannot self-assemble. They have got to be what they are through adaptive selection, shaping them to fulfil some function or purpose through adaptive processes, such as those described by Wagner [91].

As pointed out by Hartwell et al., all biology is based on purpose, and this applies right down to the molecular level [47]:

> To describe biological functions, we need a vocabulary that contains concepts such as amplification, adaptation, robustness, insulation, error correction and coincidence detection. For example, to decipher how the binding of a few molecules of an attractant to receptors on the surface of a bacterium can make the bacterium move towards the attractant (chemotaxis) will require understanding how cells robustly detect and amplify signals in a noisy environment.

[3]https://twitter.com/seanmcarroll/status/666067938311454721

> Having described such concepts, we need to explain how they arise from interactions among components in the cell. We argue here for the recognition of functional 'modules' as a critical level of biological organization.

There is a stark contrast with what is stated in the quote from Carroll given above. Physics enables this to happen, but does not by itself decide what happens [89]. Biology is not just applied physics. Rather there are biological needs that are met by physical processes.

The View from Holistic Physiology. The same theme arises at the level of systems physiology, where one studies the function of systems such as the lungs, endocrine system, immune system, heart, and so on [65, 66]. Systems level effects act down to the level of genes and molecules to determine what happens, via epigenetic processes. The logic of biology dictates which physical effects are activated (such as flows of electrons in neurons or muscles).

The View from Holistic Neuroscience. This is even more striking in the case of the brain. Just one example: memory retrieval in the hippocampus is thought to be influenced by top-down inputs from the prefrontal cortex (PFC). Relevant structures and functions have now been identified [1]:

> Anxiety-related conditions are among the most difficult neuropsychiatric diseases to treat pharmacologically, but respond to cognitive therapies. There has therefore been interest in identifying relevant top-down pathways from cognitive control regions in medial prefrontal cortex (mPFC). Identification of such pathways could contribute to our understanding of the cognitive regulation of affect, and provide pathways for intervention. Previous studies have suggested that dorsal and ventral mPFC subregions exert opposing effects on fear, as do subregions of other structures. However, precise causal targets for top-down connections among these diverse possibilities have not been established. Here we show that the basomedial amygdala (BMA) represents the major target of ventral mPFC in amygdala in mice. Moreover, BMA neurons differentiate safe and aversive environments, and BMA activation decreases fear-related freezing and high-anxiety states. Lastly, we show that the ventral mPFC-BMA projection implements top-down control of anxiety state and learned freezing, both at baseline and in stress-induced anxiety, defining a broadly relevant new top-down behavioural regulation pathway.

This shows how specific structures in the brain are constructed so as to enable top-down causation to take place, and many more examples are given in Chap. 7. Indeed paradoxically, once those structures exist, it is precisely the kinds of mechanisms described by Carroll that enable top-down effects in the brain, such as those described here, to take place. But those brain structures would not exist without the top-down effect of natural selection which enabled them to come into existence in the first place [15, 16].

The LHC. In any case, the existence and effects of the Large Hadron Collider (LHC) is a classic counterexample to Carroll's dismissal of top-down action in the quote given above. The LHC did not self-assemble in a bottom-up way. The brains of experimental physicists devised the LHC and caused it to operate, thereby causing billions of microparticles to collide. This is undeniable top-down action from the level of the brain as a whole to the level of particle physics: that is the cause of the collisions that occurred in the LHC, whereby the existence of the Higgs boson was confirmed.

So is what is claimed here based on rigorous scientific statements? My view is that it is uncontroversial, representing rigorous statements based on established science, provided one's view of science extends beyond physics to fields such as epigenetics, physiology, and neuroscience, and indeed aspects of quantum physics such as the quantum Hall effect.

1.6.3.2 Other Controversial Issues

The items mentioned above under the heading *What Is New* (Sect. 1.6.2) are all candidates for being regarded as controversial. I mention here just six particular items:

- **Multilevel selection**. This is a topic causing huge controversy in the field of evolutionary theory. In this book I defend the idea that a careful analysis of the relevant causal interrelations shows that multilevel selection must exist, and furthermore that it must have been crucial to the development of humanity and the human brain. This is discussed in Sects. 4.3.7 and 7.2.5.
- **The causal powers of abstract entities**. This concerns specifically computer algorithms, thoughts, and social constructions such as money. This is discussed in Sects. 2.7, 4.7, and 7.5.
- **Platonic entities**. The existence of, and causal effects of, Platonic entities. This is discussed in Sects. 2.7.5, 4.7, and 7.6.
- **Educational implications**. The view on how learning to read takes place, with resultant consequences for educational policy (Sect. 8.7).
- **Free will**. The multilevel argument concerning the free will debate (Sect. 7.7.4).
- **Necessity of the conclusion**. The arguments as to why the conclusion is inescapable (Sect. 1.7.2).

These are the items I would flag as being specifically in the category of speculative proposals. They are, however, supported by adequate arguments and evidence in the relevant sections.

1.7 The Necessity of the Conclusion

This book considers the nature of causation in complex systems such as living beings, and in particular in the human mind, relating this to the emergence of genuine complexity in living systems. It explores the forms of top-down causation that make this possible, and considers the implications for our understanding of the nature of causality, and hence for the nature of the scientific endeavor. I summarise here:

- The conclusion (Sect. 1.7.1).
- The necessity of the conclusion (Sect. 1.7.2).

1.7.1 The Conclusion

In brief, the conclusion is as follows:

- There are other forms of causation in the real world than those encompassed by physics and physical chemistry, for example action in the world by intelligent beings based on the outcome of a logical thought process.
- These have their own higher level logic independent of the specific nature of the lower levels of causation, e.g., an analysis of the probable outcomes of different courses of action leading to selection of the optimum action to undertake.
- These processes manifest their causal powers through top-down causation in hierarchies of structure and causation, e.g., at the level of electrons in muscles that are constructing a digital computer.
- These kinds of causation enable true complexity to emerge by constraining lower level interactions in a coordinated way so as to enable the desired higher level outputs, e.g., constructing a digital computer network that constrains electrons to flow between specific computers.
- This does not in any way override the lower level physics; rather it channels the results of physical causation at the lower levels in accordance with higher level function or purpose, e.g., creation of the internet system.

A full scientific view of the world must recognise this, or else it will ignore important aspects of causation in the real world, and so will give a causally incomplete view of things [30, 33, 35]. This has obvious implications for views regarding the relation between reductionism and emergence. Section 3.4.7 gives a brief comment on that complex debate (a good overview can be found in [7]).

1.7.2 The Necessity of the Conclusion

It is my view that there are three kinds of reasons why the outcome proposed has to be true. These are:

- The self-assembly argument.
- The cosmic context argument.
- The self-consistency argument.

1.7.2.1 The Self-Assembly Argument

It is argued in Sects. 1.3.7 and 3.5.3 that one cannot have self-assembly of living systems, or their molecular components, without prior top-down causation that stores the necessary information for this to take place in biomolecules such as RNA and DNA [15, 91]. These cannot self-assemble with the necessary level of complexity, much less with the required information needed for developmental processes to lead to functioning animals.

1.7.2.2 The Cosmic Context Argument

It is argued in Sect. 8.1 that when the cosmic context is taken properly into account, there is no way that the data needed for uniquely determining the latter development of complexity in the universe, in particular intelligent life and its outcomes such as digital computers and computer-designed aircraft, can have been written into the early universe. The standard cosmological proposal is that data is random Gaussian perturbations [27], which cannot have such outcomes coded into it. To propose otherwise is a form of intelligent design argument.

1.7.2.3 The Self-Consistency Argument

Finally, I emphasize that such a conclusion is essential to the enterprise of science. For science to take place as a human endeavour, our minds must have the power to examine the relevant arguments in a rational way and come to a conclusion based on the validity or otherwise of the rational argument. This is a higher level process of exploration that must be able to take place as a valid logical process at that level, free from any restrictions on such arguments arising from the lower level underpinnings of the operation of the brain.

If our minds are nothing but the outcome of lower level processes meaninglessly grinding away in neurons, without the higher level logical ability available to enable us to generate meaningful theories and then scientifically evaluate competing theories and make sensible choices between them shaping what takes place at the lower levels, then we do not have the capacity to undertake the scientific enterprise and produce theories such as evolutionary psychology or neuroscience or theoretical physics, which require rational consideration and choice. There is no possibility that these sophisticated outcomes could emerge in a purely bottom-up way from the meaningless operation of lower level processes, because those processes in and of themselves have no relation to the logic that is being examined by the brain. It is, however, possible if those lower level processes take place in a context of brain structure that has been adaptively shaped by our individual and collective experiences in such a way as precisely to allow such logical processes to take place.

> **The Inconsistency**. The claim that higher level processes do not exist in an autonomous way, free from determination by lower level processes, is a self-defeating claim: if it were true then making the claim has no meaning, because the mind would be unable to function sensibly and produce it as a considered outcome—if the mind is based in the brain, as current neuroscience assumes.

The Evidence. We can and do argue such issues philosophically and scientifically; this book is an example. This is already abundant evidence that the major viewpoint presented here has to be true. Certainly, the causal effectiveness of the mind is limited in a number of ways [54], but it is demonstrably real nonetheless. Philosophers and scientists who claim otherwise forget the fundamental conditions that make the very exercise of their own discipline possible.

As we are indeed pursuing understanding through science and philosophy, we must take it as a fundamental premise that higher level meaningful choice must be possible. Any theory that denies this must be rejected as an inconsistent account of the reality we experience, and so as an inadequate basis on which to conduct scientific or philosophical enquiry, for it undermines the basic necessities needed in order to carry out such an enquiry.

Meta-Conclusion. If the argument presented in this book does not work, we'll have to find another one that does, otherwise we will have shown that our very own process of scientific and philosophical investigation is impossible.

References

1. A. Adhikari, T.N. Lerner, J. Finkelstein, S. Pak, J.H. Jennings, T.J. Davidson, E. Ferenczi, L.A. Gunaydin, J.J. Mirzabekov, L. Ye, S.-Y. Kim, A. Lei, K. Deisseroth, Basomedial amygdala mediates top-down control of anxiety and fear. Nature **527**, 179–185 (2015)
2. M. Alonso, E.J. Finn, *Fundamental University Physics III: Quantum and Statistical Physics* (Addison Wesley, Reading, Mass, 1971)
3. N. Ambady, The mind in the world: culture and the brain. Assoc. Psychol. Sci. **24**(5–6), 49 (2011)
4. P.W. Anderson, More is different. Science **177**, 377 (1972). Reprinted in P.W. Anderson, *A Career in Theoretical Physics* (World Scientific, Singapore, 1994)
5. P.W. Atkins, The limitless power of science, in *Nature's Imagination: The Frontiers of Scientific Vision*, ed. by J. Cornwell (Oxford University Press, Oxford, 1995), pp. 122–132
6. G. Auletta, G.F.R. Ellis, L. Jaeger, Top-down causation: from a philosophical problem to a scientific research program. J. Roy. Soc. Interface **5**, 1159–1172 (2008). arXiv:0710.4235
7. M.A. Bedau, P. Humphreys (eds.), *Emergence: Contemporary Readings in Philosophy and Science* (MIT Press, Cambridge, Mass, 2008)
8. S. Beer, *Brain of the Firm* (Wiley, Chichester, 1981)
9. F. Bendetti, *Placebo Effects* (Oxford University Press, Oxford, 2014)
10. R.P. Benthal, *Doctoring the Mind: Why Psychiatric Treatments Fail* (Penguin, London, 2009)
11. P. Berger, T. Luckmann, *The Social Construction of Reality: A Treatise in the Sociology of Knowledge* (Anchor, New York, 1967)
12. R. Bishop, H. Atmanspacher, Contextual emergence in the description of properties. Found. Phys. **36**, 1753–1777 (2006)
13. G. Booch, *Object Oriented Analysis and Design with Applications* (Addison Wesley, New York, 1994)
14. J.T. Cacioppo, J. Decety, Social neuroscience: challenges and opportunities in the study of complex behavior. Ann. N. Y. Acad. Sci. **1224**, 162–173 (2011)
15. D.T. Campbell, Downward causation, in *Studies in the Philosophy of Biology: Reduction and Related Problems*, ed. by F.J. Ayala, T. Dobhzansky (University of California Press, Berkeley, 1974)
16. N.A. Campbell, J.B. Reece, *Biology* (Benjamin Cummings, San Francisco, 2005)
17. S. Carroll, *The Big Picture: On the Origins of Life, Meaning, and the Universe Itself* (Dutton, 2016)
18. P. Churchland, *Plato's Camera: How the Physical Brain Captures a Landscape of Abstract Universes* (MIT Press, Cambridge, Mass, 2012)
19. P.C. Clayton, P.C.W. Davies (eds.), *The Re-Emergences of Emergence: The Emergentist Hypothesis from Science to Religion* (Oxford University Press, New York, 2006)

20. S. Conway Morris, *Life's Solution: Inevitable Humans in a Lonely Universe* (Cambridge University Press, Cambridge, 2005)
21. F. Crick, *Astonishing Hypothesis: The Scientific Search for the Soul* (Scribner, 1995)
22. J.P. Crutchfield, Between order and chaos. Nat. Phys. **8**, 17–24 (2011)
23. P.C.W. Davies, The epigenome and top-down causation. Interface Focus **2**, 42–48 (2012)
24. R. Dawkins, Hierarchical organisation: a candidate principle for ethology, in *Growing Points in Ethology*, ed. by P.P.G. Bateson, R.A. Hinde (Cambridge University Press, Cambridge, 1976)
25. S. Dehaene, *Reading in the Brain: The New Science of How We Read* (Penguin, London, 2010)
26. D.C. Dennett, *Consciousness Explained* (Back Bay Books, 1992)
27. S. Dodelson, *Modern Cosmology* (Academic Press, New York, 2003)
28. M. Donald, *A Mind so Rare: The Evolution of Human Consciousness* (W W Norton, New York, 2001)
29. A. Ebe, What eye movement and miscue analysis reveals about the reading process of young bilinguals, in *Scientific Realism in Studies of Reading*, ed. by A.D. Flurkey, E.J. Paulson, K.S. Goodman (Taylor and Francis, London, 2008), pp. 131–152
30. G.F.R. Ellis, True complexity and its associated ontology, in *Science and Ultimate Reality: Quantum Theory, Cosmology and Complexity*. ed. by J.D. Barrow, P.C.W. Davies, C.L. Harper (Cambridge University Press, Cambridge, 2004), pp. 607–636
31. G.F.R. Ellis, Physics, complexity, and causality. Nature **435**, 743 (2005)
32. G.F.R. Ellis, The universe around us: an integrative view of science and cosmology (2008). http://www.mth.uct.ac.za/~ellis/cos0.html
33. G.F.R. Ellis, On the nature of causation in complex systems. Trans. Roy. Soc. S. Afr. **63**, 69–84 (2008)
34. G.F.R. Ellis, On the limits of quantum theory: contextuality and the quantum-classical cut. Ann. Phys. **327**, 1890–1932 (2012). arXiv:1108.5261v2
35. G.F.R. Ellis, Top-down causation and emergence: some comments on mechanisms. J. Roy. Soc. Interface Focus **2**, 126–140 (2012)
36. G.F.R. Ellis, D. Noble, T. O'Connor (eds.), Top-down causation: an integrating theme within and across the sciences? Roy. Soc. Interface Focus **2**, 1–140 (2012)
37. J. Emsley, *Nature's Building Blocks: An A-Z Guide to the Elements* (Oxford University Press, Oxford, 2003)
38. B. Falkenburg, M. Morrison, *Why More Is Different: Philosophical Issues in Condensed Matter Physics and Complex Systems* (Springer, Heidelberg, 2015)
39. R. Feynman, *Six Easy Pieces* (Penguin, London, 1995)
40. R.L. Flood, E.R. Carson, *Dealing with Complexity: An Introduction to the Theory and Application of Systems Science* (Plenum Press, London, 1990)
41. C.D. Frith, Free will and top-down control in the brain, in *Downward Causation and the Neurobiology of Free Will*, ed. by N. Murphy, G.F.R. Ellis, T. O'Connor (Springer, Heidelberg, 2009)
42. M. Gellman, *The Quark and the Jaguar* (Abacus, 2002)
43. S.F. Gilbert, D. Epel, *Ecological Developmental Biology* (Sinhauer, Sunderland, Mass, 2009)
44. B. Goldacres, *Bad Science* (Fourth Estate, London, 2009)
45. T. Grandin, C. Johnson, *Animals in Translation* (Scribner, New York, 2005)
46. T.N. Hanh, *The Miracle of Mindfulness: An Introduction to the Practice of Meditation* (Beacon Press, 1999)
47. L.H. Hartwell, J.J. Hopfield, S. Leibler, A.W. Murray, From molecular to modular cell biology. Nature **402**, Supplement C47–C52 (1999)
48. P.G. Hewitt, S. Lyons, J. Suchocki, J. Yeh, *Conceptual Integrated Science* (Addison Wesley, San Francisco, 2007)
49. J.H. Holland, *Adaptation in Natural and Artificial Systems* (MIT Press, Cambridge, Mass, 1992)
50. P. Hoffmann, *Life's Ratchet: How Molecular Machines Extract Order from Chaos* (Basic Books, New York, 2012)

51. L. Jaeger, E.R. Calkins, Downward causation by information control in micro-organisms. Interface Focus **2**, 26–41 (2012)
52. D. Kahneman, *Thinkng, Fast and Slow* (Farra, Straus and Giroux, New York, 2011)
53. E. Kandel, *In Search of Memory: The Emergence of a New Science of Mind* (Norton, New York, 2006)
54. E. Kandel, *The Age of Insight: The Quest to Understand the Unconscious in Art, Mind, and Brain, from Vienna 1900 to the Present* (Random House, 2012)
55. S.D. Krashen, T.D. Terrell, *The Natural Approach: Language Acquisition in the Classroom* (Alemany Press, San Francisco, 1983)
56. T. Lancaster, M. Pexton, Reduction and emergence in the fractional quantum Hall state, in *Studies in History and Philosophy of Modern Physics* (2015) (to be published)
57. R.B. Laughlin, Fractional quantisation. Rev. Mod. Phys. **71**, 863 (2000)
58. J. Le Fanu, *The Rise and Fall of Modern Medicine* (Abacus, 2011)
59. C. List, P. Menzies, Nonreductive physicalism and the limits of the exclusion principle. J. Philos. **106**, 475–502 (2009)
60. G.R. Lyon, J.M. Rumsey (eds.), *Neuroimaging: A Window to the Neurological Foundations of Learning and Behaviour in Children* (Brookes, Baltimore, 1996)
61. J. Marchant, Consider all the evidence on alternative therapies. Nature **526**, 295 (2015)
62. J. Monod, *Chance and Necessity: An Essay on the Natural Philosophy of Modern Biology* (Penguin, 1997)
63. G. McGhee, *Convergent Evolution: Limited Forms Most Beautiful* (MIT Press, Cambridge, Mass, 2011)
64. Committee on a Conceptual Framework for New K-12 Science Education Standards: *A Framework for K-12 Science Education: Practices, Crosscutting Concepts, and Core Ideas* (National Academies Press, Washington, DC, 2001)
65. D. Noble, *The Music of Life: Biology Beyond Genes* (Oxford University Press, Oxford, 2008)
66. D. Noble, A theory of biological relativity: no privileged level of causation. Interface Focus **2**, 55–64 (2012)
67. R. Oerter, *The Theory of Almost Everything: The Standard Model, the Unsung Triumph of Modern Physics* (Plume/Penguin, New York, 2006)
68. J. Panksepp, *Affective Neuroscience: The Foundations of Human and Animal Emotions* (Oxford University Press, London, UK, 1998)
69. A. Peacocke, *The Physical Chemistry of Biological Organization* (Oxford University Press, Oxford, 1989)
70. J. Pearl, *Causality: Models, Reasoning, and Inference* (Cambridge University Press, Cambridge, 2000)
71. R. Penrose, *The Emperor's New Mind: Concerning Computers, Minds and the Laws of Physics* (Oxford University Press, Oxford, 1989)
72. R. Penrose, *The Road to Reality: A Complete Guide to the Laws of the Universe* (Jonathan Cape, London, 2004)
73. R. Penrose, *Cycles of Time: An Extraordinary New View of the Universe* (Knopf, New York, 2011)
74. G.A. Petsko, D. Ringe, *Protein Structure and Function* (Oxford University Press, Oxford, 2009)
75. A.I. Rae, *Reductionism* (Oneworld, London, 2013)
76. D. Rickles, Supervenience and determination. Internet Encyclopaedia Philos. (2006)
77. A. Rosenberg, *The Atheist's Guide to Reality: Enjoying Life without Illusions* (W W Norton, 2012)
78. C. Sagan, *The Demon-Haunted World: Science as a Candle in the Dark* (Random House, New York, 1996)
79. G. Schlosser, G.P. Wagner, *Modularity in Evolution and Development* (University of Chicago Press, Chicago, 2004)
80. E. Schrödinger, *What Is Life? The Physical Aspect of the Living Cell* (The Macmillan Company, 1945)

81. A. Scott, *Stairway to the Mind* (Springer, New York, 1995)
82. W. Seager, *Natural Fabrications: Science, Emergence and Consciousness* (Springer, Heidelberg, 2012)
83. N. Shea, Distinguishing top-down from bottom-up effects, in *Perception and Its Modalities*, ed. by S. Biggs, M. Matthen, D. Stokes (Oxford University Press, Oxford, 2015), pp. 73–91
84. H.A. Simon, *The Sciences of the Artificial* (MIT Press, Cambridge, Mass, 1992)
85. E.M. Sternberg, *The Balance Within: The Science Connecting Health and Emotions* (W H Freeman, New York, 2000)
86. K. Strijkers, Y.N. Yum, J. Grainger, P.J. Holcomb, Early goal-directed top-down influences in the production of speech. Front. Psychol. **2**, 371 (2011)
87. A.S. Tanenbaum, *Structured Computer Organisation* (Prentice Hall, Englewood Cliffs, 2006)
88. R.L. Trask, *Language and Linguistics: The Key Concepts* (Routledge, Abingdon, 2007)
89. R. Van Gulick, Who's in charge here? And who's doing all the work?, in *Mental Causation*, ed. by J. Heil, A. Mele (Oxford University Press, Oxford, 1995)
90. S. Vogel, *Cats' Paws and Catapults: Mechanical Worlds of Nature and People* (W W Norton, 2000)
91. A. Wagner, *The Origins of Evolutionary Innovations* (Oxford University Press, Oxford, 2011)
92. S.I. Walker, L. Cisneros, P.C.W. Davies, Evolutionary transitions and top-down causation, in *Proceedings of Artificial Life XIII* (2012), pp. 283–290. http://arxiv.org/pdf/1207.4808v1.pdf
93. S.I. Walker, L. Cisneros, P.C.W. Davies, The algorithmic origins of life. J. Roy. Soc. Interface **10**, 0869 (2013). http://arxiv.org/pdf/1207.4803.pdf

Chapter 2
Digital Computer Systems

This chapter considers issues of emergence and causation in the case of digital computers, as a warm-up example before giving a general viewpoint on these topics in the next chapter. It will be shown that top-down causation is central to their functioning. It develops its themes as follows:

- Section 2.1 discusses the computational basics underlying the functioning of digital computers.
- Section 2.2 discusses how modular hierarchical structures enable complex higher level behaviour to emerge.
- Section 2.3 sets out the implementation and logical hierarchical structures and makes the case that software drives what happens.
- Section 2.4 discusses how both bottom-up and top-down causation happen in these hierarchies, distinguishing five types of top-down causation that have rather different dynamics.
- Section 2.5 discusses the key feature of equivalence classes that underlies the ontological nature of higher level causal elements. It characterizes in precisely what way computer programs are abstract entities.
- Section 2.6 considers the issue of clearing memory and deleting records: a selection process that leads to the irreversibility of computation. This relates to the fact that infinities cannot occur in physical reality.
- Section 2.7 looks at the nature of causation in the light of all the above, making the case for causal effectiveness of non-physical entities in digital computer systems.

2.1 Computational Basics

Digital computers are the embodiment of algorithmic operation, and are nowadays regarded as a fundamental model of causation. Many claim physics can be regarded as a computational process, and indeed that the universe is a computer [44], computational models are proposed for social life [49], and the computer is often used

G. Ellis, *How Can Physics Underlie the Mind?*, The Frontiers Collection,
DOI 10.1007/978-3-662-49809-5_2

as a model for how the mind works (some say it *is* a computer, others regard that as an analogy [59]). Accordingly, it is useful to consider how issues of emergence and causation work out in this case, so it serves as a model for effects we may see in other contexts, and in particular in the brain, as the computational metaphor does indeed seem to capture some aspects of what is going on in the brain (even though it is inadequate as a total explanation of the embodied mind).

Turing explains the basic idea as follows [64]:

> The idea behind digital computers may be explained by saying that these machines are intended to carry out any operations which could be done by a human computer. The human computer is supposed to be following fixed rules; he has no authority to deviate from them in any detail. We may suppose that these rules are supplied in a book, which is altered whenever he is put on to a new job. He has also an unlimited supply of paper on which he does his calculations. He may also do his multiplications and additions on a 'desk machine', but this is not important.

This gives the essential operational idea of algorithmic operation of a computer, also explained nicely by Hofstadter [34, pp. 33–41]. MacCormick states it thus [45, p. 3]:

> An algorithm is a precise recipe that specifies the exact sequence of steps required to solve a problem.

Turing explains how arbitrary computations can be realized if a digital computer can be regarded as consisting of three parts [64]:

1. **A Store of Information** (**Memory**). This stores data that includes an instruction table stating the rules to be obeyed by the computer (nowadays called a *program*).
2. **An Executive Unit** (**CPU**). This carries out the various individual operations involved in a calculation.
3. **Control** (**Operating System**). This sees that the instructions are obeyed correctly and in the right order.

The key feature leading to flexibility of use of the computer [21, p. 15] is the *stored program*, a set of symbolically encoded instructions in the machine's memory. By altering the program (*software*), the same physical apparatus (*hardware*) can be used to tackle many different kinds of problems. Turing demonstrated [64] that, by this means, a single machine of fixed structure is able to carry out every computation that can be carried out by any computer whatsoever. This is the special property of digital computers, namely [64]:

> They can mimic any discrete-state machine, [and this] is described by saying that they are universal machines. The existence of machines with this property has the important consequence that, considerations of speed apart, it is unnecessary to design various new machines to do various computing processes. They can all be done with one digital computer, suitably programmed for each case. It will be seen that as a consequence of this all digital computers are in a sense equivalent.

This characterizes the key property of programmable computers [14]:

> **Universal Logical Capability**. The nature of the logical operations that digital computers are able to carry out is not constrained by the specific physical implementation chosen; the underlying physics enables the chosen logic rather than controlling it.

But of course, as Turing himself showed, they are nevertheless limited in what they can do [14, 21], and furthermore, these statements do not by themselves show how to organize a computer to achieve complex behavior.

The first major question is how a combination of such simple operations can enable arbitrary complexity of behavior to emerge. We shall see below that a core requirement is:

> **Modular Hierarchical Structuring of Both Hardware and Software.** This enables structured top-down causation in the hierarchy, in particular allowing software patterns to control hardware. In this way, abstract entities have causal effects in the physical universe.

This is the first key to the emergence of true complexity, and is embodied in all current digital computers in both implementation and logical hierarchies (Sect. 2.3). In his textbook on computing, Robert Keller writes [39]:

> An abstraction is an intellectual device to simplify by eliminating factors that are irrelevant to the key idea [...] The idea of levels of abstraction is central to managing complexity of computer systems, both software and hardware. Such systems typically consist of thousands to millions of very small components (words of memory, program statements, logic gates, etc.). To design all components as a single monolith is virtually impossible intellectually. Therefore, it is common instead to view a system as being comprised of a few interacting components, each of which can be understood in terms of its components, and so forth, until the most basic level is reached.

This is particularly clear in the class/object hierarchy of object oriented languages [13].

But a further step is needed: how do we get a computer to carry out computations that are not simply logical implications of what is in the initial data? This is a core requirement on the road towards intelligence: how can we get them to learn? Turing makes a very interesting observation in this regard [64]:

> An interesting variant on the idea of a digital computer is a 'digital computer with a random element'. These have instructions involving the throwing of a die or some equivalent electronic process; one such instruction might for instance be, 'Throw the die and put the resulting number into store 1000.'

While this breaks the system out of a rigidly determined cycle of deterministic operations, by itself this won't do the job of creating intelligent behaviour. But it does open the way to programming computers to behave in an adaptive way. The second key feature needed is:

> **Adaptive Selection.** Procedures embodying adaptive selection in the manipulation of data enable the building up of meaningful information from unstructured incoming data streams or randomized internal variables. This is the basis of learning.

This is a kind of top-down causation that is crucial in enabling computers to carry out processes equivalent to learning (Sect. 2.4.4), e.g., through artificial neural networks [11] and genetic algorithms [23], and so allows local processes to flow against the stream of decay embodied in the Second Law of Thermodynamics.

Associated with this is a further crucial idea, omitted in Turing's list of operations above (because he assumed infinite memory was available):

4. **Emptying Memory**. Clearing out or overwriting short term and long term memory locations so that they can be used again.

For one thing, this enables the finite memory of the computer to act as an effectively infinite memory store, thus taming the impractical memory requirements of an infinite tape. For another, it crucially involves the element of selecting what will be kept and what discarded. As just mentioned, such selection processes are the key to building up meaningful information out of a jumble of incoming data: forgetting is the crucial counterpart of remembering! It is also where irreversibility associated with entropy production happens in computations [43].

2.2 Modular Hierarchical Structures

Digital computers involve two orthogonal but interacting hierarchies (Sect. 2.3). This is not by chance. A major theme of this book is that, as pointed out by Simon [61]:

> Genuine complexity can only emerge from networks of causation involving modular hierarchical structures.

Note that this principle applies to both physical and logical complex systems. Both kinds occur in digital computers (see the next section).

Each word is important: the physical and logical *hierarchies* (Sect. 2.3) are *structured* by means of carefully configured physical and logical connections [47, 62], and each involve interacting *modules* at many levels [30]. The system is composed of inter-related subsystems that have in turn their own subsystems, and so on, until some lowest level of component is reached where the basic work is done. This structure enables a build-up of genuine complexity if appropriately formed to fulfill some higher level function: as in biology, structure follows function. Examples are subroutines, procedures, objects. Each has a name, which identifies the specific entity, and a type, which identifies the class of entities it belongs to.

I consider in turn:

- Structures: Combination and abstraction (Sect. 2.2.1).
- Decomposition and modularity (Sect. 2.2.2).
- Encapsulation and information-hiding (Sect. 2.2.3).
- Naming, combination, and recursion (Sect. 2.2.4).
- Hierarchy: Class structure and object structure (Sect. 2.2.5).
- Evolution (Sect. 2.2.6).

2.2.1 Structures: Combination and Abstraction

In digital computer languages, as explained by Abelson and Sussman [1, p. 4], structures are formed in the following way:

> Every powerful language has three mechanisms for combining simple ideas to form more complex ideas:

- **Primitive Expressions**. These represent the simplest entities the language is concerned with.
- **Means of Combination**. These serve to build up compound elements from simpler ones.
- **Means of Abstraction**. These serve to name compound elements and are manipulated as units.

A key element here is naming compound entities, indexing them, and having rules about how they can interact:

- **Names**. Any named entity is identified as a potentially causally effective agent, whether it is physical or abstract. Indeed, any entity that is causally effective in a programme has to be given a name so that it can be referred to (Sect. 2.2.4). The name must have attributes identifying whether they refer to objects or processes or something else.
- **Indexes/Pointers**. It then also has to have an index or pointer that shows where the relevant records are stored in memory.
- **Logical Rules**. Abstract rules can then be applied to sets of named entities, these rules embodying the logic of their interactions, and which processes can interact with each object.
- **Action Rules**. Action rules can be signified by the named entity (e.g., print text.pdf).

In the end these are the abstract technologies that enable computation to function. They are causally effective because they result in the computer being able to operate. The foundational key is the ability to give names to recognisable entities—generic (hence necessarily abstract) and specific (whether physical or abstract).

Emergence. Such combinations of parts lead to the higher level functionality of a complex logical system. The behaviour of the whole is greater than the sum of its parts, and cannot even be described in terms of the language that applies to the parts. This is the phenomenon of *emergent order*: the higher levels exhibit kinds of behaviour that are more complex than those the lower level parts are capable of.

In the implementation hierarchy, much the same applies. Emergence of layers of structure and behaviour, one upon the other, lead to hierarchical structuring and this enables a build-up of higher level entities that can be characterised by abstract properties. Not only is the structure hierarchic, but the levels of this hierarchy represent different levels of abstraction, each built upon the other, and each understandable by itself (and each characterised by a different phenomenology).

Phenomenology. All parts at the same level of abstraction interact in a well-defined way, whence they have a causal reality at their own level, and each is represented in a different language describing and characterising the causal patterns at work at that level [62], which may entail their own logical hierarchies. The vocabulary to describe each of the levels in each hierarchy is different at each level, because the nature of the relevant entities at each level is quite different from that at the levels above and below.

2.2.2 Decomposition and Modularity

A hierarchy represents a decomposition of the problem into constituent parts and processes to handle those constituent parts, each requiring less data and processing, and more restricted operations than the problem as a whole [12]. The success of hierarchical structuring depends on (i) implementation of modules which handle these lower-level processes, such as the CPU and memory circuits and interconnections between them, (ii) integration of these modules into a higher-level structure, viz., the computer as a whole. The idea is to encapsulate functions in modular units with information-hiding and abstraction, so that named entities can be regarded as functional wholes whose internal functioning is hidden from the outside view. I closely follow Booch's excellent exposition of object-oriented analysis [12], together with Beer's exposition of the principles of decentralized control [7].

Modularity [12, pp. 12–13, 54–59]. The technique of mastering complexity in computer systems and in life is to decompose the problem into smaller and smaller parts, each of which we may then refine independently [12, p. 16]. The basic principle is

> **Divide and Conquer**. Divide a complex overall task into many simpler subtasks, each requiring lesser data and computational power than the whole; then integrate the results so as to attain higher level cohesive behaviour, thus creating complex outcomes.

By organising the problem into smaller parts, we break the informational bottleneck on the amount of information that has to be received, processed, and remembered at each step; and this also allows specialisation of operation. This implies the creation of a set of specialised modules to handle the smaller problems that together comprise the whole: in computer systems these will be subroutines, which Turing called 'subsidiary tables'.

According to Abelson and Sussman, one breaks up a complex problem into subproblems, each accomplished by a separate procedure. The program used can be viewed as a cluster of procedures that mirror the decomposition of the problem into subproblems [1, p. 26]:

> The importance of this decomposition strategy is not simply that one is dividing the program into parts. After all, one could take any large program and divide it into parts—the first ten lines, the next ten lines, and so on. Rather it is crucial that each procedure accomplishes an identifiable task that can be used as a module in defining other procedures.

They emphasize that when developing a program like this, we are not initially concerned with how the procedure computes its result, only with the fact that it does so. The details of how this is done can be held back until later on. Thus one actually deals with a procedural abstraction. At this level, any procedure that computes the desired output will do. This is the principle of equivalence classes, showing that top-down causation is taking place (see Sect. 2.5). Intra-component linkages are generally stronger than inter-component linkages. This fact has the effect of separating the high frequency dynamics of the components, involving their internal structure, from the low-frequency dynamics, involving interactions amongst components [61] (and it is for this reason that we can sensibly identify the components).

A further basic principle is that this allows one to:

Adapt and Re-Use [61]. In building complex systems from simple ones, or improving an already complex system, one can re-use the same modular components in new combinations, or substitute new, more efficient components, with the same functionality, for old ones.

Thus we can benefit from a library of tried and trusted components. Complex structures are made of modular units with abstraction, encapsulation, and inheritance, and this enables the modification of modules and re-use for other purposes (Sect. 2.2.6).

2.2.3 Encapsulation and Information-Hiding

Named objects carry with them expectations of behaviour that identify abstractions: specific essential characteristics of the object (ignoring other properties as inessential).

Abstraction and Labeling [12, pp. 20, 41–48]. Unable to master the entirety of a complex object, we choose to ignore its inessential details, dealing instead with a generalised idealised model of the object. An abstraction denotes the essential characteristics of an object that distinguishes it from all other kinds of objects. An abstraction focuses on the outside view of the object, and so serves to separate its essential behaviour from its implementation. It emphasises some of the system's details or properties, while suppressing others. A key feature is that compound objects can be named and treated as units (Sect. 2.2.4). This leads to the power of abstract symbolism and symbolic computation.

Encapsulation and Information-Hiding [12, pp. 49–54] **and** [57, pp. 233–234, 476–483]. In a hierarchy, through encapsulation, objects at one level of abstraction are shielded from implementation details of lower levels of abstraction. Consumers of services only specify what is to be done, leaving it to the object to decide how to do it: "No part of any complex system should depend on the internal details of any other part." Encapsulation occurs when the internal workings are hidden from the outside, so its procedures can be treated as black-box abstractions. To embody this, each class of object must have two parts: an *interface* (its outside view, encompassing an abstraction of the common behaviour of all instances of the class of objects) and an *implementation* (the internal representations and mechanisms that achieve the desired behaviour). This is formalised in declarations of public and private variables. Efficiency and usability introduce the aim of reducing the number of variables and names that are visible at the interface. This involves information-hiding, corresponding to coarse-graining in physics. The accompanying loss of detailed information is the essential source of entropy in the case of physics.

2.2.4 Naming, Combination, and Recursion

When names can be allocated to collections of names, this allows recursion, which is how real complexity gets built up (languages explicitly allow it). Indeed this is the power of *symbolic representation*: the name is a symbol for the thing it represents; and one can give names to patterns of names.

Naming. The key feature in setting up modules is first to identify them as entities by naming them: both classes, with generic features, and particular objects, with specific features, and then to refer to them by that name (an *identifier*) [57, pp. 45–46,78–80]. Associated with the name is a set of attributes that characterise the object:

- A state embodied in internal variables of specific types (the *arguments*).
- A set of characteristic behaviours that characterise how it can interact with other objects (the *methods*). These are the law-like rules of behaviour that outline the nature of the object and create ordered outcomes.
- An indexed storage location, allowing programs to access this information (involving *pointers*).

The names are referenced in an index, enabling one to locate the physical location of the items referenced by the name, and pointers enable storage in non-contiguous memory locations. Each segment of a stored item must have a clear start address and end address, as well as links to any further parts of the same stored item or memory. So objects have a state, behaviour, and identity [12, pp. 81–97].

Typing and Links. Each object has a type, that is, a precise characterisation of its structural or behavioural properties shared by a collection of entities [12, pp. 65–72]. This includes the scope of its name, i.e., whether it has global validity, or is only valid in some local context. Its possible interactions with other objects are characterised by links between objects [12, pp. 98–102]. Object diagrams show the existence of objects and their relationships in the logical design of a system [12, pp. 208–219].

Coding and Information. From a functional viewpoint, one is involved in *coding* a message from a sender to a receiver. Use of a code involves *two* pattern recognition mechanisms: one for translating an incoming message into the code, followed by some kind of transformation of the coded message, and then one further pattern recognition system for decoding the output message into a usable form that will have the desired effect. From the viewpoint of the information involved, typing also includes the rules an object has to obey, that is, the *syntax* of its allowed usage. The further aspects of a symbolic system are *semantics* (the meaning they embody) and *pragmatics* (the effect they have within the context of their usage). These aspects relate to the logical and physical effects of symbolic usage.

Collective Names. One can give a name to any pattern of symbols, including a collection of names. This is what enables one to create classification hierarchies and complex sentence structures, because one can refer to complex entities through a single name. This is also a way of reducing complexity: one does not have to deal with the details, but just with aggregate behaviour. The minimal program to solve some problem is reduced from thousands of lines of code to 'run prog.exe'. The algorithmic complexity is thereby dramatically reduced.

Combination. Given names, they can be combined in grammatical structures indicating relationships between named entities via named operations. Collections of names can be treated as single entities (phrases function as effective words). This is the key property that enables construction of hierarchical structures, i.e., structures made up of parts that are themselves made up of parts, and so on. This is a core feature of natural language [63].

Recursion. This kind of structure enables one to repeatedly call up the same named entity, nesting structures inside each other. In functional terms, the essential requirement is an operation for combining data objects such that the results of the operation can themselves be combined using the same operation. This *closure property*, for example, underlies the importance of the list structure as a representational tool in LISP [1, p. 98]. When the data object is itself an operation, this enables recursion, that is, an operation or evaluative rule that includes as one of its steps the need to invoke the rule itself [1, pp. 9, 31–42].

2.2.5 Hierarchy: Class Structure and Object Structure

The power of a class hierarchy comes from the fact that it shows the relationships between similar kinds of objects, i.e., which are generalizations of others, and which are specializations. It allows one to relate them by inheritance, a feature which often characterises the nature of the hierarchical structure (see [12, pp. 59–65] and [57, pp. 453–476, 484–494]). Thus we don't have to memorize separately all the properties of each kind of object or action: we can relate them to similar kinds of objects, remembering the class structure as a whole, and then the similarities and differences of specific members of the class (animal, mammal, dog, Dachshund, Fred). One then uses this to relate the properties of specific instances to the generic properties of a class.

To accommodate this in an object-oriented approach, objects occur in hierarchical functional classes, with inheritance of properties modified by specialization and variation. This class structure is related to the object structure because each object in the object structure is a specific instance of some class [12, pp. 14–15]. Together these form the logical model. In a list-based language, one has a hierarchy of types [1, pp. 197–199]. In both cases, crosslinks may be allowed (reflecting the fact that various hierarchies are in operation, as emphasized above).

A **class** is a set of objects that share a common structure and a common behaviour [12, pp. 103–106]. The structures chosen to define a class depend on the classification scheme used: they embody a view of the world [12, pp. 145–168]. A single object is an instance of a class. Classes are objects that can themselves be manipulated as an entity. A *metaclass* is a class whose instances are themselves classes [12, pp. 133–134], so we can have a hierarchy of classes, and class families [12, pp. 337–340]. *Class diagrams* show the existence of classes and their relationships in a logical view of a system [12, pp. 176–196], and these relationships are formalised in class specifications [12, pp. 196–199], stating their name, responsibilities, attributes, operations, and constraints. *State transition diagrams* show the state space of a class, the events that cause a state change, and the actions that result from such a change [12, pp. 199–208]. *Module diagrams* show the allocations of classes and objects to modules in the physical design of the system [12, pp. 219–223]. *Process diagrams* show the allocation of processes to processors in the physical design of the system [12, pp. 223–226].

The dual hierarchical relations are *aggregation*, denoting which whole is made of which parts [12, pp. 128–130], and *membership*, denoting which parts belong to which whole. Aggregation may or may not imply physical containment: it may just imply a conceptual whole/part relationship [12, pp. 102–103].

Inheritance [12, pp. 59–62, 107–128] **and** [57, pp. 453–476]. This is the most important feature of a classification hierarchy. It allows an object class, such as a set of modules, to inherit all the properties of its superclass, and to add further properties to them (it is a 'is a' hierarchy). This allows similarities to be described in one central place and then applied to all the objects in the class and in subclasses. It makes explicit the nature of the hierarchy of objects and classes in a system, and implements generalisation/specialisation of features (the superclass represents generalised abstractions, and subclasses represent specializations in which variables and behaviours are added, modified, or even hidden). Inheritance with exceptions enables us to understand something as a modification of something already familiar, saves unnecessary repetition of descriptions or properties, and allows nonmonotonic reasoning [48].

Patterns. Particular types of *structural patterns* recur and are worth identifying and codifying in structural classes. They include lists [68, pp. 56–75], stacks [68, pp. 75–88], queues [68, pp. 88–98] and priority queues (heaps) [68, pp. 183–224], trees [68, pp. 99–153], graphs [68, pp. 291–352], and relational databases. Similarly, particular kinds of operations often occur and are worth identifying and naming. These include *date/time operations* and *filters*, i.e., input, process, and output transformations [12, pp. 331–332], *pattern matching*, i.e., operations for searching for structured sequences within sequences [12, pp. 370–372], *searching*, i.e., operations for searching for items within structures, *sorting*, i.e., operations for ordering structures, *utilities*, i.e., common composite operations building on more primitive operations, e.g., iteration [12, pp. 355–360] and statistical analysis.

The point here is that each of these structures and operations are metaclasses that can be identified and given their own name. They then exist, in virtue of this recognition, as entities in their own right that can from now on be accorded an ontological status as effective entities. They are abstract patterns that are causally effective. They are multiply realisable at a detailed level, and hence show some form of top-down causation or influence.

At a higher level, structural patterns that often occur in modular hierarchical structures can be encoded in *design patterns* that name, explain, and evaluate important recurring designs in object-oriented systems [31, pp. 2–3]. They are:

- **Creational Patterns**. Abstract Factory, Builder, Factory Method, Prototype, and Singleton.
- **Structural Patterns**. Adapter, Bridge, Composite, Decorator, Facade, Flyweight, and Proxy.
- **Behavioural Patterns**. Chain of Responsibility, Command, Interpreter, Iterator, Mediator, Memento, Observer, State, Strategy, Template Method, and Visitor.

These are discussed in depth in [31]. They are based on foundation classes List, Iterator, ListIterator, Point, and Rectangle, with operations for construction, destruction, initialisation, and assignment of lists, and for accessing, adding, and removing elements of a list.

2.2.6 Evolution

Modularity underlies the possibility of successful development of truly complex systems [61]. One can adapt working modules for different purposes, without having to start from scratch. Selection of the most successful small variations of such classes enables incremental increase of complexity without the whole system crashing. Booch quotes Gall as follows [12, p. 13]:

> A complex system that works is inevitably found to have evolved from a simple system that worked [...] A complex system designed from scratch never works, and cannot be patched up to make it work. You have to start over, beginning with a simple system.

This is an example of adaptive selection, a crucial form of top-down causation, which is the topic of the next section.

2.3 Orthogonal Modular Hierarchical Structures

Digital computers are hierarchically structured modular systems on both the hardware and software sides. Actually, there are two orthogonal kinds of hierarchies. I discuss:

- The two kinds of hierarchies (Sect. 2.3.1).
- The implementation (vertical) hierarchies (Sect. 2.3.2).

- The logical (horizontal) hierarchies (Sect. 2.3.3).
- The relation between the two hierarchies (Sect. 2.3.4).
- Causality in the hierarchies (Sect. 2.3.5).

2.3.1 *The Two Kinds of Hierarchies*

Firstly there are *implementation hierarchies* [62], which one might call *vertical hierarchies*, whereby the logical operations of the computer are implemented. For example, digital computers are constructed of integrated circuits containing a Central Processing Unit (CPU) which in turn contains an Arithmetic Logic Unit (ALU) made of many interconnected transistors, diodes, resistors, and capacitors, each comprised of an atomic lattice infused with electrons. The higher level physical structures are emergent entities, made up of the interconnected lower level components, but each describable and functioning effectively at its own emergent level. Related to these physical components is a software hierarchy: a tower of virtual machines that implement higher level programming languages at each virtual machine level, on the basis of the underlying machine code. An example is the Java Virtual Machine (JVM). Each of these virtual machines is emergent from the one on the next lower level.

Then there are the *logical hierarchies* [12], which one might call *horizontal hierarchies*, and which exist at each higher level of the virtual machine hierarchy. High level programs contain subroutines comprised of procedures set out in program lines which relate the relevant individual operations and variables. They thus represent a hierarchical structure of operations. These programs thereby also implement hierarchical data structures, e.g., a word-processer may edit a book consisting of chapters, paragraphs, sentences, phrases, words, and letters represented as such in a word-processing program. Both the logical and the data structures cascade down the implementation hierarchy through interpreters and compilers, which translate them into combinations of lower level operations and data elements [3, 6].

At the lowest abstract implementation level, both the data and programs will be represented as strings of 0s and 1s, realised as structured electronic states in the underlying physical level. A key feature is that many different implementation hierarchies can be used to realise the same logical hierarchy. The computational process itself is indifferent as to how it is realised at the physical level. This is a core aspect of top-down causation (Sect. 2.5).

2.3.2 *The Implementation (Vertical) Hierarchies*

As regards the implementation hierarchy [62], it has hardware and software aspects. First there is the *hardware hierarchy* shown in Table 2.1. It is modular because a network of many similar identifiable lower level elements such as logic circuits and transistors underlies each of the higher level structures.

Table 2.1 The hardware hierarchy for digital computers. The computer scientist takes level 1 as the base level. However, it is based on the underlying physics hierarchy, which enables its functioning. Layers below level 0 can be taken for granted by a computer engineer: it is the base level he needs to consider

Level 7	Global network
Level 6	Local network
Level 5	Computer
Level 4	Motherboard, memory banks
Level 3	CPU, memory circuits
Level 2	ALU, primary memory, bus
Level 1	Logic circuits, registers
Level 0	Transistors, resistors, capacitors
Level −1	*Atomic physics*
Level −2	*Nuclear physics*
Level −3	*Particle physics*
Level −4	*Fundamental theory*

The lowest level, i.e., the level where the real physical work is done, is the physical base level (level −4), which is some form of fundamental physics, possibly related to quantum gravity. But we do not know what the relevant physics is, so we cannot reduce the higher level actions to lowest level actions inter alia, because the lowest level is unknown. Of necessity, for practical purposes, we have to take one of the emergent effective levels of physical operation as the base level, assuming it to exist and be real [25]. For computer scientists, this is level 1 (the gate level); for computer engineers, it is level 0 (the transistor and solid-state physics level) which can be regarded as the level where the physical work is done (see Sect. 2.7.6).[1]

However, hardware by itself will do nothing: it needs software in order to run. The *software hierarchy* is shown in Table 2.2. There is a tight logical structure at each level, governed by a set of syntactic rules for the language used at that level, and with an associated set of variables defined for that language, with typing and scoping rules. Each higher level language is emergent from the next lower level language through the way the higher level variables and operations are defined in terms of the lower level variables and operations. The magic that makes this happen is compilers and interpreters [3, 6], the foundation of truly complex functioning in computers.

The relation between level 0 and level 1 is where an appropriate physical representation of variables (in digital computers, electronic states) gives rise to a set of simple logical operations on those variables [47].

[1]For a hardcore reductionist, it is illegitimate to regard these levels as real: they are epiphenomena arising from the underlying physics. This viewpoint provides no useful understanding of the causation in action.

Table 2.2 The software hierarchy for digital computers, based on the physics of the transistors at the device level. From Tanenbaum [62]

Level 7	Applications programs	Data and operations
Level 6	Problem-oriented language level	Classes, objects
Level 5	Assembly language level	Symbolic names
Level 4	Operating system machine level	Virtual memory, paging
Level 3	Instruction set architecture level	Machine language
Level 2	Microarchitecture level	Microprograms
Level 1	Digital logic level	Gates, registers
Level 0	*Device level*	*Transistors, connectors*

Principle C1. Information is not causally effective unless it has a physical representation, and some handles whereby this representation can (i) be inserted, altered, or deleted, and (ii) be read. The relation between levels 0 and 1 is where this happens.

Virtual Machines. A key point then is that Table 2.2 represents virtual machines at every level, except the lowest (level 0). Each of them runs on top of the next lower level virtual machine [62] (see Table 2.3). The lowest level is shown as level 0, which is of a completely different character to the others: it is physically based. The relation between level 0 and level 1 is where the transition between physical and abstract causation takes place: virtual machines (level 1 up) are based on real physical entities at the bottom (level 0).

Principle C2. All the higher levels in the software hierarchy are virtual machines. They are not physical systems.

Table 2.3 A multilevel machine. From Tanenbaum [62, p. 4]

Level *n*	Virtual machine ***Mn***	Machine language ***Ln***
	⋮	⋮
Level 3	Virtual machine ***M3***	Machine language ***L3***
Level 2	Virtual machine ***M2***	Machine language ***L2***
Level 1	Virtual machine ***M1***	Machine language ***L1***
Level 0	Actual computer ***M0***	Machine language ***L0***

Table 2.4 The logical hierarchy that determines which operations happen when

Level 7	Design patterns	Data structures	Methods of argumentation
Level 6	Programs	Classes, methods	Overall purpose
Level 5	Subroutines	Objects	Steps to overall purpose
Level 4	Algorithms	Data records	Implementation methods
Level 3	Program lines	Data items	Implementation units
Level 2	Operations	Variables	Entities interacting
Level 1	Representation atoms	Entity components	Logical base level

2.3.3 The Logical (Horizontal) Hierarchies

The *logic hierarchy* (Table 2.4) structures what happens at each level of the software hierarchy, in any specific class of applications [12]. Associated with it is a *data hierarchy* as shown here, which gives the specific data related to a specific class of applications.

The key thing here is the algorithms that act on the data, specifying precisely what operation is to be performed. As explained by Turing (see the quotes in Sect. 2.1), it is these algorithms that shape the computation. What also matters then is the order in which they are implemented, and on what specific data, something which is determined by the operational context of the program as a whole, which implements the computational logic used to tackle the problem of interest.

This generic logical structure enables the logic of any specific application domain to be represented by specific variables and associated operations. Thus one might be engaged in word-processing, numerical calculations, digital image manipulation, digital sound operations, computer-aided design, and so on. A specific high level language will enable modeling of each such domain, representing the hierarchical relations of its specific structure and appropriate operations on them. For example, in the case of word-processing, one might have the data structure in Table 2.5.

The word-processor program enables insertion, edition, deletion, copying, and pasting at any level of the data hierarchy, thus enabling manipulation of the parts (words), their components (letters), and their integration into higher order entities

Table 2.5 The hierarchical structure of specific applications

Book	Specific purpose
Chapters	Major themes
Paragraphs	Subthemes
Sentences	Logical units
Phrases	Logical subunits
Words	Representational variables
Phonemes	Variable components
Letters	Logical atoms
Binary code	Digital representation

such as paragraphs and chapters. It also allows detailed formatting of the resulting text. Because of the power of language, this will enable representation of anything humans can think about, and the associated logic of that domain (science, art, philosophy, whatever).

Application Domain Logic. Word processors, music programs, image manipulation programs and so on handle general classes of data in an appropriate way: all that is required is data entry, storage, recall, editing, and deletion, with appropriate application programs and hardware to output the result. But there are many application domains with their own specific logic: mathematics, engineering, environmental issues, ecological modelling, computer-aided design, statistical data analysis, and so on. Examples chosen at random can be found in [15, 55].

These are logical hierarchies which apply generically to a class of applications. Finally, there is the *systems hierarchy* (Table 2.6) showing how these hierarchies relate to each other whenever an application program is utilised in a specific operational context. This is where *systems analysis* [9, 12] comes in: structuring the hardware, programmes, and data to suitably model some specific real world problem that needs to be solved. Design patterns [31] characterize the high level possibility structures. Thus the program structures and data model the logic of many application areas. A key issue is where these multifold logical structures come from. I consider this in Sect. 2.7.5.

2.3.4 The Relation Between the Two Hierarchies

How do the implementation and logical hierarchies relate to each other? When we load and then run a high level program, these input operations take place at the uppermost level of the implementation hierarchy (Table 2.2), representing the problem logic at that level. Compilers or interpreters [3, 6, 67] then cause all the lower implementation levels to spring into action in accord with the logic and data that has been loaded at the top level. When this occurs, the same hierarchical logical structure is represented at each of the levels of the implementation hierarchy, written in a different language at each level, using quite different kinds of commands and data representation. The operating system orchestrates the way this happens [60].

Table 2.6 The systems hierarchy: the flow of causation when tackling a specific problem. It is driven by the nature of the user's problem, which gets translated into a computer application analysing specific data according to the internal logic of the problem

User level	Specific purpose	Goal
Logical level	Problem structure	⇓
Programme level	Particular programmes	⇓
Data level	Specific data	⇓
Physics level	Hardware	Electrons

Thus the logical hierarchy of Table 2.4, as related to the specific problem at hand, recurs at each of the implementation levels of Table 2.2 in different forms. At the lower levels they are based on a set of simple logical operations, and it is very difficult to see the higher level logic which is shaping what is happening. For example:

- A word processor has high-level commands for insert, delete, copy, paste, italic, bold, change font, and so on. These are the user interface commands.
- An underlying Java Virtual Machine has instructions for the following groups of tasks: load and store, arithmetic, type conversion, object creation and manipulation, operand stack management (push/pop), control transfer (branching), method invocation and return, throwing exceptions, and monitor-based concurrency. The word-processing requirements are implemented in terms of these operations.
- These are implemented in the assembly language by commands such as MOV AL, 61h [Load AL with 97 decimal (61 hex)].
- These are implemented at the machine code level by commands such as 000000 00001 00010 00110 00000 100000 [add the contents of registers 1 and 2 and place the result in register 6].

When the machine code is run, it implements the commands at the binary level, and that induces more complex higher level commands at each higher level, thus causing the desired emergent behaviour. Why does it happen that the desired behaviour emerges? Because the system has been set up to ensure that this will be so!

Each higher level behaviour emerges from the lower level ones. But what ultimately determines what happens? The higher levels drive the lower levels. First, compilers or interpreters [3, 6] translate the higher level languages to the lower level languages. Then the lower levels implement the compiled program in a purely mechanistic bottom-up way and the desired higher level behaviour emerges from the combination of lower level operations. But those lower level states would not be there if they had not previously been determined top-down by the process of compiling a set of algorithms written in a higher level programming language. Their logic determines what happens.

Principle C3. The software drives the hardware. What specific physical interactions take place at the hardware level is controlled by specific data entered, in accord with the logic of the relevant algorithms.

It is this logical structure that is the key causal element in the sense of determining what happens next at each instant. Physical interactions in the computer are controlled by the logic of the algorithms. For example one might have an accounting system, in which case the logic of accounting systems determines what happens, or one might be modeling a chemical engineering system, in which case the logic of chemical interactions drives the system. At the lower levels, the logic of operations such as copying, deleting, and sorting determines what happens. Algorithms such as Quicksort replace physics equations as the driving logic.

The specific physical realisation is what enables this to work, but a different realisation could have been used. The essential nature of the program driving the computer is the equivalence class of all such functionally equivalent realisations (see Sect. 2.5).

2.3.5 *Causality in the Hierarchies*

Software determines what specific currents flow where and when in the hardware circuits, implementing the specific abstract logic that applies to the issue at hand. This logic is coded in a hierarchical fashion through the program and its subroutines or procedures, which embody its modular structure (Sect. 2.2).

The underlying abstract high level logic, for example, that of data compression or numerical analysis or pattern recognition, shapes the algorithms used. This determines what happens at the lower levels. This is clearly top-down causation from the higher to the lower levels. It will be explored further in Sect. 2.4. The specific outcome depends on the data supplied, which has to be hierarchically structured as required by the applications software.

Software *S* is not a physical thing, neither is data. They are realised, or instantiated, as energetic states in computer memory. The essence of software does not reside in their physical nature: it is the *patterns* of states, instantiated by electrons being in particular places at a particular time, that matters. These are not the same as those electrons themselves (just as a story is not the same as the paper on which it is written). Given the set of connections in the CPU, the pattern of electrons represents the logical structure of the program.

Programs and data together determine what specific electrical operations take place in the transistors and other physical components (level 0) in the chosen hardware, which is the context within which the software is causally effective. Thus the conclusion is:

Causal Effectiveness of Non-Physical Entities. In digital computers, non-physical entities control the behaviour of physical systems.

This will be explored further in Sects. 2.5 and 2.7.

2.4 Bottom-Up and Top-Down Causation

True complexity emerges through the interplay of bottom-up and top-down effects in the hierarchies of structure and causation [26, 27].

Bottom-Up Action. A fundamental feature of the structural hierarchy in the physical world is bottom-up action: what happens at each higher level is based on causal functioning at the level below, so what happens at the highest level is based on what happens at the bottommost level. This is the profound basis for reductionist world views. The successive levels of order entail chemistry being based on physics, material science on both physics and chemistry, geology on material science, and so on. In the case of computers, such bottom-up action is the basis of the emergence of high level languages and applications from the underlying physical and logical components. However, this only takes place once the scene has been set by processes that design the structure and so channel the lower level interactions.

Top-Down Action. The feature complementary to bottom-up action is top-down action. This occurs when the higher levels of the hierarchy direct what happens at the lower levels in a coordinated way [28]. For example, pressing a computer key leads to numerous electrons systematically flowing in specific gates and so illuminating specific photodiodes in a screen.

Generically, specifying the upper state (for example, by pressing a computer key) results in some lower level state that realises this higher level state, and then consequent lower level dynamics ensues to produce a new lower level state in a way that depends on the boundary conditions and structure of the system. The lower level action would be different if the higher level state were different. It is both convenient and causally illuminating to call this top-down action, and to represent it explicitly as an aspect of physical causation. This emphasizes how the lower level changes are constrained and guided by structures that are only meaningful in terms of a higher level description.

There are five different types of top-down causation (TD1–TD5) in the logical hierarchies, with differing characteristics. The following subsections consider them in turn. I look successively at:

- The combination of bottom-up and top-down action (Sect. 2.4.1).
- TD1: Deterministic top-down processes (Sect. 2.4.2).
- TD2: Non-adaptive feedback control systems (Sect. 2.4.3).
- TD3: Adaptive selection (Sect. 2.4.4).
- TD4: Feedback control with adaptive goals (Sect. 2.4.5).
- TD5: Adaptive selection of adaptive goals (Sect. 2.4.6).
- Goals and learning in relation to these kinds of causation (Sect. 2.4.7).

2.4.1 The Combination of Bottom-Up and Top-Down Action

In the implementation hierarchy, algorithmic processes in the bottom layers enable what happens, through suitably structured electronic interactions at the machine level combining to create the emergent stack of virtual machines (Table 2.3). But top-down control determines what happens according to the logic of the high level programs that happen to be running (music programs, image manipulation, numerical analysis, pattern recognition, or whatever). The mechanisms enabling this to happen are compilers and interpreters, as explained in Sect. 2.3.4: they transfer the application logic down from the higher to the lower implementation levels, which are all virtual machines except for the bottommost level (Table 2.2). At that level, this logic is represented as patterns of electronic excitations.

This combination of bottom-up and top-down actions enables complex higher level behaviour to emerge from simpler lower level processes, which are orchestrated from above by entering suitable data at the keyboard. That action directly alters specific memory registers, which either contain data for the program, or instructions as to what should happen next, as in Turing's description (see Sect. 2.1). Which it is

Table 2.7 The hierarchy in data communications

Level 7	Application	Message, HTTP/ SMTP /FTP
Level 6	Presentation	Data compression/encryption
Level 5	Session	Data delimitation, synchronisation
Level 4	Transport	Segments, TCP
Level 3	Network	Datagrams, IP
Level 2	Link	Frames, Ethernet/WiFi/PPP
Level 1	Physical	Individual bits, protocols

Table 2.8 Bottom-up and top-down action in the hierarchy of data communications

Source		Destination
Level 7 Application	⇒ ⇒ ⇒ ⇒ ⇒	Application
Level 6 ⇓ Presentation		⇑ Presentation
Level 5 ⇓ Session		⇑ Session
Level 4 ⇓ Transport		⇑ Transport
Level 3 ⇓ Network	Routers	⇑ Network
Level 2 ⇓ Link	Link layer switch	⇑ Link
Level 1 Physical ⇒ ⇒	⇒ Cable/wireless ⇒	Physical

depends on the context of the physical system and the pattern of excitations in all the other gates that embodies the system logic.

Emergence of Same-Level Action. The emergence of same-level action through this combination of bottom-up and top-down effects is particularly clear in the case of computer networking [40]. The internet protocol stack/OSI model is shown in Table 2.7 (levels 5 and 6 are in the OSI model). Sending a message from the source to the receiver, the process is top-down at the source: the message gets sent down from level 7 to level 1, the representation being transformed on the way down from alphabetic at level 7 to binary at level 1, and also split into packets with headers and tailers. It is sent in this form to the receiver.

A reverse bottom-up process takes place at the destination: the binary digital level 1 form gets transformed to a properly formatted output form at level 7. Encapsulation takes place: extra information is added at each level on the way down, and stripped on the way up [40]. Thus the result (Table 2.8) is effective same-level action: the message sent by the source is received in the same form at the destination. This is a good model of how same-level action emerges in general from a combination of top-down and bottom-up action.

2.4.2 TD1: Deterministic Top-Down Processes

In deterministic processes in a computer, the outcome is uniquely determined by initial and structural conditions. Data must be chosen to respect the logical conditions specifying legal data and item length limits, but it can vary arbitrarily within those

Table 2.9 The basic features of deterministic causation in a digital computer. Given the context (structural conditions and data constraints), the initial data leads to a unique final state if the calculation halts

Context	Date typing	Structural conditions
Constraints	$\Downarrow$	$\Downarrow$
Data	Initial data	Computation
Closed system	$\Downarrow$	$\Downarrow$
Outcome	Final state (deterministic)	

constraints. In fact, it varies between different runs, but is fixed for each run, and it cannot change once the run has started. That is why the dynamics is deterministic: the outcome is fixed by the input, with no uncertainty, providing the calculation halts. This is top-down causation, because the outcome depends on the context: alternative higher level states (structuring or input data) lead to different outcomes (Table 2.9).

Such computations simulate many different aspects of reality, e.g., predictions of stock control in a shop or factory, nucleosynthesis in stars, aircraft paths, weather patterns, future activity in the stock exchange. Initial data plus the algorithm determines the outcome at each time step $t_{i+1} = t_i + \Delta t$: for a system of variables $y_j(t)$,

$$y_j(t_{i+1}) = f_j\big(y_1(t_i), \ldots, y_N(t_i), \Delta t\big) , \tag{2.1}$$

there is no uncertainty in the model. But there are rounding errors affecting the outcome if it involves continuous variables, depending on the size of the step Δt. The stability of the outcome depends on whether the system modelled is stable or chaotic: attractors stabilize outcomes, strange attractors destabilize. In simulating real world systems such as stock in a shop, an aircraft in flight, or the weather, one can make the model more accurate by updating the data on an ongoing basis. Then the outcome is no longer a unique outcome of the initial data. This is the route to feedback control (TD2), discussed below (Table 2.10).

Random Initial Data. An interesting twist is to add in a random number generator to vary initial data. One uses a random seed to initialize the generator. This is a new number, unrelated to the problem domain, e.g., the time of the start of the program, or data from the weather or the atmosphere. It is chosen separately for each run and is then fixed for that run. This enables Monte Carlo simulations by choosing a whole series of runs where the seed is varied randomly but the rest of the data is fixed.

This can simulate a set of objects with varying unknown properties. Each run is deterministic, but the overall run is not. However, it shows statistical trends, which are then themselves deterministic at a higher level. These are emergent properties of

Table 2.10 Randomness and determinacy in statistical investigation

Statistical description	Statistical laws	Deterministic
$\Uparrow$	*Coarse grain* $\Uparrow$	
Ensemble	Many cases	*Random*
$\Uparrow$	*Repeat with variation* $\Uparrow$	
Individual case	Dynamics	Deterministic

an ensemble of individual lower level systems. But in the bigger scheme of things, this is still deterministic: the random number generator is not random if we take into account causal processes in the environment that determine the seed.

Indeterministic Processes. There is, however, the possibility of introducing genuine randomness into algorithmic computational systems. Here one uses quantum uncertainty to generate the seed: detection of radiation resulting from radioactive decay of atoms is used to generate a random number.[2] Then it is truly random: there is no cause for its value, provided that standard quantum theory is correct (see Sect. 6.1). The specific result of each run is not predictable from the initial data, although it must lie in the possibility space set by the algorithms. Thus it is algorithmic, but not deterministic: it is not mechanistic in the classical sense.

2.4.3 TD2: Non-adaptive Feedback Control Systems

By contrast, goals are the essence of feedback control systems. Non-adaptive control systems compare the actual present state of the system with a desired goal and feed information back to a controller to correct the system state (see Table 2.11). This is the essence of cybernetics: feedback control corrects any error in the system state (i.e., any deviation from the desired goal) by observation and measurement, continually using new data to keep it on track.

In contrast to the case just discussed (TD1), the initial data is irrelevant here. It is the full set of goals g_n that determine the outcome, through the differences $\Delta y_n(t_i)$ between the goals and the actual values. Instead of (2.1), we have

$$y_j(t_{i+1}) = f_j\big(\Delta y_1(t_i), \ldots, \Delta y_N(t_i), \Delta t\big)\,, \quad \Delta y_n(t_i) := y_n(t_i) - g_n\,. \qquad (2.2)$$

Examples are thermostats, an elevator taking one to the desired floor in a building, speed controllers in engines, fully automated electric trains, and so on. In many engineering applications, there will be computer control systems that will implement this logic of deciding what to do next on the basis of the current system state, embodied at the microscale in WHILE and IF THEN loops [14, p. 29].

Table 2.11 The basic features of a feedback control system. The goals lead to a specific final state via feedback of an error signal to an actuator. The initial state of the system is irrelevant to its final outcome, provided the system parameters are not exceeded

	Controller	$\Leftarrow$ Correction signal	
Noise $\Rightarrow$	Action $\Downarrow$	Feedback $\Uparrow$	
	State	$\Leftrightarrow$ Comparator $\Leftrightarrow$	Goal

[2]The Hotbits random number generator uses this technique: see http://www.fourmilab.ch/hotbits/.

In advanced systems (automatic pilots, control systems in chemical plants), the controller will act not on the basis of the present physical state of the system but on the basis of predicted future states as determined by the latest updates of the current system state. It is this continual updating of predictive data that gives the process its power. This is also the core principle of numerous homeostatic control systems in physiology and cell biology. This is top-down causation because the goal determines the outcome, and hence is at a causally higher level than the system controlled. It is an emergent property of the system, enabling sophisticated behaviour. But this process cannot innovate: the outcome is predictable from the outset, as it is determined by the explicit or implicit goals of the system. Like predictive algorithmic processes, non-adaptive feedback control systems cannot learn. That requires adaptive selection.[3]

2.4.4 *TD3: Adaptive Selection*

The basic feature of adaptive selection [36] is that a process of variation generates an ensemble of states, from which a best outcome is selected according to some selection criterion (see Table 2.12).[4]

The reason this is classed as a form of top-down action is that the nature of the higher level environment is crucial to using selection criteria. The outcome would be different if either the environment or the criteria were different. Its great power in evolutionary biology is due to the continued repetition of the adaptation process, with the best variant being passed on from one generation to the next by a hereditary mechanism. But that repetition is not essential to the basic process.

The basic dynamics is first a randomisation process, and then a selection process

$$y_j(t_{i+1}) = \Xi_j\big(y_1(t_i), \ldots, y_N(t_i), c_j, E\big) , \tag{2.3}$$

Table 2.12 The basic features of adaptive selection. Selection takes place from an ensemble of states, the selection being based on the outcome of some selection criteria in the context of the specific current environment. Unwanted states are discarded

System states	$\Leftarrow$	Selection agent selects state		Meta-goals
Variation $\Downarrow$		$\Uparrow$		$\Downarrow$
Ensemble of System States	$\Rightarrow$	Preferred states	$\Leftarrow$	Selection criteria
		$\Uparrow$		
		Environment		

[3] I am aware that some present day feedback control systems use principles of adaptive control. I believe they should be labeled as such, to distinguish them from the basic cybernetic processes identified by Wiener, in which the goal is fixed.

[4] This is what Penrose identifies as bottom-up organisation [53, p. 18], but this is incorrect, because he fails to recognise the top-down nature of the decision process via higher level selection criteria.

Table 2.13 The basic function of adaptive selection processes: they select what is useful or meaningful from an ensemble of mainly irrelevant stuff and reject the rest, thus creating order out of disorder by selecting states conveying meaningful information

Final data set	Meaningful information	$\Rightarrow$	Rejected set: noise
	Selection $\Uparrow$	$\Leftarrow$	Selection principle
	Varied set		
	Variation $\Uparrow$	$\Leftarrow$	Variation principle
Initial data set	Ensemble of states	Random	

where Ξ_j is a projection operator selecting one of the $y_n(t_i)$ and rejecting the rest, on the basis of the selection criterion c_j evaluated in the environmental context E. It is a non-deterministic process: because of the random element in generating the ensemble selected from, one cannot predict the outcome before the selection process takes place.

It is also for this reason that it can innovate. The process generates new information that was not there before—or rather, finds information that was hidden in noise (Table 2.13). That is the general process whereby adaptive selection generates useful information: it finds what is relevant and works from an ensemble of stuff that is mainly irrelevant or does not work, hence allowing a local flow against the general tide of increasing disorder. Inter alia, this is the process underlying learning.

Many computational processes build on this possibility. These include:

- *Artificial neural networks* [11], where selection of node weights occurs through the training process. The resulting set of node weights is not predictable. (If it were, one would not need the training process.)
- Many *optimization procedures* are of this nature, as they search the possibility space and choose the best outcome encountered. Randomness comes because one cannot explore the whole space, and we have to choose a subset of points to investigate, and steps away from these points: the result might depend on this choice, if local maxima occur.
- *Evolutionary computation* (EC) [23, 24] encompasses genetic algorithms (GA), evolution strategies (ES), evolutionary programming (EP), genetic programming (GP), and classifier systems (CS).

These are all examples of non-deterministic computing [1, pp. 412–413]:

> The key idea is that expressions in a non-deterministic language can have more than one possible value [...] our non-deterministic programme evaluator will work by automatically choosing a possible value and keeping track of the choice. If a subsequent requirement is not met the evaluator will try a different choice, and it will keep trying new choices until the evaluation succeeds, or we run out of choices. [...] the non-deterministic evaluator will free the programmer from the details of how the choice is made [...] it supports the illusion that time branches, and that our programmes can have different possible execution histories. When we reach a dead end, we can revisit a previous choice point and proceed along a different branch.

This is just a version of adaptive selection.

Table 2.14 Adaptive selection of goals

Level 3	Selection criterion	Meta-goal
	⇓	
Level 2	Goal	Adaptively selected
	⇓	
Level 1	Feedback control	⇒ Output

2.4.5 *TD4: Feedback Control with Adaptive Goals*

Higher level innovation becomes possible when one combines TD2 and TD3 to obtain TD4: feedback control with adaptive learning. Unlike TD2 where goals are fixed, these are feedback control systems that select their goals by a process of adaptive selection: equation (2.3) is applied to a set of goals g_n in (2.2) to get

$$g_j(t_{i+1}) = \Xi_j^{\mathrm{g}}\big(g_1(t_i), \ldots, g_N(t_i), c_j^{\mathrm{g}}, E\big) , \tag{2.4}$$

where c_j^{g} are criteria for feedback control goals (see Table 2.14).

This is a higher level form of top-down action, as it involves both goals in a homeostatic system (TD2) and adaptive selection criteria (TD3). It is used in engineering in adaptive forms of feedback control, which can be implemented through suitable digital computer systems.

2.4.6 *TD5: Adaptive Selection of Adaptive Goals*

One issue inevitably arises: where do the selection criteria in adaptive selection systems come from? In fact, they, too, may be adaptively selected, giving TD5: the case where adaptive selection criteria are determined by adaptive selection. Hence, (2.3) is applied to the criteria c_n [guiding selection in (2.3)] in the form

$$c_j(t_{i+1}) = \Xi_j^{\mathrm{c}}\big(c_1(t_i), \ldots, c_N(t_i), c_j^{\mathrm{c}}, E\big) , \tag{2.5}$$

where c_j^{c} are criteria for selective criteria (see Table 2.15).

This is a higher form of top-down causation, because adaptive selection is itself a form of top-down causation. It is of importance in determining strategy in every area of personal and communal life, e.g., business, education, politics, social policy. It can be exemplified by ranking systems in search engines [45], where the key element is

Table 2.15 Adaptive selection of selection criteria

Level 3	Selection criterion 2	Meta-goal
	⇓	
Level 2	Selection criterion 1	Adaptively selected
	⇓	
Level 1	Adaptive selection	⇒ Output

Table 2.16 The hierarchy of selection criteria

Level $N+1$	Selection criterion N	Non-algorithmic choice
	⇓	
Level N	Selection criterion $N-1$	Adaptively selected
	⇓	
	⋮	⋮
	⇓	
Level 3	Selection criterion 2	Adaptively selected
	⇓	
Level 2	Selection criterion 1	Adaptively selected
	⇓	
Level 1	Adaptive selection	⇒ Output

selection of criteria for ranking, and the second order adaptive outcome is successful ranking of web pages (selection of the most relevant according to the chosen criteria).

Closing the Hierarchy. Adaptive selection of adaptive criteria involves choosing a set of criteria c_j^c for suitability of adaptive criteria c_j. This appears to be the start of an infinite recursion: where do these next higher level selection criteria c_j^c come from? Are they, too, selected adaptively? How do we close the logic? (see Table 2.16).

At some point we have to stop and accept a set of highest level selection criteria as an a priori choice, otherwise we cannot close the system. (if we consider criteria for this choice and evaluate it, then through that act it is shown not to be the uppermost level). Any attempt to determine these criteria algorithmically, heuristically, or by adaptive selection will of necessity introduce a further set of selection values: it will just postpone the final decision level and choice by adding in a further level to Table 2.8. Naturally, the same issue arises in relation to adaptive selection of goals (TD4). There, too, there has to be an uppermost level which is just taken as given and sets the overall direction and purpose of the dynamics. The meta questions are:

- **Meta-analysis**. How many levels up do you go?
- **Choice**. How do you decide which criteria to use at the top?

These are philosophical issues, to be chosen according to one's philosophical position. This is where values and purpose come in: this highest level is the level of meaning ('telos'), perhaps involving ethics or aesthetics. This choice gives shape to all the rest, for it transfers down to affect choices made and outcomes at all the lower levels.

2.4.7 Goals and Learning in Relation to These Kinds of Causation

This section has looked at five distinct types of top-down causation (TD1–TD5) that can occur in computer systems. Three key points to notice are the following:

- Goals versus attractors.
- Learning and adaptive selection.
- Intelligent top-down causation.

2.4.7.1 Goals Versus Attractors

Dynamical systems with attractors (TD1) can look like feedback systems with goals (TD2) because initial conditions anywhere in a wide basin of attraction can lead to the same result. In particular this happens if there is friction (motion dies away as energy dissipates). Nevertheless, they are completely different in terms of mechanism: the second (TD2) involves active collection and use of information, while the first (TD1) does not. The second involves the causal effectiveness of goals as in (2.2), the first just the flow of the dynamical system according to the initial data as in (2.1).

2.4.7.2 Learning and Adaptive Selection

Learning, and associated collection of new information, is not possible via bottom-up action alone, or via dynamical systems (TD1) or non-adaptive feedback control (TD2). TD1 proceeds simply on the basis of information that is available at the beginning, as in (2.1), while TD2 compares updated information with goals as in (2.2). Neither generates any new information that was not there to start with. In order for new information to be acquired, and hence in order that learning can occur, one needs adaptive selection to take place, that is one needs TD3 as in (2.3), TD4 as in (2.4), or TD5 as in (2.5).

2.4.7.3 Intelligent Top-Down Causation

Intelligent top-down causation is the special case of any of TD1–TD5 where symbolic systems are used in the analysis, based on using some entity to represent something else. This is what characterizes intelligent thought: systems and situations are modeled in a symbolic way through use of language, diagrams, maps, physical models, or mathematical models. In particular higher level goals and selection criteria are analysed through use of symbolic systems and then adapted to get optimal results.

This use of symbols is an abstract technology that enables us to transcend the boundaries of what actually exists and consider what might be, what it might mean, and what methods to use when investigating these issues. The use of symbolic systems—particularly language—is a key characteristic of being human [22].

Now all digital computer systems are symbolically based—that is the core of how computers function—so their use to assist decision-making is in a sense automatically of this kind. However, sometimes computers act as explicitly symbolic computational systems, rather than just carrying out data analysis or numerical computations. Computer languages such as LISP can be used to perform logical operations and so can be used to investigate goal choice and decision-making algorithmically. Their mathematical derivatives, such as MATHEMATICA and MAPLE, are able to perform algebraic operations (solving an equation symbolically, for example) and symbolic integration and differentiation, as opposed to numerical differentiation and integration. The former hold for generic functions whereas the latter hold only for specific functions. Such languages are of considerable use in evaluating goals and adaptive criteria.

At a deeper level, computer systems act crucially as extensions of human capacity in investigating policy options symbolically: computers are used as interactive aids in decision or design systems, and it is in the human–computer interaction that the real creative capacity lies. Computer models are used to simulate reality, e.g., computer-aided design systems for houses or aircraft: the human mind intervenes and tries new options, the best one being selected. Examples are health policy, housing policy, energy policy, environmental policy. In each case examining what is possible when physical and economic constraints are taken into account can play a key role in determining what are suitable tactical and strategic goals, and indeed in working out what are the best criteria for such goals. This is particularly because of the unintended consequences that can arise in complex systems such as ecosystems: you aim for one effect, but a completely unexpected side-effect dominates the outcome. Neither the options nor the selection process can be fully algorithmic, because the former involves imagination and understanding of causal possibilities, and the latter involves decisions that cannot be reliably reduced to a numerical algorithm, for example, an architectural design involves aesthetic as well as functional features. When they are so reduced (as in the case of automated stock options), disaster may ensue.

The core causal feature is the interaction of the user and the machine, the resulting evaluations being based on models of the target area embodied in suitable symbolic systems. These evaluations then become the high level causal feature underlying our plans and consequent actions that are physically effective in the real world. One attains new patterns that were not there before by optimization and selection of goal choices, selection criteria, and methods used.

A key feature of such reasoning is that it is recursive: it can be turned on itself, to adapt the method of reasoning. An open question concerns the degree to which intelligent computer systems can capture the kind of human reasoning involved in such analysis. This is of course the contentious area of artificial intelligence [48, 56, 58]. I will not enter the fray except to give the following quote from McCarthy [48, p. 18]

> Formalizing common-sense reasoning needs contexts as objects, in order to match human ability to consider context explicitly. [...] We propose the formula *holds(p,c)* to assert that the proposition *p* holds in context *c*. It expresses explicitly how the truth of an assertion depends on context.

Thus a key to success is adapting the logic to take contextual effects into account, in line with the central argument of this book. There is, however, a specific open question as regards TD5:

Open Question. Is adaptive selection of adaptive goals *only* possible through use of symbolic systems? Or can it be possible without symbolic reasoning?

I suspect the answer is that symbolic reasoning is essential for meaningful TD5 processes. Then TD5 is necessarily a subclass of intelligent top-down causation.

2.5 The Core Feature: Equivalence Classes

The central feature of all forms of top-down causation in general is multiple realization and the associated equivalence classes [5]. This applies in particular to digital computation. I consider in turn:

- Multiple realization (Sect. 2.5.1).
- The link to top-down causation (Sect. 2.5.2).
- The ontological nature of computer programs (Sect. 2.5.3).

2.5.1 Multiple Realization

The core feature of top-down causation is the way higher level elements can emerge from many different variants of lower level ones, in both the physical and the logical context:

Multiple Realisability. Higher level structures and functions can be realised in many different ways through lower level entities and interactions.

In general many lower level states correspond to a single higher level state, because a higher level state description is arrived at by averaging over lower level states and throwing away a vast amount of lower level information (coarse-graining). Hence, specification of a higher level state determines a family of lower level states, any one of which may be implemented to obtain the higher level state (a light switch being on, for example, corresponds to many billions of alternative detailed electron configurations). The specification of structure may be loose (attainable in a very large number of ways, e.g., the state of a gas) or tight (defining a very precise structure, e.g., particular wiring of a VLSI chip in a computer). In the latter case, both description and implementation require far more information than in the former. Equivalence classes of lower level operations give the same higher level effect. Some examples in the case of digital computers are:

- At the circuit level, one can use Boolean algebra to find equivalent circuits to any circuit [47].
- At implementation level, one can compile or interpret a high level program, giving a completely different lower level process producing the same higher level outcome [67].

- One can run the same high level software on different microprocessors, using different instruction sets [47].
- One can run the same algorithms in different programming languages (Basic, Fortran, Pascal, Java, for example).
- Generic procedures can operate on data represented in different ways [1, pp. 170, 187].
- At implementation level, there is an equivalence of hardware and software. One can decide to imbed developed software in a dedicated hardware chip, giving a completely different nature of lower level physical entities for the same higher level outcome.
- At the highest level, specific tasks can be allocated either to the user or to the computer to give the desired high level output (e.g., focusing and exposure in digital cameras).

At the foundations of computing, the notion of a computable function is extremely robust and can be defined in many seemingly different, but equivalent terms [14][5]:

- One of these definitions is Turing's original definition via Turing machines that can encode numbers in the form of digits on infinite tapes that the machine can manipulate according to actions specified by its program.
- An equivalent definition is via register machines that can directly manipulate natural numbers with arithmetic operations. This is close to the (assembler) programming language.
- Another completely different but equivalent definition is purely number theoretic, avoiding reference to any kind of seemingly obscure 'machinery': a computable function is a function whose graph is a Diophantine set
- Another common characterization used in logic is via certain recursive equations, which is why the word 'recursive' is used synonymously with 'computable' in this field.

This variety of ways expresses the notion of a computable function from quite different, but nevertheless equivalent viewpoints.

2.5.2 The Link with Top-Down Causation

The connection with top-down causation is that we only normally have access to the higher level variables: these are the handles we have to affect the system state. When we change them we change numerous lower level states in accordance with the chosen higher level state, that is, we instantiate an instance of the equivalence class. It does not matter which specific one we instantiate. What matters is which equivalence class it belongs to, because this determines which higher level state it represents.

[5] I thank Vasco Brattke for these characterisations.

In the implementation hierarchy, once a particular lower level method of implementation has been chosen, that is the one that exists physically and drives the higher level dynamics.

In logical hierarchy, the higher level function drives the lower level design and hence the lower level operation. This is embodied in the nature of modularity, involving encapsulation and information hiding [57, pp. 233, 476–483]: one can change the nature of the private methods, while overall function and interface remain unchanged:

> **Equivalence Classes**. Top-down causation takes place by instantiating a specific lower level instance of an equivalence class representing a higher level variable. This happens by giving a higher level variable a specific value, an action which sets specific values for all the relevant lower level variables. Which specific such values are set is not determined by the chosen high level value.

2.5.3 The Ontological Nature of Computer Programs

Because of this multiple realisability, a higher level element is not ontologically the same as any specific lower level realization. It is the equivalence class of all of them [5]:

> **The Ontological Nature of a Computer Program**. In terms of the lower level elements that represent or instantiate it, this is nothing other than the functional equivalence class of such lower level elements that give the desired high level function.

This characterizes in precisely what way computer programs are abstract entities. They are not the same as any specific physical state: they are in essence equivalent to the set of all physical states that embodies their logic:

> **Reality of Computer Programs**. They are real and exist as higher level entities, because the equivalence class of lower level elements exists, and is causally effective. It determines uniquely what happens at the macro level.

The same is true for data: it can be represented logically in many different ways, e.g., binary or hexadecimal. It can be instantiated physically in electronic states or in printed or spoken form. The essence of the data is not any specific representation of either equivalence class: it is the equivalence class itself.

2.6 Resources: Memory and Deleting

Formal language theory proposes that there are an infinite number of possible statements in any language [38, p. 320]. This is based on the idea that statements can have an unbounded length: one can always add another clause to them. In the case of computers, the tape in a Turing machine is supposed to be infinite: it can store an infinitely long programme and an infinite amount of data. But infinities cannot occur in physical reality: resources are limited and in reality infinity is unattainable. This has important practical applications for computing. I consider in turn:

- The unphysical nature of infinity (Sect. 2.6.1).
- Deletion and garbage collection (Sect. 2.6.2).
- The memory hierarchy (Sect. 2.6.3).
- Modular hierarchical structure and scoping of variables (Sect. 2.6.4).
- Deletion, adaptive selection, and irreversibility (Sect. 2.6.5).

2.6.1 The Unphysical Nature of Infinity

Turing states [65]:

> Some years ago I was researching on what might now be described as an investigation of the theoretical possibilities and limitations of digital computing machines. I considered a type of machine which had a central mechanism, and an infinite memory which was contained on an infinite tape. This type of machine appeared to be sufficiently general [...] It was essential in these theoretical arguments that the memory should be infinite. It can easily be shown that otherwise the machine can only execute periodic operations.

But an infinite memory or an infinite tape cannot be read. Infinity is not just a very large number: it is a magnitude that is never attained. It is always beyond reach. That is its most essential feature. No matter how much has been read, there will always be more to read, because that is what infinity means—something that is never completed, it is always unattainable. David Hilbert remarked [35]:

> The infinite is nowhere to be found in reality, no matter what experiences, observations, and knowledge are appealed to.

A real computer has finite storage capacity and only survives for a finite length of time, and so can only carry out a finite number of operations in its lifetime.

One can calculate an absolute limit to what a computer can possibly read in its lifetime by estimating how many bytes can be read by a machine that reads continuously for 24 hours a day, every day for say 1200 years at a rate of say 10^9 bytes a second, giving $10^9 \times 60 \times 60 \times 24 \times 365 \times 1200 = 378\,432\,000\,000\,000\,000$ bytes: a large number but obviously not infinite. No real computer can exceed this limit in its lifetime (inter alia because it will need maintenance, and will not in fact last that length of time). Indeed the computational capacity of the entire universe is finite [44]. Hence, there is a finite limit to the length of any statement that could be read by a computer in its entire lifetime in a physically realistic setting. And anyway, sentences actually usable for computational purposes, the *raison d'être* of computers, are very much shorter:

> **Computational Finiteness.** The set of possible computable programs Ω_p and the set of potentially associated data Ω_d are both large but finite.

The implication is that there are a finite number of possible computer languages, programmes, and data, whence the possibility space for computer operations is finite. The idea of a computer that can process an infinite tape or read an infinite amount of data does not make physical sense. Formal language theory should take this into account.

What about Turing's comment that if a machine has finite memory then it can only execute periodic operations? This is in principle true, as the operation space is then compact, and if the machine continues to operate for an unlimited time, Poincaré's eternal return theorem applies: eventually all possible states will have been visited and the next and all subsequent ones will be repeats of ones already utilised, so cycles will occur. But this assumes that the machine will continue operating for an infinite time, something which cannot happen inter alia because the Earth will come to an end in a finite time, when the Sun comes to the end of its life. The alleged problem arises because of this implicit infinity, which is unphysical. Computer memories are now so large that this will not be an inevitable outcome in practice.

2.6.2 Deletion and Garbage Collection

In practical terms, this limitation on memory has important implications for how memory is handled, and leads to the need for garbage collection and the ongoing deletion of records, freeing up memory space for reuse.

Garbage Collection. During a program run [1, pp. 540–546], this is a key strategy for handling memory limits, giving the illusion of infinite memory even though in fact the memory space is finite. Memory cells used to hold intermediate results during a calculation can be cleared at the end of the calculation, freeing up memory space to be reused in the next calculation.

This is related to *persistence* [12, pp. 75–77]: keeping in memory objects and names across different contexts. Objects take up some amount of space and exist for a particular amount of time. But one has to clear them out to make room for new objects, or memory will fill up and operations will cease.

Deleting Records. As regards long term memory, deletion of records to free up memory is a key requirement, not just because storage space is limited, but also because otherwise we simply cannot handle the vast amounts of data we accumulate. We eventually forget we have stored specific data, or cannot locate the relevant records in the fog of data clogging up our machine. The key strategy here is that the user deletes all those records they don't want to keep and puts the rest into suitably formatted short term or longer term storage, depending on their usage needs. This process of sorting emails, music, digital images, and so on, deleting those that are unwanted and keeping those that are still useful, refines and organises our files into meaningful collections suited to our purposes.

Together with the organisational methods discussed in the following sections, deletion and reuse of memory is the key to handling memory limitations resulting from finite resource availability, giving an illusion of infinite memory space, despite the available space being strictly limited.

2.6.3 The Memory Hierarchy

Given the hierarchy and memory limits, one still has to handle the practical limits on memory. This is done through the memory hierarchy. Turing states the problem as follows [65]:

> A problem might easily need a storage of three million entries, and if each entry was equally likely to be the next required, the average journey up the tape would be through a million entries, and this would be intolerable. One needs some form of memory with which any required entry can be reached at short notice [...] Another desirable feature is that it should be possible to record into the memory from within the computing machine, and this should be possible whether or not the storage already contains something, i.e., the storage should be erasible.

Even with more modern forms of memory, memory bottlenecks are the key design issue for computers. This breeds the memory hierarchy of short term, medium term, and long term memory. Thus one has [39]:

- **Main memory**. DRAM semiconductor memory in which most of the program and data are stored when the program is running (short term memory).
- **Cache memory**. Very high-speed semiconductor memory that caches frequently-used programs and data from main memory (storing them in a quick access area of medium term memory).
- **Paging memory**. Slower memory, usually disk, which provides swap files as an extra area for the main memory (medium term memory not used so often).
- **Hard drives**. Disk or tape memory for files (long term memory).

There is an entire science of how to design caches [37], and special languages designed handle the memory hierarchy efficiently [29]. As is clear from the above, a key issue is what to delete and what to keep. But additionally, a suitable hierarchical structure makes a big difference.

2.6.4 Modular Hierarchical Structure and Scoping of Variables

The fundamental principle is *locality of reference*, realised in modular hierarchical structures, with related aspects of temporal locality and algorithmic locality. One limits applicability of a variable both in logical space and in time. This is done by the mechanism of *scoping*, i.e., specifying the context within which it will be valid.

Scoping as Regards Context. Algorithmic locality happens via the distinction between local and global variables, embodied in the scope of a variable. Local variables must be readily available when a module is run, but can be cleared when another one is run. Global variables must be available all the time. The existence of the modules enables this distinction and so clarifies which variables can be cleared when the active module is changed.

Scoping Variables in Time. This follows from the fact that local variables are only valid for a certain time and cease to be needed when other local variables become relevant because another module is run. But there is another aspect: a key idea is that of a function $f(t)$ with an unchanging name, which keeps its identity as we evaluate it at different times, rather than regarding each of its values as separate ontological entities $x := f(t_1)$, $y := f(t_2)$, $z := f(t_3)$, etc. This allows one to overwrite old values of the variable as new values are calculated. One can discard the old value because it is no longer needed: what matters in most cases is just the value of the function at the present time, and perhaps a few times steps before that (if we are taking numerical derivatives). Exceptions are when the records are needed in the long term (financial or medical records for example), but then they can be transferred from short term memory to long term memory and stored on hard drives for later recall if necessary. Short term memory is freed up for reuse.

Streams. A related concept is the idea of delayed evaluation of streams [1, pp. 316–330]. These are lists which can be used to represent sequences that are infinitely long (such as the set of integers), even though in fact we only compute as much of the stream as we need to access [1, p. 326]. This is done by constructing streams partially and passing the partial list to the program that uses the list. Thus one writes the program as if the entire sequence was being processed, but interleaves the construction of the stream with its use. In this case, at the end of the calculation, there is no obligation to delete the variables that are part of the list but were never needed, because they were never activated in the first place.

2.6.5 Deletion, Adaptive Selection, and Irreversibility

The big picture is that (see Table 2.13):

> **One Creates Order by Deleting**. Adaptive selection of what is meaningful, and hence creation of ordered meaningful information, is centrally based on deleting what is not wanted.

Examples are deleting old files and emails, as well as deleting old values of variables, and indeed no longer used variables themselves. This is what allows the freeing up memory for reuse, and so creates the illusion of infinite memory.

Irreversibility. As pointed out by Landauer [43], these processes are where irreversibility, associated with physical entropy production, happens in computations (quoted by Bennett [8]):

> Any logically irreversible manipulation of information, such as the erasure of a bit or the merging of two computation paths, must be accompanied by a corresponding entropy increase in non-information bearing degrees of freedom of the information processing apparatus or its environment.

One is creating logical order by deleting (see Table 2.13), as regards information locally going against the overall flow of increase of entropy. There is a consequent physical energy cost characterized by the Landauer limit: the minimum amount of energy required to change one bit of information is given by $kT \ln 2$, where $k \sim 1.38 \times 10^{-23}$ J/K is the Boltzmann constant and T is the temperature of the circuit in kelvin. This principle linking information and entropy creation has been experimentally verified by Bérut et al. [10]. Hence, there is an energy cost to generating useful information.

Ladyman et al. [41] analyse in detail what it means for a physical system to implement a logical transformation L, and make this precise by defining the notion of an L-machine. They show that logical irreversibility of L implies thermodynamic irreversibility of every corresponding L-machine. This relates in particular to the operation *Reset* which clears a logical system to its original state by replacing all variable values generated in the previous cycle with default values and so freeing it up to start a new cycle of operation. Overall, the conclusion is that dealing with logical infinity in a system of finite size is irrevocably tied to physical irreversibility.

2.7 The Outcome: Causation in Digital Computers

Even though computers are the epitome of algorithmic machine operations, they are also systems where non-physical entities (programs, algorithms, data) are causally effective, and enable symbolic operations to take place that are independent of the underlying physics. Here we consider:

- Computer programs are non-physical, but causally effective (Sect. 2.7.1).
- Computer programs embody abstract logic, and act top-down (Sect. 2.7.2).
- Room at the bottom (Sect. 2.7.3).
- Predictable explanation (Sect. 2.7.4).
- Possibility spaces and their causal effects (Sect. 2.7.5).
- Top-down action from the mind (Sect. 2.7.6).
- Genuine emergence (Sect. 2.7.7).

2.7.1 Computer Programs Are Non-physical, but Causally Effective

Virtual machines are the core of computing systems (Table 2.3), and although they do not exist as physical entities, they are real: they exist as causally effective entities.

2.7.1.1 The Non-physical Nature of Computer Programs

Computer programs are not the same as what is printed in a listing, or stored in a disc, or saved in computer memory, or presented on a blackboard, and neither are they what exists in a programmer's mind. These are all instantiations of an entity that is not itself a physical thing. It is not fully realised in any of these instantiations: precisely because it can be realised in the others. It is not the same as any of its instantiations. Rather, it is essentially equal to all of them:

- When considered in lower level terms, the real nature of a program is that it is an equivalence class of such representations (Sect. 2.5).
- When considered in higher level terms, it is an abstract entity obeying rigidly prescribed syntactic laws, and through a combination of bottom-up and top-down causation, it is causally effective at its own level.

It is not equal to any particular physical manifestation, e.g., on a CD disk or as electronic states in a computer. These are just vehicles whereby it is instantiated.

2.7.1.2 The Causal Effectiveness of Computer Programs

Given the physical computer, a loaded program, and input data, the output is uniquely determined:

$$\text{(physical structure, program, data)} \Longrightarrow \text{output}\,. \tag{2.6}$$

The first two will be fixed and unchanging in a given run (with the same high level software loaded) and can be taken for granted then. So, within this context, the given constraints imply

$$\text{(data)} \underbrace{\Longrightarrow}_{\text{program}} \text{output}\,, \tag{2.7}$$

showing that abstract information is causally effective in the given context of a specific program, which determines in a top-down way the family of results obtained from arbitrary data. But as we have seen, the program is an abstract entity. According to Abelson and Sussman [1, p. 1]:

> *Computational processes* are abstract beings that inhabit a computer. As they evolve, processes manipulate other abstract things called *data*. The evolution process is directed by a pattern of rules called a *program*. People create programs to direct processes. In effect, we conjure the spirits of the computer with our spells.

That gets it just right. Abstract entities produce concrete results. They are causally effective through the computer hardware. The ultimate reason this is so is because they were designed to do so: they are an example of the causal efficacy of the human mind. Consequently:

> **Causal Effectiveness.** Computer programs are not physical entities, but are nevertheless causally effective in numerous ways.

For example they can do engineering and calculations that result in specific physical structures such as aircraft and automobiles coming into existence. Furthermore, they facilitate economic interactions such as shopping and banking, social interaction through internet applications such as email and facebook, and education through the internet in conjunction with search engines such as Google and encyclopedias such as Wikipedia. They make a real difference in the real world. A final note for the philosophically cautious:

Existence. Because computer programs are causally effective, they clearly exist.

Here I use as a criterion that whatever is causally effective in the physical world must certainly exist (Sect. 1.3.5). If this is not true, we will have to face existence of uncaused entities or events in the physical universe.

2.7.2 Computer Programs Embody Abstract Logic, and Act Top-Down

This is possible because logical entities can cause physical effects, enabled by the interaction of bottom-up emergence and top-down causation. In particular, this happens in the interaction between the logical and physical systems. These systems are emergent systems based on the underlying physics, but then acquiring an abstract character at the higher levels.

2.7.2.1 The Implementation Hierarchy: Logical Levels and Descriptions

A series of interlocked computer programs, each representing the same logical structure, power the virtual machines at each level in the implementation hierarchy (Table 2.2). They are what give the system its dynamics. The downward link is via compilers and interpreters (Sect. 2.3.4). The upward link is via implementation, in essence according to Turing's prescription of reading a tape and performing the next logical operation specified thereon (Sect. 2.1).

The physical system is designed to embody logical relations, which are coded in a hierarchical manner through the interaction between system hardware and software. There are different layers in the description of computers and, in particular, the following[6]:

1. Digital circuits that can be directly implemented using certain physical devices.
2. Register machines that describe computation on a higher level of abstraction (in terms of very simple arithmetic operations).
3. Object-oriented programming languages that offer very abstract ways to describe data structures and operations on them.

---

[6]I am indebted to Vasco Brattke (private communication) for the following comments.

One of the questions seems to be: how is it possible to implement on one relatively primitive level a layer that seems to offer a much higher degree of abstraction?

The emergence from layer 1 to 2 above happens on the level of microcode, which is implemented in digital circuits and offers the first layer of programming. Microcode operations are very simple and they are actually used to implement assembler languages that offer pretty much the same type of instructions as register machines. On the level of microcode, one still reasons in terms of digital circuits and very elementary operations that transfer content from one position in the memory to another.

On the level of assembler languages one no longer has to think in terms of digital circuits, but the reasoning happens on the higher level of registers and certain arithmetic and logical operations. On this level one can actually implement abstract object-oriented programming languages such as Java (although in practice there are several intermediate layers, such as the operating system). In particular, all such things as indirect addressing, pointers, etc., can be implemented easily on the level of register machines.

In fact, as shown in [14], all these 'programming languages', register machines, recursive functions, Java programs, and so on, satisfy the so-called *SMN properties*(*Kleene's translation theorem*) and *UTM properties* (*Turing's universal function theorem*). Hence, it follows from the equivalence theorem of Rogers that each of them can be simulated in any of the others [14]. The level of description and abstraction is very different, but the power of expressiveness is essentially the same. Already at level 1 in Table 2.5, the zeros and ones are conceptual representations of physical states. The actual physical state is a charge or current [47]. It is conceptually referred to by binary notation: an abstraction that is the effective language of the logic that is built into the gates by their properties and connectivity in logical circuits.

Given this structure, the hierarchy of languages can be constructed, with compilers and interpreters [3, 6] acting top-down to link the levels. But they are just computer programs. Abelson and Sussman [1, p. 360] state the following:

> Metalinguistic abstraction—establishing new languages—plays an important role in all branches of engineering design. It is particularly important to computer programming, because in programming not only can we formulate new languages but we can also implement these languages by constructing evaluators. An evaluator (or interpreter) for a programming language is a procedure that, when applied to an expression of the language, performs the actions required to evaluate that expression. It is no exaggeration that this is the most fundamental idea in programming: the evaluator, which determines the meaning of expressions in a programming language, is just another program.

This enables the emergence of higher level entities such as the higher level systems programs and application programs, both realised when the low level systems programs are run. They subsequently exert top-down effects on lower level dynamics (Sect. 2.4). Universal computation is then possible, able to model arbitrarily complex systems.

2.7.2.2 The Logical Hierarchy

To enable high level computation additionally requires modular hierarchical structuring of a logical hierarchy (Table 2.4) at each level of the implementation hierarchy, enabling abstraction, information-hiding, and so on (Sect. 2.2.2). This structure enables contextual information processing. James McClelland describes it thus [46]:

> Interactive models of language processing assume that information flows both bottom-up and top-down, so that the representations formed at each level may be influenced by higher as well as lower levels. I describe a framework called the interactive activation framework that embeds this key assumption among others, including the assumption that influences from different sources are combined non-linearly. This non-linearity means information that may be decisive under some circumstances has little or no effect under other conditions. [...] feedback from higher levels is computationally desirable [because] it allows lower levels to be tuned by contextual factors so that they can supply more accurate information to higher levels.

The 5 different types of top-down causation (Sect. 2.4) can be implemented and enable complex behaviour to emerge on the basis of purely algorithmic operations at the bottom.

2.7.2.3 Symbolic Logic Independent of the Underlying Physics

It is clear from Turing's work (Sect. 2.1) that what one can do symbolically via digital computers is not in any way restricted or constrained by the lower level physical implementation [14]. It is determined by the logic of the higher level possibility space (the effective laws of logic, mathematics, and semiotic representation), not by the underlying laws of physics that enable the computer to function.

2.7.3 Room at the Bottom

How is there room at the bottom for top-down action in a mechanistic system, where the low level operations are completely deterministic?[7] The main way higher level structures exert an effect on lower levels is by setting various constraints on their functioning:

- The physical structuring of the computer (hardware) embodies patterns of connection that constrain what happens at gate level.
- The loaded high level software establishes further constraints on the logical structure of the lower level interactions.

---

[7] I only consider classical computers here, where quantum uncertainty in the underlying physics has no effect on microcomputer operations because they have been carefully designed so that this will be the case. Quantum computing raises many further possibilities I do not engage with in this text.

- Finally, the data establishes sufficient further constraints on the lower level interactions to give a unique output.

This works out in the following ways (discussed further in Sect. 5.3).

2.7.3.1 Context

Firstly, the context determines what algorithmic operations take place. The *physical context* of computer structure does not alter the lower level physics: it constrains its actions. Paradoxically, constraint creates the possibility of complexity. For example, the wiring in a computer means that a specific gate G_1 is connected only to further gates G_2 and G_3 and not to any other gates in the system, and this is what enables these three gates to produce a specific logical operation, such as AND-OR-INVERT [47]. This would not be possible if inputs from other randomly selected gates were also connected. More generally, motifs occur in complex systems and shape their behaviour by constraining interactions [2].

The *logical context* of loaded programs also constrains what happens. Gate operations at the bottom are individually identical, whether a music program, a spreadsheet, a word processor, or an image-processing program is running. The specific sequence of low level operations that takes place, and the consequent high level output, is completely different depending on the higher level context of what program is running and what data are entered.

2.7.3.2 Environment

Secondly, part of the context is the environment, which lies outside the control of the algorithmic system and exerts a causal influence on operations. In many computer applications, new data comes in during a run that was not present at the start: so the computer is not a closed system, it is influenced by the environment—a top-down effect. This happens, for example, in continually updated weather forecasting systems, online stock control systems, ATM operations, and feedback control systems.

2.7.3.3 Randomness and Adaptive Selection

Thirdly, processes of adaptive selection allow learning to take place, with new information beng garnered by selection processes whereby masses of irrelevant information are discarded as irrelevant. This is non-deterministic, and hence not uniquely implied by the initial data, because the variation processes include random elements (Sect. 5.6.6). It is top-down because the outcome depends on the choice of selection criteria at higher levels in the hierarchy of causation. It may also happen in adaptive selection processes where non-algorithmic higher level criteria are used on the fly during the selection process. This occurs, for example, in the use of spreadsheets, and all those computer-aided design processes in which the operator chooses between options.

2.7.3.4 Mutable Lower Level Elements

Fourthly and crucially, the behaviour of lower level elements is not generally immutable, but depends on context: they are adapted to their role in the hierarchy (see Sect. 5.4). Put briefly:

> **Contextually Determined Nature**. The nature of the lower level entities—the way they respond to events—is often determined by context.

In digital computers this occurs through the late time binding that enables polymorphism in object-oriented systems [57, pp. 506–531]. More generically, parameters are passed down from the higher level to set or alter the data-handling method used by modules at the lower level, thereby determining the specific outcomes. The lower class functions can in this way underlie many different higher level functions, through the setting of parameters that control function at the lower level.

2.7.3.5 The Enabling Role of Physics

One cannot derive algorithmic logic from physics: e.g., one cannot derive Quicksort either from the physical operation of electromagnetic interactions, or from the logical form of Maxwell's equations. Yet it is algorithmic logic that drives what happens at the higher levels in a computer, and hence at the lower levels.

The underlying physics enables this to happen: it dances to the tune of this abstract logic, which gets embodied in particular patterns of energy states at the micro level. They are the outcome of the logic, not its cause. The logic of the algorithms derives from the nature of what is possible in logical terms.

2.7.4 Predictable Outcome?

Computers are the epitome of algorithmic operations: is the outcome predictable? There are three ways in which the outcome may not be implied by the initial data:

1. It is not predictable because of the complexity.
2. It can have new input: data fed in during the runtime (open systems).
3. It can have a random element inserted (by a random generator or clock time or radioactive decay).

The first is non trivial, as remarked by Turing[aut] [66]:

> The view that machines cannot give rise to surprises is due, I believe, to a fallacy to which philosophers and mathematicians are particularly subject. This is the assumption that as

> soon as a fact is presented to a mind, all consequences of that fact spring into the mind simultaneously with it. It is a very useful assumption under many circumstances, but one too easily forgets that it is false. A natural consequence of doing so is that one then assumes that there is no virtue in the mere working out of consequences from data and general principles.

Indeed, if the outcome were predictable, we would not need the computer!

The second case is logically obvious, but operationally important: the cases of stock control, weather forecasting, and aircraft automatic pilots are examples.

As regards the third, unpredictable effects occur despite algorithmic operation in the case of adaptive selection, based on random lower level processes plus higher level selection effects. This results in accumulation of unpredictable information, and build-up of effective structures adapted to higher level function and environment, not uniquely determined by the initial data. Genetic algorithms and neural nets are examples. They can learn only because they get input from their environment in their training phase, enabling them to use high order selection criteria in the context of this specific environment—a form of top-down action. Then the outcome is not determined, even though the process is.

To Be Done. There is an interesting issue that arises here: such programs need a source of randomness so that the outcome is not predictable, allowing genuine learning. One can use a pseudo-random number generator, or a genuine random number generator (see the discussion in TD1 above). Both generate outcomes not implicit in the initial data, but the first is a disguised algorithmic process, while the second is not: it is truly non-deterministic. The issue is whether this makes a genuine difference to the outcome: does it really matter which choice is made? The answer is not clear.

2.7.5 Possibility Spaces and Their Causal Effects

What can be done by computers is characterized by a possibility space: the space of all possible computations Ω_c. This in turn is based on the set of all possible algorithms Ω_a, which includes the set of possible computer programs $\Omega_a(\text{prog})$.

2.7.5.1 Possible Algorithms

What is possible algorithmically is based on the space of logically possible algorithms Ω_a. This can be thought of as an eternal unchanging space of what is and what is not logically possible. We *discover* these possibilities, that is, we work out that they are indeed possible and valid first by inspiration or invention (imagining the possibilities), then by working out the details by logical argumentation (development), and then by checking that they are indeed valid (verification), again by logical argumentation).

The same algorithms are valid anywhere in the universe: near Alpha Centauri and in the Andromeda galaxy, and at any time. They were valid before humans existed and will reman valid after we are long gone. For example, there are various possible ways to sort a list: shellsort, heapsort, mergesort, bubble sort, quicksort, library sort, and so on [42, 68]. These have been discovered by human beings over the course of history, and indeed some were known long before computers existed. The corresponding subset Ω_a(sort) of Ω_a is finite (a typical list of sort algorithms will mention about 20 possibilities), as is each algorithm itself (an infinite algorithm would be of no use whatever, as discussed in Sect. 2.6.1).

The space Ω_a is hierarchically structured: more complex algorithms such as the Google search algorithm and pattern recognition algorithms [45] build on combinations of simpler ones such as quicksort. Although this logical space is progressively explored by the human mind as we discover more and more algorithms, it is independent of the mind: the logical possibility and validity of those algorithms is true independently of what we think. Like the mathematics possibility space Ω_m, the space Ω_a embodies eternal truths independent of place and time and culture, and so can be thought of as an abstract Platonic space, as is argued in the case of Ω_m by Penrose [54] and Connes [17]. In summary:

> **The Space of Algorithmic Possibilities** Ω_a. This is a hierarchically structured abstract Platonic space. We explore it through logical analysis by the action of the mind [19]. Instances of algorithms existing in Ω_a are causally effective when we implement them in computer programs [42, 45, 68].

This space is not implied by physics or physical laws, but by logic. Our understanding of this space cannot be tested by physics laboratory experiments (although these may possibly give hints as to how some algorithms operate). This understanding can, however, be tested by running computer programs embodying specific algorithms we have discovered and developed. They either work to give the desired results, or they don't!

2.7.5.2 Possible Computations: Limits of Computability and Applicability

Because computer programs are in essence just high level algorithms made by combining lower level algorithms in a structured way so as to produce a complete calculation, the space of possible computer programs is in essence a subspace Ω_a(prog) of Ω_a. But this is not the same as the space of possible computations Ω_c. Various issues intervene.

What can be computed and what cannot? There are four aspects here:

1. What kinds of problems are algorithmically expressible?
2. What algorithmic problems can be computed in principle by a physical device?
3. What is algorithmically computable by programs in a finite time?
4. What is computable in a realistic time?

These are deep issues, which I will only touch upon in the briefest of ways.

1. What kinds of problems are algorithmically expressible? How much of what humans understand can be algorithmically encoded? The brain does not naturally work in an algorithmic way, although it can be trained to do so. It operates by pattern recognition, enabled by the overall pattern of neural connections in the cortex [33], the connection weights in these neural networks (Churchland [19]), and synchronized patterns of oscillations between them (Buzsâki [16]).

These are not at all like the algorithmic operations of a digital computer, so it is not obvious that all that they can do can be represented by algorithmic processes (Penrose [52, 53]), unless those processes mimic the adaptive properties of neural networks [11], that is, they don't model the pattern of understanding attained, rather they model the process by which it is attained.

2. What kinds of algorithmic problems can be computed in principle by a physical device? This is the subject of the Church–Turing thesis, stated by Brattke [14] as follows:

> **Church–Turing Thesis (1936).** A function $f :\subseteq N^k \to N$ is computable in the formal sense if and only if it can be computed by some physical device.

This form of the thesis is not a mathematical statement since it relates the mathematically concept of computable functions to the question of what it means to compute something with a physical device. Copeland states it this way [20]:

> **Thesis M.** Whatever can be calculated by a machine (working on finite data in accordance with a finite program of instructions) is Turing-machine-computable. Thesis M itself admits of two interpretations, according to whether the phrase "can be generated by a machine" is taken in the narrow, this-worldly, sense of "can be generated by a machine that conforms to the physical laws (if not to the resource constraints) of the actual world", or in a wide sense that abstracts from the issue of whether or not the notional machine in question could exist in the actual world. Under the latter interpretation, thesis M is false. It is straightforward to describe notional machines, or 'hypercomputers' that generate functions not Turing-machine-computable. It is an open empirical question whether or not the narrow this-worldly version of thesis M is true.

The latter is the case of physical interest.

3. What is algorithmically computable by programs in a finite time? This is the issue of the *halting problem* [21]: given a valid program, will the computation come to an end in a finite time? The algorithmic structure of the program may be logically correct, but the computation may never conclude, and no algorithmic computation can determine whether this will happen or not. Chaitin states this as follows [18]:

> Turing's train of thought now takes a very dramatic turn. What, he asks, is impossible for such a machine? What can't it do? And he immediately finds a problem that no Turing machine can solve: the halting problem. This is the problem of deciding in advance whether a Turing machine (or a computer program) will eventually find its desired solution and halt.

A solution to the halting problem would determine the space of possible computations Ω_c as a subset of Ω_a(prog), but this is unsolvable by any Turing Machine.

4. What is computable in a realistic time? This is the whole subject of computational complexity and computation times. Issues occurring include time functions, complexity measures, and complexity classes [14, Sect. 3.6]. The necessary amount of auxiliary storage, stability, and effects on indexing keys are also important when comparing algorithms. Together these determine a subspace Ω_c(realisable) of Ω_c representing those possible algorithms that can be effectively implemented. This is a very context-dependent concept: as computer memory size and speed increase, what was previously impractical becomes possible. This is of great practical importance.

2.7.5.3 The Causal Effectiveness of Platonic Possibility Spaces

Overall, the key issue is the causal effectiveness of algorithms. This is what enables computer applications in engineering, science, and commerce, which cause real changes in the physical world. So where do they come from? The chain of causation is shown in Table 2.17. As explained above, algorithms ultimately originate in the Platonic space of logically possible algorithms Ω_a. Thus the conclusion is as follows:

> **Causal Effectiveness of Platonic Spaces**: The abstract possibility spaces Ω_a and Ω_c are the ultimate source of the causal powers of digital computers in the physical world.

Three-dimensional printers are able to create physical objects because the algorithms that enable this are valid algorithms, and that fact is a consequence of the nature of the Platonic space Ω_a.

Their Existence. The claim that all these spaces exist, i.e., that they are ontologically real, rests upon a philosophical analysis of what kinds of things must be recognised as existing. The view taken here (see [30] and Sect. 1.3.5) is that we must recognise the existence of any kind of entity that demonstrably has a causal influence on physical systems.

The possibility spaces discussed here are certainly causally effective, even though non-physical, so they must be realised as existing. They are the ultimate source of computational power.

Table 2.17 The origin of algorithms and programs in the abstract possibility spaces Ω_a (possible algorithms) and Ω_c (possible computations). These lead to real world effects such as 3D printing of physical objects

Level		
Level 4	Possibility space Ω_a	Possible algorithms
	$\Downarrow$	
Level 3	Possibility space Ω_c	Possible computations
	$\Downarrow$	
Level 2	Written programs p_i	Selected algorithms a_j
	$\Downarrow$	
Level 1	Computer run	Selected program and data
	$\Downarrow$	
Level 0	Output data/actions	$\Longrightarrow$ Real world effects

2.7.6 Top-Down Action from the Mind

Computer programs based in the possibility spaces Ω_a and Ω_c are not physical entities, but are nevertheless causally effective in numerous ways [45]. The final puzzle is this: how are these possibility spaces causally effective in this way? How do they influence what gets realised in computers?

The answer is through the human mind, which explores these spaces by logical reasoning. This is enabled by the ability of neural networks to learn about such abstract spaces through processes of pattern recognition based on the operation of neural networks in our brains, as explained clearly by Churchland [19]. Hence, I emphasize:

> **Causal Effectiveness of Platonic Spaces.** It is through adaptive selection processes in the mind, enabled by the neural circuits in the brain, that the possibility spaces are understood and hence causally effective.

This enables not only the existence of operational programs and algorithms, but also computers themselves: the physical entities that make this all happen. They ultimately originate, not only from our exploration of possible algorithms Ω_a, but also from our explorations of the physical possibility space Ω_{ph} restricting what is physically possible due to the nature of physical interactions (described by the laws of physics). Their development embodies the combined experience of numerous workers in aspects ranging from basic concepts to solid state physics to system design to effective algorithms to high level design patterns. This leads to the extraordinary ability of digital systems to represent language, pictures, sound, mathematical relationships, and indeed all human knowledge. Overall, this is the effect of intelligent top-down causation from the human mind to physical systems (the computer itself) and abstract systems (the set of programs that make a computer work).

At a higher level, the existence of computers is an outcome of the human drive for meaning and purpose: it is an expression of the possibility space of meanings,

the higher levels whereby we guide what actions take place. This will be discussed in Chap. 8.

2.7.7 Genuine Emergence

Although they are the ultimate in algorithmic causation, as characterized so precisely by the concept of Turing machines, digital computers embody and demonstrate the causal efficacy of various kinds of non-physical entities—algorithms, programs, data—which enable truly complex behaviour to emerge from simple constituents.

It is noteworthy here that one is able to regard level 0 in Table 2.1 as the bottommost level, the level 'where the work is really done', even though this is not in fact the case if one takes a strict reductionist viewpoint: that level emerges from lower physical levels, which are *really* where the work is done!

Why is it then legitimate to regard the emergent level 0 as real, as is taken for granted by all computer scientists and engineers? The answer is that this level does indeed do real work, as do *all* the levels in Table 2.1:

> **Genuine Emergence**. Each of the levels in Table 2.1 is a causally effective emergent level of structure. They are all equally real.

Just as in the case of neurons and the mind, and indeed biology as a whole [50], this is the only approach that makes sense. And it is valid because of the reality of top-down causation in the hierarchies, as discussed in this chapter. I revisit this issue in Sect. 8.1.

The operations at each level in both the logical and implementation hierarchies are realizations of possibilities occurring in abstract Platonic spaces such as Ω_a (Sect. 2.7.5), and these are the ultimate source of the possibility of computation. Their implementation in physical terms is possible because the human mind is able to comprehend the nature of these possibility spaces [19].

References

1. H. Abelson, G.J. Sussman, J. Sussman, *Structure and Interpretation of Computer Programs* (MIT Press, Cambridge, 1996)
2. U. Alon, *An Introduction to Systems Biology: Design Principles of Biological Circuits* (Chapman and Hall/CRC, London, 2007)
3. A.W. Appel, *Modern Compiler Implementation in Java* (Cambridge University Press, Cambridge, 2002)
4. W. Ross Ashby, *An Introduction to Cybernetics* (Chapman and Hall, London, 1957). http://pcp.lanl.gov/books/IntroCyb.pdf
5. G. Auletta, G.F.R. Ellis, L. Jaeger, Top-down causation: From a philosophical problem to a scientific research program. J. R. Soc. Interface **5**, 1159–1172 (2008). arXiv:0710.4235
6. A.V. Aho, M.S. Lam, R. Sethi, J.D. Ullman, *Compilers: Principles, Techniques, and Tools Paperback* (Pearson, 2013)
7. S. Beer, *Brain of the Firm* (Wiley, Chichester, 1981)
8. C.H. Bennett, Notes on Landauer's principle, reversible computation and Maxwell's demon. Stud. History Philos. Modern Phys. **34**, 501–510 (2003)
9. S. Bennett, S. McRobb, R. Farmer, *Object-Oriented Systems Analysis and Design* (McGraw Hill, Maidenhead, 2010)
10. A. Bérut, A. Arakelyan, A. Petrosyan, S. Ciliberto, R. Dillenschneider, E. Lutz, Experimental verification of Landauer's principle linking information and thermodynamics. Nature **483**, 187–190 (2012)
11. C.M. Bishop, *Neural Networks for Pattern Recognition* (Oxford University Press, Oxford, 1999)
12. G. Booch, *Object-Oriented Analysis and Design with Applications* (Addison Wesley, New York, 1994)
13. G. Booch, J. Rumbaugh, I. Jacobson, *The Unified Modeling Language User Guide* (Addison Wesley, New York, 1998)
14. V. Brattka, *Computability Theory* (University of Cape Town Notes, 2011)
15. V. Brilhante, Computer modelling hierarchy: the model reflects the hierarchy of the system being modelled. J. Braz. Comp. Soc. **11**(2), Campinas (2005)
16. G. Buzsáki, *Rhythms of the Brain* (Oxford University Press, Oxford, 2006)

17. J.-P. Changeux, A. Connes, *Conversations on Mind, Matter, and Mathematics* (Princeton University Press, Princeton, 1998)
18. G.J. Chaitin, Computers, paradoxes and the foundations of mathematics. Am. Sci. **90**, 164–171 (2002)
19. P. Churchland, *Plato's camera: how the physical brain captures a landscape of Abstract Universals (Cambridge)* (The MIT Press, Cambridge, 2012)
20. B.J. Copeland, The Church–Turing Thesis. The Stanford Encyclopedia of Philosophy, (Fall 2008 edition), ed. by E.N. Zalta (2002). http://plato.stanford.edu/archives/fall2008/entries/church-turing/
21. J. Copeland, *The Essential Turing* (Oxford University Press, Oxford, 2004)
22. T. Deacon, *The Symbolic Species: The Co-Evolution of Language and the Human Brain* (Penguin, London, 1997)
23. K.A. De Jong, *Evolutionary Computation: A Unified Approach* (MIT Press, Cambridge, 2006)
24. G. Dyson, *Darwin Among the Machines* (Penguin, London, 1997)
25. G.F.R. Ellis, True complexity and its associated ontology, in *Science and Ultimate Reality: Quantum Theory, Cosmology and Complexity*, ed. by J.D. Barrow, P.C.W. Davies, C.L. Harper (Cambridge University Press, Cambridge, 2004), pp. 607–636
26. G.F.R. Ellis, On the nature of causation in complex systems. Trans. R. Soc. S. Africa **63**, 69–84 (2008)
27. G.F.R. Ellis, Top-down causation and emergence: some comments on mechanisms. J. R. Soc. Interface Focus **2**, 126–140 (2012)
28. G.F.R. Ellis, D. Noble, T. O'Connor (eds.), Top-down causation: An integrating theme within and across the sciences? R. Interface Focus Spec. Issue **2**, 1–140 (2012)
29. K. Fatahalian, T.J. Knight, M. Houston, M. Erez, D.R. Horn, L. Leem, J.Y. Park, M. Ren, A. Aiken, W.J. Dally, P. Hanrahan, Sequoia: Programming the memory hierarchy, in *SC 2006 Conference, Proceedings of the ACM/IEEE* (2006)
30. R.L. Flood, E.R. Carson, *Dealing with Complexity: An Introduction to the Theory and Application of Systems Science* (Plenum Press, London, 1990)
31. E. Gamma, R. Helm, R. Johnson, J. Vlissides, *Design Patterns: Elements of Reusable Object Oriented Software* (Addison Wesley, New York, 1995)
32. P. Gray, *Psychology* (Worth Publishers, New York, 2011)
33. J. Hawkins, *On Intelligence* (Holt Paperbacks, New York, 2004)
34. D. Hofstadter, *Godel, Escher, Bach: An Eternal Golden Braid* (Penguin, London, 1980)
35. D. Hilbert, On the infinite, in *Philosophy of Mathematics*, ed. by P. Benacerraf, H. Putnam (Prentice Hall, Englewood Cliff, 1964), p. 134
36. J.H. Holland, *Adaptation in Natural and Artificial Systems* (MIT Press, Cambridge, 1992)
37. B. Jacobs, S.W. Ng, D.T. Wang, *Memory Systems: Cache, DRAM, Disk* (Elsevier, Burlington, 2008)
38. N.L. Kamorova, M.A. Nowak, Language, learning, and evolution, in *Language Evolution*, ed. by M.H Christensen, S. Kirby (Oxford University Press, Oxford, 2005), pp. 317–337
39. R.M. Keller, *Computer Science: Abstraction to Implementation*. http://www.cs.hmc.edu/~keller/cs60book/
40. J.F. Kurose, K.W. Ross, *Computer Networking: A Top-Down approach* (Addison-Wesley, New York, 2012)
41. J. Ladyman, S. Presnell, A.J. Short, B. Groisman, The connection between logical and thermodynamic irreversibility (2006). http://philsci-archive.pitt.edu/id/eprint/2689
42. R. Lafore, *Data Structures and Algorithms in Java* (SAMS, Indianapolis, 2002)
43. R. Landauer, Irreversibility and heat generation in the computing process. IBM J. Res. Dev. **5**, 183–191 (1961)
44. S. Lloyd, Computational capacity of the universe (2001). arXiv:quant-phy/0110141
45. J. MacCormack, *Nine Algorithms that Changed the Future: The Ingenious Ideas that Drive Today's Computers* (Princeton University Press, Princeton, 2012)
46. J.L. McClelland, The case for interactionism in language processing. Technical Report AIP-2 (Department of Psychology, Carnegie-Mellon University Pittsburgh, PA 15213 USA, 1987)

47. M.M. Mano, C.R. Kime, *Logic and Computer Design Fundamentals* (Pearson/Prentice Hall, 2008)
48. J. McCarthy, Artificial intelligence, logic and formalizing common sense (1990). http://www-formal.stanford.edu/jmc/
49. J.H. Miller, S.E. Page, *Complex Adaptive Systems: An Introduction to Computational Models of Social Life* (Princeton University Press, Princeton, 2007)
50. D. Noble, A theory of biological relativity: no privileged level of causation. Interface Focus **2**, 55–64 (2012)
51. Object Management Group (OMG), *OMG Unified Modeling Language (OMG UML) Superstructure Version 2.2*. http://www.omg.org/spec/UML/2.4.1/
52. R. Penrose, *The Emperor's New Mind: Concerning Computers, Minds and the Laws of Physics* (Oxford University Press, New York, 1989)
53. R. Penrose, *Shadows of the Mind: A Search for the Missing Science of Consciousness* (Oxford University Press, Oxford, 1994)
54. R. Penrose, *The Large, the Small and the Human Mind* (Cambridge University Press, Cambridge, 1997)
55. J. Porway, Q.C. Wang, S.C. Zhu, A hierarchical and contextual model for aerial image parsing. http://vcla.stat.ucla.edu/Aerial_Image_Parsing/index.html
56. S Russell and P Norvig (2009) *Artificial Intelligence: A Modern Approach* (Prentice Hall)
57. W. Savitch, *Absolute Java* (Pearson, Boston, 2010)
58. S.C. Shapiro, Artifical intelligence, in *Encyclopaedia of Artificial Intelligence*, ed. by S.C. Shapiro (Wiley, New York, 1992), pp. 54–57
59. J.R. Searle, Is the brain a digital computer? https://mywebspace.wisc.edu/lshapiro/web/Phil554_files/SEARLE-BDC.HTM
60. A. Silberschatz, P.B. Galvin, G. Gagne, *Operating System Concepts* (Wiley, New York, 2010)
61. H.A. Simon, *The Sciences of the Artificial* (MIT Press, Cambridge, 1992)
62. A.S. Tanenbaum, *Structured Computer Organisation* (Prentice Hall, Englewood Cliffs, 2006)
63. R.L. Trask, *Language and Linguistics: The Key Concepts* (Routledge, Abingdon, 2007)
64. A.M. Turing, On computable numbers, with an application to the Entscheidungsproblem, in *Proceedings of the London Mathematical Society*, vol. **42**, pp. 230–265 (1936) (Reprinted in J. Copeland, *The Essential Turing* (Oxford University Press, Oxford, 2004), p. 58)
65. A.M. Turing, Lecture on the automatic computing engine (1947) (Reprinted in J. Copeland, *The Essential Turing* (Oxford University Press, Oxford, 2004), p. 378)
66. A.M. Turing, Computing machinery and intelligence. Mind **59**, 433–460 (1950) (Reprinted in J. Copeland, *The Essential Turing* (Oxford University Press, Oxford, 2004), p. 433)
67. D.A. Watt, D.F. Brown, *Programming Language Processors in Java: Compilers and Interpreters* (Prentice Hall, Harlow, 2000)
68. M.A. Weiss, *Data Structure and Algorithm Analysis in Java* (Addison Wesley/Longman, 1999)

Chapter 3
The Basis of Complexity

This chapter looks at the basis of emergent complexity in physical systems, including life, as well as in logical systems, considering in turn:

- Section 3.1. The nature of emergence, and modular hierarchical structures.
- Section 3.2. Bottom-up effects and strong reductionism.
- Section 3.3. Emergence, higher level dynamics, and higher-level variables.
- Section 3.4. Top-down effects in physical and logical hierarchies as the key to the emergence of complexity, and the issue of supervenience.
- Section 3.5. The key enabling concept: equivalence classes of lower level states and dynamics.
- Section 3.6. Ways of demonstrating top-down causation.
- Section 3.7. Constraints on emergence.

Thus this chapter develops in more depth the ideas that were presented in outline in Chap. 1, and illustrated in some depth in Chap. 2 in the case of digital computers. The following chapter considers what different kinds of top-down causation might exist.

3.1 The Nature of Emergence

How can complex systems emerge from simple parts? As proposed in the first chapter, my main theme is this:

> Genuine complexity can only emerge from interlevel causation (both bottom-up and top-down) in modular hierarchical structures.

Each of the aspects mentioned here ('modular', 'hierarchical', and 'structure') is crucial for the emergence of complexity out of interactions between simpler units [14, 42, 48, 96, 120]. I first make some broad statements about the nature of emergence in this section, and then elaborate on them in the later sections as follows:

G. Ellis, *How Can Physics Underlie the Mind?*, The Frontiers Collection,
DOI 10.1007/978-3-662-49809-5_3

- Emergence of complexity is based on structure (Sect. 3.1.1),
- Emergence is different in different contexts (Sect. 3.1.2),
- Emergence results in a structural/functional hierarchy (Sect. 3.1.3),
- Emergence enables logical hierarchies, information flows (Sect. 3.1.4),
- Emergence has different timescales (Sect. 3.1.5),
- Emergence is based on modularity (Sect. 3.1.6).
- Emergence is based on interlevel relations (Sect. 3.1.7)

3.1.1 Emergence of Complexity Is Based on Structure

Without structure, one has random events that are chaotic at lower levels, though they may lead to order of sorts at a higher level—a gas in equilibrium is a case in point. There is indeed emergent behaviour in this case, viz., the gas laws that describe its coarse-grained behaviour, and it is not complex. The outcome is not more than the sum of the effects of the parts. One can get somewhat more interesting behaviour in the cases of sandpiles, the reaction–diffusion equation, flocks, swarms, etc. They are impressive, producing interesting spatial and temporal patterns through local interactions, but they are not by themselves capable of genuinely complex behaviour, in the end because there is no coordination of what is going on. Claims have been made of genuinely complex behaviour emerging in the case of cellular automata, indeed that they can emulate a universal Turing machine, but this does not mean they can produce the complexity of life in a viable way.[1] Reliable emergence of the complexity of life on day to day timescales requires the hierarchically structured reactions in a living cell, coordinated so as to create organised emergent behaviour that fulfils specific purposes in a living organism [63].

True complexity requires structures such as the micro-connections in a VLASIC computer chip in a computer, tissues made of cells which are in turn made of immensely complex interacting biomolecules in an animal, and neural networks made up of hierarchically interconnected neurons in a brain. Such systems are not complex merely because they are complicated: 'order' means organization, in contrast to randomness or disorder. Such structure enables the build-up of genuine complexity if it is appropriately formed to fulfil some higher level function, and this is the case in biology: structure both follows function and enables it [20]. The structure is emergent from lower level entities, but is much more than the parts. It is the patterns of structuring that count. This is a higher level property of the system: its description requires variables that relate to more than just the properties of the components.

[1] See S Weinberg, "Is the Universe a Computer?" *New York Review of Books* (October 24, 2002, available at http://www.nybooks.com/articles/archives/2002/oct/24/is-the-universe-a-computer/?pagination=false) for an illuminating discussion in a review of Steven Wolfram's book *A New Kind of Science*. Crucially, the key step is "The program for the calculation and the data to be used would be fed into a rule 110 cellular automaton as a pattern of black cells in the top row"—but who or what would do that?

Thus complexity is based on networks of interacting elements [88] which can be represented by causal diagrams [60]. They can be studied via the methods of statistical mechanics [1], identifying small-world and scale-free networks and evaluating network robustness, but that does not capture the core of their biological function, which at the microscale is embodied in the details of gene regulatory circuits and metabolic networks [128]. The network structure is an irreducible higher-level characteristic relative to the levels of the genes and molecules themselves. In addition to the properties of the units themselves, it is the set of relations between units—large scale topological relations as well as local causal motifs [2, 7, 125]—that is crucial to building up complexity. These aspects cannot be reduced to lower level variables. For example, the nature of protein function is determined on a global scale and depends on the entire connectivity pattern of the protein network [126]. Multiple functional assignments are made possible as a consequence of the existence of multiple equivalent solutions: which is the basic principle of multiple realisability discussed below (Sect. 3.5).

Higher level structural patterns channel causation at lower levels in the system, breaking symmetry and so constraining what happens at those levels. And those constraints, expressed for example in terms of effective potentials characterizing a wiring system or a neural network, lead to many different kinds of interesting behaviour by coordinating behaviour at lower levels. But structuring can take place in abstract systems as well as physical ones. In many cases it is patterns of abstract structure that determine the physical structure and behaviour that occurs, because life has needs that can be understood in logical ways, and so our brains are adapted to understand and predict logical patterns [72].

3.1.2 Emergence Is Different in Different Contexts

Emergence takes place in many different contexts, leading to many different kinds of entities:

- **Natural Objects**. Naturally occurring physical objects such as rocks, planets, stars, galaxies: no kind of purpose is evident in their nature or dynamics.
- **Life**. Bacteria, plants, animals, including intelligent beings such as humans: these are all goal-seeking (they are teleonomic).
- **Manufactured Objects and the Built Environment**. Physical artefacts, such as automobiles, aircraft, computer systems, houses, cities, bridges, water systems: these are physical entities designed to fulfil some specific purpose.
- **Organizations**. Societies, firms, armies, states, organizations: these are social constructions designed to fulfil some purpose, with both abstract and physical aspects.
- **Conceptual Structures**. Mental entities such as language, mathematics, mental models, theories, legal systems, constitutions: these are the basis of the last two.

Each of the classes mentioned has both physical and logical aspects. Physical systems are governed by law-like abstract relationships that can be expressed algorithmically. Biological development is based on the genetic code and translation of DNA sequences in an algorithmic way into sequences of amino acids. Organizations are effective because of an underlying mental model (a conceptual framework) that leads to their structure, as are manufactured entities such as aircraft and computers. This is possible because conceptual structures exist in minds that are enabled by physical brain states, and are causally effective through physical means such as speech and writing, as well as through computers.

3.1.3 Emergence Results in a Structural/Functional Hierarchy

The result of emergence in the natural and life sciences is the hierarchy of structure and causation set out in Table 3.1. A hierarchical structure will be described by a corresponding hierarchy of variables appropriate to describing the different levels of the hierarchy. This table gives a simplified representation of this hierarchy of levels of reality as characterized by corresponding academic subjects, with the natural sciences on the left [41] and the life sciences on the right [20].

Layers of emergent order and complexity build upon each other, with physics underlying chemistry, chemistry underlying biochemistry, and so on. On both sides, each lower level underlies what happens at each higher level in terms of structure and causation. There is a vast variety of existence at each higher level in the hierarchy (very large numbers of possible organic macromolecules, very many species of animals, etc.), but fewer kinds of entities at the lower levels (atoms are made of just protons, neutrons, and electrons), so complex objects with complex behaviour are made by highly structured combinations of simpler objects with simpler behaviour.

Table 3.1 The hierarchy of structure and causation for inanimate matter (*left*) and for life (*right*), as characterized by academic disciplines

Level 10	Cosmology	Sociology/economics/politics
Level 9	Astronomy	Psychology
Level 8	Space science	Physiology
Level 7	Geology, Earth science	Cell biology
Level 6	Materials science	Biochemistry
Level 5	Physical chemistry	Chemistry
Level 4	Atomic physics	Atomic physics
Level 3	Nuclear physics	Nuclear physics
Level 2	Particle physics	Particle physics
Level 1	Fundamental theory	Fundamental theory

Each level underlies what happens at the next higher level, in terms of physical causation. The existence of higher level complex behaviour, which does not occur at the lower levels, then emerges from the lower level properties both structurally and functionally (at each moment) and in evolutionary and developmental terms (over time).

Each higher level physical element, created by structured combinations of lower level elements, has different properties from the underlying lower levels—the entities at each level show behaviours characteristic of that level. Each level is described in terms of concepts relevant to that level of structure (particle physics deals with quarks and gluons, chemistry with atoms and molecules, and so on), so a different descriptive language applies at each level. Different levels of the hierarchy function according to laws of behaviour appropriate to that level, and are describable only in terms of language suited to that level. One cannot even describe higher levels in terms of lower level languages because a different phenomenological description of causation is at work at the higher levels, which may be described in terms of different causal entities. Ideas applicable to lower level causation do not by themselves succeed in explaining the higher level behaviours, for the concepts employed are simply not appropriate to the higher level kinds of causation (a motor mechanic does not talk in terms of quarks and electrons, for example).

Essential Differences Between Levels. Hierarchical structures have different kinds of order, phenomenological behaviour, and descriptive languages that characterize each level of the hierarchy.

It is sometimes queried whether these levels actually exist 'out there', or are rather impositions of the mind. My position is that different kinds of causation do indeed exist at the different levels as characterized here, and the mind recognizes these distinctions which actually exist. They are not just inventions of the mind. Atoms are different from molecules, whether characterised as such by a mind or not.

Non-Physical Entities. Note that there is no correlation between the left- and the right-hand columns above the level of chemistry, as emergence and causation is quite different in the two cases. However, the first four levels are identical (life emerges out of physics!). At the higher levels on the biology side, non-physical variables become relevant: in particular, as discussed later, thoughts and ideas and social constructions are higher level effective variables, even though they are not physical entities. Thus one can propose a 'software explanation' of behavior [30] based on overlapping hierarchies. The same is true of digital computers: computer algorithms and programs are causally effective even though they are non-physical entities (see Chap. 2). There is a link here: the existence of computers as physical entities is an example of the causative power of thoughts and ideas—computers would not exist if they had not been designed. There is no doubt whatever about the causal efficacy of thoughts.

Thus the hierarchy on the life sciences side is in terms of function and causation rather than the scale of physical entities. The hierarchy is determined by finding out what entities—physical or otherwise—exert constraints or set conditions so as to channel interactions between elements which have their own laws of interaction at their own level, for any environment acts in this way on any system it contains. Together with a careful analysis of what more complex elements emerge from simpler ones, this defines which is a higher level and which a lower level in the hierarchy.

Thus it is a hierarchy of whole–part relations, which at the bottom levels can be seen as physical (a lower level entity is physically a part of a larger one), but at higher levels is a causal hierarchy (higher level entities provide the causal context for lower level ones).

Top and Bottom. It is unclear what the topmost and bottommost levels of the hierarchy are. The topmost level is shrouded in metaphysics, and the bottommost one in the unknown nature of quantum gravity. Luckily this does not not prevent us from considering the levels in Table 3.1 and the relations between them. We would not be able to carry out scientific enquiry if this were not the case. Similarly, there may be no unique top level in the brain, where multiple interlocking hierarchies occur. That does not prevent us from considering causation in the many hierarchical structures in the brain that are easily identifiable.

Reality of Elements. Implicit in this discussion is the view that the elements at each of the levels of existence characterized by Table 3.1, except perhaps at the quantum level [61, 77], can be regarded as real [38, 40]. A table is a real table, even though it is made of atoms, which are also real, even though they are made of electrons, protons, and neutrons. And of course, the same applies to animals and people. This view, too, is needed in order that science should make sense. If a real experimenter does not exist, then experiments are not possible.

Quantum mechanics is applicable at the lower levels, but apparently not at the macrolevels except under very restricted circumstances, e.g., superconductivity, Bose–Einstein condensations, lasers, and the extraordinary recent quantum entanglement experiments over many kilometers. It does not apply under ordinary everyday circumstances at the macrolevel (which is why quantum dynamical principles are not obvious to us). Hence, experimenters talk about the classical/quantum split, or Heisenberg cut [131, p. 15], necessary for them to analyze their experiments. Things are real in the classical sense above that cut, and this includes all of the everyday world, even if it looks completely different at other levels of description [38].

> **Existence**. The different levels are all real, each existing with causal powers in its own right, because (as explained in detail in this book) they each have determinable effects on the levels above and below them. No level is more real than any other [91].

If this were not the case, we would be stuck, because we would not for example be able to treat neurons and genes as real, as Francis Crick [27] would wish us to (see the quote by Crick in Chap. 1, and response in Sect. 7.7.3). They are certainly not the bottommost level, but neither molecular biology nor neuroscience makes sense if we don't assign genes and neurons real causal powers over lower levels.

The same applies to every level: the particular science that studies that level (genetics, neuroscience, ecology, physiology, economics, for example) would not be dealing with real causal powers otherwise. Your bank manager exists with real causal powers, for example, as does the bank: this becomes very clear when you open an account or apply for a loan. It would not make sense to claim a heart attack was a cause of ill health if the level of physiology was not real. And even the quantum mechanics levels are real, if they underlie the classical levels, as they surely do.

3.1.4 Emergence Enables Logical Hierarchies, Information Flows

As well as structural hierarchies, there are logical hierarchies that are causally effective. The two may occur in a deeply entwined fashion in specific contexts, as is clear in the case of digital computers (Chap. 2): the structural/functional hierarchy is the substrate whereby the logical hierarchies are effective. They are described in terms of language, logical symbolism, and mathematics.

Language. Language is the symbolic tool by which logical thought functions [32]. It thus underlies the causal power of thought, which plays a key role in social structure and society [117]. It is based on naming entities, actions and events, and qualities, with patterns of relationships structuring their use [122]. It is hierarchically structured in two ways: first in its representation of entities and actions by words, and secondly in the way it represents references. As to the first, in the written form we have the hierarchy:

letters → words → phrases → sentences → paragraphs → chapters → books,

a hierarchical structure building up complex entities out of combinations of simple units. The base units are letters when we represent words in writing (and 0s and 1s if we represent them digitally), but the key level out of which higher order structures are built is that of words (this corresponds to the level of atoms in physics and cells in biology). The natural ordering of words is alphabetical, enabling us to store them in an ordered way that enables indexing and recall, but this ordering of words has nothing to do with their meaning, because the relation between the letters in a word and its meaning (as given in a dictionary) is arbitrary. The full set of words in use can be defined in terms of a defining vocabulary of about 3000 words which provides the referential base for the rest [92].

It is crucial here that words are understood in context. For example the word 'plane' could refer to a landscape, a vehicle, a woodworking tool, or a mathematical concept. One decides which it is from the context. The words 'they', 'it', 'then', and so on are purely contextual in nature:

> **Context and Language**. One has a top-down effect from the text as a whole to the interpretation of the words and phrases in it. Language is contextual through and through.

As to the second, given its syntactic structure, language enables reference to all our present state of knowledge, which is necessarily hierarchically classified (as in an encyclopedia) in order that we can understand it. The issue here is that one cannot understand relations between the vast variety of objects in the world without using a hierarchical characterization of properties of generic classes and specific instances. Thus for example:

animal → mammal → domestic animal → dog → guard dog → Doberman → Fred, machine → transport vehicle → automobile → sedan → Toyota → CA687-455.

Language has a hierarchical structure of meanings that is a logical hierarchy, reflecting the hierarchical structure in the world that is being represented—as is necessary in order for it to be a good semiotic representation of the system being described [33]. This hierarchy of meaning is logically largely independent of the lexical hierarchy. Dictionaries and encyclopedias give the arbitrarily assigned relation between the two.

Mathematics. This has similar hierarchies to those of language. Quantities are represented by numbers in a hierarchical way. Using a decimal basis:

millions → thousands → hundreds → tens → units,

analogously to how words are built up from letters, the units being represented in terms of the numbers 0–9 (but being represented in terms of 0s and 1s if we use a binary representation). Mathematical statements are built up hierarchically from combinations of operations on numbers and variables, leading to the whole structure of mathematical presentations in terms of lemmas, theorems, and propositions.

In addition to that, there are logical hierarchies: mathematical operations are hierarchical in that, for instance, integration is built on addition, and differentiation on subtraction and division. The former are in each case emergent from combinations of the latter. It is this kind of logical relation that is expressed in the mathematical formalism.

Computer Languages. These are hierarchically structured in two ways: there is an implementation hierarchy of languages, each level of which is structured in terms of a logical hierarchy, as described in Chap. 2. These logical layers are each causally effective in terms of determining the computer's output. One can write and run programs at any of the levels in the implementation hierarchy, it is just for human convenience that we usually use the topmost level to do so.

Information. Implementation of logical hierarchies requires *information flows*: collecting, processing, storing, and recalling logical relations. Information processing includes analyzing, filtering, and passing on the results to other modules for further processing or to initiate some form of action. Information flows can be *same level* or *interlevel*, passing messages up or down the levels in the implementation hierarchy.

As well as in spoken and written language, and in computer systems, such flows also happen in biology, for example, in physiology, and in the way genetic information is coded in DNA: it is written, stored, recalled, and translated into its biological meaning by cell processes [20]. Information flows are key causal factors in feedback control loops (Sect. 4.2), which are crucial to homeostasis throughout biology as well as in organizations [9]. They are an important feature in the functioning of society, and are of course key to the operation of the mind [59]: the brain [73] is an information-processing device par excellence (see Chap. 7). Information filtering is a key form of adaptive selection (Sects. 4.3–4.5) in all these contexts.

Table 3.2 The different timescales associated with evolution, development, and functioning

Type of system	Long term evolution	Short term evolution	Development	Function
Natural	10^9 years	10^5 years	10^5 years	Weeks to hours
Biological	10^9 years	10^5 years	20 years	Hours to msec
Artificial	10^4 years	10^2 years	10 years	Hours to μsec

In the 'natural system' row, 'function' refers to events such as volcanoes, earthquakes, typhoons, etc. In the 'biological systems' row, it refers for example to typical brain operations, while in the 'artificial systems' row it refers to a typical modern computer and its micro-operations

3.1.5 Emergence Has Different Timescales

Emergence occurs in terms of (a) evolution/coming into existence of species/type, (b) development/creation of each individual object/being, and (c) function of individual object/being, each occurring with very different timescales. The relevant timescales (Table 3.2) are related both to physical size and to degree of tightness of coupling.

Each type of emergence in biology is characterised by adaptive selection in interaction with the physical and social environment, these being the boundary conditions for the system. As life emerges, in each case there is a dramatic change from matter without complex functionality to living material.

3.1.6 Emergence Is Based on Modularity

These hierarchical structures are modular, made up of structural combinations of semi-autonomous components, each carrying out specific functions. The modules at each level will generally constitute the next lower level in the hierarchy. The structure and behaviour of modules can be studied in their own right: molecules are made of atoms, living bodies are made of cells, and so on. One can study atoms and living cells in their own right, and then see how they fit together to make molecules and bodies [20].

The key modules in physics are *atoms*: all matter is made of atoms. Feynman strongly emphasized [49] that this is probably the single most important thing we have learned in physics. These are made up of electrons and atomic nuclei, which are made up of particles (protons and neutrons), and so on, so they are just one level in the hierarchy, but with their properties as summarized in the periodic table of the elements, atoms are the key link between large and small.

Correspondingly, the key modules in biology are *cells*: all life is based on cells [20]. They are made up of nuclei, mitochondria, ribosomes, and so on, each being modules in their own right, containing biomolecules (proteins, RNA, DNA, etc.) of many kinds. They are the basic unit of life. In particular, the key modules in the brain are *neurons*: all the major brain areas are made up of networks of neurons [73]. The corresponding modules in computers are *gates*: circuits of gates create all the higher level logical operations [82].

3.1.6.1 Modules, Abstraction, and Information-Hiding

A hierarchy represents a decomposition of the problem into constituent parts and of processes into sub-processes to handle each of these sub-problems, each sub-process requiring less data and more restricted operations than the problem as a whole [120]. Modular units with abstraction, encapsulation, and inheritance handle each of these sub-processes [14]. Modules can be modified and adapted to fulfil new functions, enabling great flexibility as complex structures adapt to a changing environment. Modules at a particular level are identified by tighter binding, higher speeds of internal interaction, and higher energies than those at the next higher level in the hierarchy, and indeed it is this tighter binding that identifies them as modules. The high-frequency dynamics of the internal structures of components (relating internal variables) contrasts with the low-frequency dynamics of interactions amongst components (relating external variables). Combinations of many high-frequency lower-level interactions result in lower-frequency higher-level actions (a computer microchip may perform millions of operations per second, but the user still has to wait for the computer to do what she wants at the macro level).

The success of hierarchical structuring depends on implementing modules to handle lower-level processes and on the integration of these modules into a higher-level structure (for example, atoms comprising molecules and cells comprising a living being). This structuring enables the modification of modules and re-use for other purposes, and in addition enables fine-tuning of the internal structure of a module without affecting the large scale dynamics. It also makes the dynamics understandable. This is clear for example in complex computer programs, which may have 15 million or more lines of code: they are only understandable because they are written in a modular way with numerous separate subroutines that can each be understood on their own.

Abstraction. A key feature is that compound objects (combinations of modules) can be named and treated as single units by appropriate labeling. This leads to the power of abstract symbolism, symbolic computation, and recursion. An abstraction denotes the essential characteristics of a module that distinguishes it from all other kinds of objects. It focuses on the outside view of the module, and so serves to separate its essential behaviour from its implementation. Further, it emphasizes some of the system's details or properties, while suppressing many others.

Information is continually thrown away by the billion bits when one replaces the internal description with this external view. This is what enables modules to generate higher level structure, and it is essential to the emergence of higher level behaviour because all the micro-alternatives can neither be examined nor controlled.

Encapsulation. This occurs when the internal workings are hidden from the outside, so internal procedures can be treated as black-box abstractions. No part of any complex system should depend on the internal details of any other part: system functionality only specifies each component's function, leaving it to the object to decide how to do it.

Information-Hiding. Local variables that control the internal dynamics are invisible from the outside: access to the internal variables is only through carefully controlled interfaces. In computers, this is handled by the scoping of variables and parameter passing. In biology, it is handled by containing reactions within containers such as cell walls that allow limited access to the outside environment. Thus information-hiding is a key feature of complexity.

Inheritance. This is when specialised modules (forming a sub-class) preserve most or all of the functions of the super-class, but with extra specialisation or further properties built in. This corresponds to fine-tuning the modules to handle more specialised problems (for example generalised cells specialise to form neurons).

These issues are all discussed in greater detail in Sect. 2.2.2, in the context of digital computers and object-oriented programming [14].

3.1.7 Emergence Is Based on Interlevel Relations

Multifold causation takes place in such systems. A network of causal influences and constraints interact to produce an outcome. In order to understand such systems, we often take for granted most of these influences and concentrate on one or two of them, which we then label as 'the causes'. This has the connotation of dominant causes. But a web of influences and multiple causations is in action all the time. Nevertheless, in order to understand what is going on, it is useful to single out particular links in this causal pattern, taking all the rest for granted. I will follow this usage.

Given this understanding, the dynamics in complex systems involves the combination of bottom-up and top-down action [19] in the hierarchy of structure (see Fig. 3.1). The higher and lower levels are related to each other because the higher levels are based on the lower levels. But the higher levels set constraints on lower level dynamics in a top-down manner and this influences them in many ways [42]. It enables the activation of interlevel feedback loops which can then facilitate genuine complexity. Boundary conditions shape lower level outcomes, but these in turn influence the environment.

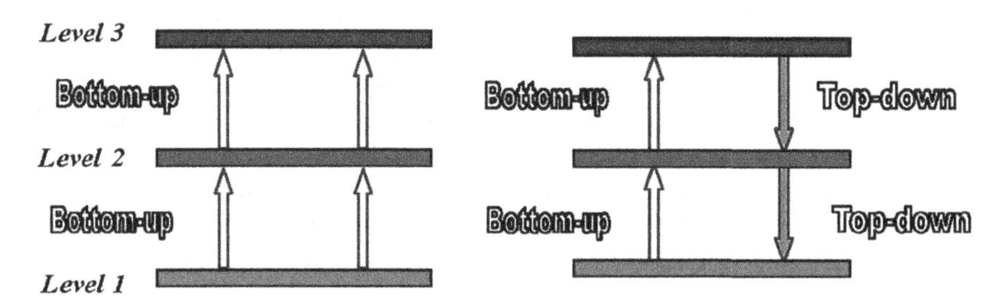

Fig. 3.1 Bottom-up and top-down causation. The fundamental importance of top-down causation is that it changes the causal relation between *higher* and *lower* levels in the hierarchy of structure and organisation. Consider the difference between bottom-up only (*left*) and the combination of bottom-up and top-down (*right*)

Bottom-up action will be reviewed in Sect. 3.2, and top-down action in Sect. 3.4. There are 5 distinct types of top-down action: they will be discussed in Chap. 4. The key idea in top-down action is equivalence classes: these are discussed in Sect. 3.5. A key question that arises in top-down causation is how there can be room for it to take place, given that physics underlies everything and is on the face of it a causally complete theory. This will be discussed in Chap. 5. The cases of top-down causation in physics and in the context of the mind will be discussed in Chaps. 6 and 7 respectively.

3.2 Bottom-Up Effects

A major theme of physics and science is that causation occurs from the lower to the higher levels of the hierarchy in Table 3.1, leading to higher level behaviour. What happens at each higher level is based on causal functioning at the level below. Hence, what happens at the highest level is based on physical functioning at the bottommost level. When I move my arm, it moves because many millions of electrons attract many millions of protons in my muscles, as described by Maxwell's equations. Thus microphysics underlies macro effects. The successive levels of order entail chemistry being based on physics, material science on physics and chemistry, geology on material science, and so on.

Bottom-Up Causation. This is the ability of lower levels of reality to have a causal effect on the higher levels which emerge from them, sometimes uniquely determining what happens at the higher levels.

This is illustrated in Fig. 3.1 (left).

3.2.1 Coarse-Graining

A feature that occurs here [3, 49] is the *coarse-graining* of lower level variables (e.g., particle states) to give higher level variables (e.g., density and pressure), accompanied by a conversion of useful energy to non-usable energy when some energy is hidden in lower level variables, and hence not available to higher levels. This is the source of entropy growth and of effective non-conservation of energy at higher levels through friction and other dissipative effects. As the lower level dynamics proceeds, for example diffusion of molecules through a gas, the corresponding coarse-grained higher level variables will change as a consequence of the lower level change. So for example a non-uniform temperature will change to a uniform temperature.

In this section, I consider in turn:

- Physics (Sect. 3.2.2).
- Biology (Sect. 3.2.3).
- Mathematics of emergence (Sect. 3.2.4).
- Strong reductionism (Sect. 3.2.5).

3.2.2 Physics

This view emerged from physics, particularly mechanics, statistical physics, and solid state physics. Its extension to understanding chemistry as arising from quantum physics was a major triumph. Examples are:

Example 3.1. Motion of Macroscopic Objects. The way an object moves, a football or steam engine for example, is due to the summation of all the forces on its component particles. Momentum conservation at the micro level leads to momentum conservation at the macro level.

Example 3.2. Gases. Determination of gas properties through the kinetic theory of gases, seen as bottom-up causation from molecular motions to gas properties. Molecular collisions lead to gas pressure, heat conduction, diffusion, and so on [3, 49].

Example 3.3. Solids. Determination of metallic properties such as thermal and electrical conductivity through the quantum theory of solids, resulting from electron motions in a lattice [22, 132].

Example 3.4. The Chemical Bond. The explanation of the chemical bond in terms of physical processes involving electron orbitals [94], with the properties of the periodic table of the elements resulting from the Pauli exclusion principle.

3.2.3 Biology

The huge change that has taken place in the last 80 years or so is the extension of this view to biology, in particular through the molecular biology revolution on the one hand [129] and the growth of understanding of neuronal processes on the other [73].

Example 3.5. Genetics. Animal development is based on the reading of genes, which determine protein synthesis and hence body structure. This is bottom-up causation from DNA to the phenotype [20].

Example 3.6. Action Potentials. Neuronal processes are based on ion transport across dendrite membranes and the resulting Hodgkin–Huxley equation and diffusion processes at synapses, leading to action potentials conveying information from one neuron to another [73, 115].

Example 3.7. The Brain. This is understood as a neural network built up from interacting neurons with outcomes determined by the integrated effects of the summation of action potentials at synapses [73, 115]. Hebbian processes ('fire together, wire together') are a bottom-up process for strengthening neural connections in a purely local way [31, pp. 281–313]. This is sometimes called unsupervised learning, but this is not a *learning* process in the sense understood in this book (see Sect. 4.3.4). It is a process of habituation independent of outcome. In a learning process, behaviour is modified according to outcome.

Example 3.7. Emergent Behaviour in Flocks and Swarms. Swarms of intelligent agents can lead to impressive kinds of patterns and emergent behaviour in a bottom-up way purely through local interactions between the agents [74]. However, it will be a theme of this book that what can happen in this way is strictly limited: it cannot produce truly complex behaviour, because it cannot lead to adaptation to an environment (this bottom-up process takes place irrespective of the nature of the environment). It cannot lead to learning, unless the agents themselves adapt to what is happening to the flock, in which case it is a top-down process from the flock level to the level of the individual agents.

3.2.4 Mathematics of Emergence

Coarse-graining of lower level variables clearly leads to effective higher level laws in simple physical cases. I will just give two examples here.

Key Example: Conservation Laws. Suppose a system S consists of N particles of mass m_i. The total mass of the system is $m := \sum_i m_i$. Then conservation of mass of the particles implies conservation of the total mass of the system:

$$\frac{\mathrm{d}m_i}{\mathrm{d}t} = 0 \quad \Longrightarrow \quad \frac{\mathrm{d}m}{\mathrm{d}t} = 0 \,. \tag{3.1}$$

The same will apply for example to electric charge. Thus microscopic conservation laws re-emerge as macroscopic conservation laws.

Classic Example: Centre of Mass Motion. The classical example of emergence of higher level linear behaviour out of lower level linear behaviour is the case of centre-of-mass motion (see [56] for a clear description). Consider a system of N point particles of mass m_i at position $\mathbf{r}_i$. Newton's law of motion for the ith particle is

$$m_i \ddot{\mathbf{r}}_i = \mathbf{F}_i^* = \mathbf{F}_i + \sum_j \mathbf{F}_{ij} \,, \tag{3.2}$$

where $\mathbf{F}_i^*$ is the total force on the ith particle, $\mathbf{F}_i$ is the external force, and $\mathbf{F}_{ij}$ is the internal force due to the jth particle (there is no self-force, i.e., $\mathbf{F}_{ii} = 0$). Newton's third law states that action and reaction are equal and opposite, i.e.,

$$\mathbf{F}_{ij} = -\mathbf{F}_{ji} \ . \tag{3.3}$$

Consequently, adding the Eq. (3.2) together for $i = 1$ to N, we find

$$\sum_i m_i \ddot{\mathbf{r}}_i = \sum_i \mathbf{F}_i^* = \sum_i \mathbf{F}_i \ . \tag{3.4}$$

Defining the total mass m, centre of mass position $\bar{\mathbf{r}}$, and total external force $\mathbf{F}$ by

$$m := \sum_i m_i \ , \quad m\dot{\bar{\mathbf{r}}} := \sum_i M_i \dot{\mathbf{r}}_i \ , \quad \mathbf{F} := \sum_i \mathbf{F}_i \ , \tag{3.5}$$

we find

$$m\ddot{\bar{\mathbf{r}}} = \mathbf{F} \ , \tag{3.6}$$

so the linear law for the individual particles [the first equality in (3.2)] is replicated by the coarse-grained variables in (3.5), and this irrespective of the nature of the internal forces.

Bottom-up action is often expressed in differential equations for the relevant variables, e.g., the evolution of quantities $\Phi_i(t)$, $(i = 1, \ldots, N)$, may be determined by

$$\left.\frac{\mathrm{d}^n \Phi_i}{\mathrm{d}t}\right|_t = f_i\left(\mathrm{d}^{n-1}\Phi_j/\mathrm{d}t(t), \ldots, \mathrm{d}\Phi_j/\mathrm{d}t(t), \Phi_j(t)\right) \ , \quad i, j = 1, \ldots, N \ . \tag{3.7}$$

The solution is determined uniquely on some interval by the initial data at an arbitrary time t_0, that is, by $\{\mathrm{d}^{n-1}\Phi_i/\mathrm{d}t(t_0), \ldots, \mathrm{d}\Phi_i/\mathrm{d}t(t_0), \Phi_i(t_0)\}$. The equation may be chaotic or unstable in some domain, in which case the solution is exquisitely sensitive to the initial data. Nevertheless, the equations are determinate in principle. Equation (3.6) is a special case, with the simple harmonic oscillator being a key example. In many other cases the relevant equations will be partial differential equations such as the wave equation or diffusion equation.

3.2.5 Strong Reductionism

The core of the reductionist view is that everything can be explained by such bottom-up mechanisms based on the laws of physics, with no remainder. This project is very successful in many cases. This is the profound basis for physicalist world views, as stated by Dirac [35]:

> The underlying physical laws necessary for the mathematical theory of a large part of physics and the whole of chemistry are thus completely known, and the difficulty is only that the exact application of these laws leads to equations much too complicated to be soluble.

The essential claim is that in the above characterization of bottom-up causation, one should delete the word 'sometimes' in the phrase 'sometimes uniquely determining what happens at the higher levels'. However, I will argue that the phrasing above is more accurate; that in fact, there is always a contextual dependence of specific outcomes. We often do not realize this feature because we take the environmental context for granted.

3.3 Emergence and Higher-Level Variables

The phenomenon of emergent order is when higher levels display new properties not evident at the lower levels. *More is different*, as famously stated by Anderson [4]. Emergence of complexity takes place where quite different laws of behaviour hold at the higher levels than at the lower levels [53, 74, 80, 113]. These properties are characterized by named higher level variables, and it is the symbolic naming of these variables that enables us to contemplate their nature. Here I consider in turn:

- Emergence of higher level structure and behavior (Sect. 3.3.1).
- Coherent higher level dynamics (Sect. 3.3.2).
- Emergent Higher level variables (Sect. 3.3.3).
- Intrinsically higher level variables (Sect. 3.3.4).

3.3.1 Emergence of Higher Level Structure and Behavior

Effective theories such as the Fermi theory of weak interactions, the gas laws, and Ohm's law give a phenomenological understanding of behaviour at higher levels [44]. The higher levels are generally more complex and less predictable than the lower levels: we have reliable phenomenological laws describing behaviour at the levels of physics and chemistry, but not at the levels of psychology and sociology. Thus this is a hierarchy of complexity. As emphasized above, one cannot even describe the higher level components or behaviour in terms of lower level language. Examples of emergence of higher level behaviour are:

- **E1 Statistical Physics**. The underlying atomic theory leads to the macroscopic gas laws, thermodynamics, and thermal properties of gases [3, pp. 434–518]. There is no similarity between the underlying theory and the emergent theory, except that concepts of mass, energy, and momentum conservation apply at both levels.
- **E2 Electrodynamics**. The process of coarse-graining leads to the polarization density of a polarized medium [123, pp. 343–349], where the electric field $\mathbf{E}$ is a coarse-grained version of the microscopic electric field $\mathbf{e}$, and the displacement vector $\mathbf{D} = \mathbf{E} + 4\pi\mathbf{P}$ includes a polarization term $\mathbf{P}$ representing coarse-grained dipole terms [68, pp. 103–108]. The fields $\mathbf{D}$ and $\mathbf{E}$ are related by a polarization tensor ϵ_{ij} such that $D_i = \epsilon_{ij} E_j$. The tensor ϵ_{ij} depends on the micro structure

of the medium. In an isotropic medium, $\epsilon_{ij} = \epsilon\delta_{ij}$ (using Cartesian tensors), while in an anisotropic medium this is not the case. The coarse-grained version of Maxwell's equations gives the divergence of **D** and curl of **E**, so a modified version of the microscopic equations emerges. The emergent theory is largely similar to the underlying theory.

- **E3 Physics to Chemistry**. The interactions of fermions lead through the Pauli exclusion principle to the nature of the hydrogen atom [3, pp. 109–148] and the electronic structure of atoms [3, pp. 158–176], and from there to the periodic table [5, 94]. The nature of the chemical bond emerges from physics [5, 94]. There is no similarity between the underlying theory and the emergent laws.
- **E4 Chemistry to Microbiology and Life**. The complex modular hierarchical structure of life emerges from the underlying physical and chemical laws [20]. There is no similarity between the underlying theory and the emergent behaviour, except that concepts of mass and energy balance apply at both levels.
- **E5 Interacting Species to Multilayered Ecological Systems**. Species interact with one another in different ways and those interactions vary spatiotemporally, leading to ecological multilayer networks [103].

In most cases, the underlying theory leads to a higher level theory characterizing quite different behaviour (after all, that is the essential content of Table 3.1). How does how higher level structure and behavior relate to lower level structure and behavior in two adjacent levels in the hierarchy of complexity?

Emergence of Structure. This occurs when higher level structure is based on lower level entities (the modules out of which the higher level entity is constructed, which may each have their own internal structure). This is represented in Table 3.3.

Examples are a crystal lattice, a star, a galaxy, a house, a computer, or a mouse. The emergence may take place spontaneously through bottom-up processes such as growth or crystallization, or it may be imposed by top-down processes such as manufacture. Either way, higher level structures are created out of lower level entities and then exist as entities in their own right. They are described by suitable higher level variables (Table 3.4).

Table 3.3 The emergence of higher level structure from lower level structure. Growth or manufacture creates a higher level entity from its parts

Level $N+1$	Structure I	Higher level entity	Whole
	$\Uparrow$	Growth/manufacture	$\Uparrow$ Emerge
Level N	Parts i	Lower level entities	Components

Table 3.4 The emergence of higher level variables from lower level variables

Level $N+1$	Variables V_I	Higher level description	Aggregated state S
	$\Uparrow$	Coarse-grain	$\Uparrow$
Level N	Variables v_i	Lower level description	Detailed states s_i

Table 3.5 The emergence of higher level behaviour from lower level dynamics

Level $N+1$	Initial state I	Higher level theory $T: \Rightarrow$	Final state F
	$\Uparrow$	Coarse-grain	$\Uparrow$
Level N	Initial state i	Lower level theory $t: \Rightarrow$	Final state f

Examples are the pressure, density, and temperature of a gas, a macroscopic magnetic field, the velocity of the centre of mass of a particle. The variables are *structural* if they are basically of a static nature—they give the higher level its identity; and they are dynamic if they are essential to its *behaviour*—they are time dependent in crucial ways. Structural variables would, for example, be those describing the shape, hardness, colour, and so on of a cricket ball; dynamic variables would be its centre-of-mass position and motion and its momentum.

Coarse-graining extracts properties of the system that characterize its higher level nature and behaviour, while throwing away large amounts of information about lower level entities. This information-hiding is a key feature of modularity (see Sect. 3.1.6) which justifies the term 'coarse-graining'. An example of coarse-graining is the averaging sums in (3.5), but it can be much more sophisticated than this, as discussed below.

Emergence of Higher Level Behaviour via Bottom-Up Causation. This occurs when higher level behavior, described in terms of higher level variables, emerges from lower level dynamics, described in terms of lower level variables (Table 3.5). The dynamics of the lower level theory maps an initial state i to a final state f. Coarse-graining the lower level variables, state i corresponds to the higher level state I and state f to the higher level state F. Hence, the lower level action $t : i \to f$ induces a higher level action $T : I \to F$. This leads to emergence $t \to T$ of higher level behaviour from the lower level. An example of this process is the steps from (3.2) to (3.6). Examples of the outcome are the perfect gas laws, black body radiation properties, and so on. One can reliably coarse-grain to get higher level variables and laws in these cases. Higher level behaviour emerges unaffected by container size or shape.

However, coarse-graining the lower level dynamics will not always result in reliable higher level dynamics: chaotic systems are a case in point. We need to consider when coherent higher level dynamics will emerge.

3.3.2 *Coherent Higher Level Dynamics*

Multiple Representation. In general, many lower level states correspond to a single higher level state (Fig. 3.2 left), because a higher level description H_1 is arrived at by ignoring the micro-differences between many lower level states L_i, and throwing away a vast amount of lower level information (coarse-graining). For example,

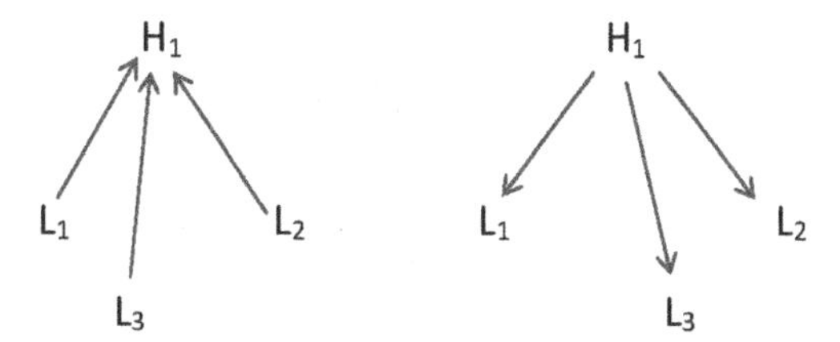

Fig. 3.2 *Left*: A set of lower level states all corresponding to the same higher level state upon coarse-graining. *Right*: Specifying a higher level state specifies an arbitrary single member of a whole family of lower level states. That specific member instantiates the higher level state

numerous microstates of particle positions and velocities correspond to a single macrostate of nitrogen gas with a pressure of one bar and a temperature of 20 K in a volume of 1 L.

The number of lower level states corresponding to a single higher level state determines the entropy of that state. This is lower level information that is hidden in that higher level view. The consequence is that specification of a higher level state H_1 determines a family of lower level states L_i, any one of which may be implemented to obtain the higher level state (Fig. 3.2 right).

Dynamics. The system dynamics (causal interactions due to physical interactions between the components) acts on each lower level state L_i to produce a new lower level state L_i'. Two major cases arise.

Incoherent Dynamics. Different lower level realisations L_i of the same higher level initial state H_1 result, through microphysical action taking each state L_i to a new state L_i', in different higher level final states H_i' (see Fig. 3.3). Here there is no coherent higher level action generated by the lower level actions and the higher level result is unpredictable. This is the case for chaotic systems with highly sensitive dependence on initial conditions, so that an arbitrarily small perturbation of the initial data may lead to vastly different future behaviour [34]. Examples are three-body systems in Newtonian mechanics and Lorenz attractors in the equations for weather.

Coherent Dynamics. Coherent higher level dynamics T emerges from the lower level action t if the same final higher level state H_1' results for all lower level states L_i that correspond to an initial higher level state H_1 [42] (see Fig. 3.4), thereby

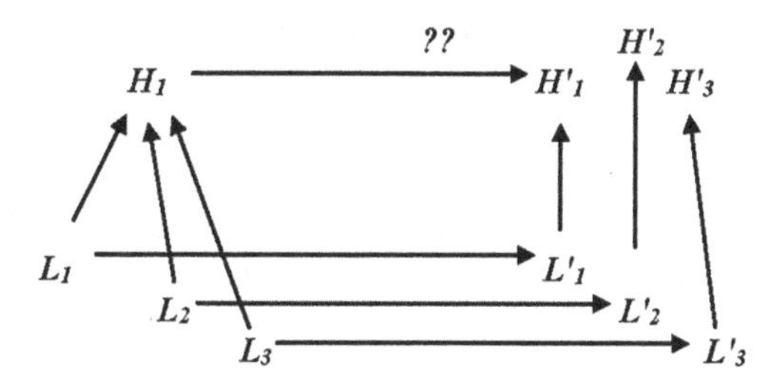

Fig. 3.3 First case: chaotic dynamics

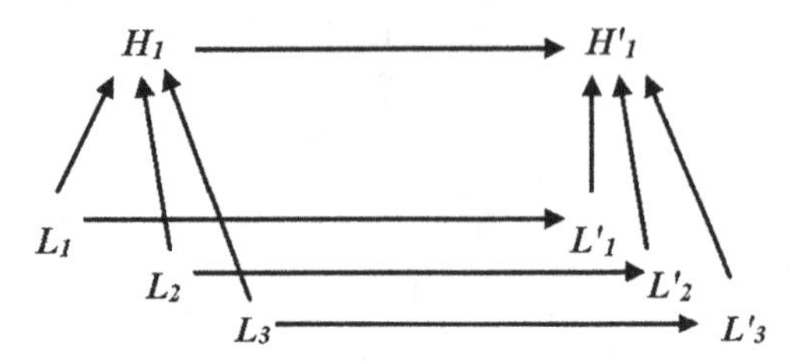

Fig. 3.4 Second case: coherent dynamics

defining an *equivalence class* of lower level states that give the same higher level action (see [6] and [8, pp. 403–407]). Different lower level realisations of the same higher level initial state result, through microphysical action, in the same higher level final state (up to the accuracy of the higher level description utilised).

It is possible that $H_1 = H_1'$, in which case we have an equilibrium state of the system. In the case of the gas, this will be so if the initial state is one of uniform temperature and density. This is also the case for structural variables in a solid: even though lower level thermal vibrations are taking place all the time, the higher level structure is a stable entity.

Effective Higher Level Dynamics. In this case, the lower level action results in a unique emergent higher level dynamics: the effective theory at the higher level. Consistent behaviour occurs at the higher level, regarded as a causal system in its own right. There is now effective higher level autonomy of action, enabled by coordinated lower level action (see Fig. 3.5):

> **Emergent Dynamics**. When coherent dynamics emerges, the resultant higher level action can be regarded as existing in its own right. It can be analysed without knowledge of the underlying lower-level interactions.

This is what enables one to talk of the existence of higher level entities in their own right. It is where the power of information-hiding arises: a coherent higher level action results from the lower level action (perhaps in a statistical sense), independent of which lower level states instantiate the higher level states.

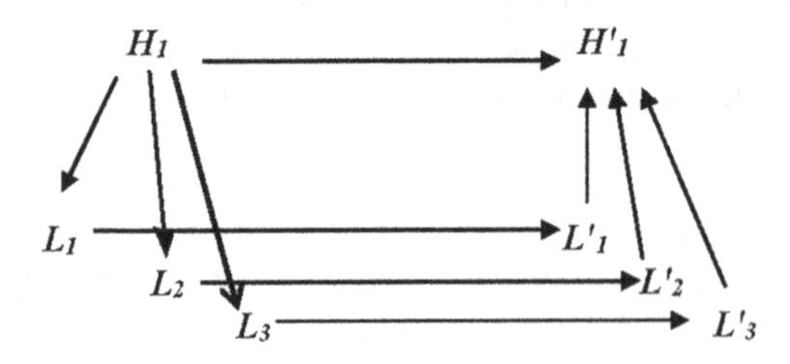

Fig. 3.5 A value of a higher level variable (such as the volume V of a gas in a cylinder) instantiates one instance of the lower level variables in an equivalence class such that the lower level dynamics leads to emergence of a coherent higher level dynamics, e.g., the perfect gas laws. Thus, for example, $\{H_1 \to L_2 \to L_2' \to H_1'\} \Rightarrow \{H_1 \to H_1'\}$

An example is gas in a container that is initially hot in one region and cooler elsewhere. Molecular diffusion will result in a final state of uniform temperature. Both the initial and final states can be realised through numerous microstates, and the higher level behaviour does not depend on which of them occurs. It is reliable behaviour at its own level, based on an equivalence class of lower level states (Sect. 3.5).

A different kind of example is pressing a key on a computer (H_1 is the computer with the key pressed), resulting in a letter being displayed on the screen (H_1' is the computer with the key pressed and the letter displayed on the screen). The higher level action is the same whatever detailed (lower level) electron motions result in the computer circuits. The lower level action and resultant final higher level state would be different if the higher level state were different (for example, if a different key were pressed).

Would the resulting higher level action be identical if the two lower level states instantiating them were identical? Not necessarily, because of quantum physics:

> **Non-Unique Development.** Different outcomes can arise from the same lower level state because of irreducible quantum uncertainty [51], which can be amplified to create macroscopic effects.

This has actually happened in two significant cases (Sect. 8.1):

- In the context of biology, cosmic rays can cause genetic variations [101], and this has made a significant difference to evolution of life on Earth
- In the case of cosmology, quantum fluctuations in the inflationary era were amplified via gravitational processes in the early universe [36], determining what astronomical structures came into existence.

In practice, there are always fluctuations at the lower level N, and these may cause significant differences in events at the next higher level $N+1$. Indeed, this randomness plays a crucial role in biology [65] and particularly in the brain [57], by allowing the existence of an ensemble of variants from which a preferred outcome can be selected by adaptive processes (see the discussion in Sect. 4.3).

3.3.3 Emergent Higher Level Variables

The essential key to understanding emergent properties is correct choice of higher-level concepts and associated variables. This enables us to identify and name the relevant causal factors at that level. Higher level variables may be emergent from the lower level variables, as in the examples discussed above. However, there are some kinds of higher level variables that are not emergent: they are intrinsically higher level variables.

3.3.3.1 Emergent Properties and Variables

Many higher level variables are functions of aggregated lower level variables, abstracting important properties of the system such as macroscopic momenta and energies, thermal conductivity, heat capacity. These higher-level variables are thus coarse-grained versions of the lower-level variables: in the sense that they represent some aspects of the system as seen from the higher-level viewpoint, with fine-grained lower-level details omitted.

Averaging. Gas pressure and density are macro-variables produced by averaging over the relevant micro-variables: numbers, masses, and momenta of constituent molecules in a given volume. A current flowing in a wire is represented at a macro-level by a number of ampere, representing the aggregate amount of charge flowing in the wire, but at the micro-level, it is described by a distribution of electrons in the wire. Stating the number of amperes flowing provides a useful coarse-grained description of the micro-situation. Together with the related resistance and energy variables, this choice gives phenomenological understanding of the higher-level behaviour (the flow of current in a wire is related to the voltage and resistance). Thus higher level variables can be considered as active agents in determining the causal outcome (a higher voltage produces a higher current, giving more heat, etc.).

The loss of lower level information associated with this coarse-graining (if we only know the current is 10 amperes, we don't know the detailed electron distribution) is the source of entropy: many lower level states correspond to the same higher-level state [100]. Consequently, the higher level states are relatively insensitive to many details of the lower level state of the system.

Effective Potentials. These are a key form of emergent property in physical systems, representing the coarse-grained effects of many interacting entities. Gemmer et al. [54, pp. 74–77] give an illuminating example of an ideal gas in a container. Other examples are the potential wells used in *nuclear shell models* [37, pp. 140–144], and the *Slater treatment of complex atoms*, explained by Pauling and Wilson thus [95, p. 230]:

> All of the methods we shall consider are based on the approximation in which the interaction of the electrons with each other has either been omitted or been replaced by a centrally symmetric force field approximately representing the average effect of all the other electrons on the one under consideration.

A similar method in astronomy is the way a coarse-grained potential energy is derived for a galaxy, and then used to find the motions of stars (see [11, pp. 67–90, 103–186] and [112, pp. 3–6]).

Note, however, that coarse-graining is not limited to simple addition or averaging, but can take many different forms. In particular there are various ways of recognizing and labeling patterns of higher level structure that are important features of the system.

Fourier Transform. This maps the detailed variables into a series of components representing coarse-graining at different wavelengths (see Bracewell [15]). Then patterns such as repetitive structures can be recognized as peaks in the Fourier transform. In a linear system, the variables at one scale interact only with other variables at the same scale, and the dynamics decouples into independent modes representing same-level dynamics at each scale corresponding to an eigenfrequency. The lower level dynamics (interactions between electrons and protons that comprise a spring, for example) then have no effect on this same-level dynamics. In nonlinear systems, the different scales interact with each other, representing bottom-up and top-down effects.

Filtering. This is a key form of coarse-graining, selecting components at some scales and ignoring others. It is of course a form of information-hiding, selecting what is relevant for some application and discarding the rest as unimportant, so it is a form of adaptive selection (see Sect. 4.3).

Crystal Structures. These are characterized by discrete symmetries represented by periodic functions, leading to coarse-grained concepts and variables such as unit cells, lattice planes, and reciprocal lattice vectors and lengths [132]. These lead to crucial concepts such as Bloch states and Fermi surfaces, and to results such as Bloch's theorem. These are all higher level properties because they depend on the crystal structure. Core electrons determine the structural variables, and valence electrons the dynamic variables such as current flows.

Molecular Structures. These are crucial in chemistry and microbiology. In simple cases they are characterised by molecular parameters and symmetries [17]. Key emergent variables are molecular potentials and binding energies, shell structures, and chemical bonds enabling emergence of complex molecules. Biomolecules have primary, secondary, tertiary structures including α helices and β sheets, and quaternary structures (the three-dimensional structure of a complex of protein molecules), leading to folded structures such as proteins [102], RNA, and DNA [20]. These in turn create structures such as vesicle walls and ion channels, leading to key variables such as action potentials. Emergent structures go all the way up through physiology [90] to the entire organism, with new emergent variables at each level that are characterized by the complex organization of their components [109].

Interaction Networks. Many kinds of these emerge [7], characterized by graphs or connection matrices [97, 98]. Examples are protein networks [70, 126], metabolic networks [107], neural nets [12], computer structures [121], and the connectome of the brain [118], characterized by the relevant wiring diagrams and by the emergence of network motifs [2]. Relevant higher level variables are currents in the case of a computer, and spiking patterns of action potentials [110] and patterns of synchronized oscillations [18] in the case of the brain connectome.

In each case one can indeed derive physical arguments for the nature and source of the higher-level properties, but only by introducing suitable higher-level concepts not implied by the underlying lower level physics [4, 78, 115].

3.3.4 *Intrinsically Higher Level Variables*

However, there are some higher level variables that are not emergent. Though they are realised in various lower level physical substrates, they are determined by higher level logic, and so are intrinsically of a higher level nature:

1. **Algorithms**. Examples are quicksort [75] or the Google search algorithm [81] (see the discussion of programs and algorithms in Chap. 2).
2. **Codified Laws of Physics**. Our mental representation of physical interactions, such as Newton's equations [49] or Maxwell's equations [50], the foundations of mechanical and electrical engineering, respectively.
3. **Social Agreements**. Examples are the rules of football, the rules of chess, legal contracts, the constitution of an organization, or exchange rates for money [117].
4. **Conceptual Plans**. Examples are the plans for a building, a town, an aircraft, or for a musical concert, a company, or a physics experiment [86].

They do not emerge from the underlying physics, but rather express (1) the logic of abstract domains, (2) the logic underlying physical behaviour, (3) the nature of social agreements reached by processes of negotiation, and (4) the plans we have individually or communally for what should happen.

Intrinsically Higher Level Variables. These are not physical variables, and there is no way to obtain them by any kind of coarse-graining process. Rather they are of a mental or abstract nature. However, they are certainly causally effective.

Each of them will have a hierarchical logical structure (which I labeled a horizontal hierarchy in Chap. 2), expressing the nature of relationships in the relevant domain. Higher level logical structures emerge from lower level ones through suitable processes of combination and naming (Sect. 3.1.4).

Biological Information. A significant question is whether biological information such as that embodied in the base sequences in DNA should be regarded as an intrinsically higher level variable or not. My provisional response is affirmative. There are two reasons. Firstly, there is no way the sequence of nucleotides in DNA can be predicted on the basis of either physics or microbiology. Rather it reflects historical effects of the environment over long periods of time [20]. Secondly, this sequence is coded in terms of the genetic code, i.e., the unique way base pair triplets are translated into amino acid sequences that generate proteins [129], which is in effect a biological agreement reached a long time ago as to how genes would function. It is analogous to translation rules from one language to another.

3.4 Top-Down Effects

This section considers top-down effects by looking in turn at the following:

- Limits to bottom-up emergence (Sect. 3.4.1).
- Top-down causation via constraints (Sect. 3.4.2).
- Top-down action via control parameters (Sect. 3.4.3).
- Top-down effects in logical hierarchies (Sect. 3.4.4).
- Top-down effects in the mind (Sect. 3.4.5).
- Top-down effects and supervenience (Sect. 3.4.6).

3.4.1 Limits to Bottom-Up Emergence

Bottom-up emergence of structure allows a certain degree of complexity to be built up spontaneously in non-equilibrium situations, often demonstrating symmetries and broken symmetries, without higher level guidance. Self-assembly and self-structuring can lead to emergence of simple structures such as those associated with energy minimization, entropy optimization, and dynamical system attractors, for example crystals and stars and galaxies. More complex patterns can occur in a bottom-up way through the reaction–diffusion equation, cellular automata, and self-organised criticality, such as sandpiles, Bénard cells, Conway's Game of Life, and biological examples such as flocks of birds. But there are limits as to how far this bottom-up process of explanation can be carried out. Even cellular automata and swarms of uncoordinated intelligent agents are limited in what they can do (unless given highly structured initial data that effectively contains coordinating information). As expressed by Campbell [19]:

> With each upward step in the hierarchy of biological order, novel properties emerge that were not present at the simpler levels of organisation. These emergent properties arise from interactions between the components [...] Unique properties of organized matter arise from how the parts are arranged and interact [...consequently] we cannot fully explain a higher level of organisation by breaking it down to its parts.

The linearity of lower level laws gets replaced by the complexity of nonlinear interactions at emergent higher levels with their own causal effectiveness, such as the networks of interacting molecules through which living cells are regulated [129]. It is the coordination of these incredibly complex interactions within their higher level contexts that enables epigenetic processes to take place [55] and so enables life to come into existence.

Not All Emergence Can Be Explained in a Bottom-Up Way. It is not possible to understand or explain the emergent properties in terms of the lower level concepts and variables alone. Superfluidity, for example, cannot be deduced from the lower level properties of electrons and atoms alone [78], and the same is true of the quantum

Hall effect [76]. The Hodgkin–Huxley equations governing membrane current propagation in neurons in the brain similarly do not follow from lower level properties alone [115, pp. 52–53]:

> The equations are not 'ordinary laws of physics' (as Schrödinger pointed out) but 'new laws' that emerge at the hierarchical level of the axon to govern the dynamics of nerve impulses. One cannot derive these new laws from physics and chemistry because they depend on the detailed organisation of the intrinsic proteins that mediate sodium and potassium current across the membrane and upon the geometric structures of the nerve fibers.

This is because in addition to bottom-up causation, *contextual effects* occur whereby the upper levels exercise crucial influences on lower level events by setting the context and boundary conditions for the lower level actions. Emergent effective laws of behaviour at higher levels play an effective role not only at their own levels, but also influence the lower levels by setting the context for their action [42, 43, 45]. To get reliable high level behaviour, one needs some kind of coordinating mechanism for lower level processes. The reliable behaviour of artificial complex systems arises from carefully designed complex structuring, where the details matter because they change the macro behaviour. A transparent example is a digital computer. One cannot coarse-grain to determine macro laws of behaviour because they are based on their own logic and act down to shape microlevel interactions (Chap. 2).

3.4.2 Top-Down Causation via Constraints

This is possible because both bottom-up (Fig. 3.1 left) and top-down causation (Fig. 3.1 right) occur in the hierarchy of structure and causation:

> **Top-Down Causation** [19, 124]. This is the ability of higher levels of reality to have a causal power over lower levels. Dynamic effects take place at some time, and the outcome would be different if the higher level context were different. Altering the high-level context alters lower level actions, which is what identifies the effect as top-down causation, where the high level context variables are not describable in lower level terms, which is what identifies them as context variables.

Top-down causation is ubiquitous in physics, chemistry, and biology, because the outcome of lower level interactions is always determined by context. For example, the motion of the Moon round the Earth causes tides locally on Earth, a top-down influence from a scale of 384,000 km to a scale of meters, and this then influences the lives of crabs. Likewise, fluctuations in the interior of the Sun cause radiation changes that alter conditions in ecosystems on Earth, influencing the distribution of micro-organisms. And so on.

This idea of top-down action in physics goes back at least to Ernst Mach in his work on Mach's principle and the origin of inertia [39, 46, 119], which strongly influenced Albert Einstein in developing general relativity theory and his static universe model. It is crucial in ideas about the origin of the arrow of time [21, 29, 39, 46, 99, 130]. Nice popular discussions of how top-down effects may take place from the universe to local physics are given in [24, 114]. Their very nature depends on the higher level structure. Top-down influences occur in convective fluid dynamics [13] and in astrophysics where, for example, star formation is suppressed by active galactic nuclei [93], and they play a key role in quantum theory as regards both decoherence and state preparation [44]. They are crucial in biology, both as regards the coding of genetic information in DNA through adaptive selection [19, 83] and as regards the reading of information through epigenetic effects [55]. And they are essential to physiology [90] and the way the mind works [52, 72].

3.4.2.1 Constraining Lower Levels

The higher level action is effective by coordinating actions at the lower levels. Whether this reliably happens may depend on the particular coarse-graining (i.e., higher level description) chosen. Describing the higher level change at the lower level is not desirable because it is not illuminating (the statement '10^{24} nuclei and associated electrons moved simultaneously in a coordinated manner so as to decrease the volume available to 10^{23} gas molecules', requiring about 10^{36} bits of information for a full description, is actually 'the piston moved and compressed the gas') and may not even be possible. Indeed this is the reason that we develop and use higher level language and mathematical descriptions. These may be employed whether or not we understand the lower level causation (see [71, p. 145] and [16, p. 89]):

> **Constraining Lower Level Interactions**. Top-down causation takes place, due to the crucial role of context in determining the outcomes of lower level causation. Higher levels of organization constrain and channel lower level interactions, paradoxically thereby increasing higher level possibilities.

These top-down effects result firstly from the fact that theoretical physics is based on partial differential equations whose solutions depend on the boundary conditions, which can be equally expressed as integral equations, whence the environment (the boundary conditions) constrains what happens locally.

Secondly, these effects exist because there are geometrical and structural relations in complex systems that dictate which components can interact with which others through which physical effects. For example the wiring in a computer channels electrons from one specific component to another and thus enables logical computations to be performed. The kind of computation performed and resultant output, and hence the detailed switching of transistors at the micro level, depends on the kind of programme loaded into the computer (word processor, music, or graphics for example), and this is a high level concept. These are constraints on the lower level dynamics and so have causal power [71]. They are causally effective only when such formal constraints from above are combined with efficient or circular causes at the same ontological level. Note that we can consider such same-level causation at each level as ontologically real: if this were not so we could not know what the ontologically real causation was, as we do not know what the fundamental bottom level of physics is.

One should note here, following Auletta's terminology [6], that there is a significant difference between having causal power (to concur in producing a certain effect) and having causal effectiveness (which, in ideal situations, would suffice to bring about an effect, given the other conditions). Formal causes, for example constraints, only have causal power but not causal effectiveness. For example, the structure of a forest (the way the trees are disposed together with other environmental items (rocks, plants, rivers, and so on) will affect (have causal power on) the way natural agents (like wind or fire) will propagate. For instance, wind may be more canalized in some parts and blocked in other ones. However, what is here causally effective is the wind (or the fire), not the structure of the forest, which would remain completely ineffective (not operative) and unable to concur in any causal process without an effective causal agent. All formal constraints have this character. Top-down causation as considered here means having causal power over lower levels, channeling causal effectiveness at those levels.

3.4.2.2 Interlevel Effects

But here's the problem: lower level (rocks, plants, wind, fire, etc.) are each made up of lower level elements, so they are not the bottom level. If we take a strict reductionist view they cannot do work either. In fact nothing does, except vibrations of superstrings, if they exist, which may or may not be the case. Lower level causality vanishes into unknown and untestable regions.

The only sensible way to handle this is to take an interlevel view, i.e., forget the bottommost level and assign real causal power to the lower level with respect to its immediate upper level, and to do this for every pair of levels [124]:

> **Interlevel Causation.** For every pair of levels $(N, N+1)$, the lower level 'does the work', but the higher level is able to influence what work is to be done by setting constraints on the lower level operations.

This is the basis for regarding every level as real: each is able to do real work. If we don't take this view, then genes and neurons are not able to do real work, as they are not the lowest level: the program of reducing brain action to that level [27] disintegrates. Every level can do work, and the implication is that the higher levels such as thoughts and imagination are also real, for they do work in changing even higher levels, such as society.

3.4.3 Top-Down Action via Control Parameters

This is the special case of top-down causation where higher level variables are purposefully manipulated so as to cause changes in lower level variables. An example is turning a light switch on. This causes electrons to flow in the wire so that the bulb lights up. The switch is a higher level entity: it cannot be described in lower level terms even though it is constructed from lower level parts. There is a same-level explanation in terms of phenomenological variables at the level of the switch and wiring. This is always what happens when reliable behaviour occurs at any level: it is described by an effective theory [62] at that level. This same-level theory is, however, enabled by the lower-level action of electrons flowing or not flowing. Whether they flow or not depends on higher level variables (in this case, the switch state) which simply cannot be meaningfully described in terms of lower level variables per se (actually because it is equal to an equivalence class of lower level variables). There is indeed a bottom-up explanation, when we are given the higher level context described by the higher level variables. Without that higher level context, there is no lower level explanation. How does one demonstrate this top-down causation in this case? By turning on the light switch and measure the resulting change in current.

But what is it that enables all this to happen? The wiring links the switch to the light, and this constrains the flow of electrons from the power grid to the light bulb. The structure of the wire and its insulating sheath channels the way they move, preventing them from moving sideways out of the wire. This constraint enables us to channel their movement to get the higher level effect we want. The bottom level physics allows this: we shape the context of the physical interactions so that they do our bidding. In mathematical terms, this happens because these physical constraints reduce partial differential equation (PDEs) to ordinary differential equations (ODEs). More generally:

> **Top-Down Action.** This is when a global variable is purposefully changed and alters local dynamics by changing the context in which the units operate, or a message is sent to the module's interface to convey a chosen new operating context.

Examples are:

- **In Organizations**. The central Government passes a law that affects all municipalities, a general sends a command to all battalions that causes them to get ready to invade France, a central office tells all the branches that from now on they will be selling at a higher price, or all stores will from now on be open from 8 am on weekdays.
- **In Engineering Systems**. A control parameter is altered that changes the rate of rotation of a turbine or the flow of a reactant, or a message is sent to a substation that turns on a generator in a power grid.
- **In Computer Systems**. A global parameter is passed to a subroutine that alters branching in the local flow of command, or a change in a class definition alters the behaviour of all objects that are members of the class.
- **In Physiology**. The brain sends action potentials to motor neurons that activate muscles, or an animal senses a threat that causes adrenaline to flow through the veins and alter the heart rate and blood vessel diameters.

In all cases the underlying physics and chemistry do not control what happens: rather they enable the desired top-down action to take place, because the control signal constrains the dynamics at the lower level so as to produce the desired outcome. The system is structured in a very precise way so that this will be the case.

3.4.4 Top-Down Effects in Logical Hierarchies

Similar effects occur in logical hierarchies (Sect. 3.1.4), based on their function of representing relationships between abstract entities and sensory experiences:

> **Contextual Dependence of Meaning.** Every logical statement depends on the context within which it is interpreted. In particular this applies to language.

Sometimes this dependence is formally prescribed in a set of rules, but more often it is implied and understood as a set of effective patterns of relationships. It is based on the fact that all logical hierarchies are 'is a' hierarchies, with a class structure and inheritance [14]. This is what enables us to understand them. We don't have to characterize all properties of entities or actions or modifiers anew in every case: we relate them in this hierarchical way to elements that are already known. Then:

- each subclass inherits most of the properties of the class,
- each instance of a class inherits the properties of the class it belongs to,

so if the class is altered, properties of both instances and subclasses change, that is, there is a top-down effect from the class definition to subclasses, and from subclasses to instances.

This is then reflected in language, because we use language to represent these logical relations, as codified in dictionaries and encyclopedias. It is also embodied in any informal or formal models we may have of these relations, such as representations in terms of computer codes or mathematical models.

Computer Languages. These are essentially hierarchically structured, as discussed in depth in Chap. 2. This is particularly clear in the case of the class structure of object-oriented languages [14] such as Java [75].

Mathematics (the Language of Patterns). This has a class structure. *Category theory* studies the relationships between classes of mathematical structures. A category consists of mathematical objects and 'morphisms', processes that are transformations between them, that can be composed with each other to give new morphisms. More specific structures inherit many properties from the higher level categories they belong to, so one can prove properties of specific structures by giving a proof in the higher level category of which they are a member. The logic chains down from the higher to the lower levels.

Natural Language: Listening and Reading. In the case of language, there is of course a hierarchical class structure in the classification of words, reflecting the logical hierarchical structuring of the concepts referred to. The mind is adapted to understanding that structure. However, given that hierarchical structuring of concepts and words, there are also further key top-down effects in the way we understand sentences and words. There is often an ambiguity of meaning if one considers only the level of words. This ambiguity is resolved where one of the possible lower level choices follows uniquely from the higher level context, viz., the phrase, sentence, or paragraph in which the word is imbedded. This is top-down effect from the context to the word meaning, and sometimes even pronunciation. An example is

> Her wound hurt her as she wound the clock.

Actually, one is always predicting what will come next on the basis of context. This is what enables understanding from a subset of the full text (you only hear a few words but still understand), or from garbled text:

> You can understand this even though words missing or spelt wrong.

Much more sophisticated contextual understanding drives language comprehension in conversation and in reading, for example when enjoying a novel: meanings are hinted at and understood on the basis of the overall social and psychological context, which is set by the text as a whole. For example:

> I attach an outline of the programme for the event. I am making arrangements for accommodation.

We effortlessly understand this in terms of previous messages (what is the event? where and when is it?) and the assumed need for someone attending the event to have accommodation (it's a multi-day event, they live out of town, everyone needs somewhere to stay at night). None of this is explicitly stated in this quote. Language is contextual through and through.

3.4.5 Top-Down Effects in the Mind

This is similar to what happens in vision, as documented for example by Chris Frith [52], Dale Purves [104], and Eric Kandel [72]. We always subconsciously interpret what we see in terms of past experience and resulting expectations [52, 72, 104]. This is evidenced by numerous visual illusions, and in particular by the way we do not normally notice the blind spot that in fact occurs in our eyes because of its physical structuring. There is a context dependence not of visual interpretation, but of vision: *what we actually see*.

This kind of contextual dependence applies to all the senses:

> **Contextual Dependence of Experience**. Every sensory input is experienced in a way that depends on the context within which it is located.

The brain is a quintessential machine for prediction on the basis of past experience [64]. All our senses interpret what they find in terms of past and present contexts which shape present expectations. This contextual dependence applies also to activities such as tasting food, watching sport, and listening to music [67, 79]. Crucially, it applies to reading [58]:

> **Contextual Dependence of Reading**. Reading is not done by reading phonemes one by one and assembling them into words that then create meaning. Rather it is carried out in a holistic way that depends on context and expectations as well as the read text, which often only hints at the intended meaning.

This is particularly important for education. It will be discussed further in Sect. 8.6. Part of this occurs through the top-down activation of symbols [87]. A key aspect of this contextual interpretation of text is:

Multiple Realizations. The same meaning can be conveyed in many different forms of the written text.

Our mind automatically regards them as equal. For example:

- It can be in large or small font size, in any number of fonts.
- It can be in English or German or Italian.
- It can use the active or the passive tense.
- It can use different word orders.
- It can state things explicitly or implicitly.
- It can use different metaphors to convey the same meaning.

It is this multiple realisability that makes language so flexible as a vehicle for expressing meaning. There is an equivalence class of sentences that convey the same meaning, namely, all the different sentences that in fact do so. This feature is a consequence of the top-down nature of language use and understanding.

3.4.6 Top-Down Effects and Supervenience

A counter to the proposal of top-down causation is sometimes claimed to arise from the idea of supervenience ([85]; see Fig. 3.6): because the higher level properties emerge from the lower level properties, the same lower level state must necessarily result in the same higher level state (see [16] and [8, pp. 81, 411]). This is expressed by Rickles [108] as follows:

Supervenience Relation. For two sets of properties, A (the supervenient set) and B (the subvenient set or supervenience base), A supervenes on B just in case there can be no difference in A without a difference in B.

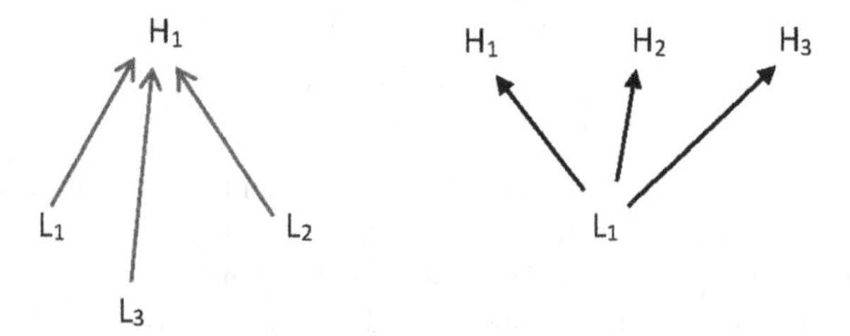

Fig. 3.6 Supervenience. *Left*: Each lower level state leads to a unique higher level state which then supervenes on the lower level state (and there can be many lower level states leading to the same higher level state, so supervenience is compatible with multiple realisation). *Right*: A lower level state leads to a number of different higher level states. There is no higher level state supervenient on the lower level state

Determination. Turning this principle on its head gives us the converse concept of determination: B determines A just in case sameness with respect to B implies sameness with respect to A. Supervenience and determination are simply two sides of the same coin.

Now this can be claimed to be the case for straightforward physical systems and biological systems, at any moment in time, that is, as a synchronic relation, and might possibly even be true for the way the micro happenings in the brain relate to the mind. However, two related issues arise:

Feedback Loops Occur from Global to Local. Higher level contextual effects act down to influence dynamics at the lower level synchronically, that is, on functional timescales. An example is epigenetic effects that decide what gene will get read next on the basis of environmental variables.

As stated by Rickles [108]:

> If the supervenient properties are understood as emergent, then it is possible that some 'global' properties, to do with a whole system, can causally effect other things, and its parts (the supervenience base). For example, a group of agents can interact to generate an economy, but the economy has properties of its own (prices, interest rates, and such like); these will be able to influence how the agents behave. In other words, there is the possibility of a 'feedback loop' from global to local.

This is of course the argument of this book. Secondly, suppose the supervenience argument is correct even in cases of complex emergence, such as digital computers (where it certainly is true) and the brain–mind relation (where it could possibly be true). That is, in these cases, the higher level outcome will be precisely the same if all lower level structures and excitations are the same. The key point as regards top-down causation is as follows:

The Origin of the State. In living systems and in digital computers, those lower level structures and states that lead to complex higher level behaviour cannot have come into being in a purely bottom-up way, because they depend on a precise conjunction of structures and energisations that reflect higher order needs and functions in a coordinated way (as reflected, for example, in non-emergent variables discussed above in Sect. 3.3.4). This requires either purposeful organisation or adaptive selection, which are both top-down processes.

That is, the supervenience argument cannot apply to diachronic emergence, because the required ongoing set of highly structured lower level states will not, in the real world, occur without the aid of top-down organisation. To put it another way, synchronic emergence of real complexity (life, the brain, digital computers) cannot occur without prior diachronic emergence.

3.4.7 Top-Down Effects and Emergence

There is a large literature on emergence and reductionism, e.g., [8, 23, 47, 106, 111, 116]. A comprehensive discussion is given in [8], which reprints many key papers surveying the field. Broadly, emergence is when phenomena arise from and depends on more basic phenomena yet are simultaneously autonomous from that base [8, pp. 1, 155–156]. It occurs as regards both entities and properties [8, p. 140]. A phenomenon is emergent if it cannot be reduced to, explained, or predicted from its constituent parts [47, p. 1].

Although there is no commonly agreed formal definition of emergence, nevertheless a series of interrelated ideas are commonly associated with it [8, pp. 9–17]:

- **Irreducibility**. The emergent phenomena are autonomous from the more basic elements that give rise to them, even though they depend on them to some extent.
- **Unpredictability**. A state or feature is emergent if it is impossible either in principle or in practice to predict on the basis of a complete theory of basic phenomena of the system.
- **New Variables Are Needed**. One needs a new conceptual or descriptive apparatus at higher levels than what is used for more basic phenomena.
- **Holism**. Some properties only arise out of wholes formed from assemblies of more basic parts. It is conceptually incoherent to conceive of them in relation to the parts alone.
- **The Whole Is More Than the Sum of the Parts**. The macro level properties cannot be obtained by simple addition of lower level properties.

It is clear that in many cases "the whole becomes not only more than, but very different from the sum of its parts" (Anderson [8, p. 226]). The key question in the end is whether the higher levels have emergent causal powers in their own right, which is strong emergence [8, pp. 141, 158–159], or the apparent higher level causal powers are in fact epiphenomena [116], which is nominal or weak emergence [8, pp. 157–160].

Reductionism. Emergence is often contrasted with reductionism. This can relate to epistemic reduction, ontological reduction, or inter-theoretic reduction [8, 23]. However, I will just refer to the discussion by Rae [106, p. 180], where reductionism is described as follows:

> First, the properties and behaviour of physical systems are controlled by the fundamental laws that apply to its components, and second, [...] genuinely new phenomena often emerge that would have been very difficult or impossible to predict from our knowledge of the components alone.

The thesis of this book is essentially that the word 'controlled' here is wrong. This is demonstrated for example by the discussion in Chap. 2 of how computers are in fact controlled by algorithms, not by physics. Furthermore, the existence of higher order causal variables such as the value of money or the rules of football demonstrates that the correct wording in the second part is 'impossible in principle'. Rae's thesis is correct insofar as it is implied by supervenience as discussed in the previous section. And as shown there, truly complex behaviour can only emerge through the top-down effects of either adaptive selection (life) [20] or deliberate design (artefacts) [14, 120]. His view denies strong emergence because it does not take top-down effects into account (interestingly, he gives superconductivity as an example of reductionism, and this is precisely the case that Laughlin uses to counter reductionism [78]).

Strong Emergence and Top-Down Causation. It is clear that strong emergence can only take place if top-down causation also takes place [8, pp. 21, 112, 139–149, 175–177, 340]. This book argues that both take place.

3.5 The Key Concept: Equivalence Classes

As in the case of computers (Sect. 2.5) and language (Sect. 3.4.5), a key feature of all top-down causation is multiple realisability of higher level functions, and the consequent existence of equivalence classes of lower level variables that correspond to the same higher level state. An equivalence class identifies all lower level states where the corresponding higher level variables are equivalent as far as the higher level behaviour is concerned, that is, they form good predictors of higher level behaviour [6]. Here, I shall look in turn at:

- Equivalence classes (Sect. 3.5.1).
- Equivalence classes and top-down causation (Sect. 3.5.2).
- Multiple realisability and supervenience (Sect. 3.5.3).

3.5.1 Equivalence Classes

The formal expression of an equivalence relation is as follows:

An *equivalence relation* is a binary relation $\sim$ satisfying three properties:

1. For every element a in X, $a \sim a$ (reflexivity).
2. For every two elements a and b in X, if $a \sim b$, then $b \sim a$ (symmetry).
3. For every three elements a, b, and c in X, if $a \sim b$ and $b \sim c$, then $a \sim c$ (transitivity).

The *equivalence class* of an element a is denoted $[a]$ and may be defined as the set of elements that are related to a by $\sim$.

The set of all equivalence classes in X given an equivalence relation $\sim$ is denoted as $X/\sim$ and called the *quotient set* of X by $\sim$. Each equivalence relation has a *canonical projection map*, the surjective function π from X to $X/\sim$ given by $\pi(x) = [x]$.

In an emergent system (see Fig. 3.2), an equivalence relation $\sim$ is defined by lower level states L_i corresponding to a high level state H_j. It satisfies the three properties above. The equivalence class $[L_i]$ is the set of all lower level states L_i that correspond to the same higher level state H_j. They are the set of all its realizations. Conceptually, H_j at the higher level is the *same thing* as $[L_i]$ at the lower level: the low level states are operationally equivalent as far as the high level description is concerned and we say that H_j emerges out of $[L_i]$. Thus the quotient set $X/\sim$ is just the set of higher level states, and the canonical projection map is the map from lower level states to the higher level state they instantiate (Fig. 3.2 left).

Equivalence Class. An equivalence class identifies all lower level states where the corresponding higher level variables are equivalent as far as the higher level behavior is concerned, i.e., they form good predictors of higher level behavior. In general, the higher level state can be realised in many different lower level states. What matters is not the specific state, but the equivalence class it belongs to.

Here are some examples.

Gas States. An equivalence class of molecular states can give the same pressure, density, and temperature of a gas [49]. It is these higher level variables that are relevant for understanding and predicting gas behavior, as for example in the ideal gas law $PV = nRT$. Entropy is a measure of the amount of variation at the lower levels that gives the same state at the higher level [100]. It characterizes how many values of hidden variables can underlie the same higher level description (they are integrated out or averaged over to attain the high level description).

System Models. Structural equation models can be observationally equivalent (see [97, Sect. 2.3] and [98]). Observations therefore determine an equivalence class of models.

Digital Computer Systems. These have numerous equivalence classes, as discussed in Sect. 2.5: higher level languages and operations can be implemented in many lower level ways.

Neural Networks. These can have different sets of link weights that give essentially the same pattern recognition properties [12]. It is the pattern recognition that matters, not the specific weights whereby they are realised.

Mathematical Relations. These can often be expressed in several different ways [99]. For example, one can use components relative to different coordinate systems to represent the same geometric structure, one can used complex variables or twice as many real variables, one can use tensors to express spinorial relations, one can use Fourier transforms to represent a function $f(x)$ in terms of frequency amplitudes $F(s)$ [15], and so on.

Physics Theories. These can be expressed in various ways: Newtonian dynamics in terms of forces, Lagrangians, or Hamiltonians, quantum physics in the Schrödinger, Heisenberg, or Dirac formalism, Maxwell's equations in 3D or 4D forms, and so on [99].

Sentences. In natural spoken or written language, these determine an equivalence class of other sentences that have the same meaning, as explained in Sect. 3.4.5. There are multiple words that label the same phenomena.

3.5.2 Equivalence Classes and Top-Down Causation

The existence of equivalence classes underlies the possibility of coherent higher level dynamics emerging from lower level dynamics (see Sect. 3.3.2). Effective same-level action occurs when top-down causation combined with bottom-up causation leads to a resulting high-level outcome that depends only on the initial high-level state. In that case, the low level dynamics commutes with coarse-graining for all low level states that correspond to each of the high level states (Fig. 3.4), and a coherent high level dynamics emerges from the lower level dynamics [42, 43]. The switching on of a light switch is an example. In this case each set of lower level states corresponding to a single higher level state forms an equivalence class as far as the higher level dynamics is concerned. The resulting same-level action allows a phenomenological description of the higher level action that is independent of the particular lower level states that realize this action. This is the basis of the independence of higher level descriptions from lower level details and the reason that we can consider same level causation at each level as ontologically real, expressed in terms of viable effective theories for the dynamics at that level [62].

If different outcomes result from different lower level realizations of top level states, we do not have reliable same level action resulting from top-down influences of the higher levels, and so do not have coherent top-down causation. Thus the possibility of coherent higher level action emerging from the lower level dynamics is based on the principle of equivalence of classes:

Principle of Equivalence Classes. The same top level state must lead to the same top level outcome, independent of which lower level states instantiates the high level state.

The high level outcome is then the same for the whole equivalence class of lower level variable values, no matter which particular one instantiates the high level state. Thus the existence of equivalence classes as discussed here is necessary for top-down causation to take place.

However, one may also ask whether the existence of such equivalence classes is *sufficient* to characterize top-down causation? The higher level variables are normally the only handles by which we can affect lower level states. For example, we can compress a gas in a cylinder by exerting a force on a piston, so changing the volume V and temperature T of the gas, and this will result in a change in the momenta p^i and positions x^i of the myriads of particle comprising the gas. But we cannot change those positions and momenta individually: apart from anything else, there are too many of them, viz., 6×10^{23} particles for 2 g of hydrogen gas. But if the principle of equivalence classes is satisfied, setting a macro variable like V and T produces some member of the equivalence class $[p^i, x^j]$ of lower level states that realizes this higher level state $\{P, T\}$ obeying its high level equation of state (the perfect gas law) in a coherent way. Thus top-down causation takes place, changing boundary conditions for the lower level states producing a coherent high level effect (Fig. 3.4).

Top-Down Causation and Equivalence Classes. Top-down causation leading to coherent higher level behavior takes place if and only if any change in the relevant higher level variables instantiate an instance of an equivalence class of lower level states that realizes the higher level dynamics. The existence of such equivalence classes is the crucial feature characterizing effects as being due to top-down causation [6].

The higher level variables drive the dynamics, while the lower level variables respond in a non-unique way, but a coherent higher level dynamics nevertheless emerges from their bottom-up action (Fig. 3.4).

3.5.3 Multiple Realisability and Supervenience

From the viewpoint of supervenience, it is multiple realisability that allows genuine emergence to occur. Rickles [108] expresses this as follows:

> Multiple realizability lies at the core of supervenience's job, namely, to describe a dependency weaker than identity and reduction. The idea is that fixing the physical properties of the work of music (the tones, durations, intensities, and so on) suffices to fix any and all aesthetic properties the piece might have. But then the idea of emergence amounts to the claim that these aesthetic properties (and similar higher-level properties) are not reducible to the physical ones, they are something 'novel' arising from the physical organization. (The distinction between physical and non-physical properties here amounts to both the fact that the latter type can be had by many objects with different natures and constitutions, and the fact that the former type obey the laws of, possibly complete, physics. However, nothing said here hinges on this distinction, one might as well say that aesthetic properties are physical too, since they occupy the world. Thus, this is just a way of speaking to label a curious fact, namely that some properties seem not to be reducible to what are standardly taken to be unproblematic 'physical' properties, such as mass, charge, spin, and so on.) Dualism and epiphenomenalism are avoided (1) because the physical facts are needed to fix the emergent facts and (2) because the emergent properties are supposed to be causally efficacious: the beauty of the Adagio from Mahler's Fifth Symphony can cause a person to cry; it isn't the durations, intensities, and pitch of sounds that is causally responsible.

He carries on to consider arguments suggesting that supervenience implies reductionism, and then counterarguments. The essence is that that properties associated with a 'special science' (for example, psychology) can be realized by a multitude of heterogeneous lower-level properties or states. Hence the lower level properties are not the essential causal factors (see [105] and [8, pp. 403–407]). This is discussed further in Sect. 7.7.2.

3.6 Demonstrating Top-Down Causation

How do you demonstrate that top-down causation is taking place? Apart from developing more formally the arguments made in Sect. 1.7 as regards the necessity of the conclusion, there are basically four options, which may overlap in practice:

- Alter context (Sect. 3.6.1).
- Identify equivalence classes (Sect. 3.6.2).
- Identify dynamics (Sect. 3.6.3).
- Computer modelling (Sect. 3.6.4).

3.6.1 Altering Context

The idea is to show that a change in high level variables results in a demonstrable change in lower level variables in a reliable way, after one has altered the high level variable. It is the reliable nature of the change that characterises it as causation and not just a random change, and this is also what leads to predictability (the result is repeatable and thus testable)

> **Change Context.** One just has to show that altering the high-level context alters the outcome in a way that depends only on the top level state, where the context variables are not describable at the lower level.

To characterise some specific causal effect as a top-down effect, we must demonstrate that a change of higher level conditions alters the way lower level actions take place. We do this by changing top level conditions and seeing what happens at the lower levels, e.g., we decrease the volume of a gas and see that it makes molecules move faster. Examples are:

- In a *control situation*, alter the parameters or the goal.
- In a *digital computer*, change the program.
- In a *manufacturing context*, change the plans.
- In an *adaptive selection context*, alter the environment.
- In *social neuroscience*, alter the social context.
- In the *brain–body relation*, study the placebo effect [10].
- In a *simulation*, alter the high level variables.

In each case one can observationally show that the higher level outcome is different, through alteration of operation of lower level mechanisms or dynamics (enabled by the underlying physics and chemistry).

3.6.2 Identifying Equivalence Classes

We should if possible demonstrate the existence of equivalence classes of lower level effects that give the same higher level outcome, as discussed in Sect. 3.5:

> **Identify Equivalence Classes.** One identifies an equivalence relation by showing that some lower level entities of mechanisms can be substituted for others and still give the same higher level outcomes.

There are various ways to do this:

- One can explicitly look for such equivalences in microbiological reactions [6, 69].
- One can identify a key role played by equivalence classes in some theory or mechanism. Crutchfield's computation mechanics [28] is an example, and Wegner proposes that a key to understanding evolution is the idea of a genotype network, defined as a set of genotypes that have the same phenotype [89, 128].

- One can regard it as evidenced by convergent evolution [25, 84], where various different mechanisms (e.g., different kinds of eyes) have evolved to solve the same higher level need (in the case of eyes, vision), showing that it is indeed the higher level need that is driving the dynamics.

Ideally, these are tested experimentally in new cases that have not yet occurred.

3.6.3 Identifying Dynamics

While all top-down causation can be characterized as due to higher level variables setting the context for lower level action, five essentially different classes of top-down causation can be identified and their existence demonstrated by many real-world examples [42, 43]. This enables one to identify mechanisms of each of these types that clearly embody top-down causation:

> **Specific Mechanisms.** One identifies a mechanism of one of the types TD1–TD5 described in Chap. 4, each of which intrinsically embodies top-down causation in its nature.

Thus one can identify:

- Specific mechanisms aimed at top-down control of lower levels, as in engineering systems or bureaucratic contexts (Sect. 4.1).
- Feedback control systems driven by goals (Sect. 4.2).
- Processes involving adaptive selection [66] according to some kind of selection criterion (Sects. 4.3–4.5).

As discussed in Chap. 4, each of these involves top-down causation: identification of the mechanism confirms the operation of top-down causation.

3.6.4 Computer Modelling

One can use computer modelling of the dynamics in order to demonstrate that altering higher level variables alters lower level outcomes. One must of course be using a multi-level model in order to do this.

An example is the way Noble has modelled the physiology of the heart in a multilevel way [90, 91], and so shown the effect in that case. Another example is the standard calculation of nucleosynthesis in the early universe in cosmology, where global cosmological parameters determine the outcome (Sect. 6.7.1). Other examples are given in Sect. 6.7.

3.7 Constraints on Emergence

All this is subject to the nature of physical processes at the bottom levels. There are various constraints from lower levels on what upper levels can accomplish. For example, one requires certain macro structures in order to walk, see, or fly. The most important such constraints are essentially because of integrated micro constraints, specifically energy and momentum conservation. Even more important, matter conservation underlies the continued existence of macro objects. Physical constraints restrict the nature of what can emerge in both living and artificial systems [127].

Constraints occur as regards:

- Matter, energy, and entropy (Sect. 3.7.1).
- Constraints on higher level possibilities (Sect. 3.7.2).
- Constraints on higher level logic (Sect. 3.7.3).

3.7.1 Matter, Energy, and Entropy

Which kinds of properties of lower levels re-emerge at higher levels in general? The basic properties of this kind are matter and energy conservation, and Newton's laws of motion [49]. These apply at the lower levels of the hierarchy of structure. On coarse-graining, they reappear at the higher levels, provided we use appropriate definitions of matter, energy, and force.

Firstly, it is these conservation laws that underlie the ongoing identity of higher level objects: the continuity of existence of macroscopic objects that we take for granted is a result of matter conservation at the lower levels [see (3.1)], together with the stability of emergent structures.

Secondly, the way macroscopic bodies can move is a result of the integration of Newton's laws of motion over all the particles in the body, with the same form of laws emerging at the higher level as effective laws [as shown in the way (3.6) emerges from (3.2)]. In particular, momentum conservation results if no external forces are applied.

Thirdly, we must distinguish energy and usable energy, matter and usable matter, the unusable energy and matter being present at the lower levels but in a disordered state that is not accessible to control or extraction by higher level variables. The second law of thermodynamics implies that the usable energy inexorably degrades to unusable energy: ordered matter states tend to become disordered. Hence, unusable energy and matter accumulates: a complex system must get rid of it. Ultimately, all this is an effect of the second law: energy is not lost, it is degraded, and the same is true of minerals and materials. New energy and materials are required on an ongoing basis to keep the system going.

Thus a consequence of lower level physics is that all complex systems need the following:

- Energy and matter input, sorting, transformation, and distribution systems.
- Heat disposal systems and waste collection and disposal systems.

These involve spreading resources to the periphery as well as the centre, and removing wastes from the periphery as well as the centre. It applies to living beings, such as cells, limbs, organisms, and communities, and to physical entities in a social environment such as automobiles, buildings, and cities as a whole.

These are profound restrictions on what is possible at the macro level. They are upwards constraints on complex systems in general and on daily life in particular.

3.7.2 Constraints on Higher Level Possibilities

As well as these constraints, other factors ensure that only certain things are attainable macroscopically in physical terms. These limitations are based on the features just mentioned (matter and energy conservation, Newton's laws of motion) together with the nature of the gravitational force [49] and electromagnetic interaction [50]. At the foundations, quantum physics and the strong force are important in enabling and constraining macroscopic possibilities [51], for instance underlying the stability of matter and the nature of the periodic table of the elements. Similarly there are possibility spaces for life [128].

A possibility space Ω_s for emergent structure characterizes these emergent possibilities. It is based on lower level properties such as:

- speed and energy of interactions,
- strength and distance of interactions and resulting bonds,
- strength and pliability of materials,
- solubility and stability of materials,
- electrical and thermal conductivity of materials,
- energy capture, storage, and transformation possibilities,
- information collection, storage, and manipulation possibilities.

These are what designers need to take into account when they design artificial systems [120], and evolution discovered as it explored the possibility space of living systems [20, 128]. As mentioned above, this leads to convergence in biology [25, 84], and even between life and engineered systems [127], because there are only a certain number of ways to sense light, to extract energy, to move, or to process information.

3.7.3 Constraints on Higher Level Logic

Intelligent systems (animals, humans, social institutions, computers) need:

- Information input, sorting, distribution systems.
- Processes to clear memory of unwanted information.

These are required so that the implementation hierarchies can support logical hierarchies, as in the case of digital computers. They are constrained by the nature of the logic operational at that level, be it mathematical, engineering, social, psychological, or whatever. For example, symbolic systems are limited by semiotic constraints on what can be done with iconic systems [32, 33].

However, there are no constraints from below on the actual logical operations that higher levels can perform. This is the essential understanding Turing gave us when he discovered the nature of universal symbolic computation [26]. This is the remarkable feature of emergence:

Logical Independence. The underlying physics does not restrict what logical operations are possible in emergent structures.

There are constraints on what is possible as regards logical operations from the nature of logic itself: that is indeed the very nature of the logical possibility spaces (Sect. 2.7.5). These limitations do not derive from lower level properties: they are independent of physics. They are immutably built into the logical structure of reality (Sects. 2.7.5, 7.6, and 8.5.4). However, those logical operations can act downwards to control physical systems, as in the case of digital computers (Chap. 2).

References

1. R. Albert, A.-L. Barabási, Statistical mechanics of complex networks. Rev. Mod. Phys. **74**, 47–97 (2002)
2. U. Alon, *An Introduction to Systems Biology: Design Principles of Biological Circuits* (Chapman and Hall/CRC, London, 2007)
3. M. Alonso, E.J. Finn, *Fundamental University Physics III: Quantum and Statistical Physics* (Addison Wesley, Reading, Mass, 1971)
4. P.W. Anderson, More is different, Science **177**, 377 (1972). Reprinted in *A Career in Theoretical Physics* (World Scientific, Singapore, 1994)
5. P.W. Atkins, *Physical Chemistry* (Oxford University Press, Oxford, 1994)
6. G. Auletta, G.F.R. Ellis, L. Jaeger, Top-down causation: from a philosophical problem to a scientific research program. J. Roy. Soc. Interface **5**, 1159–1172 (2008). arXiv:0710.4235
7. A.-L. Barabási, Z.N. Oltvai, Network biology: understanding the cell's functional organization. Nat. Rev. Genet. **5**, 101–114 (2004)
8. M.A. Bedau, P. Humphreys (eds.), *Emergence: Contemporary Readings in Philosophy and Science* (MIT Press, Cambridge, Mass, 2008)
9. S. Beer, *Brain of the Firm* (Wiley, Chichester, 1981)
10. F. Bendetti, *Placebo Effects* (Oxford University Press, Oxford, 2014)
11. J. Binney, S. Tremain, *Galactic Dynamics* (Princeton University Press, Princeton, 1987)
12. C.M. Bishop, *Neural Networks for Pattern Recognition* (Oxford University Press, Oxford, 1999)
13. R.C. Bishop, Fluid convection, constraint and causation. Interface Focus **2**, 4–12 (2012)
14. G. Booch, *Object Oriented Analysis and Design with Applications* (Addison Wesley, New York, 1994)
15. R.N. Bracewell, *The Fourier Transform and Its Applications* (McGraw Hill, New York, 1986)
16. W. Brown, N. Murphy, *Did My Neurons Make Me Do It? Philosophical and Neurobiological Perspectives on Moral Responsibility and Free Will* (Oxford University Press, New York, 2007)
17. T. Buyana, *Molecular Physics* (World Scientific, Singapore, 1997)
18. G. Buzsáki, *Rhythms of the Brain* (Oxford University Press, Oxford, 1997)

19. D.T. Campbell, Downward causation, in *Studies in the Philosophy of Biology: Reduction and Related Problems*, ed. by F.J. Ayala, T. Dobhzansky (University of California Press, Berkeley, 1974)
20. N.A. Campbell, J.B. Reece, *Biology* (Benjamin Cummings, San Francisco, 2005)
21. S. Carroll, *From Eternity to Here: The Quest for the Ultimate Arrow of Time* (Dutton, New York, 2010)
22. P.M. Chaikin, T.C. Lubensky, *Principles of Condensed Matter Physics* (Cambridge University Press, Cambridge, 2000)
23. S. Chibbaro, L. Rondoni, A. Vulpiani, *Reductionism, Emergence, and Levels of Reality* (Springer, Heidelberg, 2014)
24. M. Chown, *We Need to Talk about Kelvin* (Faber and Faber, London, 2010)
25. S. Conway Morris, *Life's Solution: Inevitable Humans in a Lonely Universe* (Cambridge University Press, Cambridge, 2005)
26. J. Copeland, *The Essential Turing* (Oxford University Press, Oxford, 2004)
27. F. Crick, *Astonishing Hypothesis: The Scientific Search for the Soul* (Scribner, 1995)
28. J.P. Crutchfield, Between order and chaos. Nat. Phys. **8**, 17–24 (2011)
29. P.C.W. Davies, *The Physics of Time Asymmetry* (Surrey University Press, 1974)
30. R. Dawkins, Hierarchical organisation: a candidate principle for ethology, in *Growing Points in Ethology*, ed. by P.P.G. Bateson, R.A. Hinde (Cambridge University Press, Cambridge, 1976)
31. P. Dayan, L. Abbot, *Theoretical Neuroscience: Computational and Mathematical Modelling of Neural Systems* (MIT Press, Cambridge, Mass, 2001)
32. T. Deacon, *The Symbolic Species: The Co-Evolution of Language and the Human Brain* (Penguin, London, 1997)
33. T. Deacon, Universal grammar and semiotic constraints, in *Language Evolution*, ed. by M. Christiansen, S. Kirby (Oxford University Press, Oxford, 2003), pp. 111–139
34. R.L. Devaney, *An Introduction to Chaotic Dynamical Systems* (Basic Books, 2003)
35. P.A.M. Dirac, Proc. R. Soc. Lond. A **123**, 714 (1929)
36. S. Dodelson, *Modern Cosmology* (Academic Press, San Diego, 2003)
37. A. Durrant, *Quantum Physics of Matter* (Institute of Physics and the Open University, Bristol, 2000)
38. A.S. Eddington, *The Nature of the Physical World* (MacMillan, London, 1928)
39. G.F.R. Ellis, Cosmology and local physics. New Astron. Rev. **46**, 645–658 (2002). http://arxiv.org/abs/gr-qc/0102017
40. G.F.R. Ellis, True complexity and its associated ontology, in *Science and Ultimate Reality: Quantum Theory, Cosmology and Complexity*, ed. by J.D. Barrow, P.C.W. Davies, C.L. Harper (Cambridge University Press, Cambridge, 2004) pp. 607–636
41. G.F.R. Ellis, *The Universe Around Us: An Integrative View of Science and Cosmology* (2004). http://www.mth.uct.ac.za/~ellis/cos0.html
42. G.F.R. Ellis, On the nature of causation in complex systems. Trans. Roy. Soc. S. Afr. **63**, 69–84 (2008)
43. G.F.R. Ellis, Top-down causation and emergence: some comments on mechanisms. J. Roy. Soc. Interface Focus **2**, 126–140 (2012)
44. G.F.R. Ellis, On the limits of quantum theory: contextuality and the quantum-classical cut. Ann. Phys. **327**, 1890–1932 (2012)
45. G.F.R. Ellis, D. Noble, T. O'Connor (eds.), *Top-down causation: an integrating theme within and across the sciences?* Roy. Soc. Interface Focus Special issue **2**, 1–140 (2012)
46. G.F.R. Ellis, D.W. Sciama, Global and non-global problems in cosmology, in *General Relativity (A Synge Festschrift)*, ed. by L. O'Raifeartaigh (Oxford University Press, Oxford, 1972) pp. 35–59
47. B. Falkenberg, M. Morrison (eds.), *Why More Is Different: Philosophical Issues in Condensed Matter Physics and Complex Systems* (Springer, Heidelberg, 2015)
48. R.L. Flood, E.R. Carson, *Dealing with Complexity: An Introduction to the Theory and Application of Systems Science* (Plenum Press, London, 1990)

49. R.P. Feynman, R.B. Leighton, M. Sands, *The Feynman Lectures on Physics: Mainly Mechanics, Radiation, and Heat* (Addison-Wesley, Reading, Mass, 1963)
50. R.P. Feynman, R.B. Leighton, M. Sands, *The Feynman Lectures on Physics: The Electromagnetic Field* (Addison-Wesley, Reading, Mass, 1963)
51. R.P. Feynman, R.B. Leighton, M. Sands, *The Feynman Lectures on Physics: Quantum Mechanics* (Addison-Wesley, Reading, Mass, 1965)
52. C. Frith, *Making up the Mind: How the Brain Creates Our Mental World* (Blackwell, Malden, 2007)
53. M. Gell-Mann, *The Quark and the Jaguar: Adventures in the Simple and the Complex* (Abacus, London, 1994)
54. J. Gemmer, M. Michel, G. Mahler, *Quantum Thermodynamics: Emergence of Thermodynamic Behaviour Within Composite Quantum Systems* (Springer, Heidelberg, 2004)
55. S. Gilbert, D. Epel, *Ecological Developmental Biology* (Sinauer, 2009)
56. M.B. Glauert, *Principles of Dynamics* (Routledge and Kegan Paul, London, 1960)
57. P.W. Glimcher, Indeterminacy in brain and behaviour. Annu. Rev. Psychol. **56**, 25 (2005)
58. K.S. Goodman, Reading: A psycholinguistic guessing game, in *Language and Literacy: The Selected Writings of Kenneth Goodman*, vol. 1, ed. by F.V. Gollaschvol (Routledge and Kegan Paul, London, 1967) pp. 33–44
59. P. Gray, *Psychology* (Worth Publishers, New York, 2011)
60. S. Greenland, J. Pearle, Causal diagrams, Technical report R0332, in *Encyclopaedia of Epidemiology* (2006)
61. G. Greenstein, A.G. Zajonc, *The Quantum Challenge: Modern Research on the Foundations of Quantum Mechanics* (Jones and Bartlett, Sudbury, Mass, 2006)
62. S. Hartmann, Effective field theories, reductionism and scientific explanation. Stud. Hist. Philos. Sci. Part B **32**, 267–304 (2001)
63. L.H. Hartwell, J.J. Hopfield, S. Leibler, A.W. Murray, From molecular to modular cell biology. Nature **402**, Supplement C47–C52 (1999)
64. J. Hawkins, *On Intelligence* (Holt Paperbacks, New York, 2004)
65. P.M. Hoffmann, *Life's Ratchets: How Molecular Machines Extract Order from Chaos* (Basic Books, New York, 2012)
66. J.H. Holland, *Adaptation in Natural and Artificial Systems* (MIT Press, Cambridge, Mass, 1992)
67. D. Huron, *Sweet Anticipation: Music and the Psychology of Expectation* (MIT Press, Cambridge, Mass, 2007)
68. J.C. Jackson, *Classical Electrodynamics* (Wiley, New York, 1967)
69. L. Jaeger, E.R. Calkins, Downward causation by information control in micro-organisms. Interface Focus **2**, 26–41 (2012)
70. H. Jeong, S.P. Mason, A. Barabasi, Z.N. Oltvai Lethality and centrality in protein networks. Nature **411**, 41–42 (2001). arXiv:cond-mat/0105306
71. A. Juarrero, *Dynamics in Action: Intentional Behaviour as a Complex System* (MIT Press, Cambridge, Mass, 2002)
72. E.R. Kandel, *The Age of Insight* (Random House, 2012)
73. E.R. Kandel, J.H. Schwartz, T.M. Jessell, *Principles of Neuroscience* (McGraw Hill, New York, 2000)
74. S.A. Kauffman, *The Origins of Order: Self-Organisation and Selection in Evolution* (Oxford, New York, 1993)
75. R. Lafore, *Data Structures and Algorithms in Java* (SAMS, 2002)
76. T. Lancaster, M. Pexton, Reduction and emergence in the fractional quantum Hall state. Stud. Hist. Philos. Mod. Phys. (2015)
77. R. Lapkiewicz, P. Li, C. Schaeff, N.K. Langford, S. Ramelow, M. Wiesniak, A. Zeilinger, Experimental non-classicality of an indivisible quantum system. Nature **474**, 490 (2011) arXiv:1106.4481v1
78. R.B. Laughlin, Fractional quantisation. Rev. Mod. Phys. **71**, 863 (2000)

79. D.J. Levitin, *This Is Your Brain on Music: The Science of a Human Obsession* (Plume, London, 2007)
80. P.L. Luisi, Emergence in chemistry: chemistry as the embodiment of emergence. Found. Chem. **4**, 183–200 (2002)
81. J. MacCormack, *9 Algorithms that Changed the Future: The Ingenious Ideas that Drive Today's Computers* (Princeton University Press, Princeton, 2012)
82. M.M. Mano, C.R. Kime, *Logic and Computer Design Fundamentals* (Pearson/Prentice Hall, 2008)
83. M Martìnez, A. Moya, Natural selection and multi-level causation. Philos. Theo. Biol. **3** (2011). http://hdl.handle.net/2027/spo.6959004.0003.002
84. G. McGhee, *Convergent Evolution: Limited Forms Most Beautiful* (MIT Press, Cambridge, Mass, 2011)
85. B. McLaughlin, K. Bennett, Supervenience, in *The Stanford Encyclopedia of Philosophy* (Winter 2011 edition), ed. by E.N. Zalta (2011). http://plato.stanford.edu/archives/win2011/entries/supervenience/
86. P. Menzies, The causal efficacy of mental states, in *Physicalism and Mental Causation*, ed. by S. Walter, H.-D. Heckmann (Imprint Academic, 2003)
87. E. Morsella, M. Lanska, C.C. Berger, A. Gazzaley, Indirect cognitive control through top-down activation of perceptual symbols. Eur. J. Soc. Psychol. **39**, 1173–1177 (2009)
88. M. Newman, A.-L. Barabási, D.J. Watts, *The Structure and Dynamics of Networks* (Princeton Unversity Press, Princeton, 2006)
89. S.A. Newman, What's new: a review of the origins of evolutionary innovations by Andreas Wegner. Philos. Theor. Biol. **4**, e304 (2012)
90. D. Noble, *The Music of Life* (Oxford University Press, Oxford, 2006)
91. D. Noble, A theory of biological relativity: no privileged level of causation. Interface Focus **2**, 55–64 (2012)
92. *Oxford Advanced Learners Dictionary* (Oxford University Press, Oxford, 2000) pp. 1414–1422
93. M.J. Page et al., The suppression of star formation by active galactic nuclei. Nature **485**, 213–216 (2012)
94. L. Pauling, *The Nature of the Chemical Bond and the Structure of Molecules and Crystals: An Introduction to Modern Structural Chemistry* (Cornell University Press, Ithaca, 1960)
95. L. Pauling, E.B. Wilson, *Introduction to Quantum Mechanics with Applications to Chemistry* (Dover, Mineola, NY, 1963)
96. A.R. Peacocke, *An Introduction to the Physical Chemistry of Biological Organization* (Oxford University Press, Oxford, 1989)
97. J. Pearl, Graphs, causality, and structural equation models. Sociol. Methods Res. **27**, 226–284 (1998)
98. J. Pearl, *Causality: Models, Reasoning, and Inference* (Cambridge University Press, Cambridge, 2000)
99. R. Penrose, *The Road to Reality: A Complete Guide to the Laws of the Universe* (Jonathan Cape, London, 2004)
100. R. Penrose, *Cycles of Time: An Extraordinary New View of the Universe* (Knopf, New York, 2011)
101. I. Percival, Schrödinger's quantum cat. Nature **351**, 357 (1991)
102. G.A. Petsko, D. Ringe, *Protein Structure and Function* (Oxford University Press, Oxford, 2009)
103. S. Pilosof, M.A. Porter, S. Kéfi, Ecological Multilayer Networks: A New Frontier for Network Ecology (2015). http://arxiv.org/abs/1511.04453
104. D. Purves, *Brains: How They Seem to Work* (FT Press Science, Upper Saddle River, 2010)
105. H. Putnam, Philosophy and our mental life, in *Mind, Language, and Reality* (Cambridge University Press, 1975)
106. A.I. Rae, *Reductionism* (Oneworld, 2013)

107. E. Ravasz, A.L. Somera, D.A. Mongru, Z.N. Oltvai, A.-L. Barabási, Hierarchical organization of modularity in metabolic networks. Science **297**, 1551–1555 (2002)
108. D. Rickles, Supervenience and determination, *Internet Encyclopedia of Philosophy (IEP)* (2013). http://www.iep.utm.edu/superven/
109. R. Rhoades, R. Pflanzer, *Human Physiology* (Saunders College Publishing, Fort Worth, 1989)
110. F. Rieke, D. Warland, R. de Ruyter van Steveninck, W. Bialek, *Spikes: Exploring the Neural Code* (MIT Press, Cambridge, Mass, 1999)
111. S. Sarkar, *Genetics and Reductionism* (Cambridge University Press, Cambridge, 1998)
112. W.C. Saslaw, *Gravitational Physics of Stellar and Galactic Systems* (Cambridge University Press, Cambridge, 1987)
113. S. Schweber, Physics, community, and the crisis in physical theory. Phys. Today 34–40 (1993)
114. D.W. Sciama, *The Unity of the Universe* (Faber and Faber, London, 1959)
115. A. Scott, *Stairway to the Mind* (Springer-Verlag, New York, 1995)
116. W. Seager, *Natural Fabrications: Science, Emergence, and Consciousness* (Springer, Heidelberg, 2012)
117. J.R. Searle, *Making the Social World: The Structure of Human Civilisation* (Oxford University Press, Oxford, 2011)
118. S. Seung, *Connectome* (Houghton Mifflin Harcourt, Boston, 2012)
119. J. Silk, *The Big Bang* (Freeman, New York, 2001)
120. H.A. Simon, *The Sciences of the Artificial* (MIT Press, Cambridge, Mass, 1992)
121. A.S. Tanenbaum, *Structured Computer Organisation* (Prentice Hall, Englewood Cliffs, 2006)
122. R.L. Trask, *Language and Linguistics: The Key Concepts* (Routledge, Abingdon, 2007)
123. K. Umashankar, *Introduction to Engineering Electromagnetic Fields* (World Scientific, Singapore, 1989)
124. R. Van Gulick, Who's in charge here? And who's doing all the work?, in *Mental Causation*, ed. by J. Heil and A. Mele (Oxford University Press, Oxford, 1995)
125. A. Vázquez, R. Dobrin, D. Sergi, J.-P. Eckmann, Z.N. Oltvai, A.-L. Barabási, The topological relationship between the large-scale attributes and local interaction patterns of complex networks. Proc. Nat. Acad. Sci. **101**, 17940–17945 (2004)
126. A. Vazquez, A. Flammini, A. Maritan, A. Vespignani, Global protein function prediction in protein–protein interaction networks. Nat. Biotech. **21**, 697–700 (2003). arXiv:con-mat/0306611
127. S. Vogel, *Cats' Paws and Catapults: Mechanical Worlds of Nature and People* (W W Norton and Company, 2000)
128. A. Wagner, *The Origins of Evolutionary Innovations* (Oxford University Press, Oxford, 2011)
129. J.D. Watson, T.A. Baker, S.P. Bell, A. Gann, M. Levine, R.M. Losick, *The Molecular Biology of the Gene* (Benjamin Cummings, 2003)
130. J.A. Wheeler, R.P. Feynman, Interaction with the absorber as the mechanism of radiation. Rev. Mod. Phys. **17**, 157–181 (1945)
131. H.M. Wiseman, G.J. Milburn, *Quantum Measurement and Control* (Cambridge University Press, Cambridge, 2010)
132. J.M. Ziman, *Principles of the Theory of Solids* (Cambridge University Press, Cambridge, 1979)

Chapter 4
Kinds of Top-Down Causation

Top-down causation is a generic concept. The previous chapters have given many examples. The overall proposal (Chap. 3) is that top-down causation takes place by higher level boundary or structural relations constraining what happens at lower levels, and thereby creating possibilities for new kinds of behaviour at the lower levels and complex emergence of higher levels. But there are various ways this can happen.

The issue that this chapter addresses is whether there are essentially different kinds of top-down causation that can be identified, with discernibly different kinds of dynamics in operation. The proposal made here (developing from [62, 64]) is that there are five different kinds of top-down causation, which I have called TD1–TD5, as indicated in Table 4.1 (with the most complex one shown at the top). To some degree the higher ones build on the lower ones.

Cases TD3–TD5 enable more complex behaviours than TD1 and TD2, as they are instances of complex adaptive systems, which allow information to be collected and learning to occur. The following sections discuss in turn,

- Section 4.1. TD1 Deterministic top-down causation.
- Section 4.2. TD2 Non-adaptive feedback control.
- Section 4.3. TD3 Adaptive selection of outcomes.
- Section 4.4. TD4 Adaptive selection of goals.
- Section 4.5. TD5 Adaptive selection of selection criteria.
- Section 4.6. Complex adaptive systems.
- Section 4.7. Intelligent top-down causation.

It will be shown that each of the five classes TD1–TD5 of top-down causation occurs in the real world, and each is essentially different from the others. Globally speaking, this chapter is an examination of causation in complex emergent systems.

G. Ellis, *How Can Physics Underlie the Mind?*, The Frontiers Collection,
DOI 10.1007/978-3-662-49809-5_4

Table 4.1 The five different kinds of top-down behaviour characterized in this chapter. Cases TD3–TD5 are all based on adaptive selection

Name	Type of top-down causation
TD5	Adaptive selection of selection criteria
TD4	Adaptive selection of goals
TD3	Adaptive selection
TD2	Non-adaptive information control
TD1	Deterministic top-down causation

4.1 Deterministic Top-Down Causation TD1

Deterministic top-down causation occurs when high-level variables have causal power over lower level dynamics through context or system structuring in such a way that the initial data uniquely determines the outcome. That is to say:

- **Determinism**. Given the higher level structural and boundary conditions, the outcome depends uniquely on the initial conditions.
- **Contextual Constraints**. If the higher level structural relations or boundary conditions are altered, the mapping from initial conditions to outcomes changes.

I discuss in turn:

- The nature of the process (Sect. 4.1.1).
- Machines (Sect. 4.1.2).
- Physical systems (Sect. 4.1.3).
- Living systems (Sect. 4.1.4).
- Logical systems (Sect. 4.1.5).
- Mathematical models: boundary conditions and constraints (Sect. 4.1.6).
- Randomness and noise (Sect. 4.1.7).

4.1.1 *The Nature of the Process*

This is the basic idea of a *machine*: it uses energy to do what you tell it to do in a reliable way. It will do so if it is a closed system, that is, if nothing outside interferes with its operation. But deterministic causation happens in many other contexts than just machines. It happens in both unstructured and structured systems (Table 4.2). The basic causal relation is that there are variables such that

Table 4.2 Deterministic determination of outcomes from initial data. Given the context (structural conditions and boundary conditions), the initial data leads to a unique final state

Context	Unstructured system	Structured system
Constraints	Boundary conditions	Structural conditions
	$\Downarrow$	$\Downarrow$
Data $\Longrightarrow$	Constrained initial data	Control parameters
Closed system	$\Downarrow$	$\Downarrow$
Outcome	Final state (deterministic)	

$$\text{(time development laws, constraints, initial conditions)} \underbrace{\Longrightarrow}_{\text{unique}} \text{(outcomes)}\,. \tag{4.1}$$

Variables. The variables here must be *effective variables*, that is they must actually affect the outcome. If they have no effect, they should be deleted from the list of variables. A challenge in any specific context is to determine a *minimal set* of effective variables, that is, to delete redundant variables.This can be done by using the constraint equations. Even when this is done, the variables will in general not be unique: one can take combinations of the variables to get a new minimal set. However, in many cases, there will be a 'best' set, in that they form a best representation of the underlying dynamics.

Time Development Laws. These determine how the relevant variables change with time. They may for example be:

- *Laws of physics*, such as Newton's laws of motion, the diffusion equation, the wave equation, Schrödinger's equation, and the Dirac equation.
- *Laws of physical chemistry*, such as the law of mass action and Fick's law of diffusion.
- *Computer algorithms*, such as quicksort, payroll procedures, and finite-element algorithms.
- *Bureaucratic processes*, such as rules as to when payments should be made, in what sequence procedures must take place to get a driver's licence, and when annual general meetings and elections must take place.
- *Rules of a game*, such as the rules of chess, contract bridge, football, tennis, and cricket.

Constraints. These are sets of time-independent relations between the variables that constrain how they relate to each other, and thereby structure what happens. They may be of many forms:

- *Boundary conditions*. A set of conditions that are the same for all members of the set of systems considered, e.g., asymptotic flatness or global topological conditions in physics.
- *Environmental conditions*. For example, the existence of a surrounding heat bath or the presence of incoming radiation from the Sun.
- *Structural relations*. For example, as the wiring in a computer system or the set of connections in a neural network.

- *Logical constraints*. For example, a constitution for an organisation, rules as to what payments are and are not allowed, rules as to what is and what is not allowed in a game.

Initial Conditions. These are the starting values of variables at some chosen time that *vary over members of the system* (molecules have different initial positions and velocities, concentrations of reactants vary in different cells, each player has a different set of cards), or *vary for the same member of the system during different runs* (we fire a canon with varying elevations, we shuffle a pack of cards for the next game, we roll a die, we choose random initial data in a Monte Carlo simulation). This is where randomness often enters: we cannot control the elevation of the gun precisely or we purposefully introduce a random element into a game or simulation.

The initial conditions for all runs must satisfy the constraints, otherwise they are not valid: either they are not possible in physical terms (you cannot have an initial speed greater then the speed of light), or are disallowed because of context (you cannot have five players in contract bridge).

Preserving the Constraints. A key feature is that the constraints must remain true at all times, therefore the dynamical evolution is required to preserve the constraints. Thus they channel the way the dynamics operates: different constraints lead to different outcomes, even though they are not operators that change the system state over time.

Causal Variables. In a given context we take for granted the items we cannot change or choose to keep fixed, and assign as causal variables those that vary due to outside causes, or that we choose to change. The outcome is then determined by the initial data, and (4.1) reduces to

$$\text{(initial conditions)} \underbrace{\Longrightarrow}_{\text{constraints}} \text{(outcomes)}\,. \tag{4.2}$$

The initial data is the 'cause', taking all the rest of the context for granted: the existence of the Universe, the existence and nature of the laws of physics, the existence of planet Earth, the existence of the experimenter, etc. These contextual features are taken as the unchanging larger context in which we consider all systems. But for any specific system, we also usually just assume the constraints specific to that system, for they are in many ways the essence of the nature of the system (a machine is characterized by its structural relations, and in biology, function is enabled by structure). They are taken to be true because it is an entity of such and such type ("It's an Apple MacBook Air", "It's a giraffe"). When we know the identity, we take the structural relations, and consequent emergent functional relations, for granted. This is an example of the power of the logical act of naming things (Sect. 2.2.1), which underlies logical hierarchies.

Deterministic Top-Down Causation (TD1). The lower level variables uniquely determine the outcome from the initial and boundary conditions, as a consequence of the system constitution and structuring. Changing these conditions leads to different lower level events and dynamical outcomes.

Provided the lower level interactions mesh together in a coherent way, the constrained operation of lower level forces, operating in a law-like/algorithmic way, leads to reliable higher level behaviour whose outcome depends on the nature of the constraints and initial conditions (Sect. 3.3.1). These constraints are often in the form of networks of interactions [10], usually including recurring network motifs [5]. These are higher level features because they cannot be described in terms of lower level concepts (the specific connections between transistors in a computer cannot be described in terms of properties of electrons) and the system ceases to function if the higher level relationships are disrupted, even though the individual lower level elements are unchanged.

This kind of deterministic causation occurs in all physical and natural systems, as well as in machines. Here are some examples.

4.1.2 Machines

Machines are the archetypal examples of deterministic causation. They are purposefully structured to attain some high level outcome, and they reliably attain that outcome by mechanistic processes at the lower levels.

4.1.2.1 Constrained Lower Level Causation

Machines achieve this by constraining lower level causation by means of specific structures such as gears and levers, hydraulic pipes and valves, electrical wires and switches, and waveguides and optical fibres. These channel the way lower level entities (water molecules, electrons, waves) flow. Such devices can be reliably controlled by setting high level variables (e.g., turning a washing machine on), which then cause lower level systems to respond appropriately (water flows into the washing machine, and the tumbler starts to rotate, because water molecules flow along pipes and electrons flow in a wire to an electric motor). The low level physics does the bidding of the person who controls the machine. Examples are:

- **Mechanical and Hydraulic Machines**. Clocks and watches (the idea of machines started with clockwork mechanisms), mechanical toys, windmills, and water turbines. We control things with levers, wires, pipes, valves, and taps.
- **Thermodynamic Machines**. Steam engines, internal combustion engines, refrigerators, and heat pumps. For example, in the case of diesel engines, compressing a gas mixture in a cylinder can result in ignition of the gas in a predictable way. A cylinder is a high level concept, as are the pressure and temperature of the gas. Low level concepts are molecules of $C_{12}H_{23}$ and O_2, and the chemical and physical reactions between them.
- **Electric Machines**. Electrical engines, electrical generators, relays, telephones. You can control things at a distance by electrical wiring and switches and relays.

- **Electronic machines** Radios, television sets, and radar. We can control things at a distance without wiring, by using radio signals.
- **Digital Computers**. An excellent example—indeed the canonical example today—is digital computers: the low level gates and transistors act in accord with the data and program loaded (word processor, music programme, image processing programme, etc.). We control things by choosing the program and the data (see Chap. 2).

Machines are carefully constructed so that any fluctuations at the lower levels, whatever their origin, will not affect higher level operational reliability.

4.1.3 Physical Systems

Top-down action occurs in natural and manufactured physical systems by:

- Setting boundary conditions for differential equations [141].
- Setting values of higher level variables having a key effect on lower level dynamics.
- Shaping effective potentials [65].

4.1.3.1 Boundary Conditions for Partial Differential Equations

The outcomes of many natural or manufactured physical systems is determined by partial differential equations or sets of ordinary differential equations, where the outcome depends on the context through boundary conditions as well as initial conditions [41]. Examples are:

Fluid Convection. This has been examined in detail as a model for downward causation in classical mechanics because of the nature of the relevant differential equations [20]. This occurs for example in Bénard convection cells, where the pattern of convection cells depends on the shape of the boundary.

Musical Instruments. Hearing the shape of a drum is a very old topic. The tone emitted results from the frequencies of its surface vibrations, which are determined by the shape of the boundary [110]. The Helmholtz equation gives the frequencies as eigenvalues of a Laplacian and the shape determines which ones occur. Essentially similar effects occur in all musical instruments, e.g., violins and pianos.

Biological Pattern Formation. Similar effects occur in biological pattern formation, for example leopard, giraffe, and zebra markings and patterns on butterfly wings ([139]:435480). The reaction–diffusion equation gives outcomes dependent on the shape of the body, and in particular its topology (closed surfaces lead to periodic boundary conditions).

4.1.3.2 Setting Contextual Variables

Lower level dynamics may also be affected by contextual variables representing the top-down effects of higher level conditions (this is similar to the way global parameters are passed to subroutines in computer programs). Examples are:

Nucleosynthesis in the Early Universe. Light elements (D, ^{3}H, ^{3}He, ^{4}He, ^{7}Li) are synthesized from hydrogen through nuclear reactions that take place in the early universe [52, 172, 188]. During the radiation-dominated early era, the Friedmann equation for the scale factor $a(t)$ is dominated by the cosmological radiation density, determining the scale factor $a(t)$ as a function of time:

$$\rho_{\text{rad}}(t) \propto 1/a^4(t) \underbrace{\Longrightarrow}_{\text{Friedmann}} a(t) \propto t^{1/2} . \tag{4.3}$$

Because $\rho_{\text{rad}} = aT^4_{\text{rad}}$, this leads to the temperature–time relation

$$T_{\text{rad}}(t) = \frac{1.5 \times 10^{10}}{t^{1/2}_{\text{sec}}} \text{ K} , \tag{4.4}$$

which then determines nuclear reaction rates that depend very sensitively on the temperature [188], and hence the way nucleosynthesis proceeds at the micro-level. The outcome thus derives from the time dependence of the macroscopic cosmological variable $a(t)$, determined by the cosmological context of the expanding early universe. Different expanding universe scenarios, for example a very anisotropic or inhomogeneous early universe, would lead to other outcomes [129, 181]. The large scale metric and density evolution set the environment (4.4) for the nuclear reactions, which determine the resulting nuclear fractions in a bottom-up manner. Hence these abundances can be used to constrain key cosmological parameters.

Phase Transitions. Changes in higher level variables cause a phase transition, representing a discontinuous alteration in the mode by which lower level interactions lead to higher level behaviour and consequent high-level equation of state [198]. The variables that cause the phase transitions are higher level (coarse-grained) variables. One cannot describe phase transitions without them. In a laboratory situation, they are manipulated by the experimenter to cause the phase transition.

4.1.3.3 Shaping Effective Potentials

In many physical systems, the lower level dynamics is governed by effective potentials that represent the summed effects of all other interacting particles [62]. The nature of these potentials will often depend on the specific higher level configuration of the other particles, for example, whether atoms are structured as a specific kind of crystal [198], or stars arranged as a particular type of galaxy [17]. They determine the lower level dynamical relations. They emerge from the lower level entities, but their nature is independent of the detailed lower level positions and velocities of particles: it depends on the patterns in which they are arranged, and associated size and energy scales.

Other examples of top-down causation in physical systems, such as the origin of the arrow of time and the use of effective potentials, are given in Chap. 5.

4.1.4 Living Systems

The molecular biology revolution led to the understanding that biology is based on molecular machines at the lower level [162], which behave in a deterministic manner and affect higher levels in a bottom-up manner. This happens through physics processes at the lower levels in the context of systems structured so as to have specific functions. The outcomes depend crucially on context. Again it happens by:

- Setting boundary conditions for differential equations.
- Setting values for contextual variables.
- Passing signals via messenger molecules.
- Constraining lower level causation through structural conditions.

4.1.4.1 Boundary Conditions for Differential Equations

Many models of biological systems consist of differential equations for the kinetics of each component [148]. These equations cannot give a solution without setting both the initial conditions (the state of the components at the time at which the simulation begins) and boundary conditions expressing what constraints are imposed on the system by its environment [141]. These lead to the kinds of patterns explored by Alan Turing, which play a role in early morphogenesis and control aspects like markings on a butterfly's wings and patterns of zebra stripings [139].

Such structured interactions occur in the form of networks of interactions [10], usually including recurring network motifs [5]. The contextual issue is that all the required reagents must occur together in a confined space, which is why cells walls and other biological membranes are so important, for example, in muscles and axons, where they control ion mobility. They provide the context enabling these networks to exist and function, leading to partitions between different functional parts of the system, for example, between shoots and roots in a plant [182, pp. 173–195]. As stated by Thornley [182, pp. 23–24]:

> A constraint may be regarded as a loss or limitation of freedom, and the general concept is crucial to all forms of modeling [...] the plant modeler seeks to construct a useful description at a particular level, and the sensitive choice of constraints (which often define the language used) is at the heart of the matter.

4.1.4.2 Setting Values for Contextual Variables

Contextual variables set by the environment must lie in a suitable range. For example, the following are crucial to life as we know it:

- The environmental temperature must lie in a very narrow band.
- Oxygen and water must be available.
- A suitable energy source must be available (sunlight for a plant, food for an animal).

Without these contextual conditions being right, much life on Earth (animals, plants, and insects) would be in trouble. Other forms of life might have different sources of energy (e.g., thermal vents), but without some energy source, they will not survive.

4.1.4.3 Contextual information

The lower level molecular machines [162] are based on physics and chemistry [190], but their outcomes depend on context communicated by 'passing parameters' from higher to lower levels via messenger molecules.

Reading Genes. This is the core of epigenetics. Thus gene expression is altered by transcriptional regulation via methylation, via the neuroendocrine system, or by microbial induction as in the vertebrate immune response [84].

Brain Function. It is a key feature of brain function. Neuronal activation is affected by neuromodulators such as dopamine, serotonin, acetylcholine, and histamine, diffused through large areas of the nervous system to affect multiple neurons [113].

Body Systems. It is also a core feature of physiology. For example, the physiology of the heart can be expressed in terms of a hierarchy of differential equations where higher level variables set the context for lower level outcomes [140].

Global Resource Cycles. At a higher level, global resource cycles such as the carbon dioxide cycle govern availability of crucial materials for plants and animals, which in turn affect global variables such as gas densities in the atmosphere [125, pp. 218–228]. This represents an interlevel feedback loop between local ecosystems and the global biogeosphere. Similar effects occur in the nitrogen and phosphorus cycles [125, pp. 229–258].

4.1.4.4 Constrained Lower Level Causation

Structured physiological systems are constructed so as to channel causation in a very precise way [156]. I give just two examples:

The Nervous System. Action potentials in an axon or dendrite in the brain travel in a mechanistic manner, based on diffusion of ions through the cell membrane [113, 166]. However, the outcome of neural network activity depends on the pattern of neuronal connections, viz., the connectome [169]. This is like the way the structured wiring of a computer determines its logical functioning (Chap. 2). At a larger scale, the entire nervous system is specifically wired to give functional outcomes. For example, the eyes are connected to specific visual areas of the cortex, and motor neurons are connected to specific muscles in our limbs.

The Cardiovascular System. This is structured so as to provide oxygen and nutrients to all the cells in the body. It routes them to each cell through a fractal-like structure, with the flow powered by the heart [156]. Because of the first and second laws of thermodynamics, we die if it fails to function.

This kind of structural determination of function occurs in each of the physiological systems in the human body [156].

4.1.5 Logical Systems

Logical systems have analogous contextual constraints to those that occur in physical systems. This involves the following:

- Constraints that restrict what may be done.
- Contextual dependence of logical flow and constraints.
- Contextual dependence of symbolic functioning, including dependence of the meaning of variables on context.

4.1.5.1 Constraints that Restrict What May Be Done

These are the rules of play for a game, or rules of logic for some enterprise, e.g., only people over 65 may live in this housing scheme, or only people with an approved educational certificate can apply for the job, and so on.

Constrained Lower Level Implementation. These are rules setting constraints on implementation of the logical system. They relate to the physical basis of emergence of the logical system, or physical aspects of how it operates:

- **Games**. The game takes place on a specific playing board or playing field, or a set of cards displays a specific set of symbols (heart and club symbols are printed on the cards).
- **Bureaucratic Rules**. This office only handles pensions, illness grants are handled in Washington. Applications must be filled in on form F101-346-1957.
- **Computer Systems**. This software is only licensed to run on brand x machines.

These are all chosen restrictions on how the logic is implemented.

4.1.5.2 Contextual Dependence of Logical Flow and Constraints

The higher level context can change the flow of the lower level dynamics. This happens by passing global variables to local domains. It can change the rules of play, or the constraints in operation. This can work by sending contextual parameters (Sect. 3.4.3) from the center to the modules to change their mode of logical operation. For example, if the age for a diver's licence has been nationally established as 17 years old, then the number 17 is such a control parameter.

Context-Dependent Rules of Play. These rules determine what action will take place when. They have the basic form:

IF X, THEN do Y to V   ELSE do Z,

where X is a higher level condition and Y, Z are operations on object V. The condition X might, for example, be a date or time, but it might be occurrence of some logical condition:

- **Bureaucratic Rules**. For example: If it is the 27th of the month, implement the payroll system.
- **Rules of Games**. For example: The referee will toss a coin. The team that wins the toss will kick the ball to start play (the coin toss sets a logical state that decides what happens next).
- **Computer Programs**. Conditional branching controlled by a global variable. For example: IF day $<$ 2 OR day $>$ 2 THEN return, IF day $=$ 2 GOTO subroutine PAY.

Contextual Constraints on Logic of Play. These are logical rules that hold all the time, and depend on the context. They may be of the form: IF X, THEN NOT Y. The context may be, for example, that one is playing a specific game (one is playing American football, not soccer), or it may be that some specific condition hold in the game (one side is on, the other side is not). The time development rules must respect these constraints:

- **Bureaucratic Rules**. In the state of Maryland, children under the age of 16 may not drive an automobile.
- **Rules of a Game**. When playing soccer, except for the goal keeper, the players must not touch the ball with their hands.
- **Computer Programs**. Contextual constraints are set by typing and scoping of variables.

4.1.5.3 Contextual Dependence of Symbolic Functioning

Top-down relationships are the key to how hierarchically structured symbolic systems work (see Sects. 3.1.6 and 3.3.3).

Class Hierarchies. The relations between classes and subclasses, and between classes and individuals, is a logical relation explicitly embodying contextual determination of lower level properties by specialization and modification of higher level properties: a form of logical top-down characterization embodied in the nature of symbolic systems such as language.

Contextual Constraints. As pointed out in Sect. 3.4.4, the meaning of words, phrases, and sentences depends on context, which is taken for granted when one reads text or listens to a story or logical argument. Just as in physical cases (Sect. 3.4.2), the higher level context constrains the lower level meanings of words, as for example when the context of an airport constrains the word 'plane' so that it primarily refers to a flying vehicle rather than a woodworking tool.

Thus the higher level context can change the logical function of lower level words, which then also alters its possible syntactical use (she *wound* the cloth around herself, her *wound* hurt). In these cases the small fragment given (a phrase) is sufficient to determine the meaning of a word, but often a larger context is needed, as in: the coach arrived late for the game. Maybe it is because it had to stop to fill up with petrol, or maybe he woke late because he was partying late the previous night.

Contextual References. All those familiar words like 'it', 'then', 'they', 'here' rely on implicit passing of a higher level variable to the lower level context ('it' was the house, 'then' was last year, 'they' were the neighbors, and so on. This is somewhat like parameter passing to local modules: the global variables are inherited by them, and they are thereby given contextual meaning. The text flows because we expect certain kinds of things to follow others:

- *contextually* in terms of specific times and places and actors involved,
- *syntactically* in terms of word patterns where collocations we have learnt enable us to read phrases as a whole [100],
- *conceptually* where we fit what we hear into larger patterns of meaning and experience, which actually shape the way we perceive things [14].

The context of decades of experience in specific cultural contexts feeds in to give meaning to the words. Those contexts shape the way we think.

4.1.6 Mathematical Models: Boundary Conditions and Constraints

The mathematics and theory underlying these effects is varied: it includes dynamical systems theory [32], partial differential equations theory [41], numerical methods such as finite elements [26], the analysis of computer algorithms [115], electronic circuit design [106], and the analysis of network motifs [5].

4.1.6.1 Partial Differential Equations (PDE)

In many cases, the relevant equations will be partial differential equations [41], such as the wave equation for a variable $\Phi_i(\mathbf{x}, t)$:

$$\frac{\partial^2 \Phi_i}{\partial t^2} - \frac{1}{c^2} \nabla^2 \Phi_i = g_i(\Phi_k) , \tag{4.5}$$

where c is the wave speed, or the diffusion equation

$$\frac{\partial \Phi_i}{\partial t} = D \nabla^2 \Phi_i , \tag{4.6}$$

where D is the diffusion coefficient [139]. The solution in each case is determined by initial data $\Phi_i(\mathbf{x}, t_0)$ given at a time t_0.

Constraints. Generically, the constraints can be expressed in the form

$$g_i\big(\Phi_j(\mathbf{x}, t)\big) = C_i = \text{constant} , \quad i, j = 1, \ldots, N . \tag{4.7}$$

These constraints may represent structural conditions, as discussed below, or they may represent boundary conditions that must be satisfied by all valid initial data. For example, models of morphogen diffusion in embryology have a source at one end and a barrier at the far end that cannot be crossed as a boundary condition [175].

Consistency with the time evolution equations gives

$$\frac{\mathrm{d}C_i}{\mathrm{d}t} = 0 \quad \Longrightarrow \quad \frac{\mathrm{d}g_i\big(\Phi_j(\mathbf{x}, t)\big)}{\mathrm{d}t} = \sum_{k=1}^{N} \frac{\partial g_i}{\partial \Phi_k} \frac{\mathrm{d}\Phi_k}{\mathrm{d}t} = 0 , \quad i, j = 1, \ldots, N , \tag{4.8}$$

which are new constraints required in order that the constraints (4.7) be conserved by the dynamics. These may be automatically satisfied in virtue of (4.7). Then the equations are consistent. If this is not the case, one must keep checking the consistency of the further constraints like (4.8) until one attains a set of equations where all such constraints are satisfied.

Where does the top-down causation come in? Basically in two ways:

- Through *boundary conditions* on the system, e.g., *asymptotic conditions* such as

$$\Phi_i(\mathbf{x}, t) \to 0 \quad \text{as} \quad |\mathbf{x}| \to \infty ,$$

  or *periodic boundary conditions* such as $\Phi_i(x, t) = \Phi_i(x + L, t)$ that must be satisfied by all initial data.
- Through *structural constraints* $g_i\big(\Phi_j(t, x)\big) = C_i = \text{constant}$ that channel the flow of causation (as in the case of electrical wiring systems and digital computers).

I will illustrate in the case of Maxwell's equations.

4.1.6.2 Maxwell's Equations

Maxwell's equations [70, 73] describe the interrelation between the electric field **E** and magnetic field **B**, with sources the charge ρ and current **J**. In differential form they consist of [70, Sect. 18-2] the time development equations

$$\nabla \times \mathbf{E} = -\frac{\partial \mathbf{B}}{\partial t}, \qquad \nabla \times \mathbf{B} = \frac{1}{c^2}\frac{\partial \mathbf{E}}{\partial t} + \frac{\mathbf{j}}{c^2}, \tag{4.9}$$

where c is the speed of light, subject to the boundary conditions

$$\nabla \cdot \mathbf{E} = \frac{\rho}{\epsilon_0}, \qquad \nabla \cdot \mathbf{B} = 0, \tag{4.10}$$

where ϵ_0 is a constant. Consistency of the time development and constraint equations gives the charge conservation equation

$$\nabla \cdot \mathbf{j} = -\frac{\partial \rho}{\partial t}. \tag{4.11}$$

The equations can also be expressed in integral forms [73, p. 130]. Combining the time derivative equations gives wave equations for **E** and **B**. The causal effectiveness of the fields derives from Maxwell's force law

$$\mathbf{F} = q(\mathbf{E} + \mathbf{v} \times \mathbf{B}), \tag{4.12}$$

which gives the force **F** experienced by a charge q moving with velocity **v**, together with Newton's force law (3.2).

4.1.6.3 Boundary Conditions

Solutions of these equations will depend on boundary conditions which may be of various kinds:

Asymptotic Conditions. If a charge is in empty space far from other charges and there is no incoming radiation, the field will die away at infinity, i.e., $\mathbf{E} \to 0, \mathbf{B} \to 0$ at infinity. Wiggling the charge emits radiation that dies away at infinity. The derivation of the radiation formula (see [70, Sect. 28-4] or [107]) assumes this condition of asymptotic decay.

Periodic Conditions. If a charge is near a plane conducting surface, the electric field due to the charge will be constrained to be normal to the surface [70, Sect. 6-9]. This constrains electrical fields in capacitors. Consequently, if the field is constrained in a container such as a resonant cavity, the wall will put conditions on the field at the boundaries leading to existence of eigenfunctions and resonant modes [70, Sect. 24-1].

Thus the local behaviour of the field is determined by non-local boundary conditions, an environmental effect acting from larger to smaller scales.

Boundary Conditions. On solutions of PDEs, these express the effect of top-down causation from global conditions to local values of the field.

4.1.6.4 Structural Constraints

When fields are constrained to flow only in one dimension, one gets guided waves, leading to the properties of transmission lines, wave guides, and optical fibres [70, Sect. 24-1]. The 3-dimensional Maxwell's equations reduce to 1-dimensional equations (see (24.1) and (24.2) in [70]), that is, PDEs get reduced to ODEs.

Similarly, when a current flows in a wire, the motion of the electrons is constrained by the non-conducting sheath of the wire: they can only move along the wire, but they cannot move out through the sheath. The potential difference V along a length of wire is then related to the current $I_{\parallel}$ flowing along the wire by Ohm's law

$$V = I_{\parallel} R \,, \qquad I_{\perp} = 0 \,, \tag{4.13}$$

where R is the resistance [69, Sect. 25-7]. These are all macroscopic variables, with their existence and values depending on the constraint that the current flows in the wire. Crucially, the perpendicular current $I_{\perp}$ vanishes due to the anisotropy of the resistance (it is effectively infinite for currents perpendicular to the length of the wire). This is what enables circuits made of wires to direct the flow of electrons. When Maxwell's equations are applied under these constrained circumstances, one obtains effective laws such as the Biot-Savart law for the magnetic field of a current in a wire [70, Sect. 14-10]. If the wire is formed onto a coil (a higher level structure), we get the formula for the magnetic field generated by a solenoid [70, Sect. 13-5].

At the next level up one gets effective laws for circuits built up out of the basic elements of a capacitor with capacitance C, resistor with resistance R, and coil with inductance L. The resulting equation for the charge q in a resonant circuit is the ordinary differential equation

$$L\frac{\mathrm{d}^2 q}{\mathrm{d}t^2} + R\frac{\mathrm{d}q}{\mathrm{d}t} + \frac{q}{C} = V(t) \,, \tag{4.14}$$

where $V(t)$ is the potential across the circuit [69, Sect. 23-6]. This equation represents the dynamics of many kinds of constrained physical systems [170].

Equations (4.13) and (4.14) are emergent laws resulting from the constraints:

Constraints. These restrict the independent variables and initial data for solutions. Top-down causation via constraints often leads to reduction of PDEs to algebraic equations, e.g., (4.13), or effective ODEs, e.g., (4.14).

The constrained operation of lower level forces in a law-like/algorithmic way leads to higher level behaviour, whose nature depends on the nature of the constraints. These are often in the form of networks of interactions [10], including recurring network motifs [5]. These are higher level features because they cannot be described in terms of lower level concepts. The high-level variables concerned are coarse-grained low level variables or their representations. It is the physical structuring and equations of state that determine the outcome resulting from particular boundary and initial conditions [170].

4.1.6.5 Ordinary Differential Equations and Dynamical Systems

The behaviour of physical systems is often described by ODEs, where evolution of quantities $\Phi_i(t)$, $i = 1, \ldots, N$, is determined by[1]

$$\left.\frac{\mathrm{d}\Phi_i}{\mathrm{d}t}\right|_t = f_i\big(\Phi_j(t), P_k(t)\big)\,, \quad i, j = 1, \ldots, N\,, \tag{4.15}$$

where $P_k(t)$ are a set of parameters determined exogenously, that is, they are unaffected by the values of the system variables $\Phi_j(t)$. They may be constants, e.g., the fundamental constants of physics, or they may be variables determined by higher level dynamics, e.g., the expansion of the universe.

These equations generically represent dynamical systems [48, 82], with the simple harmonic oscillator being a key exemplar. Typical are the equations for enzyme kinetics [139, pp. 108–118]. The solution is determined uniquely on some interval by the initial data at an arbitrary time t_0, that is, by $\{\Phi_i(t_0)\}$. The equation may be chaotic or unstable in some domain, in which case the solution is exquisitely sensitive to the initial data. Nevertheless, the equations are determinate in principle. The resulting dynamical system may have attractors, sources, sinks, and saddle points that characterize its solutions, and it may be stable or unstable, perhaps exhibiting chaotic behaviour.

The action of the system can be characterised by a mapping $\Lambda(t_0, t_1)$ from initial data $\Phi_j(t_0)$ to any later state $\Phi_j(t_1)$:

$$\Lambda(t_0, t_1) : \Phi_j(t_0) \rightarrow \Phi_j(t_1)\,. \tag{4.16}$$

If the system settles down to a final state, then $\Phi_j(t_1) \rightarrow C_j$ as $t_1 \rightarrow \infty$. In linear cases this is characterised by a transfer function.

Effective Variables. The outcome must change if the values of the variables change:

$$\forall\, j,\ \exists\, i \ \text{ such that } \{\partial f_i(t)/\partial \Phi_j\} \neq 0\,. \tag{4.17}$$

[1] The higher derivative form (3.7) can be reduced to this form of a system of first order equations by defining variables $\Phi_j^{(n)}(t) := \mathrm{d}^n \Phi_j/\mathrm{d}t(t)$.

If this is not the case, they should be deleted from the list of variables. Given a set of effective variables Φ_j, one can choose instead another non-degenerate set of variables $\Phi'_j = \Lambda_j(\Phi_i)$ where $|\partial\Lambda_j/\partial\Phi_i| \neq 0$. In general there will be a restricted set of such variables (called canonical variables) that will be simpler than other choices because they most effectively mirror the system dynamics.

Discrete Equations. Alternatively the dynamic equations may be discrete equations, where initial data plus an algorithm determines the outcome at each time step $t_{i+1} = t_i + \Delta t$. for a system of variables $y_j(t)$:

$$y_j(t_{i+1}) = f_j\big(y_1(t_i), \ldots, y_N(t_i), \Delta t\big) . \tag{4.18}$$

There is no uncertainty in the model, but it may still exhibit chaotic behavior [48]. These may be simulations of the differential equations.

Where does top-down causation come in?

Dynamical Systems Result from Constraints. The very existence of ODEs (4.15) or discrete equations (4.18) may represent contextual effects, because they often result from breaking symmetries and channeling lower level causation through structural constraints, thereby replacing fields governed by PDEs by effective variables governed by ODEs.

Shearer, Murphy, and Richardson [170] show clearly how structure governs dynamics in mechanical, hydraulic, thermodynamic, and electrical contexts (see also [111]).

In particular, what is crucial in biology is that structures such as cell walls create biological compartments that structure the kinetics of substances by controlling their flow between compartments, leading to ordinary differential equations for concentrations [158, pp. 168–220]. In plant physiology, partitioning the model produces a set of ODEs for the resulting variables [182, pp. 176–178]. A key example from neuroscience is the Hodgkin–Huxley equation for action potentials in excitable cells, resulting from the existence of ion channels in cell membranes [139, pp. 161–166]. This equation cannot be understood in a purely bottom-up way [141, 166].

4.1.7 Randomness and Noise

Randomness occurs at the bottom due to quantum fluctuations [55, 71]: in reality, the lower levels are not deterministic! Additionally, there is interlevel randomness because fluctuations in variables at level N lead to noise at the next higher level $N+1$ (Sect. 3.4.2). This is ubiquitous in physics and biology. Both effects can result in randomness at higher levels, but do not necessarily do so.

4.1.7.1 Unreliable Emergence Due to Randomness at the Bottom

Lower level randomness can get amplified to macro scales, for example, in the case of chaotic systems [48], or when catastrophes occur, leading to bifurcation of dynamical systems [180]. Such effects occur for example as regards the weather, as represented by the Lorenz equations [124], but also occur in the dynamics of the Solar System, indeed even for the gravitational 3-body problem. These are chaotic deterministic systems.

Unreliable Emergence. Deterministic lower level dynamics do not necessarily imply reliable higher level emergent behavior.

Thus these are cases where the conditions for reliable higher level emergence (Sect. 3.3.2) do not hold. Amplification of low level fluctuations can have major causal effects, as in the case of cosmology: because of the exponential nature of the inflationary era expansion, indeterministic quantum fluctuations in the very early inflationary era lead to later classical perturbations that result in large scale structures such as clusters of galaxies [52].

4.1.7.2 Reliable Emergence Despite Randomness at the Bottom

Lower level fluctuations often get washed out. This can happen in several ways.

Statistical Natural Systems. Micro-randomness gets washed out when the effects of micro-level physics are averaged to get macroscopic behavior such as the perfect gas laws. However, the randomness is apparent on closer inspection, where fluctuations occur and lead for example to the fluctuation–dissipation theorem. Their relative importance is partly just an issue of size: statistically random fluctuations in a system of N elements scale as $\sqrt{N}$, so their relative size decreases as $\sqrt{N}/N = 1/\sqrt{N}$, which is very small for everyday objects where $N \gg 10^{20}$. But whether this is important or not depends on the system structure. In everyday engineering systems such as steam engines, this is unimportant. When one pushes the system size towards nanoscales, it cannot be ignored.

Machines. These are systems where there is effectively no randomness at the macro-levels: they are designed so that this is will be the case and they function reliably (steam engines, locks, computers, etc.) This is achieved partly by having sufficient size to damp out the relative importance of micro-fluctuations, in particular washing out quantum effects, and also by error-tolerant design features that prevent the fluctuations that do occur from having a higher level effect. These include allowing for delays in arrival times of signals and materials so that variations in these times won't matter, including springs and dampers that allow for absorbtion of energy and momentum, having reservoirs of energy and essential materials that dampen fluctuations in their supply, and having activation thresholds so that lesser fluctuations have no effect. Furthermore, feedback control mechanisms (TD2) can correct for errors due to fluctuations.

Living Systems. These develop reliably most of the time. They are designed to operate in the conditions of the molecular storm at the micro-level [101], exploiting noise-tolerant design features similar to those used in machines, e.g., energy stores and activation thresholds. Stabilization also occurs via homeostatic feedback control mechanisms (TD2) that are designed to damp out fluctuations at the lower level. Paradoxically, reliability can also be the result of randomness at lower levels, utilised by adaptive selection processes (TD3) to attain desired classes of outcomes, making the higher levels effectively deterministic [101].

Reliable Emergence. Indeterministic lower level dynamics do not necessarily prevent emergence of deterministic dynamics at higher levels.

It is through the specific structuring of living entities and machines that reliable higher level behavior emerges, despite lower level randomness. Indeed this is why these structures have the form they do: they have either evolved to be reliable, or have been planned to be that way. The reliability is then the result of higher level constraints on lower level functioning.

4.1.7.3 Unreliable Emergence

When one has deterministic equations but with noise such that the lower level dynamics has a random element and higher level outcomes are not unique despite being influenced by boundary or structural conditions, we should perhaps speak of *top-down effects* rather than top-down causation. The higher level variables are still influencing the lower levels but not giving deterministic outcomes. Here is where chaotic dynamical systems play an important role, as in the case of weather patterns.

4.2 Non-Adaptive Feedback Control (TD2)

In non-adaptive feedback control:

- Goals direct what happens: higher level entities influence lower level entities so as to attain specific fixed goals through the existence of feedback control loops.
- Information flows enable this to happen: information on the difference between the actual and desired states of the system is used to lessen this discrepancy.

An example is a thermostat controlling the temperature of water in a boiler. The goal is set by setting a desired temperature on an input panel. A sensor determines the actual temperature and a controller compares this with the goal and alters the heat input so as to attain the desired temperature. A different setting results in a different temperature.

In this section I discuss in turn:

- The nature of the process (Sect. 4.2.1).
- Engineering systems (Sect. 4.2.2).
- Organisations (Sect. 4.2.3).
- Biology (Sect. 4.2.4).

Table 4.3 Basic features of a feedback control system. The goals lead to a specific final state via feedback of an error signal to an actuator. The initial state of the system is irrelevant to its final outcome, provided the system parameters are not exceeded

	Controller	⇐ Correction signal	
Noise ⇒	Action ⇓	Feedback ⇑	
	State	⇔ Comparator ⇔	Goal

- Mathematical models: control theory (Sect. 4.2.5).
- The nature of goals (Sect. 4.2.6).

4.2.1 *The Nature of the Process*

Non-adaptive control systems (see [11] and [173, Chap. 1]) compare the actual state of the system with a desired goal, then feed information on this difference back to a controller which alters conditions so as to reduce the difference between the actual system state and the desired state represented by the goal (Table 4.3).

This is the essence of cybernetics: feedback control corrects any error in the system state, i.e., any deviation from the desired goal, by observation and measurement, continually using new data to keep it on track. In contrast to the case just discussed (TD1), in this case the initial data is irrelevant. It is the full set of goals that determine the outcome, through the differences between the goals and the actual values. This involves the following features, discussed below:

- Goal-directed outcomes.
- Information flows.
- Emergent properties.
- Top-down causation.
- Ubiquitous existence in living beings.

4.2.1.1 Goal Directed Outcomes

Goals are the essence of feedback control systems. They are desired levels of significant variables. Examples are the revolutions per minute of an engine, the direction of a vehicle, the level of water in a reservoir, the temperature of a reactor, the voltage across a membrane, the amount of ATP in a cell, body temperature, blood pressure, cholesterol levels, the amount of stock in a warehouse, the number of troops ready for action, the amount of money in an account, the employment rate in a country, the pass rate at a university, customer satisfaction with a service, and so on. They embody information about a system's desired behaviour or responses. They are usually only expressible in higher level terms. The rare exception is where the goal is to specifically control lower level states, as in quantum optics. This is very difficult to achieve, and only rarely occurs.

Unlike the previous case (TD1) just considered, where the initial state plus boundary conditions determine the outcome, that outcome is not determined by the boundary or initial conditions. Rather it is determined by the goals. Indeed the whole purpose of such systems is to make initial conditions irrelevant. Thus the nature of causality is quite different when feedback control systems are guided by goals, which are higher level entities.

They are effective through specific structuring of physical systems in an implementation hierarchy. The result is

$$\text{(physics, physical structure, goals)} \implies \text{(outcomes)}\,. \tag{4.19}$$

Physics is fixed. When we consider a specific system, the physical structure is held constant and we have:

$$\text{(goals)} \underbrace{\Longrightarrow}_{\text{structure}} \text{(outcomes)}\,, \tag{4.20}$$

emphasizing that it is the specific structure that enables the goals to direct the outcomes. The series of goals in a feedback control system are causally effective. Information flows enable this, and are in general separate from the matter or energy flows that are the implementation vehicles of the system.

> **Non-Adaptive Feedback Control (TD2)**. The outcome of a feedback control system is determined by the goals rather than the initial data. The goals are causally effective through information flows from a sensor to a control element.

A different outcome will occur if the goals are changed. In general there may be multiple goals, that are themselves structured as a logical hierarchy with higher level goals constraining and directing lower level logic in a top-down fashion.

In this non-adaptive case (in contrast to the adaptive case considered in Sect. 4.4), the goals are either embodied in the system structure and so do not change with time, or are fixed by setting a control parameter, and are unchanged by the system's internal dynamics. They only change if the control parameter is externally reset. There may, however, be some associated form of information storage and retrieval, and perhaps even implicit or explicit information processing allowing the system dynamics to be based on *predictive goals*: predictions of where the system will be in the future, continually updated on the basis of incoming data in the current state of the system. Thus although the goals are fixed, complex information processing and modeling may take place in the attempt to attain those goals (Sect. 4.7).

These goals are not the same as material states, for they are desired rather than actual states, although they will be represented by material states and systems that will make them causally effective through such representations (Sect. 4.2.6). A complete causal description must necessarily take them into account. They exist as emergent properties of the system, as they are not embodied in any component on its own.

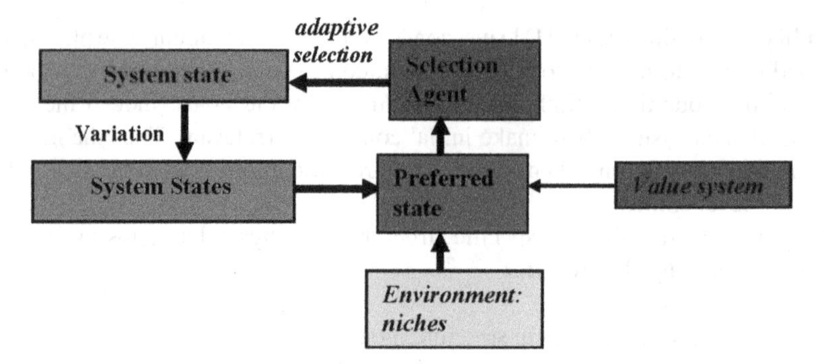

Fig. 4.1 The basic feedback control process. The goals tend to lead to a specific final state via a comparator and activator. The initial state of the system is then irrelevant to its final outcome, provided the system parameters are not exceeded. *Solid lines* are matter/energy flows. *Dotted lines* are information flows

4.2.1.2 Information Flows

Feedback control systems depend essentially on information flows from system sensors to the controller. Ensuring the goal is attained through feedback control functions by comparison of the system state and the goal by a controller. Information on the difference is fed back to the activator. These information flows are physically realised and need energy to function, but the essence of what they are is not in that energy *per se*: it is in whatever coding is used to represent the system state, for that is what is compared with a coded representation of the system goals. Information flows are distinct from the energy and material flows that make the system work. Indeed, they are represented differently from such flows in systems diagrams (see Fig. 4.1 and [132, 159]).

There is a sender and receiver, with a communication channel between them. There may be a process of encoding into a coded form—perhaps a digital code (such as in digital electronic controllers or a polypeptide sequence) or frequency format (such as spike rates in neurons)—and then decoding to an effective form (a digital code gets translated into analogue form, or a polypeptide folds into its functional three-dimensional structure). The information may be just in terms of the amplitude of a quantity, e.g., a voltage representing an engine rotation rate, or a temperature), but then that amplitude must be interrogated in a way that is not simply the transfer of energy to an object (which always in a sense conveys information, but has nothing to do with a feedback loop).

In general, it will have both syntactic and semantic aspects: it has to be coded in a specific way to be valid, e.g., DNA sequences must have start and stop codons, and has a meaning that depends on context (it is part of an engine speed controller, or part of a thermostat system, or determines amino acid sequences).

Information Flows in a Control System. These may be analog or digital, and they may be mechanical, hydraulic, electrical, chemical, or optical. They may be coded or not. In the analogue case it is the value that matters, but it is not just because it has a specific value that it is information: it is because it is deployed in a feedback control circuit with the specific purpose of controlling a particular outcome.

It can be in many forms and may be transformed between the different possible forms, for example between analog and digital (in an electronic circuit) or between electric and chemical (at a synapse). This multiple realisability is a key sign that information is a higher level entity: its specific realization does not matter (Sect. 3.5).

There are two ways information flows are directed from the source to the receiver:

- *Structured channels* that specifically link them, such as electrical wires in a thermostat or pipes in a chemical plant or axons in a brain. They physically link particular entities in order to channel information flows in specific ways. These channels may be addressable, as in landline telephone systems.
- *Broadcast messages* that are spread indiscriminately to everyone, but have some kind of addressing system that is recognized by the receiver to say this message is for me. This is the case of cellphone systems for example. It is what happens in computer bus systems and it is prevalent in biology, where the 'lock and key' mechanism implemented via molecular shapes identifies which message is meant for which receiver. Examples are hormones and neuromodulators, which activate specific receptors [190].

4.2.1.3 Emergent Properties

Feedback control is an emergent property of cybernetic systems that enables goal-seeking behaviour. Ross Ashby states this as follows [7]:

Stability and the Whole. An important feature of a system's stability (or instability) is that it is a property of the whole system and can be assigned to no part of it. The stability belongs only to the combination; it cannot be related to the parts considered separately. The fact that the stability of the system is a property of the system as a whole is related to the fact that the presence of stability always implies some co-ordination of the actions between the parts.

These control circuits are higher level entities, as they are based on higher level concepts (the high level system state and the goal). The goals are intrinsic higher level properties of the system considered and determine the outcome.

4.2.1.4 Top-Down Causation

Feedback control is a form of top-down causation for two reasons:

Effectiveness of Goals. First, because the goals determine the outcome, and hence can be regarded as being at a causally higher level than the controlled system. A goal is realised by an equivalence class of lower level states: it does not matter which particular lower level state occurs, as long as the corresponding higher level variables are in the desired range. It is this equivalence class (an irreducible high level variable) that is causally effective and determines the outcome in these systems. This underlies all homeostatic systems, and characterizes this as top-down causation (Sect. 3.5).

System Acts as a Whole. Second, the goals are implemented by higher level networks. These cannot be reduced to lower level entities, precisely because it is the relations between the parts that make the network into a feedback control system. Taking the system apart destroys those relations. The system as a whole acts down on its parts to attain the desired outcome. The way electrons flow or molecules move is a result of the system structuring—a higher level entity.

4.2.1.5 Existence and Limitations

Mechanical, chemical, and electrical engineering abound in feedback control systems [51]. The whole of biology embodies numerous genetically determined homeostatic systems based on the principle of feedback control [134]. In particular, this is true of the human body, where homeostasis is the key to physiological research [156]. And goal directed dynamics is crucial to organizations and to individuals [12].

However, such systems do not occur naturally in the physical world of materials, rocks, rivers, continents, planets, stars, galaxies. Even though feedback processes may occur in such systems, they do not occur because of information flows, but rather because of energy and matter flows *per se*. These flows are not coded to convey information about the system state to a control element. Note that I use the term 'goal' only when there is an identifiable feedback control system that leads to its reliable realization. Attractors in a dynamical system are therefore not goals, in the sense I use the term, as they do not involve control information, which is central to feedback control. Non-adaptive information control does not arise spontaneously in the natural world (in astronomy, geology, oceanography, or atmospheric physics), but it occurs in manufactured objects and is crucial to life, where it is a key aspect of complex emergence.

But this process cannot innovate: the outcome is predictable from the outset, as it is determined by the explicit or implicit goals of the system. In the biological case, the goals of physiological systems are genetically determined and are the same across a species and constant in the lifetime of an individual. Like predictive algorithmic processes, non-adaptive feedback control systems cannot learn. That requires adaptive selection (see Sect. 4.3).[2]

[2]I am aware that some present day feedback control systems use principles of adaptive control. I believe they should be labeled as such, to distinguish them from the basic cybernetic processes identified by Wiener, in which the goal is fixed.

4.2.2 Engineering Systems

Feedback control in the engineering context [51] occurs in toilet tank filling systems, automatic toasters, an elevator taking one to the desired floor in a building, voltage controllers, fully automated electric trains. In many engineering applications there will be computer control systems that will implement this logic of deciding what to do next on the basis of the current system state, which is embodied at the microscale in WHILE and IF THEN loops ([25]: 29). Examples are:

- **Steam Engine Governor**. The classic case is the control of the speed of a steam engine through a centrifugal governor. The controller is a high level device (if you disassemble it into molecules, it will no longer function). The outcome is a desired range of molecular densities and speeds in steam in a pipe (at the lower level) resulting in a desired rate of rotation of a wheel (a high level control variable).
- **Thermostat**. A commonplace example is control of water temperature in a hot water cylinder by thermostatic control of the water heater. How do we demonstrate that this is top-down causation? Change the temperature setting of the thermostat and then measure the resulting temperature changes in the water and reduced speed of motion of the water molecules. This effective action has been verified in daily life many millions of times.
- **Automatic Pilots**. In an aircraft, these control height, speed, attitude, and direction of the aircraft through engine and control surface settings. They utilize information from GPS systems, air speed indicators, compasses, altimeters, RPM indicators, etc. The ultimate goal is to arrive at a specific destination. Intermediate goals are specific heights, ground speeds, and ground paths.
- **Chemical Engineering Control Systems**. These control interface levels between two phases, the pressure of a vapor or gas, flows from storage tanks, temperatures of reactants, the composition of distillation column products, and product composition from reactors [174, pp. 268–279].

Predictive Control. In advanced systems (aircraft automatic pilots, control systems in chemical plants), the controller will act not on the basis of the present physical state of the system but on the basis of predicted future states as determined by the latest updates of the current system state. It is the continual updating of system data through incoming information that gives feedback control its power.

Adaptive Control. This occurs when a controller adapts to changes in the controlled system as its parameters vary, or has to determine parameters that are initially unknown. For example, when an aircraft proceeds on its path, its mass will decrease as fuel is burnt, and the controller must adapt to such changing conditions. However, the goals are still fixed as this happens.

4.2.3 Organisations

The very reason organisations exist is in order to attain some goal or other, otherwise they would not be there. Organisational structures and methods together form a feedback control system designed to ensure that these goals are attained. The goals structure what happens and are key to organisational effectiveness [91, pp. 554–558]. In order to attain their main goal, they have to attain a whole series of subsidiary goals, so organizational goals are hierarchically structured [91, pp. 107–135]. Tools like GANT charts help structure the attainment of successive subgoals in a coordinated way so as to achieve some main goal.

Monitoring of output and quality control relative to a desired outcome is essential to success. The essential planning cycle is to make a plan, implement it, check whether the output is as desired, and if necessary, alter conditions so that the desired effect is attained, i.e., a classic feedback process [91, pp. 522–533]. Indeed, feedback processes have become part of organizational culture: feedback forms are now commonplace at meetings, hotels, and training sessions. Feedback control is central to both management and industry: it is the core of organizational control [91, pp. 521–609] and systems approaches to industrial management [159].

- This all takes place via an *implementation hierarchy*,
- whose nature is based on a *logical hierarchy*,
- and in which there are both *bottom-up and top-down processes*.

4.2.3.1 Implementation Hierarchy

The goals are hierarchical because one splits up a complex task into simpler subtasks in a modular way, in accordance with the general principles of modular hierarchical design (Sect. 3.1.6). This leads to hierarchical organizational design (see [12] and [91, pp. 263–291]).

Work Breakdown Structure (WBS). In project management and systems engineering, this is a decomposition of a project into smaller components [23]. It groups together a project's discrete work elements which are defined in terms of outcomes or results rather than methods. It thus represents a goal structure for the organization. This ensures that the WBS is not overly prescriptive of methods, allowing for creative thinking and initiative on the part of the project participants, and so allows for multiple realisability in terms of attaining the goals. These goals usually have to be attained in a certain order. Tools such as GANT charts [136] set out the order in which these goals should be implemented. They give a project schedule, showing the start and finish dates of each part of a project. Software to create such charts are available for example through Microsoft Project, a task-planning program. The time element incorporated here is an important part of setting goals: it is often crucial that they are done on time.

Table 4.4 The hierarchy of goals in an organisation. Each level of plans is supposed to implement the next higher one, so the higher levels direct and constrain the lower ones. The lowest level is the implementation level. All the other levels are planning levels

Intention	Plan
Overall purpose	Mission statement
⇓	⇓
General approach	Strategic plan
⇓	⇓
Specific issues	Tactical plan
⇓	⇓
Specific tasks	Task goals
⇓	⇓
Subtasks	Detailed goals

Roles and Job Descriptions. Organisational roles characterise who will handle which goals [91, pp. 278–286]. You split up the task and assign the bits to different groups who assign them to specific people. This relationship is formalized in *organisational charts*, and the goals associated with each role are specified in *job descriptions*. These characterize the implementation hierarchy.

Bureaucratic and Business Procedures. In principle, these ensure that the procedures used by the organisation attain the desired goals. They specify how the logic of the goals constrains the implementation procedures that attempt to attain those goals. A crucial part of those procedures is the use of feedback processes, such as quality control procedures, to ensure that the goals are actually met as contingent events occur and upset prepared plans. As eloquently stated by Robert Burns: "The best-laid schemes o'mice an'men Gang aft agley" [31]. Feedback systems aim to correct this problem.

Resource Use Goals. These are explored using spreadsheets such as Excel, which organize information in tabular form, carrying out mathematical operations on data represented by cells in an array. They are particularly useful as regards accounting and budgeting systems, where they used for setting resource use goals. These can then be compared with the actual figures to give error messages (the difference between the two); which are the key information needed in the control of financial resources.

4.2.3.2 Logical Hierarchy

Similarly to the case of computers (Sect. 2.3), at each level of the implementation hierarchy, there is a logical hierarchy of goals. A generic structure for this hierarchy is shown in Table 4.4. In a well-functioning organization, the bottom level (the implementation level) tasks attain the detailed goals, which then combine to result in attaining the high level purpose. If this succeeds, it is because those lower level goals were specifically structured through a top-down process of planning so as to produce that desired higher level result.

4.2.3.3 Bottom-Up and Top-Down Processes

Feedback processes can occur at each level in the implementation hierarchy, as well as in terms of interlevel loops, with the higher levels acting as controllers for the lower levels. Consequently, information flows will go both up and down, while control flows go down. One must take seriously the limitations of information flows (Ross Ashby's *Principle of Sufficient Variety* [7]) as well as the issue of individual motivation and initiative: hence one should decentralise as much decision-making as possible [12], while putting in place structures and methods that ensure that lower level goals are aligned with the higher level goals.

This is of course where things tend to go wrong: there is a tension between the goals of the parts and the goals of the whole. That is why great leadership is to do with purpose-setting and communication, fixing the direction for the whole in an inspirational way [151, pp. 81–86]. Ultimately, organisations are to do with meaning-making and purpose (whether that purpose is to make a profit, to serve the community, to explore Mars, or to understand the mind).

Where do goals come from? Goals are adaptively formed in response to experience: learning takes place in particular contexts as the mind responds to the meaning of symbols in the relevant social context. This is explored in Sects. 4.4 and 4.7.

4.2.4 Biology

Living systems are goal-seeking (teleonomic) at both the physiological and behavioural levels. Homeostasis in the human body is an integrative concept that is the key to physiological research [156, pp. 22–26]. This is a core feature of physiology. Homeostatic systems maintain constant conditions in the internal environment in the human body, for example, controlling blood pressure and temperature, through integrative processes that maintain stability as external conditions vary. As stated in [92]:

> **Homeostasis**. The term homeostasis is used by physiologists to mean maintenance of static or constant conditions in the internal environment. Essentially all the organs and tissues of the body perform functions that help to maintain these constant conditions. For instance the lungs provide oxygen as required by the cells, the kidneys maintain constant ion concentrations, and the gut provides nutrients.

These systems have been built in through the adaptive process of evolution: they are constant across individuals, time, and place in specific species. They embody a hierarchy of goals at the different physical levels, from the organismal level to underlying cellular and molecular mechanisms in an integrated way. At the bottom level, the molecular basis for homeostasis is the maintenance of protein conformation [156, p. 23]. A key feature is transport across cell membranes.

Enzyme Regulation. A regulatory mechanism in cell physiology controls production of an enzyme, where the enzyme combines with the substrate to produce a product that represses the transcription of DNA to mRNA, which is the template for making the enzyme [139, p. 143–148].

Ribonuclease Action. Causal effectiveness of equivalence classes in information control in microbiology has been demonstrated in the case of structural type A and B bacterial ribonuclease P (Rnase P) RNAs [192]. These RNAs can fully replace each other in vivo despite many reported differences in their biogenesis, biochemical/biophysical properties, and enzyme function in vitro. This suggests top-down causation from these equivalence classes of RNA in vivo under standard growth conditions [9].

Axon Potential. Sodium and potassium levels in neurons are controlled by voltage-gated ion channels deployed in feedback control loops that return the membrane potential of an axon to its resting state, the resting potential [113, 166]. This is an essentially top-down affair, as Denis Noble makes clear [141]:

> The trail was blazed by Hodgkin and Huxley [99] in their Nobel prizewinning work on the nerve impulse. The ion channel proteins that sit across the cell membrane control its electrical potential by determining the quantity of charge that flows across the cell membrane to make the cell potential become negative or positive. The gating of these channels is itself in turn controlled by the cell potential. This is a multi-level loop. The potential is a cell-level parameter; the ion channel openings and closings are protein-level parameters. The loop, originally called the Hodgkin cycle, is absolutely essential to the rhythm of the heart. Breaking the feedback (downward causation) between the cell potential and the gating of the ion channels and cellular rhythm are abolished. A simple experiment on one of the cardiac cell models will demonstrate this computationally.

Organism Level. Many bodily organs exist precisely in order to enable feedback control of body conditions [156, pp. 23–26]. Each is governed by implicit goals, embodied in the physical structure of the body.
Nervous System. This includes regulatory systems for:

- *Body temperature*. Thermosensors located in the hypothalamus help maintain a goal of 37 °C [156, pp. 806–821].
- *Blood pressure*. Baroreceptors help maintain the goal of 90 mm Hg [156, pp. 343–344, 598–604].
- *Hydrogen ion concentration in the extracellular fluid*. Chemoreceptors help maintain the goal of a pH of 7.4 [156, pp. 761–781].

Endocrine System. This utilizes endocrine cells and hormones in regulatory systems for [156, pp. 372–399]:

- *Glucose concentration*. Controlled by insulin and glucagon produced by the pancreas [156, p. 447].
- *Metabolic regulation*. Inter alia via glucocorticoids such as cortisol secreted by the adrenal glands [156, pp. 427–429].
- *Electrolytes*. Calcium ion concentration in the extracellular fluid, together with phosphorus controlled by the kidneys and parathyroid glands [156, pp. 787–799], sodium [156, pp. 745–751], and potassium [156, pp. 753–755] ion concentrations, controlled by the kidneys.

Overall "the human body has literally thousands of control systems in it" [92] and the goals embodied in them are essential to life.

4.2.5 *Mathematical Models: Control Theory*

The mathematics involved in feedback control is linear and nonlinear control systems theory [51], including its applications to the biological context [134].

In contrast to the case TD1 where initial data determine the outcome, in this case the initial data is irrelevant to the final state. Consider for example the temperature in a boiler controlled by a thermostat. If you set it up as a dynamical system (4.15) and consider it as producing a transfer function from initial data to final outcome, the initial data is not relevant once the system dynamics has run its course.

It is the full set of goals g_n that determine the outcome, through the differences $\Delta y_n(t_i)$ between the goals and the actual values. Instead of (4.15), we have

$$\frac{\mathrm{d}y_j(t)}{\mathrm{d}t} = f_j\big(\Delta y_1(t), \ldots, \Delta y_N(t)\big)\,, \quad \Delta y_n(t) := y_n(t) - g_n\,. \tag{4.21}$$

Their dynamics can be described by block diagrams, Laplace transforms, and transfer functions [51, 134].

A specific case is the Goodwin's model in cell physiology representing a regulatory mechanism related to production of an enzyme E. The enzyme combines with the substrate to produce a product P that represses the transcription of DNA to mRNA M, which is the template for making the enzyme [139, pp. 143–148]. A generalised form of such equations is

$$\frac{\mathrm{d}u_1}{\mathrm{d}t} = f(u_n) - k_1 u_1\,, \quad \frac{\mathrm{d}u_r}{\mathrm{d}t} = u_{r-1} - k_r u_r\,, \tag{4.22}$$

where $k_r > 0$ and $f(u) > 0$ is the nonlinear feedback function [139, p. 144]. If $f(u)$ is a monotonic decreasing function of u: $f'(u) < 0$, this is a negative feedback loop with steady state solutions $f(u_n) = k_1 u_1$, $u_{r-1} = k_r u_r$. The goal of the system is to produce this state. Similar equations relate to regulation of protein synthesis [131, pp. 107–115], and the general equations governing homeostasis are analysed by Riggs [158, pp. 95–119].

It is true that, on the surface, dynamical systems with attractors appear to work in ways similarly to feedback control systems (with associated goals), but the mechanisms involved are completely different, and so are their outcomes in many cases. Thus for example dynamical systems include strange attractors and chaotic behaviour. Control systems will prevent this kind of behaviour. In particular they are designed to handle noise successfully, and unlike dynamical systems, their dynamics is not time reversible (we cannot tell from a later state what the initial state was).

4.2.6 The Nature of Goals

The series of goals in a feedback control system embody information about the system's desired behaviour or response. They are causally effective in the following ways:

- through structures designed to implement them (wiring of a thermostat system, computer systems programmed to attain goals on the basis of the current state), or
- through boundary conditions that create them (the structure of a cell with membranes and specific proteins and enzymes contained within a compartment so that the components needed for the feedback system to exist are there), or
- through incorporation in abstract goal systems that individuals or organisations consciously implement.

They are not material entities because they are not things as they are. Rather, they are images of how things ought to be. Thus they are abstract entities (as discussed in Sect. 7.5).

This is particularly clear in the case of plans in one's mind, but is also the case in the other examples discussed: goals are a preferred subset in the possibility space of the system. As that space is abstract, so too are goals, even when they are realised through specific physical structures or patterns. Those structures or patterns are means of attaining the goals, they are not the goals themselves.

> **Goals in Feedback Control Systems.** Goals are not the same as material states, for they are desired rather than actual states, although they will be represented by material states and systems that will make them causally effective through such representations. They are abstract entities that are causally effective through information flows enabled by system structuring and energy states.

They exist as emergent properties of the system—they are not embodied in any component on its own. A complete causal description must necessarily take them into account.

4.3 Adaptive Selection of Outcomes (TD3)

Adaptive processes, for example, those central to the Darwinian process of evolution by natural selection, take place when there is:

- **Variation of Interacting Entities.** Many entities interact, for example, the cells in a body or the individuals in a population, and variation takes place in the properties of these entities.
- **Selection of Preferred Entities.** This is followed by selection of preferred entities that are better suited to their environment or context.

However, they are not restricted to the context of Darwinian evolution: their ambit is much wider than that. They are the central way whereby order is generated from

Table 4.5 The basic features of adaptive selection. Selection takes place from an ensemble of states, the selection being based on the outcome of some selection criteria in the context of the specific current environment. The unwanted states are discarded

System states	$\Leftarrow$	Selection agent: selects state		Meta-goals
Variation $\Downarrow$		$\Uparrow$		$\Downarrow$
Ensemble of system states	$\Rightarrow$	Preferred states	$\Leftarrow$	Selection criteria
		$\Uparrow$		
		Environment		

chaos and complexity is built up [104]. In particular they are central to all learning processes. In what follows I discuss in turn:

- The nature of the process (Sect. 4.3.1).
- Physics and chemistry (Sect. 4.3.2).
- Life (Sect. 4.3.3).
- The mind: learning and perception (Sect. 4.3.4).
- Mathematical models: adaptive selection (Sect. 4.3.5).
- Multi-level selection (Sect. 4.3.6).
- The nature of selection criteria (Sect. 4.3.7).

I omit a discussion of adaptive selection and digital computers, because this was covered in Sect. 2.4.4.

4.3.1 The Nature of the Process

Adaptive processes take place when many entities interact, for example, the cells in a body or the individuals in a population, and variation takes place in the properties of these entities, followed by selection of preferred entities that are better suited to their environment or context (Table 4.5).

Higher level environments provide niches that are either favorable or unfavorable to particular kinds of lower level entities. Those variations that are better suited to the niche are preserved and the others decay away. Criteria of suitability in terms of fitting the niche can be thought of as a value system guiding adaptive selection. On this basis a selection agent or selector (the active element of the system) accepts some of the variations and rejects the rest. These selected entities then form the current system state that is the starting point for the next round of selection.

Adaptive Selection (TD3). This takes place when many entities interact, for example, the cells in a body or the individuals in a population, and variation takes place in the properties of these entities, followed by selection of preferred entities that are better suited to their environment or context.

This is the way that ever more complex structures, adaptively suited to the environment, come into existence. I discuss in turn the following features:

- The nature of the resultant causality.
- It is a form of top-down causation.
- It involves equivalence classes.
- It generates new information.
- It has predictable and unpredictable outcomes.
- It occurs widely.

4.3.1.1 The Nature of the Resultant Causality

The basic feature of adaptive selection [103] is that a process of variation generates an ensemble of states, from which a best outcome is selected according to some selection criterion, with the outcome dependent on context (see Table 4.5):

$$\text{(physics, ensemble of states, environment, selection criterion)} \Longrightarrow \text{(outcome)}. \tag{4.23}$$

In a given context we take for granted the underlying physics and accept as given some variational process that generates sufficient variety for selection to act on. The outcome is then some kind of preferred fit to the environmental niches, determined so as to adequately satisfy the selection criteria:

$$\text{(environment)} \underbrace{\Longrightarrow}_{\text{selection criteria}} \text{(outcome)}. \tag{4.24}$$

This is an adaptive process rather than a control process. It is the way new information is generated that was not present before [117], and enables emergence of complexity without dynamical attractors or specific goals guiding the process, but with an increase in complexity and embodied information, for the process searches the possible solution space in a way that is not pre-ordained and adapts to the context. Unlike feedback control, this process does not attain pre-selected goals by a specific set of mechanisms or systems. The selection criteria are meta-goals for the process: they don't prescribe specific outcomes, but rather the general direction the process should go. The system creates entities preferred by the meta-goals embodied in the fitness criteria. A classic example is weeding in a garden. This depends on one's concept of what a weed is.

Its great power in evolutionary biology is due to the continued repetition of the adaption process, with the best variant being passed on from one generation to the next by a hereditary mechanism. But that repetition is not essential to the basic process.

4.3.1.2 It Is a Form of Top-Down Causation

The reason this is classed as a form of top-down action is twofold [see (4.24)]:

Environmental Context. First, the nature of the higher level environment is crucial to using selection criteria. The lower level outcome would be different if the environment were different, resulting in different niches being available.

Selection Context. Second, the choice of selection criteria is also crucial: the outcome would be different if the criteria were changed. Thus they are at a higher level of causation than the system affected. Thus this is top-down causation from this two-fold context to the system. If the top level conditions change, the outcome will change. Pross states it thus [155, p. 146]: "In addressing the concept of fitness, context is everything."

This process usually takes place in the context of a network of interactions. A pressure of selection applied to a higher order network can constrain the evolution and adaptation of lower level constitutive networks. The idea is that the pressure of selection applied to the higher order network would lead to outcomes in the lower level constitutive networks that would be different from those observed by having the same pressure of selection applied to the lower level constitutive networks outside the context of the higher order network. In other words, the higher order network affects the way the lower level networks adapt through evolution—a form of top-down causation. Natural selection should be thought of as multilevel causation [128].

4.3.1.3 It Involves Equivalence Classes

An equivalence class of lower level variables will be favoxred by a particular niche structure in association with a specific value system (Sect. 3.5). This feature characterises adaptive selection as a form of top-down causation [9]. The essential point is that for the viewpoint of the higher level purpose embodied in the selection criteria, it does not matter which particular member of the equivalence class is realised: any one of them will do. Selection is in terms of attaining higher level function (for example, sight) rather than for the specific mechanism by which it is attained. Any mechanism which attains the goal sufficiently efficiently will suffice.

Table 4.6 Adaptive selection processes as positively construed: they select what is best suited to the environment ('survival of the fittest') according to some selection criteria

Final data set	Meaningful information	⇒	Rejected set: noise
	Selection ⇑	⇐	Selection principle
	Varied set		
	Variation ⇑	⇐	Variation principle
Initial data set	Ensemble of states	Random	

Table 4.7 The basic function of adaptive selection processes: they select what is useful or meaningful from an ensemble of mainly irrelevant stuff and reject the rest, thus creating order out of disorder by selecting states conveying meaningful information

Order	↑ ↑	Chosen
	↑ ↑	
xxxxxx	xxxxxx ↑ ↑ xxxxxx	Selective gate
	↑ ↑ ↑ ↑ ↑ ↑ ↑ ↑ ↑ ↑	
Disorder	↑ ↑ ↑ ↑ ↑ ↑ ↑ ↑ ↑ ↑	Ensemble

4.3.1.4 The Generation of New Information

The key process is deletion of what is not wanted, leaving what is meaningful. It is also for this reason that it can innovate. The process generates new information that was not there before, or rather, finds information that was hidden in noise (Table 4.6).

That is the general process whereby adaptive selection generates useful information: it seeks out what is relevant and works from an ensemble of stuff that is mainly irrelevant or does not work, hence allowing a local flow against the general tide of increasing disorder. *Inter alia*, this is the process underlying learning. This is illustrated in Table 4.7.

Through this process, the entity that is selected has incorporated into itself some kind of implicit or explicit image of the environment and available niches (specific kinds of teeth or beaks relate to specific kinds of food available in the ecosystem, and specific kinds of learnt behaviour adapt one to a specific social environment).

The information is lost once it has been erased, so erasing information is an irreversible dissipative process. It therefore increases overall entropy, as pointed out by Landauer [119] (see Bennett [13]):

> Any logically irreversible manipulation of information, such as the erasure of a bit or the merging of two computation paths, must be accompanied by a corresponding entropy increase in non-information bearing degrees of freedom of the information processing apparatus or its environment.

Thus overall entropy is increasing even as order and information in the system increases.

Selective Amplification. Paradoxically, essentially the same effect can be obtained by:

- *Selection* of preferred elements by some kind of filtering process.
- *Selective amplification* of those preferred elements by a very large amount, while the others remain unchanged.

With large enough amplification, this amounts to the same thing as deletion of rejected elements: the amplified elements win.

Kinetic Selection. This is what happens for example in *autocatalytic feedback cycles*, where chemical compounds can produce catalysts for their own production. This leads to what Pross characterizes as kinetic selection [155]:

> Kinetic selection just means 'the faster one wins'. Since the faster replicator is capable of assembling building blocks into new replicating molecules more effectively [...], the number of those faster replicators grows quickly while the number of slower replicators drops until those slower replicators die away entirely.

This is the process underlying the importance of *hypercycles*, as characterised by Eigen and Schuster [60].

Information Amplification. It also happens when, in a social context, some specific information or viewpoint is amplified a large amount and comes to dominate completely over whatever is not amplified. The latter is eventually forgotten and dies away. This is what the public relations and advertising industries are about.

4.3.1.5 Predictable and Unpredictable Outcomes

The outcome is usually not predictable either from the initial conditions or from the meta-goals, because of the random element involved, although both clearly influence the outcome. In principle, it is a non-deterministic process: the outcomes are not predicted by the initial data because of randomization that occurs before selection.

However, this depends on how specific the selection criterion is. It can produce predetermined outcomes if the selection gate is narrow enough. This is the case in nanotechnology and chemical engineering production processes and in specific state vector preparation contexts, as well as in the microbiology context of 'lock and key' mechanisms, where random lower level processes can produce reliable higher level output [101].

4.3.1.6 Occurrence

Adaptive selection underlies all life, including cells, plants, and animals. It occurs extensively in biological processes at both the micro-level and in physiology, as well as in the mind and in society. It can also occur in 'intelligent' manufactured artefacts such as computers [173, Chap. 1]. What about the natural sciences? It does not seem to occur spontaneously in the natural world at the macro-level, but does occur in physics at the micro-level and in nanotechnology. It is the key to chemistry, because purification of substances is the foundation of chemical investigation and chemical engineering. The following sections look at these various domains in turn.

4.3.2 *Physics and Chemistry*

In physics, the classic example is Maxwell's demon, selecting molecules with specific velocities from an incoming stream with random velocities [120]. It occurs in selection of specific electromagnetic frequencies from the incoming flux of radiation by specific frequency-dependent designs:

- **Photodetectors**. Photographic plates, CCDs, X-ray detectors, radio receiver antennas and dishes of specific sizes, each receiving a specific frequency range.
- **Eyes and Plant Leaves**. These select via frequency-sensitive rhodopsin and chlorophyll molecules.
- **Tuneable Frequency Filters**. As in radio circuits.

The essential process in all these cases is

$$\text{(incoming information flow)} \underbrace{\Longrightarrow}_{\text{filter}} \text{(selected flow)}\,, \qquad (4.25)$$

where the selected flow is then used for some purpose. In quantum physics, it occurs in:

- **State Vector Preparation**. Achieved using electromagnetic radiation polarisers or spin angular momentum selection apparatus, such as in the Stern–Gerlach experiment [also an example of (4.25)] [65, 105].
- **Environmental Selection of Pointer States (einselection)**. Associated with quantum decoherence [199].

It is key to

- **Nanotechnology**. Here specific desired structures are selected from an ensemble of generated entities by filtration procedures.
- **Purification Processes**. In chemistry and chemical engineering [66], achieved using physical structures such as centrifuges and distillation columns.

In all cases, since this is an irreversible process, it is intimately connected with the issue of the arrow of time [38, 150]: an unstructured set of states gets transformed to a structured set in the forward direction of time, not the reverse. The apparatus used to implement selection acts as an arrow of time detector [66].

These examples are considered in Chap. 6, but see also [65].

4.3.3 *Life*

Adaptive selection underlies all life, from cells to plants and animals and ecosystems, and is the basis for building up biological information—the foundational difference between physics and biology [95, 161]. The historical evolutionary process is a specific example. The standard story of the evolution of life is that increasingly complex structures have evolutionary advantages precisely because they constrain the lower level interactions that they are subject to [35]. Cell walls are the most striking case, but the principle applies from biopolymers to societies. It results in DNA structuring via adaptive selection over geological timescales, with the meta-goal—the higher level 'purpose' that guides the dynamics—being survival of populations of the organism (both higher level concepts).

A key feature of the biological world is that similar processes of adaptive selection take place not only on evolutionary time scales, but also on developmental and functional timescales [84]. It is central to biological functioning as well as to Darwinian evolution. Thus it occurs in both phylogeny and ontogeny. Specific cases are:

- Darwinian evolution.
- The reading of DNA codings.
- Biological mechanisms.
- The adaptive immune system.

4.3.3.1 Darwinian Evolution

Darwinian evolution results in DNA structuring through adaptive selection over geological timescales [36, 194, 196]), with the meta-goal being survival of the population of organisms (which is a higher level concept).

The Development of DNA Codings. The DNA code, or particular sequence of base pairs in the DNA, occurs through an evolutionary process of variation of phenotypes due to random gene variation, passing on to the next generation the genes that are a major factor in determining the nature of the organism, and selection of the genes of adults that are best fitted to the environmental niches. When carried out over geological timescales, the repetition of this selection process over many generations results in adaptation of organisms to ecological niches. The selector is mechanisms of death and the implicit fitness criterion is survival in that specific context.

Top-Down Causation. This is a classic case of top-down causation from the environment to detailed biological microstructure: natural selection is necessarily a form of multilevel causation [128]. Through the process of evolutionary adaptation, the environment (along with other causal factors) selects the specific DNA coding. A different niche structure results in a different set of genes. As a specific example, a polar bear *Ursus maritimus* has genes for white fur in order to adapt to the polar environment, whereas a black bear *Ursus americanus* has genes for black fur in order to be adapted to the North American forest. The detailed DNA coding differs in the two cases because of the different environments in which the respective animals live. There is no way you could predict or explain this coding on the basis of biochemistry or microphysics alone. Another example is the role of natural selection in producing the remarkable jaw structures of ants and termites, as discussed by Donald Campbell as an example of top-down causation (see [34] and [28, pp. 57–58]).

How do we demonstrate that this is top-down causation? Change the niche structure (e.g., by changing the global climate), and a different population will adapt to it. This happened in the past when bacteria changed the Earth's atmosphere to oxygen and nitrogen, and new species adapted to this change. The fact that lower level equivalence classes are selected by higher level conditions is also demonstrated by many examples of convergent evolution [40, 130], showing that it is higher level functionality that is the driving requirement.

Note that the claim is not that the environment is the only relevant factor. Rather it is that it is an important causally effective factor. Another major factor is the nature of the possibility space for animals that constrains structural, functional, and developmental possibilities [155, 186]. There will always be multiple causal factors, some bottom-up and some top-down. The final result comes from the confluence of these effects.

Gene Level Selection: Equivalence Classes. It is not the case that evolution selects for specific genes. Rather, at the gene level, evolutionary processes select for any combinatorial set of genes that is a member of an equivalence class that attains some higher level goal, such as existence of eyes (Sect. 3.5). The pressure of selection will be first on the organism as a whole rather than on the individual constitutive parts. In other words, it is for the benefit of the whole that some constitutive parts will adapt, and not for their own benefit alone [155].

This is testable, for example, by looking at the way RNA sequences evolve in vivo or in a test tube. Using RNA molecules as model systems, the type of underlying evolutionary mechanisms take advantage of 'walks' within neutral sequence networks, that is, equivalence classes that have the same higher level effect. At the level of the RNA biopolymer sequences, that will be indicative of top-down causation by adaptive selection [9, 187]. It has also been demonstrated in *Drosophila* that there can be functional genomic elements which, despite having undergone many random mutational events, have not changed in function [18]. This means that these variations form an equivalence class as regards genetic function, which is why any one of the class can be selected equally.

4.3.3.2 The Reading of DNA Codings

Developmental Biology. The reading of DNA codings in cells is not just a mechanical process, but rather is adaptive at all stages, responding to the environment as development takes place [77] and so allowing developmental plasticity [194]. For example [193]: "Light is a major environmental factor that impacts many aspects of plant development, including germination and seedling growth." This environmentally dependent plasticity is based on underlying molecular mechanisms.

4.3.3.3 Biological Mechanisms

The Lock and Key Mechanism. Molecular machines are based on a lock and key mechanism whereby the 3-dimensional shapes of molecules acts as a recognition mechanism: two molecules bind to each other if and only if they have folds and receptacles that are inverse images of each other [101]. This is a selection mechanism whereby broadcast messages are recognised as being intended for specific receptors. Thus for example, enzymes catalyze chemical reactions in a very specific way by binding with a substrate. This action is very specific, because the enzyme and the substrate possess specific complementary geometric shapes that fit exactly into one another. The substrate selects those specific enzymes that are needed for a particular purpose from all those that are available, and rejects the rest [190]. Similarly, specific neurotransmitters are recognised by specific receptors on neurons [113].

Folding of Single-Stranded DNA Sequences Following Reverse Mutations. The selection of native nucleic acid folding (an irreducible higher level variable) is an epigenetic effect, with broad implications for the evolution of plants and their viruses. The folding structure (a higher level variable) corresponds to an equivalence class of lower level sequences, and is the biologically relevant variable determining the selection that occurs. How do we demonstrate this top-down causation? This has been shown in detail experimentally by Shepherd et al. [171].

4.3.3.4 The Adaptive Immune System

Through clonal selection, the adaptive immune system functions as an adaptive system able to deal with infections never before encountered [30]. An animal randomly generates a vast diversity of lymphocytes before the body ever encounters antigens, and this enables the immune system to respond to almost any antigens it encounters. The specific lymphocytes that can react against the antigens actually encountered will be selected for and deployed against anything that expresses that antigen.

4.3.4 The Mind: Learning and Perception

Adaptive processes are key to both learning and perception. The mind works by adaptive prediction of what is likely to happen, updated on an ongoing basis [97]. This underlies most of our mental ability. It includes prediction of others' intentions, which is the basis of theories of other minds. Adaptive processes in the brain include:

- Learning.
- Perception.
- The mind in relation to society.

4.3.4.1 Learning as an Adaptive Process

Trial and error processes are the core of both learning and adaptation of the mind to its context. This basic learning procedure is crucial in personal life and business [86, 94], as well as in brain development when an infant interacts with the world around

[88]. A particular case is habituation, that is, learning to ignore a stimulus that lacks meaning [89].

Brain Plasticity. This is the underlying mechanism allowing learning and understanding [163, pp. 643–646]. Neuromodulation allows patterns of neural activity to adapt to new conditions [90]. A form of adaptive selection that has been called neural Darwinism by Gerald Edelman [56] takes place in the brain, refining neuronal connections on the basis of a higher-level 'value system' (selection criteria) that guides brain plasticity in response to environmental interactions, made effective by neurotransmitters diffused to the cortex from the limbic system. This and other processes of brain plasticity underlie learning on a minute by minute basis and enable us to adapt to whatever environment we find ourselves in, which is key to our survival [163, p. 643]: "Plasticity is the defining feature of the brain."

> **Neural Plasticity and Learning**. We are able to learn because of neural plasticity, enabling adaptive selection of specific sets of neural connections embodying best representations of the physical, social, and logical environment, determined through a process of trial and error.

It is the means by which we discover the underlying regularities in the world, finding out what works and what does not (see Chap. 7).

Competition Between Ideas. This has led to the proposal of a competition for survival between mental ideas dubbed 'memes'. This is a new twist on the old concept of cultural evolution and the adaptive nature of culture [157]. The same mechanism was labelled *evolutionary epistemology* by Donald Campbell [35]. However, the idea of a meme is a loose analogy rather than a scientific proposal: it makes no specific predictions and has no testable outcomes, *inter alia* because there is no good definition of what a meme is. In fact, it is at its core a tautological statement: *the ideas that survive are ideas that survive*, because there is no criterion for 'fitness of ideas' other than that they survive. Nevertheless, the broad idea is important: variation, selection, and retention is key to understanding [16].

4.3.4.2 Perception

The process of perception is a predictive adaptive process using Bayesian statistics to update the current perception on the basis of prediction errors. We are immersed in a sea of incoming data, and have to select what we need to pay attention to from all that is irrelevant. Boulding explains this thus [24, p. 2]:

> It is a very fundamental principle that knowledge is always gained by the orderly loss of information, that is, by condensing and abstracting and indexing the great buzzing confusion of information that comes to us from the world around into a form which we can appreciate and comprehend.

Thus this works by rejecting what is irrelevant and keeping what is relevant, as illustrated in Table 4.6. This is partly through conscious processes of discarding unwanted papers, books, files, emails, etc., and storing needed information in indexed filing systems. But additionally, this process happens unconsciously all the time. Indeed it is deeply imbedded in the very nature of perception, where we literally only see what is relevant ([79], see particularly pp. 42–43). It shapes the way we perceive the world (Sect. 7.3.3).

4.3.4.3 Organizations and Society

The mind underlies organizations and society. In order to survive, organizations must adapt to the changing social and economic environment in which they are situated. If they do not do so, they will be supplanted by other organizations. Thus learning to adapt to changing environments is a key need in organizational strategy, otherwise they will become 'dinosaur organizations'. This leads to the idea of a learning organisation: one that can adapt to changes in the environment and so survive and flourish [168].

This applies also to societies as a whole, and indeed to the whole human race. A specific issue here is global warming effects: we have to predict and adapt to changing weather patterns, whatever their cause. If it is indeed anthropogenic, it requires a global effort on the part of all humanity to avert its negative effects.

4.3.5 Mathematical Models: Adaptive Selection

One can model the selection process itself, or the statistics of the process being carried out many times over (as in Darwinian evolution).

4.3.5.1 The Process

The basic dynamics is first a randomisation process, and then a selection process

$$y_j(t_{i+1}) = \Xi_j\big(y_1(t_i), \ldots, y_N(t_i), c_j, E\big) , \tag{4.26}$$

where Ξ_j is a projection operator selecting one of the $y_n(t_i)$ and rejecting the rest, on the basis of the selection criterion c_j evaluated in the environmental context E. It is a non-deterministic process: because of the random element in generating the ensemble selected from, one cannot predict the outcome before the process of selection takes place. However, one may have probabilistic rules for the likelihood of the various possible outcomes being chosen.

In terms of decision-making, the probability of a particular outcome being chosen is based on Bayesian reasoning (see [59] and [19, pp. 18, 384–439]). The best updated estimate $P(H|D)$ of the probability of hypothesis H on the basis of new data D (the posterior probability) is given by

$$P(H|D) = \frac{P(D|H)P(H)}{P(D)}\,, \tag{4.27}$$

where the assumed prior probability for H before the data is observed is $P(H)$, $P(D|H)$ is the probability the datum D will be observed if H is true, and $P(D)$ is the unconditional probability of datum D. One can modify this to take into account probability $P(-H)$ that the hypothesis is not true to get [86, pp. 60, 98]

$$P(H|D) = \frac{P(D|H)P(H)}{P(D|H)P(D) + P(D|-H)P(-H)}\,. \tag{4.28}$$

Indeed there is evidence that the brain functions as a Bayesian predictor [45, pp. 87, 102].

The underlying neural process is sometimes characterised as *supervised learning*. The relevant equations for altering neural net connection strengths are given in [45, pp. 323–326][3] and [19]. Rejection thresholds characterise the selection function [19, p. 28].

4.3.5.2 The Statistics of the Process

While one cannot predict the specific outcome of the projection process (4.26), one can predict the statistical effects of the likely results of such choices made many times over in a population. This is the basis of population genetics studies. Maynard Smith developed the genetics of populations based on the idea of gene frequency, with random variation of genes followed by selection based on 'fitness', leading to gene ratios that could be expressed in ordinary differential equations [131, pp. 71–86].

The mathematics in general cases is the mathematics of adaptive selection examined in full generality by Holland [103], but in specific cases it results in the standard equations of population genetics [85] and molecular evolution [116]. The *Price equation* [80, 152] relates the mean and covariance of variables. It describes how the average value of any character—body mass, antler size, tendency to altruism—changes in a biological population from one generation to the next. It models the effects of gene transmission and natural selection on the proportion of genes within each new generation of a population, and is regarded as important in social evolution theory.

[3] From the viewpoint of this book, the characterisation 'supervised Hebbian learning' is a misnomer. The essence of Hebbian learning ('wire together, fire together') is that it is a local bottom-up habituation process, taking place irrespective of high level outcome. It is not a *learning* process, in the sense intended here, as the outcome is independent of context.

4.3.6 Multilevel Selection

A key question is whether or not selection is a multilevel affair: does it generally take place between multiple levels of the hierarchy? This is controversial territory, bitterly disputed by evolutionary theorists,[4] but an examination of the causal processes in action during evolutionary selection in biology, seen in the light of the hierarchical structure of living beings [36], makes a strong case that selection must indeed in general be a multilevel process. This is basically because there are very few cases where a key structure or function in higher animals corresponds to a single gene. Indeed in general there are a huge number of phenotypes corresponding to any specific genotype [187]. Selection takes place for any one of the members of this equivalence class, not for any specific individual gene (see Sect. 3.5). I discuss in turn:

- The issue of levels of selection.
- Lower level selection and aggregation.
- Higher level selection and lower level outcomes.
- Multilevel selection and the issue of group formation.
- The underlying social and biological mechanisms.
- The minimum set of levels.

4.3.6.1 The Issue of Levels of Selection

Samir Okasha's book *Evolution and the Levels of Selection* [143] gives a comprehensive discussion of the ongoing debate about levels of selection in evolutionary biology. An important feature of the book is the distinction made therein between:

- **Multilevel Selection 1 (MLS_1)**. This is concerned with the evolution of individual-level traits (level 1), e.g., being able to run fast.
- **Multilevel Selection 2 (MLS_2)**. This is concerned with the evolution of collectives and their properties (level 2), e.g., development of language.

This distinction helps considerably in clarifying some of the disputes that have arisen as regards multilevel selection processes. It is, however, useful to refine Okasha's proposal by defining individual selection properties as:

- MLS_{1E}, selection of individuals due to the environmental context E independent of the existence of the group, leading to group fitness $fitness_{2E}$.
- MLS_{1G}, selection of individuals that is essentially due to the existence of the group as an emergent entity in the environmental conext E, leading to group fitness $fitness_{2G}$.

[4] See [122] and the discussions at http://musicoflife.co.uk/ for examples.

These definitions at the individual level will be reflected by corresponding definitions at the group level:

- $\mathrm{MLS}_{2\mathrm{E}}$ is the group level selection effect due to aggregation of individual advantage characterised by $\mathrm{MLS}_{1\mathrm{E}}$. Then $fitness_{2\mathrm{E}}$ is an outcome.
- $\mathrm{MLS}_{2\mathrm{G}}$ is group level selection that occurs specifically because of the existence of the group as an emergent entity. Then $fitness_{2\mathrm{G}}$ is a cause.

4.3.6.2 Lower Level Selection and Aggregation

The combination of $\mathrm{MLS}_{1\mathrm{E}}$ and $\mathrm{MLS}_{2\mathrm{E}}$ is a form of selection that is multilevel in character, but not essentially so: properties at the individual level are selected for by $\mathrm{MLS}_{1\mathrm{E}}$ and simply lead by aggregation to group properties, i.e.,

$$\mathrm{MLS}_{1\mathrm{E}} \Longrightarrow \mathrm{MLS}_{2\mathrm{E}} . \tag{4.29}$$

This is a bottom-up process. The individual level $trait_{1\mathrm{E}}$ is an individual trait that gives an advantage to the individual in the context of the environment, and this advantage has nothing to do with the existence of the group. Examples are individuals being able to run very fast, being very strong, and so on. Selection for such traits is thus the case of selection based on individual traits alone, and can operate in an unstructured population in which there are no groups at all, but it will also operate when groups exist. This trait confers individual $fitness_{1\mathrm{E}}$ in regard to the overall environment, and is selected for by $\mathrm{MLS}_{1\mathrm{E}}$. It is not multilevel selection as regards the individuals, in that it is just selection based on properties of the individual. However, aggregation of the individual $trait_{1\mathrm{E}}$ over all the members of the group improves group survival capacity, and so underlies the group level $trait_{1\mathrm{G}}$ (the group is more likely to survive if it is made of stronger and faster members). Even though this is 'nothing but' the sum of the parts, it enhances group survival.

4.3.6.3 Higher Level Selection and Lower Level Outcomes

By contrast, the combination of $\mathrm{MLS}_{2\mathrm{G}}$ leading to $\mathrm{MLS}_{1\mathrm{G}}$ is a form of selection of individuals that is essentially multilevel in character: properties at the individual level are selected for by $\mathrm{MLS}_{1\mathrm{G}}$ because they enable desirable emergent group properties, i.e.,

$$\mathrm{MLS}_{2\mathrm{G}} \Longrightarrow \mathrm{MLS}_{1\mathrm{G}} . \tag{4.30}$$

This is a top-down process: $trait_{2\mathrm{G}}$ is a group trait that gives a selective advantage to the group as a whole because the group is an emergent entity, acting as a collective. To make the situation specific, consider two examples.

First, consider the case of why animals such as buffalo in the Kruger National Park find it important to group together in herds or tribes. The key point is that a collection of buffalo wandering around on their own is not a herd: it is an aggregation (the whole is just the sum of its parts). If the same buffalo band together as a herd interacting in a social way, they can collectively protect each other, sensing danger earlier than when alone and acting together to protect each other, and so are far more likely to survive. A herd can act together as a unit with the common purpose of warding off danger of attack by lions,[5] move together to seek water in places remembered by older animals, form a defensive ring against predators, and so on. These obviously cannot be a trait of the individuals on their own. They confer $fitness_{2G}$ due to group existence in the overall environment (which includes lions in the surroundings), and are selected for by MLS_{2G}, which in turn selects those individuals who cooperate well in the group.

Second, consider the case of bushmen in the Kalahari. Traits leading to advanced survival prospects include the use of language to communicate with each other, and group hunting of giraffe. These are, of course, impossible if the group does not exist.

Living isolated lives on their own, they are vulnerable to many dangers and will find it difficult to get food. If they band together, they can act collectively to protect young, detect dangers, ward off predators, and hunt together, and can share food, skills, resources, and information. The whole is much more than the sum of its parts and the young are much more likely to survive. As a specific example, the skills of animal tracking [121] are passed down from generation to generation through a group educational process. This collective process, impossible unless the group acts as a collective, greatly enhances the survival prospects of the group, and selects for those individuals who facilitate group processes, e.g., because they can communicate via language.

It is crucial that culture and technology can only evolve in this kind of context. Banding together into social groups probably played a key role in the evolution of language and intelligence [54, 153], which enabled human domination over all other species.

In each case, at the individual level, $trait_{1G}$ is an individual trait that gives an advantage relative to the environment because of the existence of the group. It is a capacity underlying the way the individual takes advantage of the existence of the group, which could not occur if the group did not exist. Thus cooperative individuals who are willing to learn will benefit more through the group's existence. Those who ignore group wisdom are likely to perish sooner.

4.3.6.4 Multilevel Selection and the Issue of Group Formation

The key *group level trait* underlying all these possibilities is the mere fact of group existence, that is, the tendency to live together as a cooperative group. It is the buffalo being together as a herd that allows crucial social traits to develop. The same is true for the bushmen, who through the existence of the group develop far greater survival

[5]For a graphic demonstration of the protection provided in this context by the common purpose of a collective of buffalo, see http://www.youtube.com/watch?v=xHIkUzRw2jw.

capacity, particularly through development of technology and tactics that can be taught to their children.

The key *individual level trait* underlying the individual benefit accruing from the group is the individual having a propensity to join a group. They can then benefit by learning from the group, getting its protection, and so on. It is because of this propensity for sociability at the individual level that the group emerges from the individuals that comprise it: this is the glue that holds the group together. Individuals who do not have this propensity to join a group (they like to keep away from the group and go off on their own, are uncooperative and unwilling to learn, and so on) are not so likely to survive, because they do not benefit from the group's existence. The mechanism that underlies this tendency to form groups is emotional pressure in the individual minds.

4.3.6.5 The Underlying Social and Biological Mechanisms

The key issue now is the behavioural and biological mechanisms at the individual level that underlie formation and stability of social groups, thereby leading to the group selection advantages $fitness_{2G}$.

Behavioural Systems. There are two basic behavioural needs relating the group level to the individual level in order to ensure group formation and stabilisation.

1. Group Formation: Group Level. Firstly, in order that meaningful social groups exist, cooperative behaviour between the units that make up the group is crucial [133]: "cooperation amongst lower level units is central to the emergence of new higher levels, because only cooperation can trade fitness from lower to higher levels". There are all sorts of mechanisms at the group level intended to make this happen (roles, uniforms, teaching, myths, and so on) [123, p. 39]: "each element making up a social system serves a function that assures the maintenance of the system". But that is not enough: in the case of animals and humans, the individual must be responsive to them, that is, they must want to belong to the group. There needs to be an internal mechanism to produce this response.

2. Group Stabilisation: Group Level. Generically, whenever individuals cooperate together to form groups, there must be mechanisms for regulation of conflict [143, p. 205], requiring adaptations that suppress within-group competition in order that the group emerge as a genuine whole with adaptations of its own [143, p. 221–222, 227–228]. Again there will be group level mechanisms and processes with this purpose (sanctions, teaching, a legal system, and so on), but by themselves they will not suffice: there must be internal mechanisms to help produce this outcome. How are these realised at the individual level? In the case of all mammals (including humans), there are plausible biological mechanisms at the individual level underlying both the formation and conflict regulation of such emergent groups. Social groups form because of innate tendencies of individuals to form such groups. The source for this tendency lies in innate primordial emotional systems shared by humans and all higher animals [93, 144], which are our evolutionary heritage, given to us to help guide decisions we make [67] and thereby shape intellect [43, 44]. These innate emotional systems function by giving us feelings that tend to produce specific kinds of actions that have promoted survival in our evolutionary ancestors, both human and mammalian [93, 144, 146]. In particular there are two such systems that have evolved to create and protect emergent social groups as discussed above. These are [184]:

1. Group Formation: Individual Level. An affiliation/attachment system, needed to create such groups, producing feelings of wanting to belong and loneliness when excluded, and starting with mother-child bonding [177, pp. 44–45].

2. Group Stabilisation: Individual Level. A social ranking system, needed to protect groups by regulating conflict (see [177, pp. 47–48] and [153]). This generates a dominance hierarchy which is a social ranking system (the pecking order, the alpha male, etc.). Conflict takes place to attain a place in this system, but then acceptance of one's place, and associated territorial rights, regularizes resource allocation in a largely peaceful manner.

These two systems dominate much of social life, and provide the emotional power that enables groups to function. As stated by Stephens and Price [177, p. 50]:

> In short, the evidence points to the existence of two great archetypal systems: that concerned with attachment, affiliation, care-giving, care-receiving, and altruism; and that concerned with rank, status, discipline, law and order, territory and possessions. These may well be the basic archetypal patterns on which social adjustment and maladjustment, psychiatric health and sickness depend.

These primary emotional systems have evolved over evolutionary times precisely in order to ensure that social groups will come into existence and then be stable (there can be no other reason for their existence as genetically determined systems). We know that they are selected for, because they are innate, and are shared with our ancestral relatives, human and animal [144, 146]. Thus they are key examples of individual level traits z_{1G} that have been selected for via individual selection MLS_{1G}, in order to promote group benefits Z_{1G}, as indicated by (4.21).

These systems enable the group to come into being, and would not exist as innate systems if there were no major benefit provided by existence of social groups. They have been genetically determined because they are crucial to survival.

4.3.6.6 The Minimum Set of Levels

This section has used the specific example of existence of social groups to support the usefulness of the distinction between MLS_{1G} and MLS_{1E}. Such a distinction is needed to give an evolutionary explanation both for the existence of the emotional systems that are crucial to intellectual functioning [43, 44], and also for the ascending systems and associated nuclei in the brain that are the neural bases for these emotional systems [164]. Together with the case of meiotic drive quoted by Okasha [143], this example serves as an existence proof of causal effectiveness of multilevel selection.

If correct, this analysis implies that discussion of evolutionary processes leading to innate behaviour and brain modules should take into account the key role played by emotional systems in the development of these behaviours and modules [67].

The analysis above also confirms the idea of top-down causation in the adaptive selection process [34, 128]. Indeed it supports the view that emergence of complexity such as life requires a reverse of flow of information, from bottom-up only to also including a flow from the environment down [189], where 'the environment' includes the group. This reverse flow is needed because adaptive selection causes adaptations of the organism to the environment: but this cannot happen unless information flows down from the environment into the organism, where it alters both structure and behaviour. This is what has happened in the past in the case of the coming into being of the innate primary emotional systems, which can be claimed to have played a key role in evolutionary development (if this was not the case, they would not occur as biological modules in the brain). However, in contrast to Campbell's compelling example of the jaw of a worker or termite ant [34],[6] the top-down mechanism considered here is essentially multilevel (see Table 4.8). This proposal accounts for existence both of these behavioural characteristics, and the neural systems [114, pp. 132–133] that lead to their existence. It has an essentially multilevel nature. The proposal made here suggests further useful developments:

- Seeing how this view extends to the level of cells, where the selection pressures that lead to existence of chemical synapses between neurons, rather than much faster electrical synapses, can plausibly be related to the evolutionary need to develop synaptic plasticity which can be affected by the diffusely projected neuromodulators that form Edelman's 'value system' [57].
- Seeing how this view extends even further to the underlying level of genes, where the key feature of multiple realizability, which underlies all top-down causation [64], leads to selection based on equivalence classes of sets of genes, rather than selection of individual genes [187]. Selection is not for specific genes, but for any gene in an equivalence class that produces the same higher level outcome that is selected for. It is this feature that characterizes this as top-down action.

In both these cases (selection at the neuronal level and selection at the genetic level), it is clear that the selection mechanisms in operation can only be of a multilevel kind, because there simply is no direct link from the adaptive environment to either neurons or genes. The adaptive causal link in both cases is necessarily via the survival prospects of individual animals on the one hand, and the social groups to which the individuals belong on the other, as discussed above. Multilevel selection involves more than two levels: indeed to discuss it properly, one needs to contemplate seven levels, as shown in Table 4.8. Realistic discussion and notation should refer to this full multilevel context, not just to two levels.

As a specific example, proteins such as haemoglobin, kinesin, and dynein have come about through selection pressures [187] which at the very minimum must

[6]This example is considered in depth in Brown and Murphy [28, pp. 57–58] and in Martinez and Moya [128, pp. 7/16–8/16].

Table 4.8 The levels involved in multilevel selection: direct selection effects between the environment, the collective, and the individual, and indirect effects on individuals, cells, DNA, and genes

	Level	Environmental effect	Group induced effect	Individual induced effect
7	Ecosystem	⇓		
6	Group	⇓	⇓	
5	Individual/animal	⇓	⇓	
4	Organ/systems	⇓	⇓	⇓
3	Cell			⇓
2	Genotype/DNA			⇓
1	Individual gene			⇓

involve the cellular and genotype levels, but at least in the case of haemoglobin certainly involve higher physiological levels. There is no biological mechanism that can reach down directly from the level of oxygen transport in the cardiovascular system to the level of individual genes. It has to be via the effect of oxygen in cells to the need for haemoglobin. And that oxygen is needed in muscular cells so that animals can respond to the world around them. At least all the levels 5 to 1 are involved. It is of course possible to extend the usual correlation analyses [143] to such multilevel cases.

Conclusion. Many examples occur where higher levels are selected for and then carry the lower levels along with them. In particular, once multicellular entities exist, selection cannot act directly on the gene level: there simply is no causal handle available for this to take place. It has to take place via higher levels, this selection process then acting down to the level of the genes. The point is that each cell then depends on the organism for its existence. It affects the organism's viability, but selection is via that higher level. Multilevel selection has to be the core of what is going on. A minimum for a realistic discussion of biology is the set of levels shown in Table 4.8. Any discussion based solely on correlations, rather than considering the nature of the relevant biological mechanisms, is likely to miss this point. It will not get at the essence of what is going on.

4.3.7 The Nature of Selection Criteria

Adaptive selection can be thought of as a generalised feedback loop with a meta-purpose provided by a value system, or set of selection criteria, classifying which outcomes are desirable and which are not (a higher level purpose that is not directly attained as the goals in a feedback control system are, but still effectively guides what happens by selecting preferable outcomes). The selection criteria guiding adaptive selection are not physical things. In some cases the value system may be implicit rather than explicit, being built into the way the selection agent functions rather than being a separate function. This depends on context:

Implicit. The selection criteria may be implicit, as in Darwinian evolution, where it is implied by the dynamics of the situation (death of poorly adapted animals, with their gene structure not passed on, and survival of better adapted ones, with their gene structure passed on to future generations). They may be implied by physical or biological functioning, as in electrical filters or the lock and key selection mechanism. But in each case these are the mechanisms, not the selection criteria themselves, which are abstract concepts characterising how the selection process works. It is a description of patterns of interactions that are embodied in the system dynamics through its structuring.

Explicit. The selection criteria may be explicit, as when intelligent systems select one option over another, or in intelligent machines that carry out such selection on the basis of their design and operating parameters. They are then choices we make, perhaps embedded in the logic of machines we create. In all cases they are abstract rather than physical entities.

> **Selection Criteria**. These are not physical things, they are related to processes and patterns. They may or may not be the result of conscious choice (this depends on context). However, they are indeed effective in that, given the context, they are what crucially determine the outcomes.

This is what is expressed in (4.24).

4.4 Adaptive Selection of Goals (TD4)

A key issue is where the goals in a feedback control system come from. Adaptive information control takes place when there is adaptive selection of goals:

- *Feedback control takes place*. There is a feedback control system directing dynamics according to the goals of the system.
- *Feedback control is guided by adaptive selection*. There is adaptive selection of the goals of the feedback control system.

This therefore combines both feedback control and adaptive selection. I discuss in turn:

- The nature of the process (Sect. 4.4.1).
- Evolution (Sect. 4.4.2).
- Microbiology (Sect. 4.4.3).
- Behaviour (Sect. 4.4.4).
- Engineering systems (Sect. 4.4.5).
- Mathematical models (Sect. 4.4.6).
- The nature of causality (Sect. 4.4.7).

Table 4.9 Adaptive selection of goals

Level 3	Selection criterion	Meta-goal
	⇓	
Level 2	Goal	Adaptively selected
	⇓	
Level 1	Feedback control	⇒ Output

4.4.1 The Nature of the Process

Higher level innovation becomes possible when one combines TD2 and TD3 to obtain TD4: feedback control systems with adaptive learning determining the goals. Unlike TD2 where goals are fixed, and TD3 where there are no specific goals, these are systems that select their goals by a process of adaptive selection. This is a higher level form of top-down action, as it involves both goals in a homeostatic system (TD2) and adaptive selection criteria (TD3). As expressed by MacKay [28, pp. 128–131]:

> Action loops are modulated by supervisory systems that function to set the goals of the action loops.

The goals of the feedback control system are irreducible higher level variables determining the outcome, but they are not fixed as in the case of non-adaptive feedback control. They can be adaptively changed in response to experience and information received. The overall process is guided by fitness criteria for selection of goals (Table 4.9).

This allows great flexibility of response to different environments. Indeed in conjunction with memory it enables learning and anticipation [173, Chap. 4] and underlies effective purposeful action as it enables the organism to adapt its behaviour in response to the environment in the light of past experience, and hence to build up complex levels of behaviour.

> **Adaptive Selection of Goals (TD4).** This occurs when there is adaptive selection of goals in a feedback control system, thus combining both feedback control (Sect. 4.2) and adaptive selection (Sect. 4.3).

The classical example is associative learning in animals, such as Pavlovian conditioning: animal response to a stimulus such as a sound, which is taken as a sign of something else and causes physical reactions implemented by motor neurons. The training is causally effective by top-down action from the brain to cells in muscles. The fitness criterion is avoidance of negative stimuli.

This is of course a form of top-down causation. Indeed it has two levels of top-down causation imbedded in its structure. Thus it is a second order form of top-down action, with higher levels constraining lower level actions. However, now, because the goals can be adapted through a learning process, the constraints are contextually sensitive (see [89, 109] and [28, p. 88]).

4.4.2 Evolution

Adaptive selection of feedback control system goals is fundamental to evolution: it is the process in the evolutionary history of life whereby the goals of all homeostatic systems in physiology and cell biology were determined [36] (Sect. 4.2.4). The basic Darwinian process of selection of specific successful physiological structures determines their goals [28, p. 70]. Thus they would not exist and have the values they have without this adaptive selection process.

4.4.3 Microbiology

The feedback control loops in microbiology can have an adaptive element. Indeed this will be very likely to happen wherever randomness is key to what is going on [101]. The goals of cellular feedback loops, such as those that control levels of ATP in order to provide cellular energy, alter in response to higher level conditions such as stress. The neuroendocrine system can change gene expression by transducing sensory information from the environment into the body [84, p. 46]. These are multilevel adaptive responses, altering goals of lower level feedback loops in response to environmental conditions.

Adrenalin Levels. These change if a condition of stress exists. They regulate blood vessel and air passage diameters, heart rate, and metabolic shifts in response to a 'fight or flight' situation [92, 156].

Epigenetic Processes. These determine a cell's development fate, which depends both on positional information and environmental conditions. Thus cold environmental temperatures repress gene transcription in wheat and wild mustard [84, p. 47].

4.4.4 Behaviour

This process is fundamental to animal behavior at the macro-level, based on its occurrence at the micro-level.

4.4.4.1 Animal Conditioning

Operant conditioning [89, pp.108–112] in a particular reward context is when a conditioning stimulus elicits a predictable conditioned response with some specific goal. By altering the context, one can change the goal of the conditioned response.

Associative Learning. The classic example is associative learning in animals, such as Pavlovian conditioning: animal response to a stimulus such as a sound, which is taken as a sign of something else and causes physical reactions implemented by motor neurons. It is a form of reinforcement learning [45, 89, pp. 231–358]. The training is causally effective by top-down action from the brain to cells in muscles. The fitness criterion is avoidance of negative stimuli.

Bee Responses. Bees respond to abstract classes of patterns that signify specific actions they must perform, e.g., the dances of other bees, and they can respond to symmetric and asymmetric patterns of marks in a maze [197]. The latter is learned behaviour in a particular artificial environment, reflecting top-down causation from an irreducible higher level abstract variable (a class of symmetries) to their action goals.

4.4.5 Engineering Systems

Computer-based feedback control systems can be engineered to include adaptive selection of goals. For example as an aircraft burns its fuel and loses weight, the optimum speed and height for flight may change. These new goals can be fed back to the autopilot to generate new flight patterns. The adaptive part is the optimization process that selects the new optimum height and speed. The feedback part is how they then control the aircraft movements. Implementation is via suitable software in digital computers in the aircraft control system. Similar processes will take place in chemical plant control systems and electrical grid control systems.

4.4.6 Mathematical Models

Equation (4.26) is applied to a set of goals g_n in (4.21) to get

$$g_j(t_{i+1}) = \Xi_j^{\mathrm{g}}\big(g_1(t_i), \ldots, g_N(t_i), c_j^{\mathrm{g}}, E\big) , \tag{4.31}$$

where c_j^{g} are criteria for feedback control goals (see Table 4.9).

In engineering systems, adaptive control is based on parameter estimation by methods including recursive least squares and gradient descent. These update goals in real time [8].

The evolution of eusociality has been modelled by inclusive fitness theory and Hamilton's equation:

$$R > c/b , \tag{4.32}$$

which says that cooperation is favoured by natural selection if relatedness R is greater than the cost to benefit ratio c/b. This is essentially an equation for adaptive selection of goals. It has been heavily criticised by Nowak et al., who propose alternative equations [142], leading to much controversy. While this kind of equations may describe some simple animal behaviour well, one should be very cautious in assuming that such a simple relation as (4.32) encompasses all that goes on in social contexts where memory and perception play key roles. Perhaps the point is that the cost c and benefits b, if they are indeed well defined, are extremely nonlinear functions of many other variables, so the linearity of the inequality is illusory.

It is claimed by some that the mathematics of evolutionary game theory [4] will act as an adequate basis for understanding these processes in social contexts. Personally, I have serious doubts as to how far this can succeed.

4.4.7 The Nature of Causality

The goals of the feedback control system are irreducible higher level variables determining the outcome, but they are not fixed as in the case of non-adaptive feedback control. They can be adaptively changed in response to experience and information received. The overall process is guided by fitness criteria for selection of goals, which are higher level abstract entities. This is a form of adaptive selection in which goal selection relates to future rather then present functioning of the feedback system. This allows great flexibility of response to different environments. It will by its very nature be indeterministic.

In conjunction with memory, it enables learning and anticipation and underlies effective purposeful action as it enables the organism to adapt its behaviour in response to the environment in the light of past experience, and hence to build up complex levels of behaviour.

4.5 Adaptive Selection of Selection Criteria (TD5)

Crucial to the last two forms of top-down causation(TD3 and TD4) is the question: where do the criteria of adaptive selection come from? The may just be given through the physical or biological context, but in many cases they themselves are adaptively developed:

- *Adaptive selection is taking place.* There is an adaptive selection system guided by a set of selection criteria.
- *This process of adaptive selection is guided by higher level adaptive selection.* The selection criteria are themselves adaptively selected, the outcome being shaped by higher level selection criteria.

Table 4.10 Adaptive selection of selection criteria

Level 3	Selection criterion 2	Meta-goal
	⇓	
Level 2	Selection criterion 1	Adaptively selected
	⇓	
Level 1	Adaptive selection	⇒ Output

In this section, I shall look in turn at the following:

- The nature of the process (Sect. 4.5.1).
- Evolutionary biology and animal behavior (Sect. 4.5.2).
- The mind (Sect. 4.5.3).
- Mathematical models (Sect. 4.5.4).
- Meta-causation: closing the hierarchy (Sect. 4.5.5).
- The hierarchy of goals: ethics and meaning (Sect. 4.5.6).
- Occurrence of meta-reflection (Sect. 4.5.7).

4.5.1 The Nature of the Process

The selection criteria in adaptive selection systems may prove to be unsatisfactory, and may need amendment. Thus they too may be adaptively selected on the basis of a higher set of selection criteria (see Table 4.10 and [28, p. 130]). This is a higher form of top-down causation, because adaptive selection itself is such. It is of importance in determining strategy in every area of personal and communal life: business, education, politics, and social policy, for example.

> **Adaptive Selection of Selection Criteria (TD5).** This is the case where there is an adaptive selection system, where the adaptive selection criteria are themselves determined by adaptive selection.

The outcome is not predictable from initial data: it has a double layer of random influences, modulated by selection effects. This is where the real logical depth of complex systems comes in to effect: higher level selection principles guide the creative nature of adaptive selection, searching for ways to realize their goals.

4.5.2 Evolutionary Biology and Animal Behavior

Instinctive animal behavior includes genetically determined adaptive behavior patterns, which belong in this category. As a specific example, evidence for such genetic determination of behavior has been obtained recently in the case of the burrowing behavior of *Peromyscus* mice [33, 191]. This is adaptive behavior developed as a defense mechanism against snakes invading the burrows. Because it is genetically determined, it was inbuilt through the adaptive selection process of Darwinian evolution.

4.5.3 *The Mind*

Adaptive selection of selection criteria is a key feature of how the mind works. We learn by experience that some forms of enquiry and learning are effective and others are not. Some kinds of selection criteria get us where we want, and others do not. This is a core part, not only of scientific enquiry and philosophical investigation, but also of commerce and industry, and indeed of daily life. It marks the transition from just learning, to learning how to learn. Thus it is central to all education and to all learning organisations [168]. Here are some examples:

- A strategy research project to see what kinds of objectives for an airline company (fastest flight times, cheapest fares, most comfortable seats, better entertainment systems) will generate the best passenger levels. This choice informs programs to develop those features.
- A research project to see what educational objectives work best in helping pupils become self-reliant (teach the fundamentals first and then explain the broader picture, or explain the broader picture first and then develop the fundamentals). Whichever one chooses, one then has to determine how best to achieve that objective.
- Selecting a page-ranking algorithm for internet search systems, such as that used by Google. One is then using a selection criterion for selection criteria [126].

4.5.4 *Mathematical Models*

When selection criteria area adaptively selected, (4.26) is applied to the criteria c_n [which guide selection in (4.26)] in the form

$$c_j(t_{i+1}) = \Xi_j^{\mathrm{c}}\left(c_1(t_i), \ldots, c_N(t_i), c_j^{\mathrm{c}}, E\right) , \tag{4.33}$$

where c_j^{c} are criteria for selective criteria (see Table 4.10).

Attempts to model the statistics of such processes—the selection of criteria for selection of behavioural patterns—led to mathematical models of financial markets such as the Black–Scholes formula [21]. However, these are based on specific assumptions about behavioural patterns that may or may not be true in reality. Use of such equations was a major factor leading to the global financial crisis [147]. Mathematical modelling of higher adaptive processes is a very risky enterprise.

Table 4.11 The hierarchy of selection criteria

Level $N+1$	Selection criterion N	Non-algorithmic choice
	$\Downarrow$	
Level N	Selection criterion $N-1$	Adaptively selected
	$\Downarrow$	
	⋮	⋮
	$\Downarrow$	
Level 3	Selection criterion 2	Adaptively selected
	$\Downarrow$	
Level 2	Selection criterion 1	Adaptively selected
	$\Downarrow$	
Level 1	Adaptive selection	$\Rightarrow$ Output

4.5.5 *Meta-Causation: Closing the Hierarchy*

This is a higher form of top-down causation, as it is second order adaptive selection. It has the character of meta-causation. It relates to the cause of a cause. But in principle this issue recurs: what are the criteria to use in *this* process of adaptive selection? Are these higher level selection criteria also learnt?

4.5.5.1 Closing the Hierarchy

Adaptive selection of adaptive criteria involves choosing a set of criteria c_j^c for suitability of adaptive criteria c_j. This appears to be the start of an infinite recursion: where do these next higher level selection criteria c_j^c come from? Are they too selected adaptively? How do we close the logic (see Table 4.11)?

Evolution. In the case of evolutionary biology, Darwinian processes are the topmost level. Now the key feature here is that these evolutionary processes have led not just to instinctive behaviour (Sect. 4.5.2), but also to evolution of the human mind which can indulge in meta-analysis (Sect. 4.5.3). Genes have developed to enable this, assuming intelligence has indeed been selected for, and is not just a byproduct of some other brain function.

Thus two such higher levels of selection have occurred in evolutionary history. This is probably as high as it goes: there can be just two such higher levels of selection in the standard Darwinian case.

The Mind. In the case of rational reflection and the mind, this can in principle keep on going: each higher set of criteria is chosen on some basis, that is, it is selected. At some point we have to stop and accept a set of highest level selection criteria as an *a priori* choice, otherwise we cannot close the system (if we consider the criteria for this choice and evaluate it, then through that act it is shown not to be the topmost level). Any attempt to determine these criteria algorithmically, heuristically, or by adaptive selection will implicitly introduce a further set of selection values, for one will at some point need to test whether this procedure is the best option (one could have used other algorithms or heuristics). It will just postpone the final decision level and choice by adding in a further level to Table 2.8.

The Topmost Level. This has to be just taken as given: it is either (i) determined by physical or biological processes through the nature of the interactions taking place, or (ii) a choice made on philosophical, ethical, or aesthetic grounds, expressing some viewpoint on values and meaning.

The highest level selection criteria cascades down to all lower levels. It represents a mechanism or world view that guides all lower level choices.

The same issue arises, of course, in relation to adaptive selection of goals (TD4). Here too there has to be a topmost level which is just taken as given, and sets the overall direction and purpose of the dynamics. The meta-questions in both cases are:

Meta-analysis. How many levels up do you go?

Choice. How do you decide which criteria to use at the top?

These are philosophical issues, with choices made according to one's philosophical position. This is where values and purpose come in: this highest level is the level of meaning ('telos'), perhaps involving ethics or aesthetics. This choice gives shape to all the rest, for it chains down to affect choices made and outcomes at all the lower levels (Sect. 4.5.6).

4.5.6 The Hierarchy of Goals: Ethics and Meaning

Closing the hierarchy (Sect. 4.5.5) inevitably leads to the issue of values and meaning. Thinking about these issues is an aspect of metacognition: the process of thinking reflectively about the meaning of things and the choice of goals. It involves the following inter-related trio:

- **Ethics**. What is right and wrong.
- **Aesthetics**. What is beautiful and what is ugly.
- **Meaning**. Philosophical views on ultimate purpose and what it all means.

These are the guiding principles for social and individual life.

The Topmost Level. Between them, ethics, aesthetics, and meaning form the topmost level of the hierarchy of adaptive selection criteria (Table 4.11). They are the highest level abstract principles that are causally effective in the real physical world, crucially guiding what happens in choosing goals at all levels.

4.5.6.1 Ethics: Criteria for Choice of Goals

Behavioural values, related to ethical views on what is right and what is wrong, are causally effective in a top-down way by determining the set of desirable and undesirable lower level goals. Ethics shapes goals at the highest level of the causal hierarchy, and thereby constrains the appropriate choice of lower level goals [138]. Ethical values are non-reducible higher level variables (while their choice may be affected by lower level goals, they themselves are essentially higher level entities). By determining the nature of lower level goals chosen, and hence the nature of resulting actions, ethics is a set of abstract principles that are causally effective in the real physical world. Indeed, they crucially determine what happens. This applies equally to a society and to individuals, and is a key aspect of social science when properly understood [76].

Society. Wars will be waged or not depending on the ethical stance of a society. Large-scale physical devastation of the Earth will result if thermonuclear war takes place. This ethical stance has crucial physical outcomes.

Individuals. One's goal may be to amass as much money for oneself as one can, or it may be to do what one can to help others. All the subsidiary goals in one's life—what one studies, where one goes, what jobs one does—depend on this higher level choice. Individuals such as Mahatma Gandhi and Martin Luther King spent their entire lives in the search for the right way to live.

4.5.6.2 Aesthetics and Beauty

Views on what is beautiful and what is ugly also chain down to determine many lower level decisions, establishing what is desirable and what is not in many ways.

Individuals. The individual determines a choice of design for houses and clothes, furniture and pictures, and indeed for a style of living. Exploring what is beautiful or aesthetically meaningful will be the life work of a great artist, whatever the art form (painting, sculpture, ballet, and so on). Van Gogh and Rembrandt come to mind.

Society. Society plays a great role in urban design and public architecture. In Germany, new housing estates have a percentage of the cost put aside by law to create artworks on the estate. Hence the higher level desire for art leads to specific lower level goals and expenditures.

4.5.6.3 Meaning and Purpose

Our understandings of meaning and purpose are abstract entities that constitute a high level in the hierarchy of causation in the mind. The imperative to search for meaning is a key aspect of human nature, as pointed out by Viktor Frankl [78]. Roles embody social values, which, together with individual values relating to life purpose, guide the individual and communal choice of goals and the methods used to attain these goals [76].

Society. It is embodied in rituals of social life that mark meaningful changes of status, such as initiation rites and marriage ceremonies [108]. It is also marked by public expenditure patterns, such as whether money is given to the arts and museums or to pure science for its own sake, and more generally by societal goals such as creating a great society or putting a man on the Moon.

Individual. The search for meaning and understanding can be a key feature of human life, indeed without it neither philosophy nor the entire edifice of science would exist. Literature, art, philosophy, and religion explore individual meaning and shape lives and actions by supporting understanding of the highest levels of adaptive choice criteria.

4.5.7 Occurrence of Meta-Reflection

Meta-reflection by intelligent agents is made possible through symbolic reasoning. I characterize this as *intelligent top-down causation*, which I discuss in Sect. 4.7. The main issue here is whether conscious selection of selection criteria occurs only in humans, or can occur in other animal species. The consensus seems to be that only humans (or other intelligent species elsewhere in the universe) can engage in such reflection. But perhaps one should be open-minded: maybe bonobos or parrots or dolphins can think in this way? This seems rather unlikely. This may be a key fork between *Homo sapiens* and other life forms on Earth. However, it may possibly occur in the context of computer systems that have been programmed appropriately. To what degree this is possible is a key topic of debate in the context of the study of artificial intelligence. The outcome is undecided.

4.6 Complex Adaptive Systems

Because they all have an adaptive element, the last three classes of top-down causation (TD3, TD4, and TD5) are all examples of complex adaptive systems [81]. This is the only way that biological information can be generated and incorporated into living systems, and it is also the basis of learning. In this section, I look in turn at:

- The process (Sect. 4.6.1).
- Evolutionary and developmental outcomes (Sect. 4.6.2).
- Adaptive processes and learning (Sect. 4.6.3).

4.6.1 The Process

Adaptive processes [103, 104] take place when many entities interact, for example, the cells in a body or the individuals in a population, and variation takes place in the properties of these entities, followed by selection of preferred entities that are better suited to their environment or context (Sect. 4.3.1). Higher level environments provide niches that are either favorable or unfavorable to particular kinds of lower level entities. Those variations that are better suited to the niche are preserved and the others decay away. Criteria of suitability in terms of fitting the niche can be thought of as fitness criteria guiding adaptive selection. On this basis a selection agent or selector (the active element of the system) accepts one of the states and rejects the rest. This selected state is then the current system state that forms the starting basis for the next round of selection (Table 4.6).

Thus this is top-down causation from the context to the system. An equivalence class of lower level variables will be favored by a particular niche structure in association with specific fitness criteria. Unlike feedback control, this process does not attain preselected internal goals by a specific set of mechanisms or systems. Rather it creates systems that favor the meta-goals embodied in the fitness criteria. This is an adaptive process rather than a control process.

4.6.2 Evolutionary and Developmental Outcomes

It is the way new information is generated that was not present before [116], by discarding information that is irrelevant (Table 4.7). It enables emergence of complexity without dynamical attractors or specific goals guiding the process, but with an increase in complexity and embodied information, for the process searches the possible solution space in a way that is not pre-ordained and adapts to the context. The outcome is usually not predictable either from the initial conditions or the meta-goals, because of the random element involved, although both clearly influence the outcome. This underlies all life, including cells, plants, and animals, and is the basis for building up biological information—the foundational difference between physics and biology [95, 161].

It seems that developing very complex systems such as those occurring in biology requires top-down causation, needed in order to build up the necessary biological information [116, 189]: this information cannot be derived in a bottom-up way, because it necessarily implicitly embodies information about the relevant environmental niche for the organism. It would be different in a different environment.

Adaptive Selection and Environmental Context. The importance of adaptive selection is that it can let a system adapt to ongoing changes in the environment. Indeed it is the only way of doing so [94]. Thus it is the key to genuine complexity in a biological context, not only on evolutionary timescales, but also on developmental and functional timescales.

It is also the key to the way life can apparently violate the second law of thermodynamics. Adaptive selection can accumulate structure and information by selecting a subset of entities from a vast set of many variants that explores the space of possibilities, selecting only those lower level states that correspond to a higher level selection principle, thus embodying a form of top-down action. This is an analogue of Maxwell's demon: a micro-entity that chooses, from a vast ensemble, those molecules with high energy and lets them enter a reservoir, thus violating the second law of thermodynamics locally, as negligible energy is used in the selection [120]. Maxwell's demon is envisaged as a micro-being: Darwin's demon in effect envisages a macro-demon which acts down to the molecular level to select from a vast ensemble of nucleic acids a sequence encoding specific genetic information. This again enables a local violation of the second law (although it remains globally valid because of the entropy increase in the environment).

4.6.3 Adaptive Processes and Learning

More generally, the mind works by adaptive prediction of what is likely to happen, updated on an ongoing basis [97]. This underlies most of our mental ability. For example, the process of perception is a predictive adaptive process using Bayesian statistics to update the current perception on the basis of prediction errors. This includes prediction of the intention of others, which is the basis of theories of other minds.

> **Learning and Adaptive Selection.** Learning, and associated collection of new information, is not possible via bottom-up action alone, or via dynamical systems (TD1) or non-adaptive feedback control (TD2). In order for new information to be acquired, and hence in order that learning can occur, one needs adaptive selection to take place, that is, one needs TD3, TD4, or TD5.

Dynamical systems cannot achieve this. TD1 proceeds simply on the basis of information that is available at the beginning, as in (2.1), while TD2 uses comparison of updated information with goals as in (2.2). Neither generates any new information that was not there to start with. For that one needs adaptive selection TD3 as in (2.3), TD4 as in (2.4), or TD5 as in (2.5). Hence, adaptation is the key to complex behaviour.

4.7 Intelligent Top-Down Causation

Systems may be driven by the interaction of forces and particles envisaged in physics and chemistry, or the logic of structure and function that drives biology. But some, such as computers, are driven by abstract logical structures expressed in a symbolic way. Intelligent top-down causation is the case where:

- *Top-down causation takes place.* Any of the forms of top down causation TD1–TD5 occurs.
- *Symbolic representation guides what happens.* Symbolic systems play a key role in the dynamics, based on using some entity to represent something else and so allowing abstract representation of physical, biological, and mental effects and relationships.

In this section, I look in turn at:

- The nature of the process (Sect. 4.7.1).
- Language (Sect. 4.7.2).
- Other symbolic systems (Sect. 4.7.3).
- The power of symbolic thinking (Sect. 4.7.4).
- The effectiveness of abstract variables (Sect. 4.7.5).
- The mind, intention, and goals (Sect. 4.7.6).

4.7.1 The Nature of the Process

A *symbolic system* is set of structured patterns, realised in time or space, that is arbitrarily chosen by an individual or group to represent objects, states, and relationships. It will involve combinatorial principles and hierarchical structuring, and has the potential to enable quantitative as well as qualitative investigation of outcomes. Such systems include:

- **Language (Written or Spoken).** Here metaphors are key models underlying language usage.
- **Drawings and Diagrams.** Including causal diagrams and maps.
- **Analogue Computers.** Based on electric or hydraulic circuits, with relevant variables represented by currents or flows.
- **Logic and Mathematics.** Leading to logical, geometrical, and quantitative mathematical models of systems.
- **Digital Representation.** Leading to digital computer simulations.
- **Complex Mental Models and Theories.** Built up from combinations of the above. They may be supported by physical models, such as architectural models or wind tunnel models of aircraft.

They are all representational systems for abstract information (the same information can be represented in a variety of these ways). Use of such systems enables understanding of structures, processes, and their interactions, and so enables much more complex behaviour to emerge than can be attained purely by the interactions of particles and forces.

This is what characterizes intelligent thought: systems and situations are modeled in a symbolic way through use of such representation.

Intelligent Top-Down Causation. This occurs when use of symbolic systems is an essential part of any of the forms TD1–TD5 of top-down causation, based on symbolic representation and naming of entities, actions, qualities, patterns, and combinations of symbols, allowing recursion and enabling representation and logical analysis of arbitrary relations.

In particular, higher level goals and selection criteria are analysed through use of symbolic systems and then adapted to get optimal results. This use of symbols is an abstract technology that enables us to transcend the boundaries of what actually exists and consider what might be, what it might mean, and what methods to use in investigating these issues. The use of symbolic systems, particularly language, is a key characteristic of being human [46].

The fundamental point then is this:

Abstract Causally Effective Variables. Symbolic representations are abstract entities that can be causally effective in their own right, largely independent of the physical substratum whereby they are realised.

Thus maps can guide on where to go, or the same information can be given in verbal form, a printed or spoken timetable can tell us what to do next, a mathematical model can be used in aircraft design, and computers used to create visual representations of the aircraft, and a computerised architectural model is coded in digital form and can guide construction of a building through working drawings.

Symbolic representation extends to arts such as painting and sculpture, and performing arts such as drama and mime, but these are not *systems* as intended here, which have a systemic character. They will not be considered in what follows.

4.7.2 Language

Language is a symbolic system [46] with a semiotic function [185]: its purpose is to convey meaning, facts, and concepts in a social context through systematic use of symbols [53]. It represents the world of objects, actions, and qualities, as well as relationships, ideas, and theories. This representational function involves naming, indexing, and use of metaphor [118, 185]. Facts represented are both contingent (historical, geographical, and other specific features of the world and of narratives) and generic (universal patterns characterising the way it all works in general). The relation between these two features (concrete/specific and abstract/generic) is a key aspect of thought and of language, involving development of classes of entities and classification of specific instances.

4.7.2.1 Naming and Reference

The function of language is its labeling of specific and generic objects and instances, as well as abstract entities, through systematic use of words (spoken or written) [2, 3]. Via recursion (see [47] and [185, p. 244]), this referential and representational nature can allow reference to itself, and hence disjunction from physical referents. Key design features of language are [185, p. 70]:

- *Alphabetic principle* [185, pp. 52–53].
- *Arbitrariness* [15]. This allows equivalence classes of representations to exist: the same meaning can be expressed in different symbolic forms and systems.
- *Duality of patterning*. Here a small number of meaningless units are combined to produce a large number of meaningful (semantic) units. This structure is bound by strict semiotic requirements that underlie the set of possible syntactic structures [47].
- *Stimulus-freedom*. Our ability to say anything at all in any situation, so enabling discourse that is freed from the immediate situation and stimuli. This enables us to think offline, i.e., without having to act immediately on what is thought about [15].
- *Displacement*. The ability to speak about things other than here and now, enabling us to reflect on the past and consider the future [15].
- *Open-endedness*. The ability of language to say new things never said before, virtually without limit.
- *Redundancy*. The full message is entailed by part of the given text.

A key feature of the way language functions is the use of metaphor [118, 179] in the context of conceptual schemas and cultural frames [68, pp. 135–148].

4.7.2.2 Modular Hierarchical Structures

Language needs to represent complex relations, so it has a modular hierarchical structure that enables its completely flexible representational function of similarly structured features of the world: it must represent symbolized systems in an adequate way [47].

Its *implementation hierarchy* (syntax) is enabled by the structuring of language in terms of letters, words, phrases, sentences, paragraphs, chapters, and so on, which provides rules for how the elements (nouns, verbs, adjectives, adverbs, conjunctions) may be combined. It is intimately linked to recursion [185, p. 288]: "The recognition of a suitable set of syntactic categories allows us to analyse all the sentences of a language as being built up, by means of a fairly small set of rules allowing recursion, from just these few categories."

Its *logical hierarchy* (semantics) is enabled by a class structure with inheritance, whereby members of a class inherit properties of the class, and subclasses inherit properties of superclasses [22]. Thus Mitzi is a poodle, a kind of dog, which is a type of mammal, a kind of animal, and so on. Each more specific category inherits properties of the more general category: we do not have to repeat and remember separately all those specific details of the higher level class when we consider a more specific class. We take the higher level properties for granted (all animals need to eat and drink, so Mitzi needs to eat and drink).

4.7.2.3 Recursion and Naming

Crucial aspects of language are recursion, building up patterns of patterns by imbedding, and naming: identifying compound concepts as a single unit and naming them, allowing logical structures and patterns to be recognised as effective entities in their own right.

Recursion. This is the imbedding within a syntactic category in a sentence of a smaller version of the same category. For example, a noun phrase (NP) can be imbedded as if it were a noun. Similarly, one can include verb phrases (VP) and prepositional phrases (PP). This means we can repeatedly use the same construction in a sentence to build up arbitrary complexity by imbedding with sub-sentences imbedded within sentences [2, p. 82], as indicated by tree diagrams. Thus we can say:

> (A) The theory of the interaction of electric and magnetic fields led to the development of many technological devices such as television and cell phones.

Here, the first ten words and the last nine words are noun phrases. Recursion occurs in natural languages [185] and computer languages [160, p. 546–558].

The emergence of recursion in symbolic systems [47] enables abstract thought patterns to emerge [185, p. 244]:

> Recursion is pervasive in the grammars of the languages of the world, and its presence is the chief reason we are able to produce a limitless variety of sentences of unbounded length just by combining the same few building blocks.

Its development was a key aspect in the evolution of language (see [72, 96] and [29, p. 35, 172]).

Naming. The further key development is naming of such subunits: chunking smaller units together and labelling the resulting combined entity, which can then be treated as a single entity and referred to by name. Patterns of symbols are referred to by a single signifier, which then stands for that pattern. This enables us to collapse complex sentences into simpler ones. Thus we can make the definition:

> *Maxwell's theory* is the theory of the interaction of electric and magnetic fields.

This enables us to replace (A) by the simpler sentence (B):

> (B) Maxwell's theory led to the development of many technological devices such as television and cell phones.

This chunking and naming is the key to abstract thinking: it builds up a hierarchy of concepts that can be referred to and studied as entities in their own right, e.g., a triangle is a three-sided polygon, the sum of the interior angles of a triangle is 180°, and so on.

> **Recursion and Naming Is Possible**. A symbol can represent collections of symbols. This enables the power of meta-analysis and associated higher level kinds of causation.

It must be based on specific aspects of local neural connections, involving specific kinds of connections enabling naming procedures to be applied to names themselves. This type of neural connectivity, presumably involving links from higher levels of structure to lower levels, should be characterisable in the same kind of way that Hawkins [97] and Churchland [39] identify the neural bases of naming.

4.7.2.4 Multiple Realisability

At the core of language is the arbitrary nature of symbolic choice. This occurs in both the implementation and the logical hierarchies. This multiple realisability is a key feature of top-down causation: the higher logical levels are driving the lower physical and logical levels (Sect. 3.5).

Equivalence Class of Representations and Embodiment. Language is embodied via an equivalence class of physical representations. In particular it has spoken and written forms. Physical realisation of language can be neural (in an individual's brain), spoken (sound), written (visual), electronic (digital), or occur in the form of visually transmitted sign patterns (sign languages). The same logical patterns are embodied in these different representations. Meaning is embodied in an equivalence class of such surface representations: it is independent of whether language is spoken, written or signed, and independent of dialect/pronunciation and font.

They are all enabled by the physical structure of the brain, which is hierarchically structured so as to enable an interplay of sensory interpretation and prediction, based on pattern recognition, classification, memory, and extrapolation [97]. A profound ability of the mind, underlying the flexibility of language usage and representation, is to recognize them all as functionally equivalent.

Equivalence Class of Logical Structures. The same entity or action can be represented by different words in the same language, or in different languages ('dog', 'hound', 'chien', 'hund', etc.). The same concepts can be represented by different phrases ("Albert drove the car", "Albert was the car's driver", and so on). The essential concept being communicated is an abstract entity: the equivalence class of all such representations. Again the mind has the extraordinary capacity to recognise these equivalences and respond to the concepts rather than the specific representation. Features represented are recognized as entities that exist in their own right, which can be labelled and represented in many different ways.

4.7.2.5 Contextual Dependence

When used for communication purposes (the *raison d'être* of language), everything is context dependent. The parts obtain their meaning by being imbedded in the whole, which sets the overall context and reference frame for the parts, and determines what is meant by words such as 'it' and 'there' and 'then' (a restaurant in Paris in 1930). Often the meaning and even the pronunciation of words depends on context (e.g., the various meanings of the words 'plane' and 'wound').

Thus reading and language understanding are contextual processes, involving a psycholinguistic guessing game [87] that works when we share a common culture [14]. Reading and listening cannot be successfully undertaken in a purely bottom-up way. Rather (as in the case of vision [79]) expectations of what we are likely to read or hear shape to a considerable extent what we actually do read or hear. This can be shown by miscue experiments [75]. *Inter alia*, this is the reason why it is so difficult to proofread an article that one has written oneself. These matters are all pursued in further depth in Chap. 7.

4.7.3 Other Symbolic Systems

The same kinds of principles occur in all the other forms of symbolic systems we use to understand and control the world. As to the last, some symbol systems are indeed rule-based, while others are not. The mind adapts more easily to the more flexible pattern-based systems (think of the frustration of dealing with 'syntax error 147' when interacting with a computer, where replacing a comma by a full stop or misspelling a word can cause everything to grind to a halt). Other symbolic systems include:

- Diagrams and maps.
- Mathematics.
- Formal logical systems.
- Computer programs.

All these use a common set of symbolic principles.

4.7.3.1 Diagrams and Maps

Diagrams and maps can indicate various kinds of relationships:

- *Spatial relationships* by geometric drawings, construction diagrams, maps.
- *Temporal relationships* by calendars, diaries, and timelines.
- *Numerical relationships* by graphs, pie charts, bar charts.
- *Structural relationships* by structural diagrams showing networks of connections.
- *Causal relationships* by causal diagrams characterising networks of interactions.

Each has a hierarchical modular structure and can be represented in multiple physical ways and by multiple symbolic choices. Because visual thinking is a powerful form of thought [6], these are all effective ways of presenting information to the user.

4.7.3.2 Mathematics

Mathematics is a key symbolic system for developing science and technology. It has been characterised as the 'science of patterns' [49, 176], representing quantitative relations by equations and inequalities. Thus it involves the following:

- *Quantities*. Numbers (magnitude), vectors (directions), tensors, dimensions.
- *Variables*. Quantities that can have arbitrary values and vary in space and time.
- *Functions of variables*. Relations between them.
- *Operations on variables and functions*. Addition, multiplication, division, differentiation, integration, determining new variables and functions from old.
- *Equations and inequalities*. Specifying relations between variables and functions.
- *Spaces of variables and functions with specified properties*.

This creates hierarchical modular structures and enables mathematical models using equations to represent the dynamics of physical entities. Numerical analysis can be used to determine outcomes of equations by repeated numerical operations.

4.7.3.3 Formal Logical Systems

Abstract logical notation and argumentation gives a formal kind of reasoning that is strictly logical. It is based on the use of logical variables (A, B, etc.), combinatorial operations (AND, OR, NOT) and demarkers (brackets of various kinds), equality and implication symbols ($=$, $\Rightarrow$), truth values (TRUE, FALSE), and existential qualifications ($\exists$, $\forall$). In combination, they lead to new identities and truth tables, with the brackets playing a crucial role in terms of denoting operator range and chunking things together.

Logic can be used to represent and examine causal relations in general. For example, Boolean logic is used in the design of computer logic circuits [127, pp. 53–166].

4.7.3.4 Computer Programs

Computer programs are formal symbolic systems powering digital computers. The procedures described in the program will run algorithmically on the basis of the initial data (entered as initial values of the variables) to produce output data (the final values of the variables) by repeated application of logical operations on variables (Chap. 2).

The fundamentals of a computer language are [1, p. 4]:

- *Primitive expressions* which represent the simplest entities with which the language is concerned.
- *Means of combination* by which compound expressions are built from simpler ones.

- *Means of abstraction* by which compound expressions can be named and treated as a unit.
- *Rules for procedures* by which data are manipulated.

The programs will have a hierarchical modular structure, as will the data, and will be realisable in multiple ways. This is discussed in depth in Chap. 2.

4.7.3.5 The Common Principles

In addition to the basic features of symbolic systems of representing one entity via an arbitrarily assigned other entity (Sect. 4.7.1), which implies the multiple realisability of symbolic systems (as in the case of language, see Sect. 4.7.2), two major principles apply to all these examples:

- **Combination and Naming**. Rules of combination enable one to build up new logical entities labelled by some name, say E, out of logical components labelled by their names, say e_i. Because the new entity can be identified as such by its name, it can be referred to by that name and treated as a logical entity in its own right, with its own structure and rules of behaviour that result from the lower level structures and rules of behaviour.

This is the basic principle by which complex logic can be built up, leading to modular hierarchical structures (as in the case of language, see Sect. 4.7.2). It occurs in all the examples above.

- **Top-Down Effects**. Contextual dependence of behaviour of lower level entities takes place either via explicit or implicit parameter or variable passing from a higher logical level to a lower level, thus affecting the lower level referential meaning and dynamics, or by 'IF ...THEN ...ELSE' and 'WHILE ...THEN ...' logic which controls the branching of lower level dynamics.

These are ways that contextual effects occur in logical systems. In particular, they allow adaptive selection to occur, for that comes about via logical branching taking place in the context of causally effective contextual variables.

4.7.4 *The Power of Symbolic Thinking*

Symbolic top-down causation from the mind to the world underlies all human planning and action (Sect. 4.7.6). It is indeed the way the human mind alters the world, and has led to all the achievements of technology and engineering that underlie the rise of civilisation [27].

The key feature of this higher level of causation is its use of language and abstract symbolism [46]. These are irreducible higher level variables (of an abstract nature, since they form equivalence classes of representations), but are causally effective by top-down action from the brain to cells in our muscles [137]. This enables information to be stored and retrieved, classified and selected as relevant or discarded, processed in the light of other information, and used to make qualitative and quantitative projections of outcomes and plan future actions in a rational way (see [98] and [173, Chap. 4]), altering goals according to an intelligent understanding of past experiences and future expectations. Intentional action [109] then enables one to implement the resulting plans, and so change the physical world.

Symbolic thinking enables offline planning, symbolic representation of concrete situations allowing analysis of underlying patterns and structures, and rule-based procedures such as algorithms. But particularly it allows us to unpick the causal features of physics and chemistry that underlie the way things work. This is what has led to the power of technology, which has transformed the world.

But where does all this success derive from? It comes from the effectiveness of such symbolic systems, with the properties outlined above, in capturing the essence of causal processes that shape what happens. This has been much discussed in the context of mathematics, in the context of Wigner's famous question [195] as to why mathematics is so effective for understanding physical processes. But this power of symbolic systems extends much more widely to symbolic systems in general, as indicated in this chapter, through the interaction between mind, matter, and mathematics [149], extended to generic symbolic systems, enabling human organisation and attainment of desired goals.

The Causal Power of Symbolic Systems. Images and formal and informal causal models of the natural and social worlds, ranging from mental images of what might happen to elaborate quantitative models of physical entities and societies, derive from their success as models of what happens. Through their structure they enable us to understand and predict with success because these abstract structures mirror key aspects of the world around us with great precision.

These abstract entities (which are shared among many minds) play a large part in formulating our understanding and consequent actions, and hence are causally effective in the real world as they help us to attain our goals.

4.7.5 The Effectiveness of Abstract Variables

There are various kinds of abstract entities that are causally effective in top-down causation, both in the case of mental causation and more generally. They include goals in feedback control systems and selection criteria in adaptive systems. In the case of the mind, they include conscious goals and plans, abstract theories, social constructions, and ethical values.

Intrinsically Higher Level Variables. The key element here is that these high level variables cannot even in principle be determined by coarse-graining of lower level variables. As a simple example, Maxwell's theory of electromagnetism is a human construction that is causally effective because it accurately models the nature of reality. Historically, it has led to the existence of radio, TV, cellphones, and so on, an undeniable demonstration of causal efficacy. It is inconceivable that somehow coarse-graining or partitioning of any lower level variables whatever will lead to this theory, or indeed that any bottom-up process can account for its existence or nature. It cannot emerge spontaneously from atoms and molecules, *inter alia* because its ultimate source is an abstract Platonic space of possibilities, apprehended by the brain [39].

4.7.6 The Mind, Intention, and Goals

Similarly to an organization, goals guide action in an individual, and hence in society. A great deal of individual and social activity is to act so as to attain those goals. They are the core of intentional activity, for they are what lead to specific actions ("I'm going to make a cup of tea", "I'm going to take a holiday in China", "We are going to sending an American safely to the Moon before the end of the decade").

Conscious Goals in Human Activity. Our actions are governed by hierarchically structured goals. They occur at all structural levels in society (individuals, families, groups, a society as a whole). They may be explicit or implicit, qualitative or quantitative.

These goals are structured as a logical hierarchy, effective through the mind and brain, which form the implementation hierarchy. This dual system is similar to the case of computers (Sect. 2.3). They are not physical quantities (see Sect. 4.2.6), but can be represented in many ways, so they effectively constitute an equivalence class of representations, reflecting the fact that they enable top-down causation (Sect. 3.5). What happens is based on the following:

- The causal efficacy of mental goals.
- A logical hierarchy.
- An implementation hierarchy.

4.7.6.1 Causal Efficacy of Mental Goals

Our goals cause real physical change in the world. I will give just one example:

Aircraft Design. Plans for a Jumbo Jet aircraft result in billions of atoms being deployed to create the aircraft in accordance with those plans. This is a non-trivial example: it costs a great deal of money to employ experts in aerodynamics, structures, materials, fuels, lubrication, controls, etc., to design and then to manufacture the aircraft in accordance with those plans, which comprise a set of goals for what will be made.

The plan itself is not equivalent to any single person's brain state. It is an abstract hierarchically structured equivalence class of representations which can be represented in many ways, viz., spoken, drawn, or given in abstract specifications, digitally in computers, in brains, etc. It is the equivalence class of all such representations together that comprise the design.

This logical structure is hierarchically organised. The plane consists of a fuselage, engines, control systems, etc. Each of those consists of parts: rotors, fuel systems, lubrication systems, etc. Each of those consists of parts: electric engines, wiring, computers, hydraulic links, etc. And so it goes. The logical plan specifies what each of them should be. The implementation plan specifies how, where, and when each of them will be made and by whom.

The entire plan is a set of intricately linked logic and implementation goals. It is clearly causally effective: the aircraft would not exist without it. It in effect acts top-down from the mind to the level of atoms and particles via all the intermediate levels of engines and wings, compressor blades and electrical motors, materials and screws, and all the other elements that make up an aircraft. There is no way they can spontaneously assemble to make an aircraft by any bottom-up process. I return to this in Sect. 7.5.3.

4.7.6.2 The Logical Hierarchy

The logical hierarchy sets out the lower level goals that must be met to attain each higher level goal: to make a cake I need a recipe with a list of ingredients, to get the missing ingredients I need to go to the shop, to go to the shop I need to catch the bus, to catch the bus I need to go to the bus stop, and so on. Similar to the case of organisations, at the highest level is a purpose that drives the rest: ("I want to lift up the poor", "I want to make lots of money"). This higher level purpose cascades down to shape and constrain all lower level goals.

Their Origin. They are adaptively formed by choice in response to experience: learning takes place in particular physical and social contexts where the mind responds to events and to the meaning of symbols. This is discussed in Sects. 4.4–4.7.

4.7.6.3 Implementation Hierarchy

The brain is the physical system that implements the decisions of the mind, making intentional action possible. The goals we have result in decisions being taken that lead to action potentials in motor neurons flowing from the motor cortex to muscles [89, p. 139]. Regulatory drives help preserve homeostasis [89, p. 187], while purposive activity seeks to attain conscious goals through a feedback process of comparing goals with images of the expected future based on current actuality, and adjusting our actions to increase the chances of attaining our goals. It is these abstract images that are causally effective [24, pp. 3, 61]:

> It is images of the future which determine present behaviour through the process of decision [...] what all decision-makers are deciding about are alternative images of the future in their own minds.

So the situation is:

$$(\text{goals}) \underbrace{\Longrightarrow}_{\text{images of future}} (\text{outcomes})\,. \tag{4.34}$$

Thus both the goals and the images are causally effective. The way this happens is the topic of Chap. 8.

Tools we use to help organize our actions via a structured system of goals are less formal versions of those used in organisations, as discussed above: diaries, calendars, to-do lists, organizers, household budgets, etc. These are physical extensions of our minds that increase our effectiveness in attaining chosen goals.

4.7.6.4 Conscious Goals and Plans

Through the mind, abstract entities have causal effects by its goal choices (Sect. 4.7.6). The symbolic representation and choice of goals entails the causal efficacy of abstract entities such as action plans, the theory of the laser, social agreements such as the value of money, and ethical value systems. They are implemented through social institutions, which play a key role in society [167].

Plans of What to Do. Ideas in the mind that lead to action plans and associated goals are causally effective. For example, recipes for a cake, plans for a fete, or plans for an aircraft (see Sect. 4.7.6). When a human being has a plan in mind (say a proposal for a bridge being built) and this is implemented, then enormous numbers of microparticles (comprising the protons, neutrons, and electrons in the sand, concrete, bricks, etc. that become the bridge) are moved around as a consequence of this plan and in conformity with it. Thus in the real world, the detailed micro-configurations of many objects (which electrons and protons go where) is determined by the plans humans have for what will happen, and the way they implement them. Human choices based on self-reflective intelligence and imagination are thus causally effective, allowing anticipation and intelligent planning and design based on abstract models of reality [173, Chaps. 5and6].

Example: Timetables. A timetable for an airline determines when the aircraft fly, in a more or less reliable way. It results in the aircraft flying on a particular path at a particular time, resulting in particular passengers arriving at particular destinations st specified times, courtesy of the many billions of atoms that comprise the structure of the aircraft. How do we demonstrate top-down causation? Change the timetable and different patterns of travel will result. The timetable is not a physical thing: it is an abstract structure that can be represented in many forms. It can be spoken about, printed on paper, displayed on an airport screen, or stored in digital form in a computer.

Physics can describe the material out of which the printed representation of the timetable is made and the ink markings on the paper, but it cannot comprehend the causal chain by which this leads to particular aircraft flying at particular times. The relevant variables (the entries in the timetable) belong to an irreducible equivalence class of abstract entities coding information that controls what happens in the real world. When you choose a flight, you adaptively select the one that works best for you.

Example: Physics Experiments. Every physics experiment is an example of intelligent top-down causation. The experimenter manipulates the world so as to create a repeatable experiment, e.g., collisions between subatomic particles that are then measured with precision. These collisions would not take place without the active intervention of the experimenter in accordance with his mental plans, so this is top-down action from the human brain to sub-microscopic scales.

4.7.6.5 Social Agreements and Understandings

Social agreements are abstract entities that govern social life. They come into being by a complex process of negotiation, and thereafter structure what happens. They include regulations and laws, the constitution of a voluntary society, an employment contract, and so on.

Example: The Value of Money. Physically, money is just coins or pieces of paper with patterned marks on them. This does not explain its causal significance. The effectiveness of money, which can cause physical change in the world such as the construction of buildings, roads, bridges, and so on, by top-down action of the mind on material objects, is based on social agreements that lead to the value of money (pricing systems) and exchange rates. These are abstract entities arising from social interaction over an extended period of time, and are neither the same as individual brain states, nor equivalent to an aggregate of current values of any lower level variables (although they are causally effective through such states and variables).

Roles, Frames, and Expectations. Social roles are socially determined abstract entities that are causally effective in structuring society. They are a key aspect of the causal power of social structures [61]. Roles are developed by an adaptive process which is a combination of bottom-up and top-down interaction between society and the individuals who make up the society. They are then inculcated into the individual by top-down social processes [14], whereafter they become a core feature of individual psychology in relation to society, together with expectations guiding the choice of goals and actions and hence being causally effective in a top-down way from the mind to the body.

Expectations arise both from the nature of roles and from social frames where a set pattern of interactions is expected. For example, entering a restaurant as a customer, one expects the waiter to produce a menu, take an order, bring the food, and so on. These are informal patterns of behaviour that guide our actions and make life predictable.

4.7.6.6 Abstract Theories

Abstract theories are non-physical entities that can have enormous causal power.

Physics Theories. Maxwell's theory of electromagnetism (an abstract entity, described by Maxwell's equations) led to the development of radio, cell phones, TV, and so on. It can be represented in many different ways: as 3D vector equations, 4D tensor equations, via variational principles, and so on. These are all causally efficient: they affect the nature of physical objects in the world. It is shown to be true by experiment and by its technological outcomes. Maxwell's theory is not the same as any single person's brain state: individuals can die but the theory lives on in books, in other peoples' brains, and in computer programs. It is an irreducible higher level causal factor (it cannot be derived by coarse-graining any lower level variables). The abstract theory has altered physical configurations in the real world, and hence is causally effective, through being realised in neuronal structures. The origin of such theories will be considered in Chap. 8.

4.7.6.7 Values

Values are abstract entities that are causally effective by shaping all the other goals of an individual, selecting those that are desirable from those that are not. They are what ultimately shape our lives, as discussed in Sect. 4.5.6.

In society, roles and practices embody social values, which, together with individual values relating to life purpose, guide the individual and communal choice of goals and the methods used to attain these goals. They are the ultimate adaptive selection criteria that form the framework for all the rest (Sect. 4.5.5).

4.7.6.8 Physicalism and Causation

These examples show that it is not the case that the only entities with causal powers are physical things such as particles, forces, and physical fields. A variety of non-physical entities shape what happens in complex systems. They have a causal effect on outcomes, as can be demonstrated by changing their nature and hence altering outcomes:

- Change the exchange rate of money and different things will happen in the economy.
- Change the building byelaws and shopping centres will arise next to ecologically important wetlands.
- Change the airline timetable and aircraft will fly at different times.

Physics enables all this to happen, but does not determine the outcome. That is shaped by the various forms of top-down causation discussed in this chapter, each of which acts in a rather different way to the others.

References

1. H. Abelson, G.J. Sussman, J. Sussman, *Structure and Interpretation of Computer Programs* (MIT Press, 1996)
2. J. Aitchison, *Linguistics* (Hodder and Stoughton, Seven Oaks, 1987)
3. J. Aitchison, *Words in the Mind* (Wiley-Blackwell, Chichester, 2012)
4. J.M. Alexander, Evolutionary game theory, in *The Stanford Encyclopedia of Philosophy*, Fall 2009 edn. ed. by E.N. Zalta (2009). http://plato.stanford.edu/archives/fall2009/entries/game-evolutionary/
5. U. Alon, *An Introduction to Systems Biology: Design Principles of Biological Circuits* (Chapman and Hall /CRC, London, 2007)
6. R. Arnheim, *Visual Thinking* (University of California Press, Berkeley, 1969)
7. W.R. Ashby, *Design for a Brain*. See *Feedback, Adaptation and Stability: Selected Passages from Design for a Brain* (1960). http://www.panarchy.org/ashby/adaptation.1960.html
8. K. Astrom, *Adaptive Control* (Dover, 2008)
9. G. Auletta, G.F.R. Ellis, L. Jaeger, Top-down causation: from a philosophical problem to a scientific research program. J. R. Soc. Interface **5**, 1159–1172 (2008). arXiv:0710.4235
10. A.-L. Barabási, Z.N. Oltvai, Network biology: understanding the cell's functional organization. Nat. Rev. Genet. **5**, 101–114 (2004)
11. S. Beer, *Decision and Control* (Wiley, New York, 1966)
12. S. Beer, *Brain of the Firm* (Wiley, Chichester, 1981)
13. C.H. Bennett, Notes on Landauer's principle, reversible computation and Maxwell's demon. Stud. Hist. Philos. Mod. Phys. **34**, 501–510 (2003)
14. P. Berger, T. Luckmann, *The Social Construction of Reality: A Treatise in the Sociology of Knowledge* (Anchor, New York, 1967)
15. D. Bickerton, *Language and Human Behaviour* (University of Washington Press, Seattle, 2001)
16. M.H. Bickhard, D.T. Campbell, Variations in variation and selection: the ubiquity of the variation-and-selective-retention ratchet in emergent organizational complexity. Found. Sci. **8**, 215–282 (2003)
17. J. Binney, S. Tremain, *Galactic Dynamics* (Princeton University Press, Princeton, 1987)
18. E. Birney, Come fly with us. Nature **450**, 184–185 (2007)
19. C.M. Bishop, *Neural Networks for Pattern Recognition* (Oxford University Press, Oxford, 1999)
20. R.C. Bishop, Fluid convection, constraint and causation. Interface Focus **2**, 4–12 (2012)
21. F. Black, M. Scholes, The pricing of options and corporate liabilities. J. Polit. Econ. **81**, 637–654 (1973)
22. G. Booch, *Object-Oriented Analysis and Design with Applications* (Addison-Wesley, Menlo Park, 2007)
23. Booz, Allen, and Hamilton: Earned Value Management Tutorial Module 2: Work Breakdown Structure, US Department of Energy: Office of Science, Tools and Resources for Project Management. http://science.energy.gov/opa/project-management/tools-and-resources/
24. K.E. Boulding, *Economics as a Science* (McGraw Hill, New York, 1970)
25. V. Brattka, *Computability Theory* (University of Cape Town Notes, 2011)
26. S.C. Brenner, L.R. Scott, *The Mathematical Theory of Finite Element Methods* (Springer, Heidelberg, 2007)
27. J. Bronowski, *The Ascent of Man* (London: BBC Books, 1973). (reprint 2011)
28. W. Brown, N. Murphy, *Did My Neurons Make Me Do it? Philosophical and Neurobiological Perspectives on Moral Responsibility and Free Will* (Oxford University Press, New York, 2007)
29. R. Burling, *The Talking Ape: How Language Evolved* (Oxford University Press, Oxford, 2007)
30. F.M. Burnet, *The Clonal Selection Theory of Acquired Immunity* (Cambridge University Press, Cambridge, 1959)

31. R. Burns, To a mouse, on turning her up in her nest with the plough, in *Kilmarnock Volume* (John Wilson, Kilmarnock, 1785)
32. J.R. Busemeyer, Dynamic systems, in *Encyclopedia of Cognitive Science* (Macmillan, 2011). http://ebookbrowse.com/busemeyer-03-pdf-d282315897
33. E. Callaway, Behaviour genes unearthed: speedy sequencing underpins genetic analysis of burrowing in wild oldfield mice. Nature **493**, 284 (2013)
34. D.T. Campbell, Downward Causation, in *Studies in the Philosophy of Biology: Reduction and Related Problems*, ed. by F.J. Ayala, T. Dobhzansky (University of California Press, Berkeley, 1974)
35. D.T. Campbell, Evolutionary epistemology, in *Evolutionary Epistemology, Rationality, and the Sociology of Knowledge* (Open Court Publishing, 1987), pp. 47–89
36. N.A. Campbell, J.B. Reece, *Biology* (Benjamin Cummings, San Francisco, 2005)
37. G.A. Carpenter, S. Grossberg, *Pattern Recognition by Self-Organising Neural Networks* (MIT Press, Cambridge, Mass, 1991)
38. S. Carroll, *From Eternity to Here: The Quest for the Ultimate Arrow of Time* (Dutton, New York, 2010)
39. P. Churchland, *Plato's Camera* (MIT Press, Cambridge, Mass, 2012)
40. S. Conway Morris, *Life's Solution: Inevitable Humans in a Lonely Universe* (Cambridge University Press, Cambridge, 2005)
41. R. Courant, D. Hilbert, *Methods of Mathematical Physics II* (Wiley Interscience, New York, 1962)
42. F. Crick, *Astonishing Hypothesis: The Scientific Search for the Soul* (Scribner, 1995)
43. A. Damasio, *Descartes' Error: Emotion, Reason, and the Human Brain* (Avon Books, New York, 1995)
44. A. Damasio, *The Feeling of What Happens: Body, Emotion, and the Making of Consciousness* (Harcourt, New York, 1999)
45. P. Dayan, L. Abbot, *Theoretical Neuroscience: Computational and Mathematical Modelling of Neural Systems* (MIT Press, Cambridge, Mass, 2001)
46. T. Deacon, *The Symbolic Species: The Co-Evolution of Language and the Human Brain* (Penguin, London, 1997)
47. T. Deacon, Universal grammar and semiotic constraints, in *Language Evolution*, ed. by M. Christiansen, S. Kirby (Oxford University Press, Oxford, 2003), pp. 111–139
48. R.L. Devaney, *An Introduction to Chaotic Dynamical Systems* (Basic Books, 2003)
49. K. Devlin, *Mathematics: The Science of Patterns: The Search for Order in Life, Mind and the Universe* (Scientific American Paperback Library, 1996)
50. E. de Waal, C. Borgeaud, A. White, Potent social learning and conformity shape a wild primate's foraging decisions. Science **26**, 483–485 (2013)
51. J.J. DiStefano, A.R. Stubberud, I.J. Williams, *Feedback and Control Systems (Schaum's Outlines)* (McGraw Hill, New York, 1995)
52. S. Dodelson, *Modern Cosmology* (Academic Press, San Diego, 2003)
53. M. Donald, *Origins of the Modern Mind* (Harvard University Press, 1991)
54. R. Dunbar, *Human Evolution* (Pelican Books, London, 2014)
55. A. Durrant, *Quantum Physics of Matter* (Institute of Physics and the Open University, Bristol, 2000)
56. G.M. Edelman, *Neural Darwinism: The Theory of Group Neuronal Selection* (Oxford University Press, Oxford, 1989)
57. G.M. Edelman, *Brilliant Air, Brilliant Fire: On the Matter of Mind* (Basic Books, New York, 1992)
58. G.M. Edelman, G. Tononi, *Consciousness: How Matter Becomes Imagination* (Penguin Books, London, 2001)
59. W. Edwards, Conservatism in human information processing, in *Judgement under Uncertainty: Heuristics and Biases*, ed. by D. Kahneman, P. Slovic, A. Tversky (Cambridge University Press, Cambridge, 1982), pp. 359–361

60. M. Eigen, P. Schuster, *The Hypercycle: A Principle of Natural Organisation* (Springer, Berlin, 1979)
61. D. Elder-Vass, *The Causal Power of Social Structures: Emergence, Structure and Agency* (Cambridge University Press, Cambridge, 2010)
62. G.F.R. Ellis, On the nature of causation in complex systems. Trans. R. Soc. S. Afr. **63**, 69–84 (2008)
63. G.F.R. Ellis, Commentary on 'An Evolutionarily Informed Education Science' by David C Geary. Educ. Psychol. **43**, 206–213 (2008)
64. G.F.R. Ellis, Top-down causation and emergence: Some comments on mechanisms. J. R. Soc. Interface Focus **2**, 126–140 (2012). http://rsfs.royalsocietypublishing.org/content/2/1/126.full.pdf+html
65. G.F.R. Ellis, On the limits of quantum theory: contextuality and the quantum-classical cut. Ann. Phys. **327**, 1890–1932 (2012)
66. G.F.R. Ellis, The arrow of time and the nature of spacetime. Stud. Hist. Philos. Sci. Part B Stud. Hist. Philos. Mod. Phys. **44**, 242–262 (2013). arXiv:1302.7291, http://www.mth.uct.ac.za/~ellis/Quantum_arrowoftime_gfre.pdf
67. G.F.R. Ellis, J. Toronchuk, Neural development: affective and immune system influences, in *Consciousness and Emotion: Agency, Conscious Choice, and Selective Perception*, ed. by R.D. Ellis, N. Newton (John Benjamins, 2005), pp. 81–119
68. J. Feldmann, *From Molecule to Metaphor: A Neural Theory of Language* (MIT Press, Cambridge, Mass, 2006)
69. R.P. Feynman, R.B. Leighton, M. Sands, *The Feynman Lectures on Physics: Mainly Mechanics, Radiation, and Heat* (Addison-Wesley, Reading, Mass, 1963)
70. R.P. Feynman, R.B. Leighton, M. Sands, *The Feynman Lectures on Physics: The Electromagnetic Field* (Addison-Wesley, Reading, Mass, 1963)
71. R.P. Feynman, R.B. Leighton, M. Sands, *The Feynman Lectures on Physics: Quantum Mechanics* (Addison-Wesley, Reading, Mass, 1965)
72. W.T. Fitch, M.D. Hauser, N. Chomsky, The evolution of the language faculty: clarifications and implications. Cognition **97**, 179–210 (2005)
73. D. Fleisch, *A Student's Guide to Maxwell's Equations* (Cambridge University Press, Cambridge, 2008)
74. R.L. Flood, E.R. Carson, *Dealing with Complexity: An Introduction to the Theory and Application of Systems Science* (Plenum Press, London, 1990)
75. A.D. Flurkey, E.J. Paulsen, K.S. Goodman, *Scientific Realism in Studies of Reading* (Laurence Erlbaum, New York, 2008)
76. B. Flyvbjerg, T. Landman, S. Schram, *Real Social Science: Applied Phronesis* (Cambridge University Press, Cambridge, 2012)
77. E. Fox Keller, *The Century of the Gene* (Harvard University Press, Cambridge, Mass, 2000)
78. V. Frankl, *Man's Search for Meaning* (Beacon Press, 2006)
79. C. Frith, *Making up the Mind: How the Brain Creates Our Mental World* (Blackwell, Malden, 2007)
80. A. Gardner, The Price equation. Curr. Biol. **18**, R198 (2008)
81. M. Gell-Mann, *The Quark and the Jaguar: Adventures in the Simple and the Complex* (Abacus, London, 1994)
82. P. Glendinning, *Stability, Instability and Chaos: An Introduction to the Theory of Non-Linear Differential Equations* (Cambridge University Press, Cambridge, 1996)
83. S.F. Gilbert, *Developmental Biology* (Sinauer Associates, Sunderland, MA, 2006)
84. S. Gilbert, D. Epel, *Ecological Developmental Biology* (Sinauer, 2009)
85. J.H. Gillespie, *Population Genetics* (Johns Hopkins University Press, Baltimore, 2004)
86. G. Gigerenzer, *Adaptive Thinking: Rationality in the Real World* (Oxford University Press, Oxford, 2000)
87. K.S. Goodman, Reading: a psycholinguistic guessing game, in *Language and Literacy: The Selected Writiings of Kenneth Goodman*, vol. 1, ed. by F.V. Gollasch (Routledge and Kegan Paul, London, 1967), pp. 33–44

88. A. Gopnik, A.N. Meltzhoff, P.K. Kuhl, *The Scientist in the Crib* (Harper Collins, New York, 1999)
89. P. Gray, *Psychology* (Worth Publishers, New York, 2011)
90. R.J. Greenspan, *An Introduction to Nervous Systems* (Cold Spring Harbor Laboratory Press, Cold Spring Harbor, 2007)
91. R.W. Griffin, *Management* (Houghton Mifflin, Boston, 1987)
92. A.C. Guyton, *Basic Human Physiology: Normal Function and Mechanisms of Disease* (W B Saunders, Philadelphia, 1977), pp. 4–5
93. J. Haidt, S. Kesebir, Morality, in *Handbook of Social Psychology*, ed. by S. Fiske, D. Gilbert, G. Lindzey (Wiley: Hoboken, NJ, 2010)
94. T. Harford, *Adapt: Why Success Always Starts with Failure* (Abacus, London, 2011)
95. L.H. Hartwell, J.J. Hopfeld, A.W. Murray, From molecular to systems biology. Nature **402**, C47–C52 (1999)
96. M.D. Hauser, N. Chomsky, W.T. Fitch, The faculty of language: what is it, who has it, and how did it evolve? Science **298**, 1569–1579 (2002)
97. J. Hawkins, *On Intelligence* (Holt Paperbacks, New York, 2004)
98. J.C. Higgins, *Information Systems for Planning and Control: Concepts and Cases* (Edward Arnold, London, 1976)
99. A.L. Hodgkin, A.F. Huxley, A quantitative description of membrane current and its application to conduction and excitation in nerve. J. Physiol. **117**, 500–544 (1952)
100. M. Hoey, *Lexical Priming: A New Theory of Words and Language* (Routledge, London, 2005)
101. P.M. Hoffmann, *Life's Ratchets: How Molecular Machines Extract Order from Chaos* (Basic Books, New York, 2012)
102. D.R. Hofstadter, *Godel, Escher, Bach: An Eternal Golden Braid* (Penguin books, Harmandsworth, 1980)
103. J.H. Holland, *Adaptation in Natural and Artificial Systems* (MIT Press, Cambridge, Mass, 1992)
104. J.H. Holland, *Hidden Order: How Adaptation Builds Complexity* (Basic Books, New York, 1995)
105. C.J. Isham, *Lectures on Quantum Theory: Mathematical and Structural Foundations* (Imperial College Press, London, 1995)
106. R. Jaeger, *Microelectronic Circuit Design* (McGraw-Hill, 1997)
107. J.C. Jackson, *Classical Electrodynamics* (Wiley, New York, 1967)
108. D. Jones, The ritual animal. Nature **493**, 470–472 (2013)
109. A. Juarrero, *Dynamics in Action: Intentional Behaviour as a Complex System* (MIT Press, Cambridge, Mass, 2002)
110. M. Kac, Can one hear the shape of a drum? Am. Math. Mon. **73**, 1–23 (1966)
111. D. Karnopp, R. Rosenberg, *Systems Dynamics: A Unified Approach* (Wiley Interscience, New York, 1975)
112. S.A. Kauffman, *The Origins of Order: Self-Organisation and Selection in Evolution* (Oxford, New York, 1993)
113. E.R. Kandel, J.H. Schwartz, T.M. Jessell, *Principles of Neuroscience* (McGraw Hill, New York, 2000)
114. R.E. Kingsley, *Concise Text of Neuroscience* (Lippincot Williams and Wilkins, 1996)
115. D.E. Knuth, *Selected Papers on Design of Algorithms* (Center for the Study of Language and Information, Stanford, California, 2010)
116. B.-O. Kuppers, *Molecular Theory of Evolution* (Springer, Berlin, 1985)
117. B.-O. Kuppers, *Information and the Origin of Life* (MIT Press, Cambridge, Mass, 1994)
118. G. Lakoff, M. Johnson, *Metaphors We Live by* (University of Chicago Press, Chicago, 1980)
119. R. Landauer, Irreversibility and heat generation in the computing process. IBM J. Res. Dev. **5**, 183–191 (1961)
120. H.S. Leff, A.F. Rex (eds.), *Maxwell's Demon: Entropy, Information, Computing* (Adam Hilger, Bristol, 1990)

121. L. Liebenberg, *The art of tracking: The origin of science* (David Philip, Cape Town, 2001) http://www.cybertracker.org/downloads/tracking/The-Art-of-Tracking-The-Origin-of-Science-Louis-Liebenberg.pdf
122. E. Lloyd, Units and levels of selection, in *Stanford Encyclopedia of Philosophy* (2005). http://plato.stanford.edu/archives/fall2005/entries/selection-units/
123. J.F. Longres, *Human Behavior in the Social Environment* (F E Peacock Publishers, Itasca, Ill, 1990)
124. E.N. Lorenz, Deterministic nonperiodic flow. J. Atmos. Sci. **20**, 130–141 (1963)
125. A.J. Lotka, *Elements of Mathematical Biology* (Dover, New York, 1956)
126. J. MacCormack, *9 Algorithms that Changed the Future: The Ingenious Ideas that Drive Today's Computers* (Princeton University Press, Princeton, 2012)
127. M.M. Mano, C.R. Kime, *Logic and Computer Design Fundamentals* (Pearson/Prentice Hall, 2008)
128. M. Martinez, A. Moya, Natural selection and multi-level causation. Philos. Theory Biol. **3**(2), 1/16–16/16 (2011)
129. R. Matzner, T. Rothman, G.F.R. Ellis, Conjecture on isotope production in the Bianchi cosmologies. Phys. Rev. D **34**, 2926–2933 (1986)
130. G. McGhee, *Convergent Evolution: Limited Forms Most Beautiful* (MIT Press, Cambridge, Mass, 2011)
131. J. Maynard Smith, *Mathematical Ideas in Biology* (Cambridge University Press, Cambridge, 1968)
132. D. Meadows, *Thinking in Systems: A Primer* (Chelsea Green, White River Junction, 2008)
133. R.E. Michod, *Darwinian Dynamics: Evolutionary Transitions in Fitness and Individuality* (Princeton University Press, 1999)
134. J.H. Milsum, *Biological Control Systems Analysis* (McGraw Hill, 1966)
135. M. Mitchell, *An Introduction to Genetic Algorithms* (MIT Press, Cambridge, Mass, 1998)
136. P.W.G. Morris, *The Management of Projects* (Thomas Telford, 1994)
137. N. Murphy, Emergence and mental causation, in *The Re-Emergence of Emergence*, ed. by P. Clayton, P. Davies (Oxford University Press, Oxford, 2006), pp. 227–243
138. N. Murphy, G.F.R. Ellis, *On the Moral Nature of the Universe* (Fortress Press, Minneapolis, 1995)
139. J.D. Murray, *Mathematical Biology* (Springer, 1990)
140. D. Noble, From the Hodgkin-Huxley axon to the virtual heart. J. Physiol. **580**, 15–22 (2007)
141. D. Noble, A theory of biological relativity: no privileged level of causation. Interface Focus **2**, 55–64 (2012)
142. M.A. Nowak, C.E. Tarnita, E.O. Wilson, The evolution of eusociality. Nature, **466**, 1057–1062 (2010). For further information http://www.ped.fas.harvard.edu/SI.pdf
143. S. Okasha, *Evolution and the Levels of Selection: Toward a Broader Conception of Theoretical Biology* (Oxford University Press, 2006)
144. J. Panksepp, *Affective Neuroscience: The Foundations of Human and Animal Emotions* (Oxford University Press, 1998)
145. J. Panksepp, The neuro-evolutionary cusp between emotions and cognitions: implications for understanding consciousness and the emergence of a unified mind science. Evol. Cogn. **7**, 141–149 (2001)
146. J. Panksepp, L. Biven, *The Archaeology of Mind: Neuroevolutionary Origins of Human Emotions* (W W Norton and Company, 2012)
147. S. Patterson, *The Quants: How a New Breed of Math Whizzes Conquered Wall Street and Nearly Destroyed it* (Crown Business, New York, 2010)
148. A.R. Peacocke, *An Introduction to the Physical Chemistry of Biological Organization* (Oxford University Press, Oxford, 1989)
149. R. Penrose, *The Large, the Small and the Human Mind* (Cambridge University Press, Cambridge, 1997)
150. R. Penrose, *Cycles of Time: An Extraordinary New View of the Universe* (Knopf, New York, 2011)

151. T.J. Peters, R.H. Waterman, *In Search of Excellence* (Warner Books, New York, 1982)
152. G.R. Price, Selection and covariance. Nature **227**, 520–521 (1970)
153. J.S. Price, Hypothesis: the dominance hierarchy and the evolution of mental illness. Lancet **2**, 243–246 (1967)
154. H. Pringle, The roots of human genius are deeper than expected, in *Scientific American* (2013). http://www.scientificamerican.com/article.cfm?id=the-origin-human-creativity-suprisingly-complex
155. A. Pross, *What Is Life? How Chemistry Becomes Biology* (Oxford University Press, Oxford, 2012)
156. R. Rhoades, R. Pflanzer, *Human Physiology* (Saunders College Publishing, Fort Worth, 1989)
157. P.J. Richerson, R. Boyd, *Not by Genes Alone: How Culture Transformed Human Evolution* (University of Chicago Press, Chicago, 2005)
158. D.S. Riggs, *The Mathematical Approach to Physiological problems* (MIT Press, Cambridge, Mass, 1972)
159. E.B. Roberts, *Managerial Application of System Dynamics* (MIT Press, Cambridge, Mass, 1991)
160. E.S. Roberts, *Java: An Introduction to Computer Science* (Addison Wesley, Boston, 2009)
161. J.G. Roederer, *Information and Its Role in Nature* (Springer, Berlin, 2005)
162. B. Roux, *Molecular Machines* (World Scientific, Singapore, 2011)
163. M. Rosenzweig, S.M. Breedlove, A. Leiman, *Biological Psychology* (Sinauer, Sunderland, Mass, 2002)
164. E. Scarr, A.S. Gibbons, J. Neo, M. Udawela, B. Dean, Cholinergic connectivity: its implications for psychiatric disorders. Front. Cell. Neurosci. **7**, 55 (2013)
165. G. Schlosser, G.P. Wagner, *Modularity in Evolution and Development* (University of Chicago Press, Chicago, 2004)
166. A. Scott, *Stairway to the Mind* (Springer, New York, 1995)
167. J.R. Searle, *Making the Social World: The Structure of Human Civilisation* (Oxford University Press, Oxford, 2011)
168. P.M. Senge, *The Fifth Discipline* (Century Business, London, 1990)
169. S. Seung, *Connectome: How the Brain's Wiring Makes Us Who We Are* (Houghton Mifflin Harcourt, Boston, 2012)
170. J.L. Shearer, A.T. Murphy, H.H. Richardson, *System Dynamics* (Addison Wesley, Reading, Mass, 1971)
171. D.M. Shepherd, D.P. Martin, A. Varsani, J.A. Thomson, E.P. Rybicki, H.H. Klump, Restoration of native folding of single stranded DNA sequences through reverse mutations: An indication of a new epigenetic mechanism. Arch. Biochem. Biophys. **453**, 108–122 (2006)
172. J. Silk, *The Big Bang* (Freeman, New York, 2001)
173. H.A. Simon, *The Sciences of the Artificial* (MIT Press, Cambridge, Mass, 1992)
174. R. Sinnott, G. Towler, *Chemical Engineering Design* (Elsevier, Amsterdam, 2009)
175. J.M.W. Slack, *From Egg to Embryo: Regional Specification in Early Development* (Cambridge University Press, Cambridge, 1991)
176. L.A. Steen, The science of patterns. Science **240**, 611–616 (1988)
177. A. Stevens, J. Price, *Evolutionary Psychiatry: A New Beginning* (Routledge, 2000)
178. G.F. Striedter, *Principles of Brain Evolution* (Sinauer Associates, Sunderland, Mass, 2005)
179. J. Sutter, Multimodality: moving beyond 'language'/rethinking 'meaning'. Lang. Issues **20**, 53–60 (2008)
180. R. Thom, *Structural Stability and Morphogenesis: An Outline of a General Theory of Models* (Addison-Wesley, Reading, MA, 1989)
181. K.S. Thorne, Primordial element formation, primordial magnetic fields, and the isotropy of the universe. Astrophys. J. **148**, 51–68 (1967)
182. J.H.M. Thornley, *Mathematical Models in Plant Physiology* (Academic Press, London, 1975)
183. J.A. Toronchuk, G.F.R. Ellis, Disgust: sensory affect or primary emotional system? Cogn. Emot. **21**, 1799–1818 (2007)

184. J.A. Toronchuk, G.F.R. Ellis, Affective neuronal selection: the nature of the primordial emotion systems. Front. Psychol. **3**, 589 (2013)
185. R.L. Trask, *Language and Linguistics: The Key Concepts* (Routledge, Abingdon, 2007)
186. S. Vogel, *Cats' Paws and Catapults: Mechanical Worlds of Nature and People* (W W Norton and Company, 2000)
187. A. Wagner, *The Origins of Evolutionary Innovations* (Oxford University Press, Oxford, 2011)
188. R.V. Wagoner, W. Fowler, F. Hoyle, On the synthesis of elements at very high temperatures. Astrophys. J. **148**, 3–49 (1967)
189. S.I. Walker, L. Cisneros, P.C.W. Davies, Evolutionary transitions and top-down causation, in *Proceedings of Artificial Life* **XIII**, 283–290 (2012). arXiv:1207.4808
190. J.D. Watson, T.A. Baker, S.P. Bell, A. Gann, M. Levine, R. Losick, *The Molecular Biology of the Gene* (Benjamin Cummings, 2003)
191. J.N. Weber, B.K. Peterson, H.E. Hoekstra, Discrete genetic modules are responsible for complex burrow evolution in Peromyscus mice. Nature **493**, 402–405 (2013)
192. B. Wegscheid, C. Condon, R.K. Hartmann, Type A and type B RNase P RNAs are interchangeable in vivo despite substantial biophysical differences. EMBO Rep. **7**, 411–417 (2006)
193. D. Weigel, C. Dean, Development, evolution and adaptation. Curr. Opin. Plant Biol. **5**, 11–13 (2002)
194. M.J. West-Eberhard, *Developmental Plasticity and Evolution* (Oxford University Press, Oxford, 2003)
195. E. Wigner, The unreasonable effectiveness of mathematics in the natural sciences. Commun. Pure Appl. Math. **13**, 1–14 (1960)
196. G.C. Williams, *Adaptation and Natural Selection* (Princeton University Press, Princeton, 1992)
197. S. Zhang, M. Srinivasan, Exploration of cognitive capacity in honeybees: higher functions emerge from a small brain, in *Complex Worlds from Simpler Nervous Systems*, ed. by F.R. Prete (MIT Press, Cambridge, Mass, 2004)
198. J.M. Ziman, *Principles of the Theory of Solids* (Cambridge University Press, Cambridge, 1979)
199. W.H. Zurek, Decoherence, einselection, and the quantum origins of the classical. Rev. Mod. Phys. **75**, 715 (2003). http://lanl.arxiv.org/abs/quant-ph/0105127

Chapter 5
Room at the Bottom?

The previous chapters have given numerous examples of top-down causation. They appear to make the case for existence of top-down causation unshakeable.

However, lower level causation seems to give a complete account of what happens [4]. Where does the causal slack lie, enabling top-down action to take place? Massimo Pigliucci reports a discussion on the issue as follows [66]:

> Steven Weinberg played what he thought was a trump card in favor of reductionism 'all the way down': he mentioned the causal completeness of the laws of physics. I asked him to elaborate on the point, and he said that the laws of Newtonian mechanics, for instance, are causally complete in the sense that there is no room within the equations for any unaccounted parameters. It follows, according to Weinberg, that those equations are a complete description of the causality of the system, leaving no room for emergent properties.

This is the 'exclusion' argument [4, pp. 111–120, 159, 177]. How is there freedom for higher level causation to be efficacious? In this chapter, I consider in turn:

- Section 5.1. Room at the bottom: over-determination?
- Section 5.2. Contextual constraints.
- Section 5.3. Structure and constraints.
- Section 5.4. Changing the nature of constituent entities.
- Section 5.5. Leading to existence of the elements.
- Section 5.6. Deleting lower level elements.
- Section 5.7. Queries.

I do not revisit here the supervenience perspective, which was covered in Sect. 3.5.3, nor the argument for the necessity of the conclusion, given in Sect. 1.7.

5.1 Room at the Bottom: Over-Determination?

The key problem is how, if physics is causally closed as proposed by Putnam, Papineau, Weinberg, Rosenberg, and others [62, 66, 75], physics can allow top-down causation in the structure/function hierarchy characterised in Sect. 3.1.3.

G. Ellis, *How Can Physics Underlie the Mind?*, The Frontiers Collection,
DOI 10.1007/978-3-662-49809-5_5

> **Overdeterminism**. How can top-down causation be possible in the case of the implementation hierarchy, if the physics at the bottom is a causally closed system, determining all that happens through interactions of particles and fields mediated by forces and potentials? Isn't the system already fully determined, so there is no room for any kind of top-down causation?

Similar issues arise in the case of the logical hierarchy, characterised in Sect. 3.1.4:

> **Logical Space**. Higher level structures in the logical hierarchy emerge from combinations of lower level logical elements. How can the higher levels affect what logical behavior occurs at a lower level, when the behaviour of elements at that level is fully specified?

In either case, the issue does not necessarily concern the topmost and bottommost levels (if we can identify them). It arises between any two adjacent levels. Each lower level may appear to fully determine what happens at each next higher level. How is there any space for contextual effects and top-down causation?

In particular, the issue arises in relation to conscious activity. The mind is based on brain operations (electrons flowing in dendrites and axons, for example), and these are based on the underlying physics (the forces and fields described by Maxwell's equations, for example) [71]. Therefore the claim can be made that physics does not just constrain what happens, it uniquely determines what happens in the brain. If basic physics determines all, the situation is causally closed and there is no room for higher level influences. Despite appearances, the operations of the mind are epi-phenomena.

This chapter will make the case that there is no problem here. In essence, physical causes are not the only one that affect lower level dynamics and outcomes. Rather, they form a vehicle for other kinds of causality to operate. In the case of the implementation hierarchy, the underlying physics establishes the set of possibilities that can happen, but not the specific events that actually happen. It does the work, but does not choose what work will be done [10, 74]. The higher levels constrain the lower level interactions, thereby creating possibilities of complex behaviour, and selecting from all the possibilities the behaviours that actually happen in specific contexts. Similar effects happen in the logical hierarchy.

There are five different ways these top-down effects can be efficacious. Firstly, higher levels channel and constrain lower level dynamics:

- **Contextual Constraints**. Higher level boundary conditions or variables can constrain lower level outcomes (Sect. 5.2).
- **Constraining Structures**. Higher level structures can constrain lower level outcomes (Sect. 5.3).

Second, in many cases there is not just a fixed given set of lower level elements that interact with each other. Rather higher level processes modify and select the lower level elements:

- **Changing the Nature of Lower Level Elements**. Higher level effects can change the nature of lower level elements (Sect. 5.4).
- **Existence of Lower Level Elements**. Higher level structures or dynamics can lead to the existence of lower level elements (Sect. 5.5).

- **Deleting Lower Level Elements**. Adaptive selection selectively deletes lower level entities in accordance with higher level selection criteria, thereby creating order out of chaos (Sect. 5.6), and the outcome is not uniquely determined from initial data because of lower level random processes (Sect. 5.6.6).

These are not mutually exclusive options: some of them can occur simultaneously. Together these possibilities resolve the puzzle of overdetermination of the lower levels.

In all these cases it is equivalence classes of lower level variables, corresponding to specific higher level contexts and structures, that are the real causal variables when top-down influences occur (Sect. 3.5). The specific lower level states that instantiate an equivalence class are inconsequential. Higher levels act down, not by constraining specific lower level states, but by constraining an equivalence class of states.

5.2 Contextual Constraints

The first way top-down effects are efficacious is through contextual constraints on lower level dynamics, where:

- higher levels set boundary conditions (Sect. 5.2.1), or
- pass higher level variables to local systems (Sect. 5.2.2).

Multiple realisability at the lower levels enables the same higher level function to be realized in many different ways at the lower levels. Together these features set the environment in which the lower level components operate, and so determine their outcomes [59].

5.2.1 Boundary Conditions

When considering specific physical and biological systems, contextual effects can occur via the boundary conditions, which affect local outcomes. This was discussed in Sect. 4.1.3. Here I will just give a few examples from physics.

5.2.1.1 Global Topology

Global topology of spaces affects families of solutions of differential equations, which can have physical outcomes.

Particle Properties. In the M-Theory approach to a unified theory of fundamental physics,[1] the topology of Calabi–Yau spaces constrains low-energy string vibrational

[1]An approach to understanding the most fundamental level of physics. For an introduction, see http://superstringtheory.com/.

patterns and hence determines the families of elementary particles that can occur [15]: different topologies give different classes of particles.

CMB Anisotropies. In cosmology, the patterns of microwave background radiation anisotropies that can occur in universes with closed space sections is determined by the spatial topology [51, 69].

5.2.1.2 Asymptotic Boundary Conditions

By contrast, one can have unbounded spaces and conditions at infinity such as asymptotic flatness of spacetime and associated outgoing radiation conditions. These are the usual boundary conditions for isolated systems, such as a binary pulsar. They are a family of conditions, specified flexibly enough to allow variation in terms of time-dependent multipoles of the central object and associated emission of gravitational radiation [17]. This contrasts with exact spherical symmetry, where this is not possible.

5.2.1.3 Shape and Geometry of the Boundary

The size and shape of a boundary selects specific physical outcomes from all the possibilities.

Eigenfunctions. The shape and nature of a boundary affects eigenfunctions and eigenvalues of vibrating systems. The boundary may be open (an organ) or closed (a guitar). It determines the possible motions of air molecules in a tube and of the atoms making up strings, so affecting the tone and quality of musical instruments such as horns, organs, saxophones, guitars, violins, pianos, and drums [30].

The Reaction–Diffusion Equation. Patterning of animal coats, like leopard spots and zebra stripes, is determined as a result of instabilities in the diffusion of morphogenetic chemicals during the embryonic stage of animal development. Body shape (the morphogen concentration must take the same value at the same point on going once round a closed curve) and boundary conditions (there is no flow across the boundary) determine the outcome, resulting in specific patterns of skin stripings in zebras and spots on a leopard, and the specific wing patterns of butterflies [57].

5.2.1.4 Determination of the Arrow of Time

Which is the future direction of time in which entropy should increase? Fundamental microphysics cannot tell, as it is time symmetric, so time's arrow must come from global boundary conditions: the universe must have been very smooth on large scales at early times [16, 63]. This is discussed further in Chap. 6.

5.2.2 *Passing Higher Level Variables or Parameters*

Lower level systems function in ways that depend on contextual variables, passed from the global to the local context.

5.2.2.1 Physical Systems

This was discussed in Sect. 4.1.3, considering for example the case of nucleosynthesis in stars and in the early universe. Here I will just give two further examples.

Thermodynamics. A core idea in thermodynamics is a heat bath that surrounds a system and provides its environment (the heat bath is so large that it is unaffected by the system). The temperature of the heat bath is a key variable in determining the system state and so the motions of its constituent molecules. This is the basis of the analysis of thermodynamic machines such as characterised by a Carnot cycle [45]. The contextual variable is the temperature of the heat bath.

The Solar System. The key elements in the Earth's environment are the Sun (which provides the Earth with high grade thermal radiation) and the dark sky (which provides a heat sink for low grade radiation emitted from the Earth). Together these determine the Earth's heat budget [56], which controls the temperature and so in particular influences agriculture. Alteration in solar radiation can cause climate change on Earth. This is top-down causation from the local context (the Solar System) to the system (the Earth). The contextual variable is the temperature of the Sun's radiation.

5.2.2.2 Structured Systems

Control parameters and variables determine outcomes of behaviour in structured systems [73]. They are passed down from the higher levels to the modules.

Feedback Control. The setting of goals in a feedback control system alters lower level behaviour, as discussed in Sect. 3.4.3. For example, setting the desired temperature in a thermostat controls the lower level dynamics in such a way as to reliably attain the desired temperature. The contextual variable is the desired temperature. Changing the desired destination in an aircraft automatic pilot results in different control surface movements that lead the aircraft to the new destination.

Computational Systems. As discussed in Sect. 2.7.3, the choice of control parameters determines the mode of operation of logical systems and hence determines their outcome through branching logic of the form IF $X = X_0$ THEN Y ELSE Z, where X is a contextual variable. For example, a computerised payroll system may have a variable PAYCLASS with values PC = 1 corresponding to STATUS = CEO and PC = 5 corresponding to STATUS = MACHINE OPERATOR. Setting PC = 1 rather than PC = 5 results in different flows of electrons in computer gates, and hence different outcomes in terms of payments made to a bank account. The variable PAYCLASS is a contextual variable that is set by a data base to control the subroutine function, and changes the detailed logical flow and output of the program.

Organisations. These have equivalent parameter passing mechanisms. For example, the HR section passes a message to Payroll: "Joseph Wing has been appointed Director of Communications". This causes Payroll to alter settings in their employee database, whereupon Wing occupies a different payclass and receives different payments.

5.3 Structure and Constraints

The second way top-down effects are efficacious is when emergent structures (Sect. 3.1.1) act as constraints on lower level dynamics (Sect. 3.4.2), thus channeling the way they function [42]. The constraints change the dynamics by breaking symmetries [2] and so create more general possibilities than are available to unstructured systems. This happens in:

- Physical systems (Sect. 5.3.1).
- Artefacts (Sect. 5.3.2).
- Biology (Sect. 5.3.3).
- The Brain (Sect. 5.3.4).
- Organisations (Sect. 5.3.5).

5.3.1 Physical Systems

Crystal Structures. These are characterised by symmetry groups, and they govern material properties and behaviour [77]. The periodic crystal structure in a metal breaks symmetry [2] and leads to lattice waves and an electronic band structure depending on the particular solid involved [77], and this in turn results in the physical behaviour of the material. Crystal structures set up the conditions for electron flows that lead to properties such as heat conductivity and electrical conductivity, with a key feature being the introduction of impurities, leading to the existence of different types of transistors [47, pp. 123–136]. A crystal doped with boron results in a p-type semiconductor, whereas doping with phosphorus creates an n-type semiconductor. The very concept of doping (introducing an impurity) only makes sense in the context of the background silicon crystal. It is a contextual concept.

Thus the properties at the crystal level, only describable at that level, act down to the level of electrons. It is the higher level patterns that are the essential causal variable in solid state physics, by creating specific band structures in solids (hence, for example, the search for materials that will permit high temperature superconductivity). Which specific lower level entities create them is irrelevant: you can move around specific protons and electrons while leaving the band structure unchanged.

Effective Potentials. This multiple realisability is always the case where effective potentials occur due to local matter distribution. Top-down causation takes place, for example, through the way the effective potential representing the combined effects of many stars in a galaxy controls the motions of stars in the galaxy [7]. The specific detailed distribution of stars leading to the effective potential is irrelevant: it is the large scale pattern that determines the lower scale motions. For instance, the outcome is quite different if it is a spiral or elliptical galaxy.

Gemmer et al. give an illuminating example [32, pp. 74–77] of an ideal gas in a container, where the way the container wall (a higher level construct) constrains lower level entities (the gas molecules) can be described by a coarse-graining process. But this property of containment would not occur if the higher level structure (the wall) did not exist, with the specific size and shape it has. This is discussed in detail, with many other examples, in [26].

5.3.2 Artefacts

Artefacts such as cameras and computers have been structured precisely in order to attain desired high level outcomes by constraining lower level physics in an appropriate way [73]. This is achieved by channeling the operational logic of the system.

Physical Constraints. The specific connections in an electric motor or a digital computer act as constraints on lower level dynamics, thus channeling the way they function. This functioning is enabled by structure: the constraints imposed by the existence of wires channels electron flows (Sect. 4.1.6). For example the specific connections in a computer (which could have been different) create logic gates which are combined to give a CPU and memory banks, structured to realise specific logical functions [52] (as discussed in Chap. 2).

Logical Constraints. A digital computer is a vehicle for logical operations. It can be operated as music system or word processor, or in many other ways. Which occurs depends on the high level software loaded, which logically constrains lower level logical and physical operations by imposing a higher level logical structure on them (Sect. 2.4.1).

5.3.3 Biology

Plant and animal physiology describes a set of higher level biological structures that constrain and shape lower level dynamics (Sect. 4.2.4). For example, in all the physiological systems in the human body [68], the specific functioning (according to the laws of physics) of the component parts, given their nature, is determined by their physiological context [59]. In particular there are numerous physiological feedback systems (Sect. 4.2) that channel lower level actions so as to attain higher level goals, such as a constant body temperature.

5.3.4 *The Brain*

Structure. The specific wiring of the brain [72] shapes the way action potentials flow between neurons, thereby enabling the logical functioning of the mind [58]. Different neural network connections will lead to different detailed functioning and outputs. This logical functioning is based on the way that the structure of dendrites and axons channels signals between neurons, this channeling being caused by their physical membrane structure which underlies the propagation of action potentials in the brain. This in turn is based on the properties of voltage-gated ion channels in the membrane [71].

Function. The same muscle cells can be used in football or in playing music. Different action programs set different logical structures in action (scoring a goal, playing Mozart) that use the same physical components to attain different outcomes. In effect, different software changes the detailed dynamics of the hardware. In general, goals and choices shape lower level function in a top-down way by structuring the logic of mental processes [31].

5.3.5 *Organisations*

The structure of an organisation is laid down in a constitution and an organisation chart. Its functioning is spelt out in a mission statement, organisational goals, operating procedures, ethical guidelines, and so on. The two are linked by job descriptions and directives that specify who will do what and when. Taken together, this is an abstract structure that guides and often determines what happens in a top-down way. Organisational departments respond by structuring more detailed section activities in the department. Sectional goals and deadlines structure individual activities. The trick in management is to get these lower level processes and goals aligned with the overall organisational purpose [5].

5.4 Changing the Nature of Constituent Entities

The third way top-down effects are efficacious is by changing the nature of lower level entities. The standard bottom-up view is based on a billiard ball model of unchanging lower level entities underlying higher level structure. The key point is the implicit assumption that lower level entities are independent of higher level context. But that is often wrong. Hydrogen in a water molecule has completely different properties than when free, for electrons bound in atom interact with radiation quite differently than when free. The higher level context has changed the nature of the underlying components, because the nature of an entity is characterised by the way it interacts with other entities, which is how we recognize it for what it is.

> **Mutable Elements**. There need not be a situation of invariant lower level elements obeying fixed physical laws. Higher level context can change the nature of the lower level elements [50]. Thus the nature of micro-causation is changed by top-down processes, profoundly altering the mechanistic view of how things work.

This often ensures that the lower level elements function in such a way as to fulfil higher level purposes: this is a commonplace aspect of adaptive selection in biology. The entire discussion is different when this crucial feature is taken into account. Top-down causation is then seen as not just the shaper of bottom level activity, but also the shaper of bottom level properties. This happens in:

- Physics and chemistry (Sect. 5.4.1).
- Biology (Sect. 5.4.2).
- The Brain (Sect. 5.4.3).
- Society (Sect. 5.4.4).
- Logic (Sect. 5.4.5).

5.4.1 Physics and Chemistry

5.4.1.1 Molecules

Water. Water has an essentially different nature in the form of liquid water, steam, and crystalline ice, with quite different properties in each case. Which state occurs depends on environmental variables (temperature and pressure) according to the phase diagram. Isolated water molecules do not occur in practice.

Hydrogen Atoms. Hydrogen has an essentially different nature in different contexts. Buchanan expresses this as follows [11]:

> We tend to think that the character of the hydrogen atom follows from the laws of particle physics (quantum electrodynamics). But a hydrogen atom in relative isolation in the interstellar medium has very different properties from one trapped in a dense liquid of hydrogen under high pressure; the 'normal' radiative spectrum of hydrogen alters radically. Which is the 'true' hydrogen? There's obviously no answer. The nature of hydrogen depends on context.

5.4.1.2 Particle Properties

Neutrons. The neutron has a half life of 11 min in isolation, decaying to form a proton, an electron, and a neutrino. It is stable with a half-life of billions of years when bound in a nucleus. Its properties are therefore dramatically different in these different contexts.

Chameleon Particles in Cosmology. This is a postulated scalar particle in cosmology with an effective mass that depends on its environment because of a non-linear self-interaction. It is a possible candidate for both dark energy and dark matter in cosmology because of this environmental dependence [44].

Particle Properties. In the string theory approach to quantum gravity, particles are realised as low-energy string vibrational patterns. The topology of Calabi–Yau spaces constrains low-energy string vibrational patterns and hence determines the families of elementary particles that can occur [15]: different topologies give different classes of particles. Then particle properties are determined not by the laws of physics but by the contingent nature of which specific string theory false vacuum occurs. The specific kinds of particles that occur in the universe are environmentally determined.

5.4.1.3 Chemical Bonding

Chemical bonding radically changes the nature of elements. Sodium and chlorine have completely different properties from sodium chloride (common salt). When bound, the atoms no longer have the properties they had when free. A hydrogen atom in isolation is quite different from one covalently bonded with oxygen to form water [37]. The orbital structure is quite different and the hydrogen atom thus has different properties than in isolation. Indeed it is really to some degree a misuse of language to call it hydrogen when bonded.

5.4.2 Biology

Cell Differentiation. Cells in a living body start off identical and then are each fitted to their specific role in the body by a developmental process of specialization guided in a top-down way by morphogens. Through the processes of developmental biology, cells get differentiated to perform specific functions. This changes their nature in an adaptive way [34]. Cells differentiate into neurons that get adapted to their location in the brain, into muscle cells adapted to their role in the heart, and so on. They each develop so as to fit into their allotted role in the body, and are then fine-tuned for their function.

5.4.3 The Brain

This top-down adaptive influence shapes lower level entities at all scales in the brain.

Neurons. A particular case of the specialisation of cells to fit their future functions (just discussed) is the specialisation of cells to become motor neurons, sensory neurons, or interneurons in the nervous system [36, pp. 138–140].

Neuronal Connections. Memory is stored by changes in connections between nerve cells. The patterns of neural connections and weights in a neural net get adapted to the patterns it has learned to recognise, resulting in long-term memory encoding in a neural net fitted to that specific purpose. For example, particular facial features are learnt and hence patterns of synaptic connections are structured, through experience, to recognise specific individuals. The neural net is shaped by that particular social context.

Humans in Society. Individual minds develop in the context of their interactions with other minds, and brain development cannot be understood outside this context [22]. Individuals are shaped by society so that they fit into that society, for example, learning a specific language and a variety of societal roles and expectations [6]. This is top-down causation from the society to the individual, and indeed to their synaptic connections: their brain is adapted to fit into the society in which they live [1]. Thus the detailed nature of micro causation in the brain is changed by these top-down processes, profoundly altering the mechanistic view of how things work. This is a key feature of the adaptive selection processes TD3, TD4, and TD5 discussed in Chap. 4.

5.4.4 *Society*

Social Organisations. These usually have mechanisms to ensure that their members are suited to the needs of the organisation. This leads to the 'the basketball team model' of downward causation described by Vicente [75]:

> In a basketball team, the players are effective causes of what the team is able to do. However, the behaviour of the players cannot be understood if we forget that they are playing for and in the team. Teams 'selectively activate' the causal powers of the players, and it can even be said that teams 'recruit' players, i.e., that the players are there because they have the powers that the team requires from them. Teams, then, are self-preserving self-organized entities which constrain and partly explain the behaviour of their players.

This applies to most organisations: you only belong to an organisation if you satisfy its membership criteria or are selected as suitable by a selection committee. Thus you are a 'member' only because the organisation makes you one. Once you belong, the organisation fits you to your role by training programmes, and has rules of procedure and codes of conduct that constrain the way you behave.

5.4.5 *Logic*

Essentially the same thing happens in the case of logical hierarchies, with the nature of lower level elements often being adapted to their higher level contexts (Sects. 3.4.4 and 4.1.5).

Language. The majority of words in a dictionary have multiple meanings [61]. An example is the word 'plane', which may be a noun (an aircraft or flat piece of land or woodworking tool), a verb (in carpentry or in the case of the motion of a boat), or an adjective (in geometry or geography). The context determines what part of speech it is and what it means, i.e., context changes its logical properties.

5.5 Leading to Existence of the Elements

The fourth way top-down effects are efficacious is by creating the possibility of existence of lower level entities. This is the case whenever the parts cannot survive on their own: the higher level context is essential to the existence of the constituent entities.

> **Contextually Dependent Existence.** In many cases the lower level entities would not exist without the higher level structure.

In these situations, context is a creator as well as a modifier. This occurs in:

- Physics (Sect. 5.5.1).
- Biology (Sect. 5.5.2).
- Society (Sect. 5.5.3).
- Logical hierarchies (Sect. 5.5.4).

5.5.1 Physics

Emergence of higher level entities has clearly occurred when lower level entities cannot exist outside their higher level context.

Phonons. Phonons are quasi-particles that play an important role in the physical properties of solids, such as thermal and electrical conductivity. The very possibility of the existence of phonons is a result of the physical structure of specific materials [48]. The periodic crystal structure in a metal leads (via Bloch's theorem) to lattice waves, and the existence of quasiparticles such as phonons results from vibrations of the lattice structure [77]. The entire machinery for describing the lattice periodicity refers to a scale much larger than that of the electron, and hence is not describable in terms appropriate to that scale. This structure is at a higher level of description than that of electrons.

Cooper Pairs. Quantum cooperative effects occur in superconductivity, superfluidity, and the quantum Hall effect. In superconductivity, the electrons, despite their repulsion for each other, form pairs called Cooper pairs which are the basic entities of the superconducting state. This happens by a cooperative process: the negatively charged electrons cause distortions of the lattice of positive ions in which they move, and the real attraction occurs between these distortions. Thus these effects all exist because of the macro-level properties of the solid, i.e., the crystal structure depending on the particular solid involved, and hence represent top-down causation from that structure to the electron states.

Because these are all based on top-down action, they are emergent phenomena in the sense that they simply would not exist if the macro-structure did not exist, and hence cannot be understood by a purely bottom-up analysis, as emphasized strongly by Laughlin in his Nobel lecture [48]:

> One of my favourite times in the academic year occurs in early spring when I give my class of extremely bright graduate students, who have mastered quantum mechanics but are otherwise unsuspecting and innocent, a take-home exam in which they are asked to deduce superfluidity from first principles. There is no doubt a very special place in hell being reserved for me at this very moment for this mean trick, for the task is impossible. Superfluidity, like the fractional Hall effect, is an emergent phenomenon, a low-energy collective effect of huge numbers of particles that cannot be deduced from the microscopic equations of motion in a rigorous way, and that disappears completely when the system is taken apart [2].

This collective dynamics is only possible because of top-down causation; and that is why it cannot be derived in a bottom-up way. The relevant physics is discussed in more detail in Sect. 6.2.3. It is key that quasi-particles should be regarded as real [28, pp. 227–250].

Topological effects. Another key example is topological effects in physics [76].

5.5.2 *Biology*

Cells in Multicellular Animals. These can only exist in the context of a live animal body, once they have differentiated. The body provides them with nutrition and energy via the blood stream, and removes waste [68]. They die if that context fails, for example if the heart stops beating. Their existence is dependent on their context.

Excitable Cells. Action potentials in excitable cells such as neurons and muscle cells are generated by voltage-gated ion channels embedded in a cell membrane. The spiking patterns of signals conveyed by action potentials would not exist if the membrane structure did not exist [71].

Symbiotic Pairs. Symbiosis is rife in biology, where animals depend on each other for life, or animals and plants need each other for their survival. An example is mycorrhiza, which is a fungus living in symbiosis with the roots of a vascular plant [33]. The higher level entity is the symbiotic pair, and it is its existence that makes the existence of each of the symbiotic partners possible.

Ecosystems. The animals in an ecosystem cannot exist unless the ecosystem itself exists. We are all interrelated [14] and if the system fails, we die, for we depend on it for our supply of food: we are all part of the food chain.

5.5.3 *Society*

Membership of social organisations only has meaning because the organisation exists. The membership categories derive from the structure of the organisation, as set out in its constitution. Thus the categories 'Member', 'Chairman', 'Secretary', etc., derive their existence from the existence of the organisation. Once in existence, they derive bottom-up causal powers from the powers assigned to that membership category by the constitution.

5.5.4 *Logical Hierarchies*

Essentially the same effect occurs in the logical hierarchies of language and computation. Some lower level elements do not exist in a logical sense independently of their context.

5.5.4.1 Contextual Variables

Referential Meaning. In natural language, the referential meaning of some lower level entities depends on the current higher level context. Thus the words 'then', 'there', 'he', 'she', 'it', and so on, have no meaning by themselves. Their reference is set in some higher level context than the sentence in which they occur. The same applies to phrases such as 'at that time' or 'in that place'. Hence they do not exist in a logical sense independently of this context, which continually changes.

5.5.4.2 Semantic Meaning

Class Structures in Computer Programs. In object-oriented computer languages, objects that are members of a class derive their nature from the class they belong to [9]. They only have meaning within the context of that class, which determines their methods and hence their logical nature. This is not just a case of the meaning varying according to context, as in Sect. 5.4. It is the base meaning that is provided by the higher level context. If the class is not defined, members of the class do not have known properties; if one compiles a Java program where an object is assigned to a class that does not exist, the compiler will give an error message and grind to a halt.

Natural Language. The same applies to members of the classification hierarchies in ordinary language. The word 'dog' does not derive its meaning from a sentence where it occurs ('the dog ate its meal'). It brings that meaning to the sentence from its taken-for-granted semantic context, a socially determined relatively fixed logical environment with a class structure that brings with it expectations of how the word will function in a semantic sense (we don't expect 'the dog drove the car' because that syntactically valid sentence does not correspond to events that occur on the real world).

5.6 Deleting Lower Level Elements

The fifth way top-down effects are efficacious is through selective deletion of lower level elements, thereby shaping the nature of the population of constituent entities.

> **Selection of Elements.** Generically, there is no fixed set of lower level entities. The core of adaptive selection is when selection processes create order out of disorder by deleting unwanted lower level entities or states according to some higher level selection criteria, which selects what lower elements will survive.

This is the process whereby useful information is garnered and order is created out of disorder (Sect. 4.3). The selection criteria are regarded as being at a higher causal level, because they shape the outcomes: if you alter the criteria, you get different outcomes. This is a key effect in:

- Biology (Sect. 5.6.1).
- Computing (Sect. 5.6.2).
- The mind (Sect. 5.6.3).
- Organisations (Sect. 5.6.4).
- Physics and chemistry (Sect. 5.6.5).

Together with the way top-down processes alter lower level elements (Sect. 5.4), and even create them (Sect. 5.5), this shows how higher causal levels can adapt lower level entities so as to shape lower level dynamics in accord with desired higher level outcomes. The required freedom from bottom-up determinism lies in micro-indeterminism (Sect. 5.6.6):

> **Indeterminate Outcomes.** Micro indeterminism provides the space for adaptive selection processes to generate new outcomes that were not implied by the initial state. Statistical variation or quantum indeterminacy at the micro-level provide a repertoire of variant systems that are then subject to processes of adaptive selection. The outcome cannot be uniquely predicted from the initial state.

5.6.1 Biology

5.6.1.1 Darwinian Evolution

Selective deletion is crucial in evolutionary biology, as is made clear in the slogan 'survival of the fittest', resulting in selection of genes of individuals best adapted to specific niches [14]. This is a top-down process, selecting which species survive and which do not [13].

5.6.1.2 Developmental Biology

Selection occurs in developmental biology, where plants and animals adapt to local conditions via many epigenetic mechanisms that select one developmental path from all those that are possible [34]. This dynamic is based on the underlying molecular machinery.

5.6.1.3 Molecular Machines

This flexibility at the macro-level is enabled by a corresponding flexibility at the micro-level. Molecular machines work in a noisy environment because of the confluence of thermal, chemical, mechanical, and electrostatic energies at the scale of biological macromolecules [65]. There is a molecular storm at the cellular level [40, p. 72]: in a cellular context, the average molecule undergoes ten billion collisions with water molecules every millisecond [40, p. 150]. Selection of the relevant molecules in this turbulent environment is provided by the lock and key mechanism, which recognizes some enzymes and ignores others. This variability is crucial to molecular machinery, which extracts order from chaos [40].

5.6.1.4 Protein Folding

One of the key events in molecular biology processes is protein folding. Hoffmann explains how this works in a very clear way [40, p. 115]:

> Protein folding is possibly the best example of how physical laws, randomness, and information—provided by evolution—work together to create life's complexities. The amino acid sequence of a protein is determined by the cell's DNA, according to the genetic code. This information evolved over billions of years. But the amino acid sequence in our DNA only encodes the amino acid sequence: it does not encode the final 3D shape of the protein. The 3D shape is the result of the energy landscape, which is determined by physical forces (hydrophobic forces, electrostatic forces, binding energies, etc.) acting on the particular sequence of amino acids. This shape also depends on external conditions (pH, temperature, ion concentration). Thus a large part of the necessary information to form a protein is not contained in DNA, but rather in physical laws governing charges, thermodynamics, and mechanics. And finally, randomness is needed to allow the amino acid chain to search the space of possible shapes and to find its optimal shape.

Thus global parameters are passed down and affect the local dynamics, while randomness provides a repertoire of alternatives from which a best one is selected. The physics enables this, but the information in the DNA is the ultimate cause of the 3D shape, which is the biologically effective feature. Physics opens up the opportunity space, and provides the means for change to take place, but biological variables are the causally effective agents determining what happens both at lower and higher levels, according to the logic of biological processes.

5.6.1.5 Adaptive Immune System

Through clonal selection, the adaptive immune system functions as an adaptive system able to deal with infections never before encountered [12].

5.6.1.6 Genetic Circuits

Variable components and random lower level processes occur in genetic circuits. According to Eldar and Elowitz [24]:

> The genetic circuits that regulate cellular functions are subject to stochastic fluctuations, or 'noise', in the levels of their components. Noise, far from just a nuisance, has begun to be appreciated for its essential role in key cellular activities. Noise functions in both microbial and eukaryotic cells, in multicellular development, and in evolution. It enables coordination of gene expression across large regions, as well as probabilistic differentiation strategies that function across cell populations. At the longest timescales, noise may facilitate evolutionary transitions [...] Emerging principles connect noise, the architecture of the gene circuits in which it is present, and the biological functions it enables.

The function of noise is to create space within which selection processes can choose the outcome.

5.6.1.7 Randomness

Chance (statistics associated with coarse-graining, and random boundary conditions) means that physical outcomes in biological systems are not uniquely determined by physics alone (Sect. 4.3.3). This provides the openness needed for Darwinian selection processes to choose outcomes that satisfy higher level goals and values. This occurs in developmental and functional contexts as well as the Darwinian evolutionary context.

5.6.2 Computers

Computer memory is finite, so if unwanted items are not deleted, memory banks will eventually fill up and the system will slow down and then grind to a halt (Sect. 2.6). Hence, to keep the system working, garbage collection is needed in working memory, and deletion of records (old data files, draft texts, unwanted emails, poor quality photographs, unused programs, etc.) in long term memory (Sect. 2.6.2). The key issue is the criteria used to decide what to remove and what to leave behind. These criteria are the factors that create order and adapt the remaining files to suit our purpose (Sect. 2.4.4).

Deletion is a two stage process: removing data files from an index of files in use, and overwriting them with new data. The first step is reversible, the second is not.

The first step orders data according to our needs, but does not save memory space. The irreversible second step is needed to keep the system working.

5.6.3 *The Mind*

Adaptive processes are crucial in the mind [43], allowing humans to continuously adapt to local conditions. This takes place at multiple levels (Chap. 7).

5.6.3.1 Effective Level

Adaptive processes take place all the time as the mind senses the environment, interprets it, and learns from it (Sect. 4.3.4). In particular, consider the following.

Perception. This is based on paying attention to the incoming data that really matters, and ignoring the rest [31]. We could not function if we had to pay detailed attention to all incoming sensory data.

Learning. This is based on formulating hypotheses, testing them, and keeping the ones that work while discarding the rest. This may take place subconsciously as well as consciously, as our mind sorts through the world for meaningful predictive patterns [36, pp. 93–135]. The inability to discard failed hypotheses is the recipe for failure. So the theoretician's most powerful tool is her wastepaper basket, or, in the present day context, the delete key.

Remembering. This is based on forgetting, because the mind subconsciously selects what to remember and what to forget, transferring only some items to long-term memory and clearing out the rest [36, pp. 310–345]. As in the case of computers, this has to be the case: if we did not subconsciously select what to store in long-term memory, our minds would be jammed full of irrelevant memories that would crowd out what is important. The same is true of short term memory: we must continually clear it of old data that has become irrelevant (e.g., Where did I leave my keys?), because of the famous limit of seven plus or minus two items that can be stored in this fast-access memory [54].

Learning and remembering is based on the underlying feature of synaptic plasticity [58, pp. 227–242], which allows the brain to adapt to the world around.

5.6.3.2 Network and Neuron Level

These behaviors are based on the adaptive properties of neuronal nets, which change link weights as the network learns from the environment [19]. According to the famous aphorism, "Wire together, fire together", unused connections wither away, the selection criterion being disuse. At the neuron level, synapse properties are altered by processes of neuronal group selection characterized by Edelman as neural Darwinism [23].

5.6.4 Society

Social organisations have selection mechanisms to get rid of members not suited to the organisation's function or needs (see Sect. 5.4.4). Complex disciplinary procedures lead to firing of employees in a company or ejection of members from an organisation. This is a process of adaptively selecting the membership to suit the organisation's explicit or implicit goals (which may be in conflict with each other).

5.6.5 Physics and Chemistry

Adaptive selection effects in physics and chemistry are discussed in Sect. 4.3.2. I briefly mention them here.

5.6.5.1 Purification Processes

Purification processes isolating specific elements and compounds are the foundation of the possibility of doing chemistry and chemical engineering. These are adaptive selection processes of various kinds, as described in [38]. Nanotechnology is a case in point: self-assembly processes create a variety of nanostructures. One needs to select the ones one wants, and discard the rest. This is discussed in Sect. 6.5.

5.6.5.2 State Vector Preparation

State vector preparation is key to experimental setups in quantum physics, and is a non-unitary process because it can produce particles in a specific eigenstate from a stream of particles that are not in such a state. Isham [41] points out that the outcome states are drawn from some collection E_i of initial states by being selected by some suitable apparatus, for example, being chosen to have some specific spin state in the Stern–Gerlach experiment. The other states are discarded. Selection takes place from a (statistical) ensemble of initial states according to some higher level selection criterion, which is a form of top-down causation from the apparatus to the particles [26]. The apparatus is specifically designed to have this non-unitary effect on the lower level (Sect. 6.6.5).

5.6.6 *Micro Indeterminism and Adaptive Selection*

Because of the existence of random processes at the bottom, there is sufficient causal slack to allow this kind of top-down effect to occur without violation of physical causation. In general, the outcome is not uniquely determined from initial data because of lower level random processes. Physical processes do not determine a unique outcome.

A reductionist response might be that, in principle, the lower levels are determined: we call it random just because we don't have enough information, i.e., the randomness is not real, it's just ignorance! There is a double reply to this. First, this is an interlevel effect. This happens because of effective interlevel randomness between adjacent levels. As such it is indeed a real effect in the relations between any two levels. But what about a possible bottom level, where the real causal work is done and there is no underlying level, so the interlevel argument cannot apply? Well any such level is not deterministic! Quantum uncertainty applies at the bottom. Here I look at:

- Interlevel randomness.
- Quantum randomness.

5.6.6.1 Interlevel Randomness

Random outcomes at the next lower level $N-1$ allow variation at any level N, which then leads to selection at the micro-level $N-1$, but based on macro-level properties and meaning. Statistical variation provides a repertoire of variant systems that are then subject to processes of Darwinian section, based on higher level qualities of the overall system.

The randomness is real because the higher level does not have access to the relevant lower level variables. It only has access to the equivalence class of lower level variables that can be controlled by altering higher level variables (Sect. 3.5), so from a higher level viewpoint these lower level fluctuations are indeed random.

Internal Random Variables. There can exist lower level fluctuations which cannot be manipulated via higher level variables, but have higher level causal effects. They give a random input to the dynamics.

This is the feature of information-hiding that is a key aspect of modular hierarchical structures (Sect. 3.1.6). The effectiveness of this randomness in causal terms is recognized in Monod's classic book *Chance and Necessity* [55]. Whatever the philosophical issues may be, it is an effect that must be taken into account in studies of causation.

In biological cases, developmental biology processes [34] amplify molecular level variation to system level changes. That these random processes do indeed occur in biology is indicated by many kinds of evidence [18, 24, 35, 40]. This mechanism can only work because of the huge number of micro-components involved: atoms in a cell, cells in a human body, etc. [40]. Emergence of genuine complexity requires the vast numbers of micro components entailed in biological reality.

5.6.6.2 Quantum Randomness

Because of the existence of quantum processes at the bottom, physics is *not* deterministic, despite the way many writers represent the situation as if it is. It seems that the profound nature of the quantum revolution has still not permeated the consciousness of many physicists and biologists, who present the situation as if physics were deterministic all the way down. This is not the case: the bottom level is not deterministic [26, 29]!

At its base level, the universe is indeterministic, allowing the needed causal slack freeing higher levels from lower level causal determinism. And this quantum indeterminism can affect biological processes [3, 46, 49]. Some processes of molecular biology, e.g., involving replication of mutated molecules, act as amplifiers, even allowing quantum effects to change evolutionary outcomes [64, 70]. By itself, that randomness does not lead to emergence of higher level order, but it can provide a basis for this to occur through the process of adaptive selection (Sect. 8.1.2).

If those levels cannot be reduced to lower levels (interactions of electrons and quarks for example), as assumed by Crick in his book *The Astonishing Hypothesis* [20], then the principle of irreducible intermediate levels is established. If this was not the case we'd be in deep trouble as we don't in fact know what the bottom level is. We'd have to suspend neuroscience and genetics research until the theory of quantum gravity has been sorted out. In reality, no particular level is privileged in causal terms: they all have causal power [60].

5.7 Queries

Finally, I turn to some comments opposing or querying the view presented here, and suggest some answers.

Question. How can top-down action take place without violating the causal closure of physics: if everything is physically determined from a micro-level, how can genuine top-down action be possible?

Answer. Physics creates a possibility space of a variety of physical states and lays down constraints on how changes between them may take place. Top-down action chooses which actually occur.

Question. Vicent says [75] there are two influential reductive views as to what causation is: one is, causation is nothing but the action of forces; the other is, causation is the transference, transmission, or exchange of conserved quantities. Are these the only forms of causality?

Answer. No, they are not the only form of causation in town. Information is causally effective, as in the case of DNA. Abstract algorithms are causally effective, as in the case of digital computers. Human theories and concepts have causal effects as in the design and manufacture of a jumbo jet aircraft, through an understanding of

Newton's laws of motion, the theory of control systems, aerodynamics, and so on. Such abstract theories in the engineer's minds are causally effective. Physics per se has no explanation of how the plane came into existence: there is no Jumbo jet potential, no minimum action principle, no conjunction of forces, and no conserved quantity that can account for its coming into being or the specific nature of its existence (i.e., all its design features).

What leads to an aircraft flying at a particular time from London to Berlin? The bottom-up explanation is in terms of Bernoulli's law in relation to molecules impinging on its wings. The top-down explanation is that it was, at great expense, designed to fly. The same level explanation is that the pilot is operating the controls to make it do so. And the topmost explanation is that someone believes it will make money for them if it does so. *All these explanations are simultaneously necessary in order that it fly on schedule.* Only the bottom-up explanation relates to physical forces and conserved quantities. The plane cannot exist without the causal efficacy of mental states [53].

5.7.1 Criticism and Response

In response to my argument above about the multiple causes leading to an aircraft flying, Tim O'Connor (private communication) has commented as follows[2]:

> Note that the higher-level explanations appeal to (intentional) states long prior to the plane's flying. That is, the explanation works by setting the event to be explained in a larger spatiotemporal context. The reductionist might retort: if those prior intentional states are themselves wholly fixed by more fundamental physical facts that compose them, we could have in principle a completely physical (bottom-up) explanation spanning each step of the larger context to which you point. This would be an explanation wholly independent of high-level intentional explanations—and appeal to facts that are themselves collectively responsible for their being a co-existing intentional level of explanation. Taking the widest scope possible (the universe as a whole), the fundamental physical facts and the laws that directly govern them asymmetrically determine the existence of higher-level systems and the forms of explanation they make possible. Or, at any rate, it is not clear that anything to which you appeal conflicts with this assertion. And if that is correct, then why is there not a perfectly good sense in which all action takes place down below?

My response is as follows. As I understand it, this proposes that the larger spatiotemporal context of cosmology sets initial data that completely determines the present day situation and so explains all current lower and higher levels. The answer is twofold.

Firstly, because of quantum uncertainty, such a proposal to explain present day details in terms of cosmological initial data cannot work even in principle, as explained in Sect. 8.1. Quantum fluctuations can change the genetic inheritance, and hence existence of animals. If our own existence cannot uniquely follow from that initial data, neither can any specific thoughts or intentions.

---

[2]This is in effect also the burden of a paper by Purves, Wojtach, and Lotto [67].

Secondly, if we disregard this impossibility, we are in effect faced with the proposal that the future specific events and outcomes, such Darwin's development of the theory of evolution, or Witten's development of M-Theory, are specifically written into the fluctuations on the last scattering surface in the early universe that we now observe through the WMAP satellite [39]. This would require firstly, a unique dynamical mapping from the data on that surface to these present day outcomes. There would have to be some kind of coding for this mapping from then to the present day (what configurations of atoms would uniquely imply that Einstein would say the words "God does not play dice"?). Given such a coding, there would also have to be some process that could have fine-tuned that initial data to give this specific result. But the standard view is that what is present on this surface is not some form of subtle structure that codes the fact these specific unique apparently intelligent outcomes will occur, but rather that they are random Gaussian fluctuations [21], which contain no such intelligent statements. There is no mechanism in sight that could have written such coding in to the data on the last scattering surface.

Consequently, the only plausible way these outcomes could have happened is for that initial data to be such as to lead to the development of complex entities such as ourselves capable of the kind of logical thought that causally leads to these outcomes. That is, genuine higher level causal powers have to come into being with their own inherent logic, these then leading to these extraordinary outcomes, *inter alia* causing billions of electrons to move in brains in ways essentially determined by higher level causal factors such as the logic of general relativity theory. There is no way that they could be implied by physics per se.

Finally, O'Connor suggests that, although it has to be conceded that biological top-down causation in cell differentiation does show that some non-fundamental levels of causes and explanations are not independent of those above them, the reductionist will claim that whatever the fundamental physical facts and laws turn out to be will be independent of higher-level entities that they make possible.

Perhaps this is so, but their specific outcomes will not. Physics *per se* will never be able to predict that either a teapot or a Jumbo jet aircraft or a giraffe will exist [25]. And it may be that the supposition itself is incorrect. The quantum process that determines specific outcomes from an initial quantum state is at least partially environmentally dependent, because decoherence is a selection process determined by the environment [78, 79]. We cannot pronounce on the measurement process itself, because quantum physics is still unable to explain how this happens. This, too, may be contextually dependent. What is clear is that the local context (such as what type of experimental apparatus is used) influences quantum measurement outcomes [26]. For example, if we measure spin, the resulting final state is different than if we measure momentum. The lower level physics is not immune to higher level influences.

5.7.1.1 The Outcome

Putting this all together: the big picture is as follows:

> **Interlevel Causation**. Bottom-up effects do indeed determine the higher level outcomes, given the specific initial dispositions of particles, fields, and energetic states on the lower level. But those dispositions would not be what they are if it were not for top-down effects. Furthermore, the outcome is not uniquely determined by the lower level initial states because of both interlevel statistical fluctuations and quantum uncertainty at the bottom. This opens the space for higher level selection criteria to crucially shape the outcome.

The final point is that top-down causation also occurs in physics itself. That is the topic of the next chapter. So even if we could reduce everything to physical causation (which we cannot, see Sect. 5.6.1 for specific examples), that would not eliminate top-down causation.

References

1. N. Ambady, The mind in the world: culture and the brain. Assoc. Psychol. Sci. **24**(5–6), 49 (2011)
2. P.W. Anderson, More is different. Science **177**, 377 (1972). Reprinted in *A Career in Theoretical Physics* (World Scientific, Singapore, 1994)
3. P. Ball, The dawn of quantum biology. Nature **474**, 272–274 (2011)
4. M.A. Bedau, P. Humphreys (eds.), *Emergence: Contemporary Readings in Philosophy and Science* (MIT Press, Cambridge, Mass, 2008)
5. S. Beer, *Brain of the Firm* (Wiley, Chichester, 1981)
6. P. Berger, T. Luckmann, *The Social Construction of Reality: A Treatise in the Sociology of Knowledge* (Anchor, New York, 1967)
7. J. Binney, S. Tremaine, *Galactic Dynamics* (Princeton University Press, Princeton, 1987)
8. R.C. Bishop, Fluid convection, constraint and causation. Interface Focus **2**, 4–12 (2012)
9. G. Booch, *Object Oriented Analysis and Design with Applications* (Addison Wesley, New York, 1994)
10. W. Brown, N. Murphy, *Did My Neurons Make Me Do It? Philosophical and Neurobiological Perspectives on Moral Responsibility and Free Will* (Oxford University Press, New York, 2007)
11. M. Buchanan, Going up, going down. Nat. Phys. **9**, 63 (2013)
12. F.M. Burnet, *The Clonal Selection Theory of Acquired Immunity* (Cambridge University Press, Cambridge, 1959)
13. D.T. Campbell, Downward causation, in *Studies in the Philosophy of Biology: Reduction and Related Problems*, ed. by F.J. Ayala, T. Dobhzansky (University of California Press, Berkeley, 1974)
14. N.A. Campbell, J.B. Reece, *Biology* (Benjamin Cummings, San Francisco, 2005)
15. P. Candelas, G. Horowitz, A. Strominger, E. Witten, Vacuum configurations for superstrings. Nucl. Phys. B **258**, 46–74 (1985)
16. S. Carroll, *From Eternity to Here: The Quest for the Ultimate Arrow of Time* (Dutton, New York, 2010)
17. J.M. Centrella, Resource letter GrW-1: Gravitational waves. Amer. J. Phys. **71**, 520–525 (2003). arXiv:gr-qc/0211084
18. T. Chouard, Breaking the protein rules: if dogma dictates that proteins need a structure to function, then why do so many of them live in a state of disorder? Nature **471**, 151 (2011)
19. P. Churchland, *Plato's Camera* (MIT Press, Cambridge, Mass, 2012)

20. F. Crick, *Astonishing Hypothesis: The Scientific Search for the Soul* (Scribner, 1995)
21. S. Dodelson, *Modern Cosmology* (Academic Press, San Diego, 2003)
22. M. Donald, *A Mind so Rare: The Evolution of Human Consciousness* (W W Norton, 2001)
23. G.M. Edelman, *Neural Darwinism: The Theory of Group Neuronal Selection* (Oxford University Press, Oxford, 1989)
24. A. Eldar, M.B. Elowitz, Functional roles for noise in genetic circuits. Nature **467**, 167–173 (2010)
25. G.F.R. Ellis, Physics, complexity, and causality. Nature **435**, 743 (2005)
26. G.F.R. Ellis, On the limits of quantum theory: contextuality and the quantum–classical cut. Ann. Phys. **327**, 1890–1932 (2001). http://lanl.arxiv.org/abs/1108.5261
27. G.F.R. Ellis, Top-down causation and emergence: some comments on mechanisms. J. Roy. Soc. Interface Focus **2**, 126–140 (2012)
28. B. Falkenberg, M. Morrison (eds.), *Why More Is Different: Philosophical Issues in Condensed Matter Physics and Complex Systems* (Springer, Heidelberg, 2015)
29. R.P. Feynman, R.B. Leighton, M. Sands, *The Feynman Lectures on Physics: Quantum Mechanics* (Addison-Wesley, Reading, Mass, 1965)
30. N.H. Fletcher, T.D. Rossing, *The Physics of Musical Instruments* (Springer, New York, 2010)
31. C. Frith, Free will and top-down control in the brain, in *Downward Causation and the Neurobiology of Free Will*, ed. by N. Murphy, G.F.R. Ellis, T. O'Connor (Springer, Heidelberg, 2009)
32. J. Gemmer, M. Michel, G. Mahler, *Quantum Thermodynamics: Emergence of Thermodynamic Behaviour within Composite Quantum Systems* (Springer, Heidelberg, 2004)
33. J.W. Gerdemann, Vesicular-arbuscular mycorrhiza and plant growth. Annu. Rev. Phytopathol. **6**, 397–418 (1968)
34. S.F. Gilbert, *Developmental Biology* (Sunderland Sinauer Associates, MA, 2006)
35. P.W. Glimcher, Indeterminacy in brain and behaviour. Annu. Rev. Psychol. **56**, 25 (2005)
36. P. Gray, *Psychology* (Worth Publishers, New York, 2011)
37. H.B. Grey, *Chemical Bonds: An Introduction to Atomic and Molecular Structure* (Benjamin Cummings, Menlo Park, 1973)
38. E.J. Henley, J.D. Seader, D.K. Roper, *Separation Processes and Principles* (Wiley, Asia, 2011)
39. G. Hinshaw, WMAP data put cosmic inflation to the test. Physics World **19**, 16–19 (2006)
40. P.M. Hoffmann, *Life's Ratchets: How Molecular Machines Extract Order from Chaos* (Basic Books, New York, 2012)
41. C.J. Isham, *Lectures on Quantum Theory* (Imperial College Press, London, 1995)
42. A. Juarrero, *Dynamics in Action: Intentional Behaviour as a Complex System* (MIT Press, Cambridge, Mass, 2002)
43. E. Kandel, *The Age of Insight: The Quest to Understand the Unconscious in Art, Mind, and Brain, from Vienna 1900 to the Present.* (Random House, 2012)
44. J. Khoury, A. Weltman, Chameleon cosmology. Phys. Rev. C **69**, 044026 (2004). arXiv:astro-ph/0309411
45. C. Kittel and H. Kroemer, *Thermal Physics* (W H Freeman Company, 1980)
46. N. Lambert, Y.-N. Chen, Y.-C. Cheng, C.-M li, G.-Y Chen, F. Nori, Quantum biology. Nat. Phys. **9**, 10–18 (2013)
47. P. Landshoff, A. Metherall, *Simple Quantum Physics* (Cambridge University Press, Cambridge, 1979)
48. R.B. Laughlin, Fractional quantisation. Rev. Mod. Phys. **71**, 863 (2000)
49. S. Lloyd, A bit of quantum hanky-panky. Phys. World 26–29 (2011)
50. P.L. Luisi, Emergence in chemistry: Chemistry as the embodiment of emergence. Found. Chem. **4**, 183–200 (2002)
51. J.-P. Luminet, J. Weeks, A. Riazuelo, R. Lehoucq, J.-P. Uzan, Dodecahedral space topology as an explanation for weak wide-angle temperature correlations in the cosmic microwave background. Nature **425**, 593 (2003). arXiv:astro-ph/0310253
52. M.M. Mano, C.R. Kime, *Logic and Computer Design Fundamentals* (Pearson/Prentice Hall, 2008)

53. P. Menzies, The causal efficacy of mental states, In *Physicalism and Mental Causation*, ed. by S. Walter, H.-D Heckmann (Imprint Academic, 2003)
54. G.A. Miller, The magical number seven, plus or minus two: some limits on our capacity for processing information. Psychol. Rev. **63**, 81–97 (1956)
55. J. Monod, *Chance and Necessity: Essay on the Natural Philosophy of Modern Biology* (Alfred A Knopf, New York, 1971)
56. J.L. Monteith, *Principles of Environmental Physics* (Edwin Arnold, London, 1973)
57. J.D. Murray, *Mathematical Biology* (Springer, 1990)
58. J.G. Nicholls, A.R. Martin, B.G. Wallace, P.A. Fuchs, *From Neuron to Brain* (Sunderland Sinauer, Mass, 2001)
59. D. Noble, *The Music of Life* (Oxford University Press, Oxford, 2006)
60. D. Noble, A theory of biological relativity: no privileged level of causation. Interface Focus **2**, 55–64 (2012)
61. *Oxford Advanced Learner's Dictionary* (Oxford University Press, Oxford, 2000)
62. D. Papineau, The rise of physicalism, in *Physicalism and Its Discontents*, ed. by C. Gillet, B. Loewer (Cambridge University Press, Cambridge, 2001)
63. R. Penrose, *Cycles of Time: An Extraordinary New View of the Universe* (Knopf, New York, 2011)
64. I. Percival, Schrödinger's quantum cat. Nature **351**, 357 (1991)
65. R. Phillips, S.R. Quale, The biological frontier of physics. Phys. Today 38–43 (2006)
66. M. Pigliucci, On the causal completeness of physics, Rationally Speaking blog (2013), http://rationallyspeaking.blogspot.com/2013/02/on-causal-completeness-of-physics-part-i.html. Accessed 27 Feb 2013
67. D. Purves, W.T. Wojtach, R.B. Lotto, Understanding vision in wholly empirical terms (2011). PNAS Early Edition, www.pnas.org/cgi/doi/10.1073/pnas.1012178108
68. R. Rhoades, R. Pflanzer, *Human Physiology* (Saunders College Publishing, Fort Worth, 1989)
69. A. Riazuelo, J. Weeks, J.-P. Uzan, R. Lehoucq, J.-P. Luminet, Cosmic microwave background anisotropies in multi-connected flat spaces. Phys. Rev. D **69**, 103518 (2004). arXiv:astro-ph/0311314
70. J. Scalo, J.C. Wheeler, P. Williams, Intermittent jolts of galactic UV radiation: mutagenetic effects, in *Frontiers of Life, 12ième Rencontres de Blois*, ed. by L.M. Celnikier (2001). arxiv:0104209
71. A. Scott, *Stairway to the Mind* (Springer Verlag, New York, 1995)
72. S. Seung, *Connectome* (Houghton Mifflin Harcourt, Boston, 2012)
73. H.A. Simon, *The Sciences of the Artificial* (MIT Press, Cambridge, Mass, 1992)
74. R. Van Gulick, Who's in charge here? And who's doing all the work?, in *Mental Causation* ed. by J. Heil, A. Mele (Oxford University Press, Oxford, 1995)
75. A. Vicente, On the causal completeness of physics. Int. Stud. Philos. Sci. **20**, 149–171 (2006)
76. S. Zhang, Topological states of quantum matter. Physics **1**, 6 (2008)
77. J.M. Ziman, *Principles of the Theory of Solids* (Cambridge University Press, Cambridge, 1979)
78. W.H. Zurek, Decoherence, einselection, and the quantum origins of the classical. Rev. Mod. Phys. **75**, 715 (2003). http://lanl.arxiv.org/abs/quant-ph/0105127
79. W.H. Zurek, Quantum Darwinism and invariance, in *Science and Ultimate Reality: Quantum Theory, Cosmology, and Complexity*, ed. by J. Barrow, P.C.W. Davies, C. Harper (Cambridge University Press, Cambridge, 2004), pp. 121–134

Chapter 6
The Foundations: Physics and Top-Down Causation

At the bottom level, what happens is based on physics: it enables the emergence of higher level entities, which then in turn act down on the lower level components. Hence top-down causation takes place also in the context of physics. That is the topic of this chapter, which is more technical than the others, and may be skipped by those wishing to move on to the next chapter (on the brain). This chapter discusses:

- Section 6.1. The bottom level: quantum physics.
- Section 6.2. Emergence of higher level behaviour from the lower levels.
- Section 6.3. Top-down action in physics in general.
- Section 6.4. Deterministic top-down effects in physics.
- Section 6.5. Adaptive selection in physics and chemistry.
- Section 6.6. Top-down effects in micro physics.
- Section 6.7. Top-down effects in cosmology.

The basic assumption that will be made here as to how physics underlies complexity is as follows [37]:

The Nature of Physical Reality

1. **Combinatorial Structure**. Physical reality is made of linearly behaving components combined in linear and non-linear ways.
2. **Emergence**. Higher level structure and behaviour emerges from these combinations of lower level elements, leading to a hierarchy of causality and complexity. The nature of this emergent behaviour depends on the way the lower level elements are combined.
3. **Contextuality**. The way the lower level elements behave, and the specific outcomes of their interactions, depend on the context in which they are imbedded, including the nature of relevant emergent structures.
4. **Quantum Foundations**. Quantum theory is the universal foundation of what happens, applying locally to the lower level (very small scale) entities at all times and places. It may or may not apply at higher levels.

G. Ellis, *How Can Physics Underlie the Mind?*, The Frontiers Collection,
DOI 10.1007/978-3-662-49809-5_6

Item (3) is where top-down constraints come in, determining what specific outcomes occur in specific physical contexts. There is no overriding of the lower level physics. Rather, there is a channelling of its effects by structural conditions, or determination of effects through boundary conditions.

6.1 The Bottom Level: Quantum Dynamics

The micro-level dynamics is governed by quantum physics. I look in turn at:

- The basic dynamics (Sect. 6.1.1).
- Alternative possibilities (Sect. 6.1.2).
- The outcome (Sect. 6.1.3).
- Particle–wave duality (Sect. 6.1.4).

6.1.1 The Basic Dynamics

The basic expansion postulate of quantum mechanics [55, 62, 84, 97] is that, before a measurement is made, the state vector $|\psi\rangle$, which lives in a Hilbert space H, can be written as a linear combination of unit orthogonal basis vectors, viz.,

$$|\psi_1\rangle = \sum_n c_n |u_n(x)\rangle \ , \tag{6.1}$$

where u_n is an eigenstate of some observable $\hat{A}$ [62, pp. 5–7]. The evolution of the system can be completely described by a unitary operator $\widehat{U}(t_2, t_1)$, and so evolves as

$$|\psi_2\rangle = \widehat{U}(t_2, t_1)|\psi_1\rangle \ . \tag{6.2}$$

Here $\widehat{U}(t_2, t_1)$ is the standard evolution operator, determined by the evolution equation

$$\mathrm{i}\hbar \frac{\mathrm{d}}{\mathrm{d}t}|\psi_t\rangle = \widehat{H}|\psi_t\rangle \ . \tag{6.3}$$

When the Hamiltonian $\widehat{H}$ is time-independent, $\widehat{U}$ has the form [62, pp. 102–103]

$$\widehat{U}(t_2, t_1) = \mathrm{e}^{-\mathrm{i}\widehat{H}(t_2 - t_1)/\hbar} \ , \tag{6.4}$$

which is unitary [62, pp. 109–113], i.e., it has the property

$$\widehat{U}\widehat{U}^{\dagger} = 1 \ . \tag{6.5}$$

Applying this to (6.1) with $\widehat{U}(t_2, t_1)|u_n(x)\rangle = |u_n(x)\rangle$ (an invariant basis) gives

$$|\psi_2\rangle = \sum_n C_n|u_n(x)\rangle \ , \quad C_n := \widehat{U}(t_2, t_1)c_n \ . \tag{6.6}$$

Immediately after a measurement is made at a time $t = t^*$, however, the relevant part of the wavefunction is found to be in one of the eigenstates:

$$|\psi_2\rangle = c_N|u_N(x)\rangle \ , \tag{6.7}$$

for some specific index N.

This is where the quantization of entities and energy comes from (the discreteness principle): only eigenstates can result from a measurement. The eigenvalue c_N is determined by the operator representing the relevant physical variables, and hence is unrelated to the initial wave function (6.1). The data for $t < t^*$ do not determine either N or c_N. They merely determine a probability for each possible outcome (6.7), labelled by N, through the fundamental equation

$$p_N = c_N^2 = |\langle e_N|\psi_1\rangle|^2 \ , \tag{6.8}$$

which is known as the Born rule. One can think of this projection process as due to the probabilistic time-irreversible reduction of the wave function

$$\begin{array}{ccc} |\psi_1\rangle = \sum_n c_n|u_n(x)\rangle & \longrightarrow & |\psi_2\rangle = c_N u_N(x) \\ \text{indeterminate} & \text{transition} & \text{determinate} \end{array} \tag{6.9}$$

This is the event where the uncertainties of quantum theory become manifest (up to this time the evolution is determinate and time reversible). It will not be a unitary transformation (6.6) unless the initial state was already an eigenstate of $\hat{A}$, in which case we have the identity projection

$$|\psi_1\rangle = c_N u_N(x) \longrightarrow |\psi_2\rangle = c_N u_N(x) \ . \tag{6.10}$$

6.1.1.1 Unpredictability

Thus there is a deterministic prescription for evolution of the quantum state determining probabilities of outcomes of measurements, but indeterminacy of the specific outcomes of those measurements, even if the quantum state is fully known. Examples are:

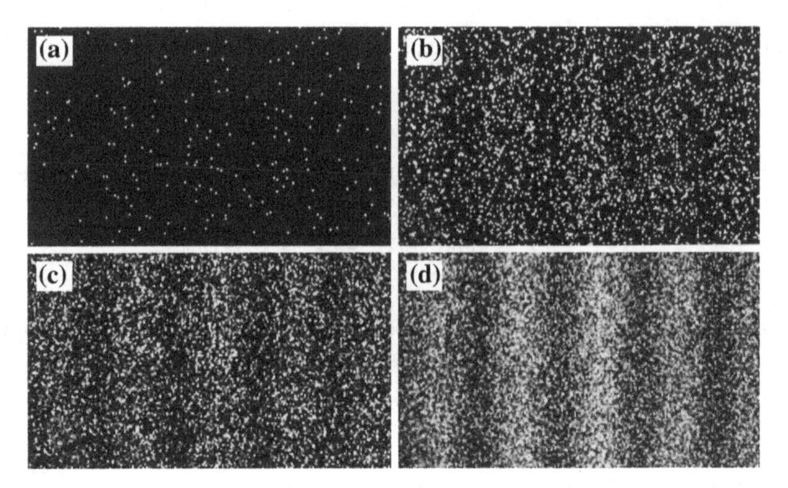

Fig. 6.1 Quantum uncertainty. Double-slit-experiment performed by Tonomura, showing the build-up of an interference pattern of single electrons. The numbers of electrons are **a** 200, **b** 6 000, **c** 40 000, **d** 140 000. Credit: Wikimedia Commons (Belsazar)

Radioactive Decay. We cannot predict when an excited nucleus will decay, or what the velocities of the resultant particles will be.

The Foundational Two-Slit Experiments. We cannot predict precisely where a photon, electron, neutron, or atom will end up on a screen after passing through close parallel slits in a barrier between the source and the screen. The quantum mechanical interference pattern is built up photon by photon as individual photons arrive at a detector in an indeterministic way. Each individual photon arrives at an unpredictable place, but the predicted overall interference pattern gradually builds up over time (see Fig. 6.1).

This discussion presents the simplest idealized case of a measurement [89, pp. 542–549]. More generally, one has projection into a subspace of eigenvectors (see [62, p. 136] or [118, pp. 10–12] or a transformation of density matrices [62, p. 137], or any other of a large set of possibilities [118, pp. 8–42], but the essential feature of non-unitary evolution remains the core of the process.

Thus there is a deterministic prescription for the evolution of the quantum state determining probabilities of outcomes of measurements [48], but indeterminacy of the specific outcomes of those measurements, even if the quantum state is fully known. The fact that such unpredictable measurement events happen at the quantum level does not prevent them from having macro-level effects. Many systems can act to amplify them to macro-levels, including photomultipliers (whose output can be used in computers or electronic control systems). Quantum fluctuations can change the genetic inheritance of animals [91] and so influence the course of evolutionary history on Earth, and they have changed the course of structure formation in the universe [35]. Thus quantum implications are not confined to the micro realm.

6.1.1.2 Uncertainty of Outcomes

It is a fundamental aspect of quantum theory that the uncertainty of measurement outcomes is unresolvable: it is not even in principle possible to obtain enough data to determine a unique outcome of quantum events [50, 62, 88]. This unpredictability is not a result of a lack of information: it is the very nature of the underlying physics. This uncertainty is made manifest when a measurement takes place, and only then. Without measurements, there is no uncertainty in quantum processes.

Here by a *measurement*, we mean a process whereby quantum uncertainty is changed into a definite classical outcome that can be recorded and examined as evidence of what has happened. It is not necessary for an observer to actually make any measurements. For example, it happens when a photon falls on a physical object such as a screen, a photographic plate, or the leaf of a plant, and deposits energy in a particular spot on the object at a particular time and place. In more technical terms, it occurs generically when some component of a general wavefunction collapses to an eigenstate of an operator [see (6.9)].

This is not a side-effect in quantum theory: it is absolutely central to its real world applications. As stated by Leggett [75, p. 87]:

> [...] it is the act of measurement that is the bridge between the microworld, which does not by itself possess definite properties, and the macroworld, which does. [...] the concept of measurement, prima facie at least, is absolutely central to the interpretation of the quantum mechanical formalism.

In addition, the uncertainty principle tells us we cannot simultaneously make accurate measurements of two non-commuting variables, for example position and momentum, so we cannot measure the initial data that would lead to unique results in the first place. Thus *irreducible randomness occurs in physics at the quantum level* [50, 55].

6.1.2 Alternative Possibilities

The above is the standard view: according to Heisenberg, Dirac, von Neumann, Feynman and others: irreducible randomness occurs in quantum theory. Determinism does not hold in the real world, at the micro-level. This was very worrying to many people, in particular Albert Einstein, and all possible alternatives have been carefully explored:

- **Hidden Variables**. Many investigations have tried to see if physicists have somehow missed some hidden variables that underlie this randomness. This involved the Bohr–Einstein debate [115], a famous paper by Einstein et al. [32], and a set of inequalities by Bell [14, 15]. These have shown that hidden variables are incompatible with the usual concepts of realism and locality. Locality means no instantaneous ('spooky') action at a distance, while realism means objects are there even when not observed. Bell's inequalities are violated due to quantum entanglement, as has been demonstrated experimentally.

- **Pilot Wave Theory**. This is a realistic and deterministic hidden variable theory which is manifestly nonlocal and gives exactly the same results as ordinary quantum mechanics [17, 18]. However, there can exist wave functions propagating in space and time carrying no energy or momentum, and not associated with a particle. As it makes no difference to the experienced outcome, it does not change the experimentally determined phenomena outlined above.
- **Many Worlds**. Everett [46] proposed that there is no collapse of the wave function. Everything that is possible occurs, as possibilities multiply and the wave function splits into innumerable branches, see [62, 101] for summaries of the various proposals as to how this happens. But this has no cash value: it does not change the experimental situation described above, which is what we experience in the real world. Any number of hypothetical other worlds that are supposed to be realised somewhere else make no difference to this well tested outcome. And this is not a testable proposal.
- **Decoherence**. This effectively removes entanglement by diagonalising the density matrix, and so some have suggested that the measurement problem is solved by environmental decoherence [55, 62, 125, 126]. However, while this diagonalizes the density matrix, it leaves a superposition of states, and so does not lead to a specific classical outcome. It does not predict where the individual spots in Fig. 6.1 will occur, and neither does any other result from quantum physics.

All these options are discussed by Isham [62], and many of the original papers have been presented with commentaries by Wheeler and Zurek [115].

6.1.3 The Outcome

There are various alternatives to the standard view, but in the end they amount to proposing some kind of machinery hidden behind the scenes that makes no difference to the practical outcomes described by (6.1)–(6.8). You have no ensemble to which you can apply statistics unless you have the individual events that make up the ensemble, and those are what quantum physics is unable to predict.

The irreducible uncertainty of specific events, as shown in the two-slit experiment in Fig. 6.1, is what we have to deal with in all experienced quantum phenomena [107]. There is indeed genuine unpredictability in the real world, even though we can predict statistics of microevents with precision.

6.1.4 Particle–Wave Duality

This is a further key aspect of quantum physics. Whether an entity acts as a particle or a wave is context dependent: this is the heart of particle–wave duality, where one can determine whether particles going through a slit should behave as particles or waves by the way one carries out the experiment [50, Sects. 1–1 to 1–7]. This has now been realised experimentally in the case of a version of Wheeler's delayed choice experiment [114], where the which-way choice is made after the particle has passed the slits [64]: a case of top-down causation from the apparatus to the very nature of the particle/wave at the time it passed through the slits.

6.2 The Emergence of Higher Level Behavior

Next, we consider how higher level behavior emerges from lower level behavior in two adjacent levels in the hierarchy of complexity (Table 6.1). As stated above, the fundamental viewpoint will be that the higher level behavior emerges from that at the lower levels.

The dynamics of the lower level theory maps an initial state i to a final state f. Choose a set of higher level effective variables that arise from the lower level variables in the context of the emergent structure. I will refer to this as a coarse-graining of the lower level variables because it leaves out all the details of the micro-states that underlie the macro states. Then on coarse-graining the lower level variables, state i corresponds to the higher level state I and state f to the higher level state F. Hence, the lower level action $t : i \rightarrow f$ induces a higher level action $T : I \rightarrow F$. A *coherent higher level dynamics* T emerges from the lower level action t if the same higher level action T results for all lower level states i that correspond to the same higher level state I [36], so defining an *equivalence class* of lower level states that give the same higher level action [9]. (If this is not the case, the lower level dynamics does not induce a coherent higher level dynamics, as for example in the case of a chaotic system.) Then on coarse-graining, e.g., integrating out fine scale degrees of freedom so as to give only macro degrees of freedom, the lower level action results in an emergent higher level dynamics: the effective theory at the higher level. Two key points follow:

EM1: Emergent Dynamics. The effective higher level dynamics will in general not be the same as the lower level dynamics [5].

Among other things, this is because a great deal of information is hidden in the higher level view. The second point is:

Table 6.1 The emergence of higher level behaviour from lower level theory. Coarse-graining the action of the lower level theory results in an effective higher level theory

Level $N+1$	Initial state I	Higher level theory $T \Rightarrow$	Final state F
	$\Uparrow$	Coarse-grain	$\Uparrow$
Level N	Initial state i	Lower level theory $t \Rightarrow$	Final state f

EM2: Essential Higher Level Variables. Not all effective higher level variables can be derived by coarse-graining in the sense of just integrating out lower level degrees of freedom. They may depend on detailed features of the emergent structures, and hence be essentially higher level variables [36].

In this section, I shall look in turn at the following:

- Examples of emergence (Sect. 6.2.1).
- Statistical mechanics (Sect. 6.2.2).
- Condensed matter physics (Sect. 6.2.3).
- Chemistry and biology (Sect. 6.2.4).
- Bottom-up effects: cosmology (Sect. 6.2.5).

6.2.1 Examples of Emergence

Here are some examples:

- **E1: Statistical Physics**. The underlying atomic theory leads to the macroscopic gas laws, thermodynamics, and thermal properties of gases [4, pp. 434–518]. The underlying theory and the emergent theory are quite different. This is discussed further below (Sect. 6.2.2).
- **E2: Electrodynamics**. The process of coarse-graining leads to the polarization density of a polarized medium [109, pp. 343–349], where the electric field **E** is a coarse-grained version of the microscopic electric field **e**, and the displacement vector $\mathbf{D} = \mathbf{E} + 4\pi\mathbf{P}$ includes a polarization term **P** representing coarse-grained dipole terms [63, pp. 103–108]. The fields **D** and **E** are related by a polarization tensor ϵ_{ij} such that $D_i = \epsilon_{ij} E_j$. The tensor ϵ_{ij} depends on the micro structure of the medium. In an isotropic medium, $\epsilon_{ij} = \epsilon\delta_{ij}$ (using Cartesian tensors). In an anisotropic medium this is not the case. The coarse-grained version of Maxwell's equations gives the divergence of **D** and curl of **E**, so a modified version of the microscopic equations emerges. The emergent theory is largely similar to the underlying theory.
- **E3: Physics to Chemistry**. The interactions of fermions lead through the Fermi exclusion principle to the nature of the hydrogen atom [4, pp. 109–148] and the electronic structure of atoms [4, pp. 158–176], and hence also to the periodic table [8, 86]. The nature of the chemical bond emerges from physics [8, 86]. There is no similarity between the underlying theory and the emergent laws.
- **E4: Chemistry to Microbiology and Life**. The complex modular hierarchical structure of life emerges from the underlying physical and chemical laws [23]. There is no similarity between the underlying theory and the emergent behaviour, except that, with suitable definitions, concepts of mass and energy balance apply at both levels.

In most cases, the underlying theory leads to a higher level theory characterizing quite different behaviour (after all, that is the essential content of Table 6.1). However, sometimes linear higher level behaviour will result from the locally linear lower level behaviour.

Classical to Classical Example: Geometric Optics. In the high frequency limit, Maxwell's equations for the electromagnetic field lead to geometric optics [59, 63, 77], with light propagating in a way described by Hamiltonian dynamics. The different wavelengths do not interfere with each other because the system is linear, whence spectral decomposition makes sense.

Classical to Classical Example: Engineering and Natural Systems. As emphasized by Bracewell [20], many manufactured and engineering systems have a linear dynamics that leads to periodic behaviour and the suitability of Fourier analysis. This occurs particularly when the system is engineered to have linear modes, for example organ pipes, guitars, linear electrical and electronic circuits, and so on. However, there may be such modes in other cases, for example, wave modes in suspension bridges and torsional oscillations of buildings. There are also similar instances in the natural world, for example propagation of water waves and sound waves—indeed anywhere where Fourier Analysis applies, linearity of the relevant degrees of freedom leading to the splitting of the system into normal modes with different frequencies that do not interfere with each other.

But these examples, although ubiquitous, are also limited: the engineering examples are carefully tailored to behave in this way, often at considerable expense, and they have frequency limits beyond which the linear behaviour ceases. Similarly, the linear behaviour of natural systems is very limited in time and space. Non-linearities intrude when we examine behaviour beyond these limits.

Quantum to Classical Example: Ehrenfest's Theorem. As a consequence of the Schrödinger equation (6.3), the time derivative of the expectation value for a quantum mechanical operator is determined by the commutator of the operator with the Hamiltonian of the system:

$$\frac{\mathrm{d}}{\mathrm{d}t}\langle A\rangle = \frac{1}{\mathrm{i}\hbar}\langle [A, H]\rangle + \left\langle \frac{\partial A}{\partial t} \right\rangle . \tag{6.11}$$

Applying this to the case of a particle of mass m and momentum p moving in a potential V so that $H = p^2/2m + V$, and defining $\langle F\rangle = -\langle \nabla V\rangle$, one finds

$$\frac{\Delta\langle p\rangle}{\Delta t} = \langle F\rangle , \qquad \frac{\Delta^2\langle x\rangle}{\Delta t^2} = \frac{1}{m}\langle F\rangle , \tag{6.12}$$

in agreement with the classical equation. Hence, the linearity of (6.3) results in the linearity of the relations (6.12), which are not however quantum relations (they have a classical form).

Quantum to Quantum: Renormalization Group. In some cases one can prove that coarse-graining a Hamiltonian system leads to another Hamiltonian system with the same Hamiltonian but different values of the constants. One example is the Wilson approach to renormalization theory, where the high momentum degrees of freedom in the generating functional $Z[J]$ are integrated out, leading to the renormalization group relating parameters of the original Lagrangian to the new Lagrangian (see [93, pp. 394–409] or [121, pp. 341–345]). However, this is possible only in restricted circumstances [93, pp. 402–403].

Another example is the Kadanoff construction, explicitly coarse-graining an Ising model, and thereby defining a coarse-grained lattice and block spin variables. The coarse-grained dynamics are governed by a Hamiltonian that is a function of the coarse-grained variables on the coarse-grained lattice [25, pp. 237–242]. Indeed, the block spins interact via the same Hamiltonian as the original spins, leading to a scaling of free energy and applicability of the Wilson renormalization group (see [117] and [25, pp. 245–248]).

Quantum to Quantum: Effective Theories. In some cases, coarse-graining will result in a Hamiltonian theory at the higher level, but with a Hamiltonian that has a different form. This is the case of *effective field theories* that emerge at higher level from the underlying physics (see [58] and [121, pp. 437–440]: an effective Lagrangian or Hamiltonian governs the higher level dynamics, but it is different from the original one. One cannot always derive this higher level effective action by explicit coarse-graining, but one can often determine the form the effective action should take by symmetry and conservation principles. The classic example [121, p. 441] is Fermi's β-decay theory [116], now embodied in *Fermi's Golden Rule* [100, p. 332], which is of wide application (see, e.g., [27, pp. 84–86] or [54, pp. 20, 165–166]).

Other examples are effective field theories of a Hall fluid [121, pp. 302–303] and of proton decay [121, pp. 440–441]. A more recent application relates to gravitational theory and the early universe. When one treats cosmological inflation in the early universe as being due to an effective theory, integrating out physics above some energy scale Λ induces non-renormalizable operators in the effective theory. This can also lead to corrections to the kinetic terms which contain higher powers of derivatives. The effects on the early universe are different than in the standard theory (see [51, 52] and references therein).

Quantum to Quantum: Long Range Order. The electron system in superconductors can exhibit *long range order*, with strong correlations in the wave functions of pairs of particles over distances longer than the coherence length [124, pp. 402–403]. Hence, one can introduce a macroscopic wave function $\Psi(r)$ (the Ginzburg–Landau order parameter) for the superfluid component of the electron density, leading to flux quantization [124, pp. 404–405] as a macroscopic manifestation of quantum mechanics. $\Psi(r)$ obeys a time dependent Schrödinger equation [124, (11.87)], which underlies the Josephson effect [124, pp. 405–410].

This is possible only in the context of metals with a periodic lattice structure, or other materials that allow superconductivity [124, pp. 396, 410–414]. The restricted nature of the contexts that allow this emergence of higher level effective quantum equations is shown by the great difficulty in identifying superconductors other than metals. In the case of metals, it is only possible when the temperature is exceedingly low, so that the non-linear interactions that would occur at higher temperatures are suppressed.

6.2.2 *Statistical Mechanics*

Statistical mechanics characterises the way higher level variables emerge from statistical disorder at the lower levels.

6.2.2.1 The Classical Case

In the classical case, coarse-graining the underlying atomic theory leads to the macroscopic gas laws, thermodynamics, and thermal properties of gases [4, pp. 434–518]. There is no similarity between the underlying theory and the emergent theory, except for the constraints that suitable mass, energy, and momentum conservation laws apply at both levels.

In the kinetic theory of gases, on coarse-graining, the pressure P exerted by a gas of molecules of mass m and number density n is determined by their average velocity as follows:

$$P = \frac{nm\overline{v^2}}{3} . \tag{6.13}$$

Thus the macroscopic pressure P is an emergent property, given by the average of the kinetic energy per molecule $m\overline{v^2}/2$, which is a microscopic property. The temperature T is given by

$$T = \frac{m\overline{v^2}}{3k_{\mathrm{B}}} , \tag{6.14}$$

where k_{B} is the Boltzmann constant. The macroscopic variables are related by the ideal gas law

$$PV = Nk_{\mathrm{B}}T , \tag{6.15}$$

where N is the number of molecules, each of mass m, enclosed in a container of volume V, and k_{B} is the Boltzmann constant. Similarly all the effective high level variables (P, T, etc.) are formed of equivalence classes of lower level variables: huge numbers of lower level states correspond to the same coarse-grained higher level state, with the entropy of the state being a measure of this multiplicity. As explained by Penrose [90, pp. 25–34], Boltzmann's entropy formula can be written

$$S = k \log V_{\mathrm{m}} , \tag{6.16}$$

where V_{m} is the volume of the coarse-graining region in phase space that has a macro property p (it defines the equivalence class of micro-states giving that macro state). The huge size of these volumes leads to the second law of thermodynamics [90]. It is the very existence of these equivalence classes that characterizes the existence of top-down causation: one only has a handle on the macro properties, not the micro ones. Altering a macro variable, e.g., changing the volume from $V_{\mathrm{m}}(t_1)$ to a new value $V_{\mathrm{m}}(t_2)$, results in any one of an equivalence class of micro-states corresponding to the new macro state. This change of the macro state alters the micro-states so that other macro variables vary according to the emergent macro relation (6.15), but it does not determine the specific micro outcome.

6.2.2.2 The Quantum Case

In the quantum case, new features come in because of the indistinguishability of particles, leading to new ways of counting states that result in Fermi–Dirac and Bose–Einstein statistics.

Boson Gas. In the case of a boson gas, the wave function at the quantum level is symmetric [29, pp. 205–211], resulting in the Bose–Einstein distribution law [4, pp. 528–530] on coarse-graining. Non-linear macroscopic laws of behaviour emerge, describable in purely classical terms. For example, in the case of photons, one obtains the blackbody spectrum for radiation [4, pp. 7–11, 531–532], and the associated formula for energy density and pressure of a photon gas:

$$\rho(T) = \frac{8\pi h}{c^3} \int_0^\infty \frac{\nu^3 d\nu}{e^{h\nu/kT} - 1}\,, \qquad p(T) = \frac{\rho(T)}{3c^2}\,. \tag{6.17}$$

The key point is that *these are relations for classical variables*: there is nothing in the behaviour at this higher level corresponding to superposition of states or entanglement. The situation is shown in Table 6.2.

Similarly, one obtains the macro formula for the pressure and density of a gas of molecules with zero integral spin [4, (13.32)]. In a metal, a phonon gas leads to a formula for the heat capacity C_V of a solid [4, (13.28)]. These are all emergent classical properties, as in the case of the energy density and pressure in (6.17).

Fermi–Dirac Gas. In the case of an electron gas, the wave function at the quantum level is antisymmetric [29, pp. 205–211], resulting in Fermi–Dirac statistics [4, pp. 519–522]. This again results in higher level non-linear behaviour describable in

Table 6.2 The emergence of higher level effective classical variables from the underlying quantum theory

Classical level	Gas laws	Temperature T, density ρ, pressure p
	Coarse grain ⤊	⤊ Bose–Einstein statistics
Quantum level	Photon gas	Symmetric wave function

purely classical terms, e.g., the thermoelectric current density coming from a metal surface in terms of the temperature of the metal [4, (13.11)].
Overall, the emergence of these classical levels from the underlying quantum theory is in accord with the view put forward in the previous chapters:

> **Higher Level Emergence.** Each of the higher levels of the hierarchy of complexity is real in its own right, described by relevant variables for that level, and laws of behaviour that are effective at that level. These variables and interactions emerge from the underlying quantum variables, and in many if not most cases, are classical variables.

But the word 'effective' sounds pejorative: they are *the* laws of behaviour applicable at that level. When equilibrium occurs, classical higher level thermodynamic behaviour emerges from the underlying quantum structure and reliably characterises what happens at that level, as for example in (6.17). The way this happens is presented by Gemmer et al. [54].

The essential point is that the statistical interactions between the components that lead to equilibrium destroy any coherence among the higher level variables: they do not display either constructive interference or destructive interference. An example is that the transition to equilibrium in a crystal relies on the Umklapp process [54, p. 223], which does not preserve momentum, and so is not a unitary process. Presumably, this corresponds to frequent collapse of the wave function at the micro-level: for if that does not take place, the necessary interactions between the components for thermalisation will not have occurred, and they can be expected to occur very frequently.

The key feature of such emergence is as follows:

> **Bottom-Up Emergence.** The macro laws (6.15) and (6.17) hold independently of the context. For example, the size and shape of the container, what it is made of, and the history of the gas are irrelevant. This is the hallmark of pure bottom-up effects.

Thus blackbody radiation is the same in the very early universe and in a laboratory today. This great difference in context is irrelevant.

6.2.3 Condensed Matter Physics

Gases are disordered, so straightforward statistical physics determines what happens. However, solids are a different matter: they are often comprised of highly ordered emergent structures. Specifically, crystal structures are highly ordered because energy minimisation leads to molecules forming symmetric patterns characterised by discrete symmetry groups. They may occur naturally, as in the case of felspar and quartz. However, often they can only occur through careful manufacturing processes, as in the case of sheet glass, metals such as in copper wires, semiconductors (where a high degree of chemical purity and crystalline perfection is required), superconductors, and so on. These emergent physical structures, with their specific symmetries, lead to particular electronic band structures and optical properties, which then act down to determine electronic properties.

6.2.3.1 Lattice Waves and Quasiparticles

The periodic crystal structure in a metal leads via Bloch's theorem [124, pp. 16–20] to lattice waves [124, pp. 27–75], and an electronic band structure depending on the particular solid involved [124, pp. 93–94, 119–128], resulting in all the associated phenomena deriving from the band structure. The entire machinery for describing the lattice periodicity refers to a scale much larger than that of the electron, and hence is not describable in terms appropriate to that scale. Thus these effects are all macro properties of the solid, i.e., the crystal structure, which emerges from the lower level interactions.

This can then lead to the existence of quasiparticles such as *phonons* [124, pp. 59–62] that result from vibrations of the lattice structure, and hence associated phenomena such as the *U-process*, whereby momentum in electron scattering processes is transferred to the system as a whole. It also leads to *Cooper pairs*, produced by the exchange of phonons between electrons [124, pp. 382–386] and hence to superconductivity [124, pp. 386–394] and associated phenomena such as superfluidity in metals [124, pp. 394–396]. Other examples are *holes*, conduction electrons with negative effective mass as determined by the energy surface $\mathcal{E}(k)$ [124, pp. 182–186], which are central to the physics of semiconductors [124, pp. 59–62], and *plasmons*, particles derived from plasma oscillations. The quantum Hall effect is a result of the existence of composite fermions, realised in the interface between two semiconductors [65]. The basic dynamics in each case is as shown in Table 6.3.

These are emergent phenomena arising from the coming into being of specific higher level atomic structures, with those crystal structures arising either spontaneously, or through purposeful design.

Emergent Effects. Because the electronic band structures and the resultant lower level entities such as phonons are based on the higher level crystal structure rather than simply being based solely on properties of the lower level constituent, they are both *emergent phenomena*. They simply would not exist if the macro-structure did not exist.

This implies, as discussed below, that they cannot be understood by a purely bottom-up analysis, as emphasized strongly by Laughlin [73]. This is the key difference relative to the cases that can be analysed purely by statistical mechanics. In addition, the very existence of many of these materials cannot be explained in a bottom-up way: they sometimes only occur in manufactured form. This applies for example to transistors, most superconducting materials, and even electric wires, and of course also to the laboratories and complex apparatus by which these properties are tested.

Table 6.3 Emergence leads to a higher level context that then results in the existence of emergent lower level entities

Context	Emergent structure	Effect	Contextual effect
Higher level	Lattice structure	$\Rightarrow$	Band structure, collective oscillations
	Emergence $\Uparrow$		Context $\Downarrow$
Lower level	Basic constituents		New entities

6.2.4 *Chemistry and Biology*

Simple molecules can emerge spontaneously through bottom-up processes, but there are limits as to how far this can go.

6.2.4.1 Inorganic Chemistry

Basic inorganic chemical properties emerge in a bottom-up way through the various forces that bind atoms together to form molecules, settling to a most probable state by minimising energy [8, 72]. The shell structure of the atoms, governed by the Pauli exclusion principle [29], determines the nature of the binding [86]. One can understand such molecules by physical principles alone and they can in principle come into existence spontaneously in a purely bottom-up way, although this depends on the availability of sufficient supplies of the requisite chemical elements in pure enough form at the right time and place, which may not easily occur naturally in some cases.

6.2.4.2 Organic Chemistry and Biology

Much more complex properties occur in microbiology and macrobiology, based on organic chemistry properties. These are again enabled by the various forces that bind atoms together to form molecules, settling to a most probable state by minimising energy. Various kinds of forces occur, as outlined by Watson in his classic book on molecular biology [111], enabling simple organic molecules to form spontaneously. Thus molecules such as formaldehyde, polycyclic aromatic hydrocarbons, amino acids, glycolaldehyde, glycine, fullerenes, and many others have been detected in interstellar space [68]. Also phospholipid structures can self-assemble.

However, the case will be made below that emergence of the genuinely complex molecules needed for life, involving primary, secondary, tertiary, and quaternary structures [6], is only possible when top-down causation is taken into account. This is because while simple biomolecules can form spontaneously, complex biological molecules such as haemoglobin, chlorophyll, rhodopsin, kinesin, and dynesin, and many proteins needed for cellular functioning can only get to be what they are through natural selection, which is a top-down process based on the local environment provided by cells situated in living beings. This applies particularly to information carrying molecules such as RNA and DNA. As stated by Lodish et al. [78, Sect. 1.2]:

> Macromolecules, though, are the most interesting and characteristic molecules of living systems; in a true sense the evolution of life as we know it is the evolution of macromolecular structures.

The same is true *a fortiori* for biological cells, and all living beings made out of them: they too can only come into being through natural selection processes [23]. In the real universe, with its finite lifespan, they cannot come into being by statistical mechanical processes.

6.2.5 Bottom-Up Effects: Cosmology

Turning from the small to the very large, the major thrust of present day scientific cosmology [30] is the way the structure of the universe emerges in a bottom-up way from gravitational interactions between its constituent entities, following Einstein's application of the local law of gravitation to determine spacetime structure in the large. This represents the effect of local physical laws on the large-scale structure of the cosmos, determining the evolution of the cosmic scale factor $S(t)$ according to the Friedmann equation

$$3\frac{\dot{S}^2}{S^2} - \kappa\rho - \Lambda = -\frac{3k}{S^2}\,, \tag{6.18}$$

where $\rho(t)$ is the total density of matter in the universe and Λ the cosmological constant. The spatial curvature is k/S^2, where k may be 0 (flat spatial sections), $+1$ (spherical space sections), or -1 (hyperbolic space sections). The gravitational equations, however, do not determine k or the spatial topology of universe. The matter present determines the effective equation of state for $\rho(t)$ and so determines the evolution of the universe. During the radiation dominated era in the early universe, the energy density is given by

$$\rho = \rho_{\nu_e} + \rho_{\bar{\nu}_e} + \rho_{\nu_\mu} + \rho_{\bar{\nu}_\mu} + \rho_\gamma = aT_\gamma^4 + \frac{7}{4}aT_\nu^4 = 1.45aT_\gamma^4\,, \tag{6.19}$$

so that we have a hot state evolving as

$$S(t) \propto t^{2/3}\,, \quad t = \left(\frac{c^2}{15.5\pi GaT^4}\right)^{1/2} \implies \frac{T}{10^{10}\,\mathrm{K}} = \left(\frac{t}{1.92\ \mathrm{sec}}\right)^2. \tag{6.20}$$

This is the context in which nucleosynthesis takes place, as discussed below. At later times,

$$\rho = \rho_{\mathrm{bar}} + \rho_{\mathrm{CDM}} + \rho_{\mathrm{DE}} = \frac{\rho_{\mathrm{bar}}(t_0)}{S^3} + \frac{\rho_{\mathrm{CDM}}(t_0)}{S^3} + \rho_{\mathrm{DE}}\,, \tag{6.21}$$

and that is the background for structure formation.

Overall, in a magnificent extension of our understanding of the way apples fall to the surface of the Earth, the Moon circles the Earth, and the Earth circles the Sun, we now understand the rate of expansion of the universe as being controlled in a bottom-up way by the cumulative effect of the gravitational force exerted by all the particles in it on each other. And that extrapolation from laboratory scale to the universe as a whole seems to work. The dynamics of the universe as a whole emerges in a bottom-up way from the cumulative effect of the dynamics of the particles that make it up.

6.3 Top-Down Causation

We now turn to the other half of the causal story: namely top-down effects. As explained in the previous chapters, the higher levels of the hierarchy of complexity and causation (Table 6.1) provide the context within which the lower level actions take place. By setting the context in terms of initial conditions, boundary conditions, and structural relations, the higher levels determine the way the lower level actions occur (Sect. 3.4). The general picture is shown in Table 6.4.

The lower levels do the work, but the higher levels decide what is to be done.

This can be regarded as top-down causation in the hierarchy of complexity. Such causation, in conjunction with bottom-up action, is the key to the emergence of complexity from the underlying physics [36, 43]. The fundamental importance of top-down causation is that it changes the causal relation between upper and lower levels in the hierarchy, in particular enabling inter-level feedback loops.

Proving Top-Down Causation. How do we prove that top-down effects are occurring? One has to show that changing some higher level condition changes lower level dynamics or behaviour. For example, changing the length of an organ pipe changes the wavelengths of possible standing waves, so the sound it emits depends on its size. Similarly, changing the shape of a drum changes the sounds it emits. By contrast, the blackbody spectrum (6.17) is independent of the size and shape of an oven that emits blackbody radiation. It is determined by purely local effects.

In this section, I shall look in turn at the following:

- Equivalence classes (Sect. 6.3.1).
- Changing or creating the basic elements (Sect. 6.3.2).
- Types of top-down causation in physics (Sect. 6.3.3).

Table 6.4 The emergent effective higher level theory exerts contextual effects on the operation of the underlying quantum theory

Level $N+1$	Higher level theory	Effective Theory
	Emergence $\Uparrow$	Top-down effects $\Downarrow$
Level N	Lower level theory	Contextual effects

6.3.1 *Equivalence Classes*

Technically, the way this works is that equivalence classes of lower level states correspond to a single higher level state [9]. For example, in the case of a gas in a cylinder, a myriad of lower level molecular states s_i will correspond to a specific higher level state $\mathcal{S}$ characterized by a temperature T, volume V, and pressure p, which are the effective macroscopic variables. The number of such lower level states that correspond to the higher level state determines the entropy of that state [90]. One can only access the equivalence class by manipulating higher level variables rather than the detailed lower level variables, so one cannot determine by higher level action which specific lower level state s_i realizes the higher level state $\mathcal{S}$. (But there is a proviso: one can design the kind of apparatus that occurs in a quantum optics laboratory so that some higher level variables access specific lower level states. However, these are exceptional situations.) Philosophers characterise this existence of equivalence classes through the phrase 'multiple realization' of the higher level state.

6.3.2 *Changing or Creating the Basic Elements*

One further point of importance is that it is not necessarily the case that one always has unchanging lower level elements being combined in different ways to form higher level complex structures. It may occur that the higher level context actually changes the very nature of the lower level entities that are combined to make the whole. An example from physics is that a free neutron has completely different behaviour than one bound in a nucleus: the former decays with a half life of 11 min, the latter last billions of years. Therefore, its essential nature is changed by context. An example from chemistry is that a free hydrogen is quite different than a hydrogen atom incorporated in a water molecule. It is an essentially different entity. More than that, top-down effects may even create the lower level elements (see Sect. 6.2.3.1) or delete them (see Sect. 6.5.1 on adaptive selection below).

Thus the idea of higher level causation being due to interactions between invariant lower level elements is crucially wrong. The existence of the lower level elements, and the nature of the interactions between them, can be contextually dependent.

6.3.3 *Types of Top-Down Causation in Physics*

The different classes of top-down causation have been discussed in previous chapters, but some of them (TD4 in Sect. 4.4 and TD5 in Sect. 4.5) only occur in the context of biological systems. In the case of purely physical systems, we have:

- **Deterministic Top-Down Causation (TD1)**. This is ubiquitous (Sect. 4.1). The way boundary conditions affect outcomes is a standard part of physical understanding, leading to effects such as shaping the sound of a violin and hearing the shape of a drum. Furthermore, contextual variables control lower level systems in cases such as nucleosynthesis in the early universe and in stars.
- **Adaptive Selection (TD3)**. Selection of a preferred outcome from an ensemble of entities or states (Sect. 4.3), which also occurs in physical systems, for example, in state vector preparation, nanotechnology selection procedures, and purification procedures which are critical to the very existence of the study of chemistry and solid state physics.

By contrast, feedback control (TD2) (Sect. 4.2) probably does not occur in natural physical systems, but only in engineered systems and biology (feedback processes occur in astrophysics and geophysics, but they are not control processes governed by goals.)

6.4 Deterministic Top-Down Effects in Physics (TD1)

Examples of deterministic top-down causation (TD1) in physics include:

- The effects of contextual variables (Sect. 6.4.1).
- The effects of boundary conditions (Sect. 6.4.2).
- Effective potentials and structural conditions (Sect. 6.4.3).
- Binding energies (Sect. 6.4.4).
- Features of computational mechanics (Sect. 6.4.5).

6.4.1 Contextual Variables

Outcomes in physics depend on the values of macro variables which are the context of the micro-state, hence they influence micro variables in a top-down way. This is quite non-controversial, it is just not usually expressed in this way.

A simple example is provided by the ideal gas laws (6.15) with macro variables (6.13), (6.14), and micro variables the particle positions and velocities. For gas constrained in a cylinder by a piston, we can change the micro-states by altering macro variables, e.g., by compressing the gas, that is, by changing V, which then changes lower level states (speeding up the molecules and so changing their positions). This is the top-down effect of the higher level variables on lower level states. The universal constant R is the link between the micro and macro states, because it relates the energy of micro-states to the values observed at the bulk level. The relation between the macroscopic variables pressure P, volume V, amount of gas n, and temperature T does not depend on detailed microscopic variables such as velocities, positions, and masses of any specific molecules. Indeed, we don't know those values. The gas is simply constrained by the cylinder and piston.

Equivalence Classes of Lower Level States. The variables at the macro-level are the only handle we have on lower level states: we cannot (except in unusual circumstances) manipulate the micro-level variables directly. The set of all lower states corresponding to a single higher level state form an equivalence class as far as the higher level variables and behaviour are concerned. They are characterised by corresponding subspaces of the particle phase space [90], and are the effective variables that matter in terms of controlling the gas behaviour, rather than the specific lower level state that instantiates the higher level one. This is why lower level equivalence classes, rather than individual lower level states, are the key to understanding the dynamics: engineering design will specify the higher level state required, and the engineer has no interest in which specific lower level states instantiate the specified higher level variables as an engine or refrigerator performs its duty cycle.

6.4.2 *Effect of Boundary Conditions*

Solutions of partial differential equations (PDEs) depend on sources and boundary conditions, either at infinity or on some finite boundary. This is standard fare in theoretical physics [26] and examples abound:

- A standard example is boundary conditions at infinity (a macro state) governing outgoing radiation from an antenna, and hence influencing local electron movements and field configurations [63].
- More complex is the reaction–diffusion equation

$$\partial_t q = \underline{\underline{D}} \nabla^2 q + R(q) \ , \tag{6.22}$$

  which creates spatial patterns that depend on the boundaries set by context, with results that are significant in biological pattern formation [85]. The local distribution of molecules is shaped by these larger scale conditions.
- A classic example is hearing the shape of a drum [66]. The frequencies at which a drumhead can vibrate, and hence the positions of the atoms that make up the drumhead, depend on its shape. The Helmholtz equation tells us the frequencies if we know the shape. These frequencies are the eigenvalues of the Laplacian in the region.
- The distinctive sound of a particular violin is the result of interactions between its component parts when a bow is drawn across the strings, causing them to vibrate and transmit the vibration to the body of the violin. The tension and type of strings, the structure of the bow, and the shape and construction of the body will all affect the tonal quality of the sound by selecting which atoms move where and when [120].

6.4.3 Structural Conditions and Effective Potentials

6.4.3.1 Structural Conditions

An important form of top-down causation in complex systems is through structural conditions, such as the electrical wiring which channels currents in electric circuits. These circuits are an emergent higher level entity arising out of the configurations of the atoms that make up the wiring. One does not need to include a representation of each individual interacting atom to characterise the wiring diagram or function. It is the specific nature of the connections that matters: which relay, transistor, or other component is connected to which through the specific wiring pattern used. This connectivity pattern cannot be reduced to a description simply in terms of the properties of atoms or electrons, even though it is made up of them. It is a higher level of structure, described at a different scale.

The key feature here is not just that an electric wire allows currents to flow along the wire; it is that it prevents them flowing orthogonally to the wire, because the resistance of the insulation surrounding the wire is effectively infinite. This feature can be represented by a square well potential. Thus electrical wiring systems can be represented through an effective potential system which channels the flow of electrons. Other examples range from integrated circuits to split-gate devices used in nanotechnology [83, pp. 96, 104, 112], to telephone systems, chemical plants, and neuronal connections via dendrites and axons in a brain.

> **Constraints Create Possibilities.** In each case it is the constraints created by the structure that channel lower level causation and so create possibilities that are not there when this structure is not present, as in a gas.

6.4.3.2 Effective Potentials

One can use effective potentials to represent such higher level structures emerging from the underlying physical levels, and then acting down on the lower level components to channel the way they interact with each other. One does not need to include a representation of each individual interacting atom in the structure. Rather one represents the interactions between many atoms in terms of an effective overall potential, representing the contextual system, i.e., the network of interconnections, as a single functional entity.

These are examples of the method of *mean field theory* [25, pp. 198–208], which can be applied in many other contexts as well as representing structural conditions. Gemmer et al. give an illuminating example [54, pp. 74–77] in there discussion of an ideal gas in a container. The container provides the environment for the gas and is made up of an interacting set of particles [54, Fig. 7.2]. Starting with a standard interaction Hamiltonian, coarse-graining leads to an effective 'box' potential $\hat{V}^g$ for each gas particle, comprising the mean effect of all the atoms in the container walls. This mean potential is then the higher level context within which the gas particle moves. It can be represented [54, Fig. 7.3] by a smooth set of equipotential lines, the transition from Figs. 7.2 and 7.3 being a classic illustration of the coarse-graining process. One can regard the result as top-down action by the potential (regarded as an entity in its own right) on the gas particles. The underlying equivalence classes are all the different configurations of particles making up the container that lead to the same effective potential. It is the equivalence class, equivalent to a macro state, that is the significant causal entity, rather than any detailed particle configuration that leads to the potential. It is precisely because this approach makes sense that it is meaningful to refer to a cylinder containing the gas in an engine. The cylinder is given functionally by the effective potential that contains the gas particles within a specific volume. This is the higher level reality that engineers deal with.

Similar examples are the potential wells used in nuclear shell models [31, pp. 140–144], and the Slater treatment of complex atoms, explained by Pauling and Wilson in the following terms [87, p. 230]:

> All of the methods we shall consider are based on the approximation in which the interaction of the electrons with each other has either been omitted or been replaced by a centrally symmetric force field approximately representing the average effect of all the other electrons on the one under consideration.

6.4.4 Binding Energies and Altered Properties

When there are such terms in the interaction representing the overall context, this will result in changes in energies. An important example is *nuclear binding energies*, the cost of putting emergent nuclear structures together, which can be reclaimed on dismantling the structure [70]. These energies would not be there if the structure (a nucleus) was not there, so it is a direct result of the existence of the higher level structure. Nucleons on their own have no such energy term. These are of course of crucial importance in nuclear physics, leading both to the stability of matter under ordinary circumstances and the possibility of extracting nuclear energy is a suitably hot environment. Similarly, there are atomic binding energies, key in atomic physics, and molecular binding energies, of crucial importance in chemistry. Each is associated with a threshold above which entities are dissociated and release energy (e.g., burning of a material) and below which they are stable. In each case the energy is associated with the emergent structure, not with the parts that make it up.

Associated with this is the fact that bound entities may interact in a different way with external particles or fields than when they are free. Thus electrons bound in an atom interact with light to give spectra characteristic of the atom, while free electrons just scatter light. The nature of the photon–electron interaction is quite different in these two contexts. As mentioned before, neutrons decay by β-decay when free, with a half-life of about 10 min, emitting an electron and an electron anti-neutrino, but they last billions of years when bound in a stable nucleus because of the Pauli exclusion principle. The same fundamental particle interactions give quite different outcomes in different settings. But it is the way an entity interacts with others that characterises what it is. These entities have changed their nature due to their local context (Sect. 6.3.2).

6.4.5 *Computational Mechanics*

This book has emphasized the way top-down effects are characterised by the existence of equivalence classes of lower level entities as the key to a dynamic process. Shalizi and Crutchfield [106] have developed computational mechanics, an approach to structural complexity, that is based on causal states of a process being represented in such a way. Hence it is in fact a representation of top-down effects. They call the dynamics of this representation an ε-machine, and show that it is the minimal one consistent with accurate prediction.

The core of this approach is to focus attention on patterns within a statistical ensemble and their possible representations. Using ideas from information theory, they define causal states that are equivalence classes of behaviors. The structure of transitions between causal states is the ε-machine. They show that the causal states are ideal from the point of view of Occam's razor, being the simplest way of attaining the maximum possible predictive power. And the causal states are uniquely optimal. Computational mechanics is not characterised as such, but in fact seems an intriguing way of developing aspects of top-down causation in physical systems.

6.5 Adaptive Selection in Physics and Chemistry (TD3)

The process of adaptive selection (see Sect. 4.3) [53, 67] is ubiquitous in biology [23] and the way the brain functions (Chap. 7). It is perhaps something of a surprise that it also occurs in the contexts of physics and chemistry. Here I consider in turn:

- Adaptive selection (Sect. 6.5.1).
- Maxwell's demon (Sect. 6.5.2).
- Separation and purification processes (Sect. 6.5.3).

6.5.1 *Adaptive Selection*

The basic process in adaptive selection is that selection takes place from an ensemble of initial states to produce a restricted set of final states that satisfy some given selection criterion. The process is summarized in Table 6.5.

Table 6.5 The basic features of adaptive selection

System state	⇐ Selection agent: selects state	
Variation ⇓	⇑	Meta-goals
Ensemble of system states	⇒ Preferred state in ensemble ⇐	Selection criteria
	⇑	
	Environment	

Selection takes place from an ensemble of states, the selection being based on the action of some selection criteria in the context of the specific current environment

In effect, in a selection event, the selection agent compares the entities available in the initial ensemble to determine the best candidates on the basis of the preset selection criteria, evaluated in the current environmental context. The best candidates are selected and retained as the outcome of the event and the rest are discarded. The meta-goals embodied in the selection criteria do not necessarily lead to a specific final state (although they may do in some restricted circumstances): rather they lead to any one of a class of states that tends to promote the meta-goals. Thus the final state is not uniquely determined by the initial data. Random variation influences the outcome by leading to a suite of states from which an adaptive selection is made in the context of both the selection criteria and the environment. Note that it can take place in a one-off form: in biology it gains its enormous strength because it is repeated so many times, but that repetition is not essential to the concept of selection. One could call it simply *selection*, but I prefer *adaptive selection*, to emphasize that it always takes place as a consequence of the existence of selection criteria, which are higher level entities in the hierarchy of causation. Hence this is another form of top-down action [36]. Examples are Maxwell's demon (Sect. 6.5.2), purification processes (Sect. 6.5.3), decoherence (Sect. 6.6.2), and state vector preparation (Sect. 6.6.5).

6.5.2 *Maxwell's Demon*

Selection is what enables an apparent local violation of the second law of thermodynamics, as in the case of *Maxwell's demon* (see [2, pp. 4–6], [24, pp. 186–189, 196–199], [49, Sect. 46–5], [74]). This is indeed an example of an adaptive selection agent, acting against the local stream of entropy growth by selecting high-energy molecules from those with random velocities approaching a trap-door between two compartments. The selection criterion is the threshold velocity v_{c}, deciding whether a molecule will be admitted into the other partition or not. It is significant that Maxwell's demon type devices can be created in the lab [95, 96, 99, 102], explicitly demonstrating that adaptive selection can arise in a quantum physics context. It occurs also in microbiology, where active transport systems are enabled by voltage-gated ion channels [76, pp. 191–206].

6.5.3 Separation and Purification Processes

Particularly important is the way separation and purification processes underlie our technological capabilities by enabling us to obtain specific chemical elements and compounds as needed. This is another case of adaptive selection, locally going against the grain of the second law of thermodynamics at the expense of the environment.

The specific processes that enable these non-unitary effects are detailed in [7, 60]. Methods used include adsorption, centrifugation, chromatography, crystallization, decantation, distillation, electrophoresis, evaporation, leaching, flotation, flocculation, filtration, magnetic separation, precipitation, sedimentation, sieving, sublimation, and winnowing. Indeed, one can use almost any physical or chemical difference between components as the basis of separation and purification, for example, differences in size, shape, mass, density, electric or magnetic properties, or chemical affinity. Chemistry and chemical engineering would be impossible if it were not for this capability, as well as most of civil, electrical, mechanical engineering, an manufacturing, which all depend on using the right components made of the right materials for specific purposes.

6.6 Top-Down Effects: Micro Physics

Quantum theory underlies all physics, so it is of interest to see where top-down effects can occur in this domain. It happens in terms of the following:

- Open systems and their environment (Sect. 6.6.1).
- Decoherence (Sect. 6.6.2).
- Lattice waves and quasiparticles (Sect. 6.6.3).
- Topological effects (Sect. 6.6.4).
- State preparation (Sect. 6.6.5).
- Measurement (Sect. 6.6.6).

6.6.1 Open Systems and Their Environment

Effect of the Environment on the System. Following Breuer and Petruccione, consider an open quantum system $\mathcal{S}$ ('the system') coupled to another quantum system $\mathcal{B}$ ('the environment'), with respective Hilbert spaces $\mathcal{H}_{\mathcal{S}}$ and $\mathcal{H}_{\mathcal{B}}$ [21, pp. 110–120]. The Hilbert space $\mathcal{H}$ of the combined system $\mathcal{T} = \mathcal{S}+\mathcal{B}$ is $\mathcal{H} = \mathcal{H}_{\mathcal{S}}\otimes\mathcal{H}_{\mathcal{B}}$. The total Hamiltonian $H_{\mathcal{T}}$ is taken to be of the form

$$H_{\mathcal{T}} = H_{\mathcal{S}} \otimes I_{\mathcal{B}} + I_{\mathcal{S}} \otimes H_{\mathcal{B}} + \widehat{H}_I(t) , \tag{6.23}$$

where $H_{\mathcal{S}}$ is the self-Hamiltonian of the open system, $H_{\mathcal{B}}$ the free Hamiltonian of the environment, and $\widehat{H}_{\rm I}(t)$ the Hamiltonian describing the interaction between the system and the environment. Now an ensemble $\mathcal{E}$ of pure ensembles $\mathcal{E}_\alpha$ for the total system $\mathcal{S}$ with weights w_α has a density matrix

$$\rho = \sum_\alpha w_\alpha |\psi_\alpha\rangle\langle\psi_\alpha| . \tag{6.24}$$

The reduced density matrix for the system $\mathcal{S}$, given by tracing out the environment, is

$$\rho_{\mathcal{S}} = tr_{\mathcal{S}}\rho . \tag{6.25}$$

It follows from the unitary evolution of the total density matrix ρ that the reduced density matrix evolves according to the Lindblad master equation

$$\frac{\mathrm{d}}{\mathrm{d}t}\rho_{\mathcal{S}}(t) = -\mathrm{i}\big[H, \rho_{\mathcal{S}}(t)\big] + \mathcal{D}\big(\rho_{\mathcal{S}}(t)\big) , \tag{6.26}$$

where the unitary part of the dynamics is generated by the new Hamiltonian H and the dissipator $\mathcal{D}(\rho_{\mathcal{S}})$ is determined by the spectral decomposition of the density matrix $\rho_{\mathcal{B}}$ of the environment [21, pp. 103–119]. The viewpoint here is that shown in Table 6.6.

The two key points then are that (i) in general $H \neq H_{\mathcal{S}}$—this is what opens the way to the renormalization group and higher level effective Hamiltonian theories—and (ii) generically $\mathcal{D}(\rho_{\mathcal{S}}) \neq 0$, i.e., the higher level system is not Hamiltonian, and hence (6.26) is associated with the generation of entropy (see [21, pp. 123–125] and [122]). This carries through to all the other versions of the master equation, e.g., the interaction picture master equation [21, p. 130] and the quantum optical master equation [21, pp. 140–149].

6.6.1.1 The Caldeira–Leggett Model

This is a system plus heat reservoir model for the description of dissipation phenomena in solid state physics (see [21, pp. 166–172] and [22]). Here the Lagrangian of the composite system T consisting of the system S of interest and a heat reservoir R takes the form

$$L_{\rm T} = L_{\rm S} + L_{\rm R} + L_{\rm I} + L_{\rm CT} , \tag{6.27}$$

Table 6.6 The system plus environment evolve in a Hamiltonian way, and interact with each other. When the environment is traced over, the system evolves in a non-Hamiltonian way

(Hamiltonian)	System plus environment $\mathcal{T}$	
	$\Downarrow\Downarrow$	(Coarse-grained)
System $\mathcal{S}$	$\Longleftrightarrow$ components $\Longleftrightarrow$	Environment $\mathcal{B}$
(Non-Hamiltonian)		

where

$$L_S = \frac{1}{2}M\dot{q}^2 - V(q) , \tag{6.28}$$

$$L_I = q \sum_k C_k q_k , \tag{6.29}$$

$$L_R = \sum_k \frac{1}{2} m_k \left(\frac{dq_k}{dt} \right)^2 - \sum_k \frac{1}{2} m_k \omega_k 2 q_k^2 , \tag{6.30}$$

$$L_{CT} = -q^2 \sum_k \frac{1}{2} C_k^2 / m_k \omega_k^2 , \tag{6.31}$$

are respectively the Lagrangians of the system of interest, interaction, reservoir, and counterterm (see below). The reservoir consists of a set of non-interacting harmonic oscillators with coordinates q_k, masses m_k, and natural frequencies ω_k. Each one of them is coupled to the system of interest by a coupling constant C_k. The counterterm L_{CT} is introduced to cancel an extra harmonic contribution that would come from the coupling to the environmental oscillators. This term represents a top-down effect from the environment to the system, because L_I completely represents the lower level interactions between the system and the environment. The effect of the heat bath is more than the sum of its parts when $L_{CT} \neq 0$, because the summed effect of those parts is given by L_I. The term L_{CT} is a contextual term, because it would not be there if there was no heat bath.

6.6.2 *Decoherence*

Decoherence is the process whereby the environment (a macro context) decoheres the wave function and selects preferred pointer states, thus crucially determining the nature of micro outcomes (see [21, pp. 212–270], [62, p. 155], and [118, pp. 121–141]). Zurek argues that this can be seen as a Darwinian-like process he calls environmental selection (*Einselection*) [125, 126]. This can therefore be seen as a case of top-down causation by adaptive selection (Sect. 6.5.1): the lower level dynamics does not by itself determine the outcome, which is shaped by the higher level context of the environment.

6.6.3 *Lattice Waves and Quasiparticles*

The basic idea was discussed in Sect. 6.2.3.1. Higher level emergent structures lead to the existence of lower level entities that then cause interesting physical effects.

6.6.3.1 Band Structure and Quasiparticles

The periodic crystal structure in a metal leads via Bloch's theorem [124, pp. 16–20] to lattice waves [124, pp. 27–75] and an electronic band structure depending on the particular solid involved [124, pp. 93–94, 119–128], resulting in all the phenomena associated with the band structure. The entire machinery for describing the lattice periodicity refers to a scale much larger than that of the electron, and hence is not describable in terms appropriate to that scale. Thus these effects all exist because of the macro properties of the solid, i.e., the crystal structure, and hence represent top-down causation from that structure to the electron states.

For example, this can lead to the existence of quasiparticles such as *phonons* [124, pp. 59–62] that result from vibrations of the lattice structure, and hence associated phenomena such as the *U-process* whereby momentum in electron scattering processes is transferred to the system as a whole. Other examples are *holes*, conduction electrons with negative effective mass as determined by the energy surface $\mathcal{E}(k)$ [124, pp. 182–186], which are central to the physics of semiconductors [124, pp. 59–62], and *plasmons* (particles derived from plasma oscillations). The quantum Hall effect is a result of the existence of composite fermions, realised in the interface between two semiconductors [65]. In all cases, it is the higher level context that leads to their existence, because it determines the form of $\mathcal{E}(k)$. This represents the effective result of the existence of the macro structure, similarly to the way effective potentials do (Sect. 6.4.3).

6.6.3.2 Superconductivity

In superconductivity, despite their repulsion for each other, the electrons form pairs (called Cooper pairs), which are the basic entities of the superconducting state. This happens by a cooperative process: the negatively charged electrons cause distortions of the lattice of positive ions in which they move, and the real attraction occurs between these distortions. The Cooper pairs are produced by the exchange of phonons between electrons [124, pp. 382–386] and lead to superconductivity [124, pp. 386–394] and associated phenomena such as superfluidity in metals [124, pp. 394–396]. The Nobel lecture by Laughlin [73] discusses the implications (see Sect. 5.5.1).

The claim made here is that this dynamics is possible because of top-down causation. They are *emergent phenomena* in the sense that they simply would not exist if the macro-structure did not exist (Sect. 6.3), and hence cannot be understood by a purely bottom-up analysis.

6.6.4 Topological Effects

Some quantum effects, however, are topological effects [16, 119], and as such dependent on non-local emergent properties.[1] They are therefore top-down effects.

6.6.4.1 The Quantum Hall Effect

The fractional quantum Hall (FQH) effect is an emergent phenomenon which is not based on symmetry-breaking, often seen as the basic means by which interesting emergent structures arise in physics. Instead, FQH states are part of a distinct class of ordered matter that is defined topologically [71]. Topologically ordered states result from complex long-ranged correlations between their constituent parts, such that the system displays strongly irreducible, qualitatively novel properties. This is clearly a top-down effect from the collective level to the level of the 2D system of electrons.

6.6.4.2 The Aharanov–Bohm Effect

The Aharanov–Bohm effect [1, 92] is a nonlocal effect in quantum physics when described in terms of fields, in which an electrically charged particle is affected by an electromagnetic field (**E**, **H**) despite moving in a domain in which both **E** and **H** are zero. However, the paths of the relevant particles circle a domain where **E** and **H** are nonzero. The effect has been observed experimentally [13, 108]. It implies that holonomy is key to quantum physics.

6.6.5 State Preparation

State vector preparation is key to experimental setups in quantum physics, and is a non-unitary process, because it can produce particles in a specific eigenstate. Indeed, it acts just like state vector reduction (6.1), being a non-unitary transition that maps a mixed state to a pure state. How can this happen in a way compatible with quantum theoretical dynamics?

The crucial feature of quantum state preparation is pointed out by Isham [62, pp. 74, 134] as follows: selected states are drawn from some collection $\mathcal{E}_i$ of initial states by some suitable apparatus, for example to have some specific spin state, as in the Stern–Gerlach experiment, and the other states are discarded. This is another case of adaptive selection (see Sect. 6.5.1): selection takes place from a (statistical) variety of initial states according to some higher level selection criterion. As explained in Sect. 6.5.1, this is the characteristic way one can generate order out of a disordered

[1] See https://physics.aps.org/articles/v1/6 for a brief discussion.

set of states by a process of selection from an ensemble of systems, and so generate useful information [98], just as in the case of Maxwell's demon.

This top-down effect from the apparatus to the particles causes an effective non-unitary dynamics at the lower levels, which cannot therefore be described by the Schrödinger or Dirac equations. The apparatus is specifically designed to have this non-unitary effect on the lower level. This happens in two basic ways:

- separation and selection, which is unitary up to the moment of selection, or
- selective absorption, which absorbs energy and so is non-unitary all the time.

Both are examples of emergence leading to top-down action.

6.6.5.1 Separation and Selection

This is a very general basis for state selection. In the case of the *Stern-Gerlach experiment* [50, Sects. 5-1–5-9], collimation of an incoming stream of atoms by some slits is followed by deflection in a non-uniform magnetic field, which separates the initial beam into final beams according to their spin. Each final beam is then a polarized beam in a prepared spin state. Thus when we choose to examine a particular spin by selecting one of these beams, one set of incoming states is selected and the other discarded. A *mass spectrometer* works on the same principle, separating out particle masses, as does a *spectrograph*, where a prism or diffraction grating sorts out light by wavelength (so one can select a specific pure colour by using a slit to collimate the light after it has passed through the prism).

Another example is a *Nicol prism*, used to generate a beam of polarized light [77, p. 132]. A crystal of Iceland spar is cut diagonally, the two parts being joined by Canada balsam. When unpolarized light enters the crystal, it is split into two polarized rays by *birefringence* (see [59, pp. 111–118] and [77, p. 131]), the decomposition of a light ray into two rays by an anisotropic crystal. The crystal is shaped so that one beam is totally internally reflected and lost, while the other emerges parallel to the incidence direction. Birefringence is caused by electromagnetic polarization in an anisotropic medium with dielectric tensor ϵ_{ij} resulting from the coarse-graining of the dipole contributions to the electric field (Sect. 6.2) [63, pp. 116–122].

Polarization is also caused by *reflection of light* at less than the critical angle at a surface separating two transparent media. Then partial reflection and partial transmission take place (see [59, pp. 40–41, 108–109] and [77, pp. 109–110]), again separating the initial beam into two polarized beams. So this can also be used to prepare polarized states. The anisotropy in this case is caused by the layer separating the two media. The reflected light is polarized normal to the incidence plane.

6.6.5.2 Selective Absorption

Dichroism is the selective absorption of one polarization state due to a linear structure in a polarizer, which therefore selects a specific spin state from a beam of incoming photons, thereby rejecting the other states. This may be realised by a *wire grid polarizer* [59, pp. 105–106]: a set of closely spaced fine conducting wires. If a wave interacts with these wires, the electric field component parallel to the wires drives electrons along the wire, generating an alternating current which encounters resistance. This absorbs energy from this component of the incoming field, heating the material. The electrons re-radiate a wave which further tends to cancel this component of the incident wave, while the transverse component is not so affected. Hence the transmitted wave is linearly polarized. The same effect occurs in a *polaroid polarizer*, consisting of many parallel microscopic crystals embedded in a transparent polymer film (see [59, p. 105] and [77, pp. 132–133]). Similarly, a spin-polarized current in a metal can be generated by passing the current through a ferromagnetic material.

A different example is a *filter* that absorbs some wavelengths of light and transmits others, because of the molecular structure of the glass, hence selecting a particular frequency range by adaptive absorption.

6.6.5.3 Emergence Leading to Top-Down Action

In each case, the underlying unitary quantum electrodynamics leads to emergence of higher level classical structures (wires, crystals, and so on) that can then act down to the particle level to cause non-unitary transformations which can change a mixed incoming beam to a pure state (Table 6.7).

As in the case of the band structures of metals, this top-down action depends on the physical structure of the polarizing material or device as indicated in the above examples, and so is a case of top-down causation by adaptive selection in the context of the structure of the material. In the case of separation and selection, the lower level evolution is unitary until selection takes place. In the case of selective absorption, the ongoing non-unitary nature of the resulting higher level effective action is reflected in an energy loss and heating associated with the process.

6.6.6 Measurement

Measurement is a process with significant parallels to the process of state preparation, in that both can change a wave function that is a superposition of states to an eigenfunction. Hence they are non-unitary processes that are not equivalent to action by the Schrödinger equation. The experimental viewpoint is that the macro observer and apparatus have an existence as macro entities that can be taken for granted, and

Table 6.7 The postulated contextual view of state vector preparation

Classical apparatus	Non-linear system	Non-unitary
Emergence ⇑	Contextual effects ⇓	Adaptive selection
Quantum systems	State vector selection	Non-unitary

that can influence quantum states both in terms of enabling state vector preparation, and in terms of determining the context for outcomes of a measurement, for example, by determining the axes along which spin will be measured. These are of course both cases of top-down causation.

Some of the more advanced measurement techniques seem to directly involve adaptive selection. For example, this occurs in *weak measurements*, which are based on post-selection [2, pp. 225–227, 230–235]. This way of selecting some outcomes and discarding others is also central to the *generalized theory of quantum measurement* characterized by Breuer and Petruccione [21, pp. 83–85]. It may well be worth pursuing the idea that adaptive selection is the heart of the measurement process [37].

6.7 Top-Down Effects: Cosmology

The major thrust of present day scientific cosmology is that of examining the effect of local physical laws on the large-scale structure of the cosmos (Sect. 6.2.5). However, from the earliest times there has also been a counter-theme: the study of the way that global properties of the universe can influence its local properties. Writers such as Sciama consistently emphasized the *interconnectedness of the universe* [104]: each part interacting with each other part, and with very distant parts being as important as local regions in these interactions, a prime example being Olbers' paradox [19]. Because of this interconnection, in principle one can obtain some understanding of the whole from any part. Indeed, if one were clever enough and understood enough physics, one could in principle completely deduce the nature of the whole from a sufficient study of its parts. An example of this line of argument is Bondi's suggestion that one could in principle deduce the expansion of the universe, and even the approximate value of the Hubble constant, from the existence of bus tickets.

Cosmology provides the environment for local physics and acts top-down in two ways. First, it determines contextual variables for local physics (Sect. 6.4.1). This occurs in particular for the following:

- Element formation (Sect. 6.7.1).
- Structure formation (Sect. 6.7.2).

Second, it sets boundary conditions for isolated systems (Sect. 6.4.2), a form of top-down constraint. This occurs in particular as regards the following:

- Mach's principle (Sect. 6.7.4).
- Olbers' paradox (Sect. 6.7.3).
- The arrow of time (Sect. 6.7.5).
- Existence of isolated systems (Sect. 6.7.6).

Together these show how local systems are indeed influenced by the universe at large [34, 45].

6.7.1 Element Formation

The way cosmology determines the rate of change of temperature with time in the early universe is discussed in Sect. 6.2.5. In this *hot big bang* epoch in the early universe, we can use standard physical laws to examine the processes going on in the expanding mixture of matter and radiation [30, 94, 112]. At very early times and high temperatures, only elementary particles can survive and even neutrinos had a very small mean free path. But as the universe cooled down, neutrinos decoupled from the matter and streamed freely through space. At these times the expansion of the universe was radiation-dominated, and we can approximate the universe then by models with $k = 0$, $p = \rho/3$, and $\Lambda = 0$, the resulting simple solution of the Friedmann equation uniquely relating time to temperature, see (6.20) (there are no free constants in this equation).

Nucleosynthesis refers to the formation of the light elements. Above about 10^9 K, nuclei could not exist because the radiation was so energetic that as fast as they formed, they were disrupted into their constituent parts (protons and neutrons). However, below this temperature, if particles collided with each other with sufficient energy for nuclear reactions to take place, the resultant nuclei remained intact (the radiation being less energetic than their binding energy and hence unable to disrupt them). Thus the nuclei of the light elements—deuterium, tritium, helium, and lithium—were created by neutron capture. This process ceased when the temperature dropped below about 10^8 K (the nuclear reaction threshold). In this way, the proportions of these light elements at the end of nucleosynthesis were determined and they have remained virtually unchanged since. The rate of reaction was extremely high. All this took place within the first three minutes of the expansion of the universe. Theory and observations agree extremely well, and this is one of the major triumphs of the big bang theory:

Nucleosynthesis. Theory and observation are in excellent agreement provided the density of baryons is low, i.e., $\Omega_{\text{bar}\,0} \approx 0.044$. Then the predicted abundances of these elements (25 % helium and 75 % hydrogen by mass, the others being less than 1 %) agrees very closely with the observed abundances.

Thus the standard cosmological model explains the origin of the light elements in terms of known nuclear reactions taking place in the early universe [103, 112]. However, heavier elements could not have formed in the time available (about 3 min).

Because the global expansion rate and resulting temperature–time relation (6.20) determines the outcome of the local physical interactions at the time of nucleosynthesis, one can use measurements of element abundances to constrain the dynamics of the early universe and hence to help fix cosmological parameters such as the average density of baryons. This has an important outcome: it tells us dark matter is non-baryonic [30, 94].

6.7.2 Structure Formation

In a similar way, structure formation in the expanding universe is controlled initially by contextual variables related to the expansion rate of the universe. This is true for

- the initiation of perturbations during inflation,
- their evolution as acoustic waves during the hot big bang era,
- the subsequent growth of perturbations due to gravitational attraction in the early matter-dominated era.

(see [30, 94] for discussion). I will illustrate the point just by considering the latter era, when in terms of the gauge invariant and covariant fractional density $\mathcal{D}_a$, the linear perturbation growth equation is [40]

$$(\mathcal{D}_a)^{\cdot\cdot} + \frac{2}{3}\Theta(\mathcal{D}_a)^{\cdot} - \frac{1}{2}\kappa\rho\mathcal{D}_a = 0\,, \tag{6.32}$$

where Θ is the background expansion rate and ρ the background energy density. These are determined by the Friedmann equation and conservation equation, giving the background dynamics at this time as:

$$a(t) = Ct^{2/3} \quad \Longrightarrow \quad \Theta = 3\frac{\dot{a}}{a} = \frac{2}{t}\,, \quad \kappa\rho = \frac{1}{3}\Theta^2 = \frac{4}{3}\left(\frac{1}{t}\right)^2\,. \tag{6.33}$$

The second order perturbation equation (6.32) becomes

$$(\mathcal{D}_a)^{\cdot\cdot} + \frac{4}{3}\frac{1}{t}(\mathcal{D}_a)^{\cdot} - \frac{2}{3}\left(\frac{1}{t}\right)^2\mathcal{D}_a = 0\,. \tag{6.34}$$

The solution is

$$\mathcal{D}_a = t^{2/3}C_{1a} + t^{-1}C_{2a}\,, \tag{6.35}$$

showing how there are power law growing and decaying modes. Thus cosmological contextual variables determine the rates of growth and decay of inhomogeneities due to gravitational attraction. One can make the point forcefully by noting that, if the inhomogeneity were to occur in a static universe (which would require a cosmological constant), then we would have $\Theta = 0$, $\rho =$ constant in (6.32) and the inhomogeneity growth would be exponential rather than a power law.

Putting together the various points mentioned above and working out the effects on the cosmic microwave background interacting with dark matter and baryons during the hot big bang era [30], one can use observations of the angular power spectra of the microwave background and its polarisation to put useful constraints on inflationary universe models [80, 81]. This is illustrated in Fig. 6.2. The galaxy cluster and WMAP limits are derived by studying the effect of the background model on structure formation in the expanding universe. They are much tighter than those derived from direct measurement of the background geometry by using supernova data alone.

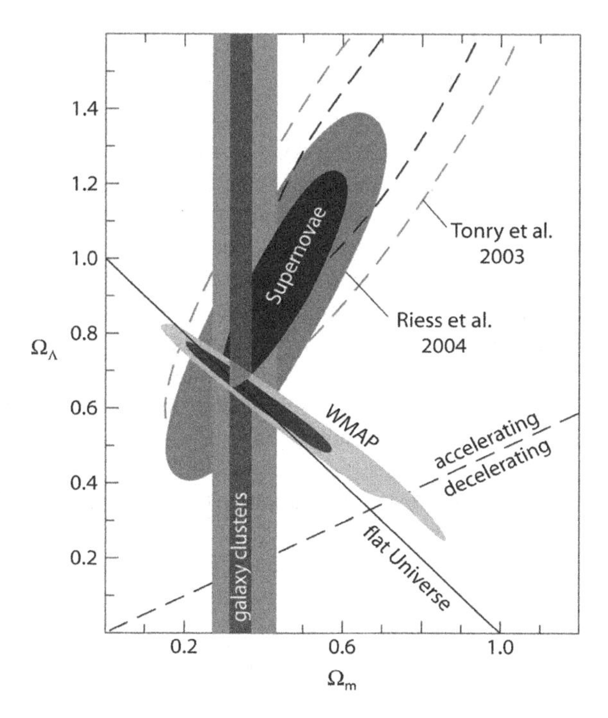

Fig. 6.2 Cosmological observations determine the present day values of dark energy density Ω_Λ and matter density Ω_m. The supernova data (*green*) is based on observing the background model geometry directly. The galaxy cluster data (*red*) reflects how the background model has affected structure formation. The WMAP data (*blue*) represents how this structure affects the observed CBR temperature power spectrum. It is the latter two—top-down effects from the cosmological background to smaller scales—that give us the best estimates of the cosmolgical parameters Ω_Λ, Ω_m. Taken from [69]. Reproduced by permission of the AAS (color figure online)

The possibility of determining these limits arises precisely because global variables have significant effects on local physical processes. This confirms the importance of top-down effects in cosmology, and the way we can use these effects to determine the structure of the universe.

> **Cosmological Tests.** The most sensitive limits on inflationary cosmological models come, not from direct measurement of spacetime curvature, for example by supernova observations, but from observations of structures that have formed and their effects on CMB anisotropies [80]: that is, the top-down effect of global cosmological conditions (which we want to measure) on structure formation.

On a smaller scale, astrophysical structure formation at the local level is crucially affected by environmental effects [56, 110]: the evolution of galaxies is a strong function of environment, because galaxies entering high density regions like clusters are prone to processes like ram pressure stripping, starvation, strangulation, and tidal stripping. These physical processes are significant because they influence the star formation rates and colours of galaxies as a function of environment.

6.7.3 *Olbers' Paradox*

Probably the oldest scientific investigation of top-down effects in the cosmological context was Olbers' paradox (why is the sky dark at night?) [19, 57, 104], actually developed by Halley. It arises from integrating over all very distant radiation sources in a static universe, supposing they extend to infinity and shine forever. In flat space-time, the radiation from such sources diverges, because their number goes up as r^2 and this compensates for the decrease in flux from each source, which goes down as $1/r^2$, where r is their radial distance from the observer. However, one can then take into account the fact that more distant sources would be occluded by closer ones: as each line of sight would intersect the surface of some star or other in the 'forest of stars' [57], the night sky in this context (and the day sky, for that matter) should be as bright as the surface of the Sun. Doing the calculation in static Robertson–Walker universes with r chosen as the area distance gives the same outcome, because the total intensity I of radiation received from a source (that is, flux per unit solid angle) is independent of the source's area distance.

However, in any evolving universe, the reciprocity theorem [33] shows that the total intensity of radiation received from a source is

$$I = \frac{1}{(1+z)^4} I_G \,, \tag{6.36}$$

where I_G its bolometric surface brightness and z its redshift, while the pointwise specific intensity I_ν of radiation received (the intensity per unit frequency range) in direction θ, ϕ is

$$I_\nu(\theta,\phi) = \frac{\mathcal{I}(\nu(1+z))}{(1+z)^3} I_G(\theta,\phi), \tag{6.37}$$

where $\mathcal{I}(\nu)$ is the source spectrum and $I_G(\theta,\phi)$ its surface brightness in the direction of observation. Hence, in an expanding universe, very distant sources at high z appear much fainter than nearby ones, and so the sky can appear dark even though every line of sight eventually intersects a source of some kind. This in principle resolves the paradox: we live in an expanding universe where intensity is diminished by redshift as in (6.36) and (6.37).

However, as emphasized by Ted Harrison [57], that is not the whole story: in the end, the dark night sky is the result of the fact that there is not enough radiation in the universe to generate a bright night sky. Most of the lines of sight from the earth effectively end up on the surface of last scattering of radiation after the hot big bang, the effect of numerous intervening galaxies being small.

6.7.3.1 Discrete Sources and Background Radiation

Thus the present day astronomical version of the Olbers' studies comprises two parts. The first is the study of the integrated radiation from all individually unresolvable sources, resulting in a predicted spectrum of integrated background radiation at all wavelengths coming from these sources, in particular, the optical, radio, and X-ray backgrounds [61]. The detailed theory of this background then resides in detailed astrophysical speculations about these sources and their evolution [79].

The second part is a dominant theme of present day theoretical and observational cosmology: namely, prediction and observation of the cosmic blackbody background radiation spectrum and anisotropies. This radiation is the relic radiation from the hot big bang era in the early universe, when matter and radiation were tightly coupled together. They decoupled, at the last scattering surface, when the temperature of this plasma dropped below its ionisation temperature of about $T_e = 4\,000\,\mathrm{K}$. It follows from (6.37) that the blackbody radiation emitted at that temperature is received as blackbody radiation at a temperature

$$T_o = T_e(1+z)^{-1}, \tag{6.38}$$

where z is the redshift of the emission. The redshift factor is about 1 100, so we receive cosmic relic blackbody radiation at about $3\,000\,\mathrm{K}/1100 = 2.73\,\mathrm{K}$ from that surface. There are small variations in this temperature over the sky due to gravitational field fluctuations and velocity effects, and study of these anisotropies is a major part of present day cosmology via the COBE, WMAP, and Planck satellites, as well as many ground-based experiments [30, 94].

Thus this can legitimately be regarded as the culmination of Olbers' investigations into the integrated radiation emitted by all sources in the universe, because this radiation dominates over all the radiation from discrete sources. The source is the uniformly distributed matter at the surface of last scattering, for we cannot receive electromagnetic radiation from more distant matter by any kind of radiation because the universe would have been opaque at earlier times. Hence, this matter comprises the visual horizon [42], and is the proper endpoint of the integral in the Olbers' calculation. Thus in the real universe, Olbers' integral does not extend to infinity, and 2.73 K is the temperature of the night sky (and for the day sky as well, away from the Sun). That temperature is not due to discrete sources, but rather is due to the primeval radiation from the hot big bang.

6.7.3.2 The Anthropic Version of Olbers' Paradox

This calculation is of considerable importance for daily life on Earth. It implies that the Earth is in contact with a thermal reservoir at a temperature of 3 K, which is the sink into which we dispose of our excess entropy, generated by all the processes of life. That sink is essential to the thermodynamic functioning of the biosphere, and hence to the existence of life on earth, for the biosphere functions by receiving high grade blackbody radiation from the Sun, using it to run biological and atmospheric processes, and then radiating the low grade waste heat away to the sky [90]. If the temperature of the sky were much higher, and certainly if it were above 300 K, life like ours on Earth would not be possible. Thus from this viewpoint, the reason we observe the night sky to be dark is that if that were not true, we would not be here to see it!

This illustrates the kind of 'anthropic' argument whereby cosmological conditions in the universe are important for local biology [10, 12].

6.7.4 Mach's Principle

Largely motivated by his position in the philosophical debates about relative versus absolute motion, Ernst Mach conjectured that the origin of inertia is interactions of local matter with very distant matter [19, 105]. The question relates particularly to what determines the nature of local inertial rest frames, and Newton's 'bucket experiment' suggest that it relates to the rest frame defined by very distant matter in the universe [105]. To some degree this principle seems to be embodied in general relativity, for example, through frame dragging (Lense–Thirring) and gravitoelectromagnetic effects [82], but the issue remains controversial [11]. However, it played a key role in motivating Einstein's thoughts as he developed general relativity theory.

6.7.5 The Arrow of Time

A fundamental problem in physics is the relation between reversible microphysics and irreversible emergent macrophysics [3, 19, 24, 28, 45, 90, 104]. This is a major gulf between the time-reversible microphysics that underlies all physical processes, where neither direction of time is preferred, and time irreversible macrophysics characterised by the second law of thermodynamics and the various arrows of time (radiation, thermodynamical, quantum mechanical, chemical, biological, and psychological) that are dominant in the macro world, all indicating the same future direction, characterising the passage of time.

The relation of a phenomenological (macro) definition to microscopic properties is often (Sect. 6.2) given via a process of coarse-graining [90], through which the macro description (given only in terms of macroscopic variables) explicitly loses information that is available in the detailed micro description (given in terms of microscopic variables). Then the basic quantitative question is how many different micro-states correspond to the same macro-state. A macro-state is more probable if it corresponds to a greater number of different micro-states, and time evolution will tend to go from a less probable to a more probable state, defined in this way [90]. This appears at first sight to solve the issue of how irreversibility emerges from underlying reversible dynamics.

6.7.5.1 The Classical H-Theorem

Boltzmann's H-theorem [123, pp. 43–48] makes this explicit. It shows that, in the case of interactions between particles in Newtonian theory, coarse-graining from micro to macro scales results in entropy increasing with time, because random motions in phase space take one from less probable smaller to more probable larger regions of phase space (see [24, pp. 172–174], [54, pp. 43–47], [89, pp. 686–696], and [90, pp. 9–56]). Indeed, statistical mechanics shows that the entropy

$$S = -k \sum_i p_i \ln p_i \tag{6.39}$$

of an isolated system, where k is Boltzmann's constant and p_i is the probability of the system being in a coarse-grained state i, increases in the future direction of time:

$$\frac{\mathrm{d}S}{\mathrm{d}t} \geq 0\,. \tag{6.40}$$

Hence, one can show that entropy increases to the future. The second law of thermodynamics (6.40) at the macro-level emerges from the coarse-grained underlying micro theory. This appears to show the origin of the arrow of time, as it shows that entropy increases towards the future, in accordance with the second law.

But this appearance of an arrow of time arising from the underlying theory is an illusion, as the underlying theory is time-symmetric, so there is no way an arrow of time can emerge only by a local coarse-graining procedure. Indeed Boltzmann's derivation of the increase of entropy applies equally to both directions of time! Let $t \to t' := -t$. Then the same proof that leads to (6.40) shows, step by step, that

$$\frac{\mathrm{d}S}{\mathrm{d}t'} \geq 0\,. \tag{6.41}$$

This is Loschmidt's paradox (see [88, Fig. 7.6], [89, pp. 696–699], and [90]):

Loschmidt's Paradox. The classical H-theorem predicts that entropy will increase to both the future and the past.

It has to do so: there is no direction of time in the microphysics. It does not therefore lead to an arrow of time.

6.7.5.2 The Quantum H-Theorem

Does quantum physics change this? Weinberg's derivation of the H-theorem on the basis of quantum field theory [113, pp. 150–151] depends only on unitarity. However, unitary transformations are time reversible. There is therefore nothing in the dynamics that can choose one time direction against the other. What works one way will work equally the other way. In more detail, entropy is once again defined by (6.39). Differentiating with respect to time gives

$$\frac{\mathrm{d}S}{\mathrm{d}t} = -k\sum_i \left(\frac{\mathrm{d}p_i}{\mathrm{d}t}\ln p_i + \frac{\mathrm{d}p_i}{\mathrm{d}t}\right) = -k\sum_i \frac{\mathrm{d}p_i}{\mathrm{d}t}\ln p_i\,, \tag{6.42}$$

using the fact that $\sum \mathrm{d}p_i/\mathrm{d}t = 0$, since $\sum p_i = 1$. Now Fermi's golden rule gives a master equation for the average rate of quantum jumps from state α to β, and from state β to α. For an isolated system, the jumps will make contributions

$$\frac{\mathrm{d}p_\alpha}{\mathrm{d}t} = \sum_\beta \nu_{\alpha\beta}(p_\beta - p_\alpha)\,, \quad \frac{\mathrm{d}p_\beta}{\mathrm{d}t} = \sum_\alpha \nu_{\alpha\beta}(p_\alpha - p_\beta)\,, \tag{6.43}$$

where the reversibility of the dynamics ensures that the same transition constant $\nu_{\alpha\beta}$ appears in both expressions. So

$$\frac{\mathrm{d}S}{\mathrm{d}t} = \frac{1}{2}k\sum_{\alpha\beta} \nu_{\alpha\beta}(p_\beta - p_\alpha)(\ln p_\beta - \ln p_\alpha)\,. \tag{6.44}$$

But the two brackets will have the same sign, so no contribution to $\mathrm{d}S/\mathrm{d}t$ can be negative. Therefore $\mathrm{d}S/\mathrm{d}t \geq 0$ for an isolated system: the second law (6.40) holds in the + direction of time given by the coordinate t.

But now choose the opposite direction of time. Define

$$t' = -t\,, \tag{6.45}$$

and relabel the articles $\alpha \to \beta$ and $\beta \to \alpha$ in (6.43) (this is allowed as there is no intrinsic difference between them). Then the proof goes through unchanged! Therefore $\mathrm{d}S/\mathrm{d}t' > 0$ [see (6.41)] holds also for an isolated system, as in the classical case:

> **Loschmidt's Paradox Extended.** The quantum H-theorem predicts entropy will increase to both the future and the past.

Loschmidt's paradox holds in this case too, and the H-theorem cannot provide the direction of time. The same will apply to the quantum theoretical derivation of an increase in entropy through evolution of the density matrix (see [21, pp. 123–125] and [54, pp. 38–42, 53–58]). It cannot resolve where the arrow of time comes from, or indeed why it is the same everywhere.

6.7.5.3 The Past Hypothesis

Given that the arrow of time cannot be derived in a bottom-up way from the microphysics, the only viable option for explaining it and the associated increase in entropy seems to be in terms of cosmological conditions [90, 104], that is, through some kind of *past hypothesis* (see [2, p. 176] and [3]):

> **The Past Hypothesis.** The direction of time must be derived by a top-down process from cosmological to local scales.

This is strongly supported by the fact that the entropy of the universe could have started off in a maximal state if initial conditions were different (see [24, pp. 345–346] and [88]), because black holes have much more entropy than standard homogeneous cosmologies (see [24, pp. 299–302] and [89, pp. 728–731]). Entropy is able to increase because the universe started off in a very special initial state, characterized by the Weyl curvature hypothesis: it was asymptotically conformally flat at the big bang [89, pp. 765–769]. This is the reason why the second law holds: entropy can increase to the future direction of time because it was small to start with. If it had started off in a maximally entropic state, this would not have been possible.

Thus the solution to the arrow of time problem seems to be that special boundary conditions existed on a cosmological scale at the beginning of spacetime: additional conditions are needed to get the second law of thermodynamics [47, pp. 41–54]. This then makes a fundamentally important difference to local physical behaviour, and so is a crucial form of top-down causation from the whole universe to local systems whose behaviour is not determined on the basis of local physical laws alone [3, 24, 88, 89]. It also depends in an essential way on these initial conditions.

6.7.5.4 The Direction of Time and Arrows of Time

However, this is not the whole story. We should distinguish two related concepts [38, 39]:

- **The Direction of Time**. This derives non-locally from the passage of time as the universe evolves. Thus it points from the start of the universe, which is a fixed time in the past, to the present day, which is at a continually increasing proper time from the start of the universe (it is currently 13.7 billion years form the start). This probably makes best sense in the context of an evolving block universe [35, 41].
- **Arrows of Time**. These determine the future locally as characterised by different physical and biological effects (thermodynamics, fluid dynamics, electrodynamics, chemistry, biology, psychology, and so on): entropy increases, waves arrive after they were sent, birth precedes death, and so on.

The direction of time is determined by cosmology, and the arrows of time are determined by the way local physics works out in the context given by cosmology. It is then a key feature of physics that these arrows of time all point in the same direction as the direction of time.

Table 6.8 Contextual determination of the arrow of time cascades down from cosmology to the underlying micro processes, on the natural sciences side, and then up to the brain and society, on the human sciences side

The arrow of time		
Cosmology		Brain, society
Top-down effects ⇓		⇑ Bottom-up effects
Non-equilibrium environment	⇒	Molecular processes
Top-down effects ⇓		⇑ Bottom-up effects
Quantum theory	⇒	Quantum theory

6.7.5.5 The Arrow of Time Cascade

The way this happens is by a top-down cascade from cosmological conditions to fundamental physics, followed by a bottom-up cascade through emergent structures [39]. The picture that emerges is shown in Table 6.8. In summary, this view proposes that:

- The fundamental direction of time from the start of the universe till today was set at the start of the universe. This probably makes best sense if spacetime is an evolving block universe, which grows as time evolves [35, 38, 44].
- The observable part of the universe started off in a special state which allowed structure formation to take place and entropy to grow [2, 3, 24].
- The arrow of time cascades down from cosmology to the quantum level (top-down effects) and then cascades up in biological systems (emergence effects) [39]. This is enabled by the expanding universe context which leads to a dark night sky allowing local non-equilibrium processes to occur [90].
- The arrow of time parameter t in the basic equations of physics (the Dirac and Schrödinger equations, Maxwell's equations and Einstein's equations in the 1+3 covariant formulation [42]) determines the rate of physical processes and hence the way time emerges in relation to physical objects.
- Each of these processes is enabled by top-down action taking place in suitable emergent local structural contexts, provided by molecular or solid-state structures [37]. These effects could not occur in a purely bottom-up way.

The detailed argument is in [38, 39, 41].

6.7.6 *Existence of Isolated Systems*

The final effect to be discussed here is the existence of isolated systems in the expanding universe. Paradoxically, an isolated system is a special case of top-down causation: it is the special case where there is no interference on the system by the outside environment, enabling the system to evolve solely under the influence of its own internal dynamics.

6.7.6.1 Isolated Systems and Life

An isolated system is only possible in a subset of cosmologies where tidal gravitational forces and gravitational waves are small, and likewise for incoming electromagnetic radiation. In most cosmologies, this will not be true, and the universe will interfere with local systems [34]. In those cases laboratory determination of the laws of physics will be difficult if not impossible. More than that, it is unlikely that life (and hence an experimenter) will come into existence in a universe where there are no stable environments in which evolutionary development of life can take place over billions of years. Thus lack of such interference is a particularly fortuitous form of top-down causation from the global context: it is an anthropic requirement that the universe allows local systems to evolve on their own, without significant interference.

This is not true in the early universe, where there are no isolated systems. The evolution of the universe can be regarded as *evolution from connected to more or less isolated systems* (inter alia, where the virial theorem holds). The initial conditions set by the almost spatially homogeneous (almost Friedmann–Lemaître–Robertson–Walker, or FLRW) universe are indeed such that locally independent systems can evolve more or less freely, which is a key requirement for biological complexity and life to exist. Hence, *it would seem that the universe has to be almost FLRW for the existence of life to occur*. This is discussed further in [34].

6.7.6.2 Possibility of Newtonian Physics

A further interesting point is what kinds of conditions on a surface $\mathcal{F}$ surrounding a local physical system will be required in order that system can validly be described in terms of Newtonian physics? Too much interference from the outside will prevent a good Newtonian limit existing, e.g., a local system imbedded in a universe with high-intensity gravitational waves will not have a good Newtonian description. There will be limits on the particles and gravitational waves crossing any bounding surface like $\mathcal{F}$, in order that such a description be possible.

It seems likely that the existence of a surface $\mathcal{F}$ in an almost flat region at a finite distance may be necessary for Newtonian-like behaviour [34]. It will not guarantee it, since black holes may form inside such a surface.

6.7.6.3 Conclusion

As well as the major way that microphysics affects macrophysics in a bottom-up way, there are many themes whereby there is a top-down action of the cosmos as a whole on local physical systems. This plays a key role in local physics, particularly as regards the question of the arrow of time. Many of the themes discussed here have practical applications in terms of being related to tests of cosmological theories, precisely because, if the universe has an influence on local systems, then observing local systems, e.g., determining the abundance of elements, tells us something about the universe as a whole.

References

1. Y. Aharonov, D. Bohm, Significance of electromagnetic potentials in quantum theory. Phys. Rev. **115**, 485–491 (1959)
2. Y. Aharonov, D. Rohrlich, *Quantum Paradoxes* (Wiley-VCH, Weinheim, 2005)
3. D. Albert, *Time and Chance* (Harvard University Press, Harvard, 2003)
4. M. Alonso, E.J. Finn, *Fundamental University Phyiscs III: Quantum and Statistical Physics* (Addison Wesley, Reading, 1971)
5. P.W. Anderson, More is different. Science **177**, 377 (1972). Reprinted in *P W Anderson: A Career in Theoretical Physics* (World Scientific, Singapore, 1994)
6. K. Ariga, T. Kunitake, *Supramolecular Chemistry: Fundamentals and Applications* (Springer, Berlin, 2006)
7. W.L.F. Armarego, C. Chai, *Purification of Laboratory Chemicals* (Butterworth, 2013)
8. P.W. Atkins, *Physical Chemistry* (Oxford University Press, Oxford, 1994)
9. G. Auletta, G. Ellis, L. Jaeger, Top-down causation: from a philosophical problem to a scientific research program. J. R. Soc. Interface **5**, 1159–1172 (2008), arXiv:0710.4235
10. Y. Balashov, Resource letter AP-1: the anthropic principle. Am. J. Phys. **54**, 1069 (1991)
11. J. Barbour, H. Pfister (eds.), *Mach's Principle: From Newton's Bucket to Quantum Gravity* (Birkhäuser, 1995)
12. J. Barrow, F. Tipler, *The Cosmological Anthropic Principle* (Oxford University Press, Oxford, 1984)
13. H. Batelaan, A. Tonomura, The Aharonov-Bohm effects: variations on a subtle theme. Phys. Today **62**, 38–43 (2009)
14. J.S. Bell, On the Einstein-Poldolsky-Rosen paradox. Physics **1**, 195–200 (1964)
15. J.S. Bell, *Speakable and Unspeakable in Quantum Mechanics* (Cambridge University Press, Cambridge, 1987)
16. Y. Ben-Aryeh, Geometric phases and topological effects in quantum mechanics, in *Quantum Mechanics*, ed. by J.P. Groffe (Nova Science Publishers, 2012)
17. D. Bohm, A suggested interpretation of the quantum theory in terms of hidden variables. I. Phys. Rev. **85**, 166–179 (1952)
18. D. Bohm, A suggested interpretation of the quantum theory in terms of hidden variables. II. Phys. Rev. **85**, 180–193 (1952)
19. H. Bondi, *Cosmology* (Cambridge University Press, Cambridge, 1960)
20. R.N. Bracewell, *The Fourier Transform and its Applications* (McGraw Hill, New York, 1986)
21. H.-P. Breuer, F. Petruccione, *The Theory of Open Quantum Systems* (Clarendon Press, Oxford, 2006)
22. A.O. Caldeira, Caldeira–Leggett model. Scholarpedia article (2010)
23. N.A. Campbell, J.B. Reece, *Biology* (Benjamin Cummings, 2005)
24. S. Carroll, *From Eternity to Here: The Quest for the Ultimate Arrow of Time* (Dutton, New York, 2010)
25. P.M. Chaikin, T.C. Lubensky, *Principles of Condensed Matter Physics* (Cambridge University Press, Cambridge, 2000)
26. R. Courant, D. Hilbert, *Methods of Mathematical Physics*, vol. II (Wiley-Interscience, New York, 1962)
27. D.P. Craig, T. Thirunamachandran, *Molecular Quantum Electrodynamics: An Introduction to Radiation-Molecule Interactions* (Dover, New York, 1984)

28. P.C.W. Davies, *The Physics of Time Asymmetry* (University of California Press, 1977)
29. P.A.M. Dirac, *The Principles of Quantum Mechanics* (Oxford University Press, Oxford, 1958)
30. S. Dodelson, *Modern Cosmology* (Academic Press, New York, 2003)
31. A. Durrant, *Quantum Physics of Matter* (Institute of Physics and The Open University, Bristol, 2000)
32. A. Einstein, B. Podolsky, N. Rosen, Can quantum-mechanical description of physical reality be considered complete? Phys. Rev. **47**, 777–780 (1935)
33. G.F.R. Ellis, Relativistic cosmology, in *General Relativity and Cosmology, Proceedings of the XLVII Enrico Fermi Summer School*, ed. by R.K. Sachs (Academic Press, New York, 1971). Reprinted as GRG Golden Oldie, Gen. Relativ. Gravit. **41**, 581–660 (2009)
34. G.F.R. Ellis, Cosmology and local physics. New Astron. Rev. **46**, 645–658 (2002), arXiv:gr-qc/0102017
35. G.F.R. Ellis, Physics in the real universe: time and spacetime. GRG **38**, 1797–1824 (2006), arXiv:gr-qc/0605049
36. G.F.R. Ellis, On the nature of causation in complex systems. Trans. R. Soc. South Africa **63**, 69–84 (2008)
37. G.F.R. Ellis, On the limits of quantum theory: contextuality and the quantum-classical cut. Ann. Phys. **327**, 1890–1932 (2012), arXiv:1108.5261
38. G.F.R. Ellis, The arrow of time and the nature of spacetime. Stud. Hist. Philos. Modern Phys. **44**, 242–262 (2013), arXiv:1302.7291
39. G.F.R. Ellis, The evolving block universe and the meshing together of times. Ann. N. Y. Acad. Sci. (2014), arXiv:1407.7243
40. G.F.R. Ellis, M. Bruni, Covariant and gauge-invariant approach to cosmological density fluctuations. Phys. Rev. D **40**, 1804 (1989)
41. G.F.R. Ellis, R. Goswami, Space time and the passage of time, in *Springer Handbook of Spacetime*, eds. by A. Ashtekar, V. Petkov (Springer, Heidelberg, 2014) Chap. 13, arXiv:1208.2611
42. G.F.R. Ellis, R. Maartens, M.A.H. MacCallum, *Relativistic Cosmology* (Cambridge University Press, Cambridge, 2012)
43. G.F.R. Ellis, D. Noble, T. O'Connor, Top-down causation: an integrating theme within and across the sciences. J. R. Soc. Interface Focus **2**, 1–3 (2011)
44. G.F.R. Ellis, T. Rothman, Crystallizing block universes. Int. J. Theoret. Phys. **49**, 988 (2010), arXiv:0912.0808
45. G.F.R. Ellis, D.W. Sciama, Global and non-global problems in cosmology, in *General Relativity: Papers in Honour of J L Synge*, ed. by L. O'Raifertaigh (Oxford University Press, Oxford, 1972), p. 35
46. H. Everett, Relative state formulation of quantum mechanics. Rev. Mod. Phys. **29**, 454–462 (1957)
47. B. Falkenberg, M. Morrison (eds.), *Why More Is Different: Philosophical Issues in Condensed Matter Physics and Complex Systems* (Springer, Heidelberg, 2015)
48. R.P. Feynman, A.R. Hibbs, *Quantum Mechanics and Path Integrals*, ed. by D.F. Styer (Dover, New York, 1965)
49. R.P. Feynman, R.B. Leighton, M. Sands, *The Feynman Lectures on Physics I: Mainly Mechanics, Radiation, and Heat* (Addison-Wesley, Reading, 1963)
50. R.P. Feynman, R.B. Leighton, M. Sands, *The Feynman Lectures on Physics III: Quantum Mechanics* (Addison-Wesley, Reading, 1965)
51. P. Franche, R. Gwyn, B. Underwood, A. Wissanji, Attractive Lagrangians for non-canonical inflation. Phys. Rev. D **81**, 123526 (2009), arXiv:0912.1857v3
52. P. Franche, R. Gwyn, B. Underwood, A. Wissanji, Initial conditions for non-canonical inflation. Phys. Rev. D **82**, 063528 (2010), arXiv:1002.2639v1
53. M. Gell-Mann, *The Quark and the Jaguar: Adventures in the Simple and the Complex* (Abacus, London, 1994)
54. J. Gemmer, M. Michel, G. Mahler, *Quantum Thermodynamics: Emergence of Thermodynamic Behaviour Within Composite Quantum Systems* (Springer, Heidelberg, 2004)

55. G. Greenstein, A.G. Zajonc, *The Quantum Challenge: Modern Research on the Foundations of Quantum Mechanics* (Jones and Bartlett, Sudbury, 2006)
56. M.R. Haas, J. Schaye, A. Jeeson-Daniel, Disentangling galaxy environment and host halo mass. MNRAS **419**, 2133 (2012), arXiv:1103.0547
57. E.R. Harrison, *Darkness at Night: A Riddle of the Universe* (Harvard University Press, 1987)
58. S. Hartmann, Effective field theories, reductionism, and scientific explanation. Stud. Hist. Phil. Mod. Phys. **32**, 267 (2001)
59. E. Hecht, *Optics* (McGraw Hill, Schaum, 1975)
60. E.J. Henley, J.D. Seader, D.K. Roper, *Separation Processes and Principles* (Wiley Asia, 2011)
61. R.C. Henry, Diffuse background radiation. ApJ **516**, L49 (1999), arXiv:astro-ph/9903294
62. C.J. Isham, *Lectures on Quantum Theory: Mathematical and Structural Foundations* (Imperial College Press, London, 1995)
63. J.C. Jackson, *Classical Electrodynamics* (Wiley, New York, 1967)
64. V. Jacques, E. Wu, F. Grosshans, F. Treussart, P. Grangier, A. Aspect, J.-F. Roch, Experimental realization of Wheeler's delayed-choice Gedanken experiment. Science **315**, 5814 (2007), arXiv:quant-ph/0610241v1
65. J.K. Jain, The composite fermion: a quantum particle and its quantum fluids. Phys. Today 39–45 (2000)
66. M. Kac, Can one hear the shape of a drum? Am. Math. Monthly **73**, 4 Part 2 (1966)
67. S.A. Kauffman, *The Origins of Order: Self-Organisation and Selection in Evolution* (Oxford, New York, 1993)
68. Fachgruppe Physik, Universität zu Köln, Molecules in space (2013), http://www.astro.uni-koeln.de/cdms/molecules/
69. M. Kowalski et al., Improved cosmological constrains from new, old, and combined supernova data sets. Astrophys. J. **686**, 749–778 (2008), http://iopscience.iop.org/article/10.1086/589937/pdf
70. K.S. Krane, *Introductory Nuclear Physics* (Wiley-VCH, 1987)
71. T. Lancaster, M. Pexton, Reduction and emergence in the fractional quantum Hall state. Stud. Hist Philos Modern Phys. **52** (Part B), 343–357 (2015)
72. K.J. Laidler, J.H. Meiser, B.C. Sanctuary, *Physical Chemistry* (Houghton Mifflin, 2002)
73. R.B. Laughlin, Fractional quantisation. Rev. Mod. Phys. **71**, 863 (2000)
74. H.S. Leff, A.F. Rex (eds.), *Maxwell's Demon: Entropy, Information, Computing* (Adam Hilger, Bristol, 1990)
75. A.J. Leggett, Reflections on the quantum measurement paradox, in *Quantum Implications: Essays in Honour of David Bohm*, ed. by B.J. Hiley, F.D. Peat (Routledge, London, 1991), pp. 85–104
76. A.L. Lehninger, *Bioenergetics* (W A Benjamin, Menlo Park, 1973)
77. S.G. Lipson, H. Lipson, *Optical Physics* (Cambridge University Press, Cambridge, 1969)
78. H. Lodish, A. Berk, S.L. Zipursky et al., *Molecular Cell Biology* (W H Freeman, New York, 2000), http://www.ncbi.nlm.nih.gov/books/NBK21473/
79. M. Longair, The physics of background radiation, in *The Deep Universe: Saas-Fee Advanced Course 23*, A.R. Sandage, R.G. Kron, M.S. Longair (Springer, 1993, 1995)
80. J. Martin, C. Ringeval, V. Vennin, Encyclopaedia Inflationaris (2013), arXiv:1303.3787
81. J. Martin, C. Ringeval, V. Vennin, How well can future CMB missions constrain cosmic inflation? (2014), arXiv:1407.4034
82. B. Mashhoon, Gravito-electromagnetism: a brief review (2014), arXiv:1407.4034
83. G.J. Milburn, *Schrödinger's Machines: The Quantum Technology Reshaping Everyday Life* (W H Freeman, New York, 1997)
84. M.A. Morrison, *Understanding Quantum Physics: A User's Manual* (Prentice Hall International, Englewood Ciffs, 1990)
85. J.D. Murray, *Mathematical Biology II* (Springer, 2003)
86. L. Pauling, *The Nature of the Chemical Bond and the Structure of Molecules and Crystals: An Introduction to Modern Structural Chemistry* (Cornell University Press, Ithaca, 1960)

87. L. Pauling, E.B. Wilson, *Introduction to Quantum Mechanics with Applications to Chemistry* (Dover, Mineola, 1963)
88. R. Penrose, *The Emperor's New Mind: Concerning Computers, Minds and the Laws of Physics* (Oxford University Press, Oxford, 1989)
89. R. Penrose, *The Road to Reality: A Complete Guide to the Laws of the Universe* (Jonathan Cape, London, 2004)
90. R. Penrose, *Cycles of Time* (Vintage, 2011)
91. I. Percival, Schrödinger's quantum cat. Nature **351**, 357 (1991)
92. M. Peshkin, A. Tonomura, *The Aharonov–Bohm Effect* (Springer, 1989)
93. M.E. Peskin, D.V. Schroeder, *An Introduction to Quantum Field Theory* (Perseus, Reading, 1995)
94. P. Peter, J.-P. Uzan, *Primordial Cosmology* (Oxford University Press, Oxford, 2013)
95. G.N. Price, S.T. Bannerman, E. Narevicius, M.G. Raizen, Single-photon atomic cooling. Laser Phys. **17**, 1–4 (2007)
96. G.N. Price, S.T. Bannerman, K. Viering, E. Narevicius, M.G. Raizen, Single-photon atomic cooling. Phys. Rev. Lett. **100**, 093004 (2008)
97. A. Rae, *Quantum Physics: Illusion or Reality?* (Cambridge University Press, Cambridge, 1994)
98. J.G. Roederer, *Information and Its Role in Nature* (Springer, Heidelberg, 2005)
99. A. Ruschhaupt, J.G. Muga, M.G. Raizen, One-photon atomic cooling with an optical Maxwell demon valve. J. Phys. B: At. Mol. Opt. Phys. **39**, 3833–3838 (2006)
100. J.J. Sakurai, *Modern Quantum Mechanics* (Addison Wesley Longman, Reading, 1994)
101. S. Saunders, J. Barrett, A. Kent, D. Wallace, *Many Worlds: Everett, Quantum Theory and Reality* (Oxford University Press, Oxford, 2011)
102. G. Schaller, C. Emary, G. Kiesslich, T. Brandes, Probing the power of an electronic Maxwell demon (2011), arXiv:1106.4670v2
103. D.N. Schramm, M.S. Turner, Big-bang nucleosynthesis enters the precision era. Rev. Mod. Phys. **70**, 303–318 (1998)
104. D.W. Sciama, *The Unity of the Universe* (Faber and Faber, 1959)
105. D.W. Sciama, *The Physical Foundations of General Relativity* (Doubleday, 1969)
106. C.R. Shalizi, J.P. Crutchfield, Computational mechanics: pattern and prediction, structure and simplicity. J. Stat. Phys. **104**(3/4) (2001)
107. L. Susskind, A. Friedman, *Quantum Mechanics: The Theoretical Minimum* (Basic Books, 2014)
108. A. Tonomura, N. Osakabe, T. Matsuda, T. Kawasaki, J. Endo, Evidence for Aharonov-Bohm effect with magnetic field completely shielded from electron wave. Phys. Rev. Lett. **56**, 792–795 (1986)
109. K. Umashankar, *Introduction to Engineering Electromagnetic Fields* (World Scientific, Singapore, 1989)
110. M. Vogelsberger, S. Genel, V. Springel, P. Torrey, D. Sijacki, D. Xu, G.F. Snyder, D. Nelson, L. Hernquist, Introducing the Illustris project: simulating the coevolution of dark and visible matter in the universe (2014), arXiv:1405.2921v1
111. J.D. Watson, *Molecular Biology of the Gene* (W A Benjamin, 1970)
112. S.W. Weinberg, *Gravitation and Cosmology* (Wiley, New York, 1972)
113. S. Weinberg, *The Quantum Theory of Fields, Volume 1: Foundations* (Cambridge University Press, Cambridge, 2005)
114. J.A. Wheeler, The 'past' and the 'delayed-choice double-slit experiment', in *Mathematical Foundations of Quantum Theory*, ed. by A.R. Marlow (Academic Press, New York, 1978), pp. 9–48
115. J.A. Wheeler, W.H. Zurek, *Quantum Theory and Measurement* (Princeton University Press, Princeton, 1983)
116. F.L. Wilson, Fermi's theory of β-decay. Am. J. Phys. **36**, 1150–1160 (1968)
117. K.G. Wilson, The renormalization group: critical phenomena and the Kondo problem. Rev. Mod. Phys. **47**, 773–840 (1975)

118. H.M. Wiseman, G.J. Milburn, *Quantum Measurement and Control* (Cambridge University Press, Cambridge, 2010)
119. E. Witten, Three lectures on topological phases of matter (2015), arXiv:1510.07698
120. J. Wolfe, Violin acoustics: an introduction, http://www.phys.unsw.edu.au/jw/violintro.html
121. A. Zee, *Quantum Field Theory in a Nutshell* (Princeton University Press, Princeton, 2003)
122. H.-D. Zeh, Quantum measurement and entropy, in *Complexity, Entropy and the Physics of Information*, ed. by W.H. Zurek (Addison Wesley, Redwood City, 1990), pp. 405–421
123. H.-D. Zeh, *The Physical Basis of the Direction of Time* (Springer, Berlin, 2007)
124. J.M. Ziman, *Principles of the Theory of Solids* (Cambridge University Press, Cambridge, 1979)
125. W.H. Zurek, Decoherence, einselection, and the quantum origins of the classical. Rev. Mod. Phys. **75**, 715 (2003), http://lanl.arxiv.org/abs/quant-ph/0105127
126. W.H. Zurek, Quantum Darwinism and invariance, in *Science and Ultimate Reality: Quantum theory, Cosmology, and Complexity*, ed. by J. Barrow, P.C.W. Davies, C. Harper (Cambridge University Press, Cambridge, 2004), pp. 121–134

Chapter 7
The Mind and the Brain

This chapter looks broadly at how the overall thesis of the book relates to the human mind and brain, which develop to be what they are because they are located in a specific society. The main claim will be that one cannot begin to understand the brain properly without taking top-down causation into account. An example is that you are able to read this book, written in English, because your neuronal connections have been adapted to understanding that language as you interacted with members of the society in which you grew up (your parents, siblings, and school mates if it is your home language). This is top-down causation from the social milieu to detailed aspects of brain structure. There is no way this can be understood purely on the basis of the underlying physics.

It is obviously not possible to give a detailed study of the brain in the space available, and nor do I have the competence to do so. Rather the point of this chapter, the culmination of what has gone before, is to point out firstly that the combination of bottom-up and top-down causation certainly occurs in the brain. And secondly that it is a crucial aspect of the way physics can underlie the extraordinary nature of the functioning of the mind. This viewpoint at least partly helps to see how the higher levels can have genuine causal powers in their own right, not determined uniquely by the underlying physics, but rather shaping the contexts of those physical events so that they can enable higher level functioning to take place meaningfully as events at their own level of causation.

While I will make the case for top-down causation in the brain, I do not claim this solves the hard problem of consciousness or even the issue of free will, but rather that this sets in place some understandings that provide a platform towards tackling those issues adequately. Any approach which ignores the top-down aspect of causation in the brain will be bound to go wrong (as will any approach that ignores the bottom-up aspects). It is a key aspect of how our mental abilities can emerge on the basis of the underlying physics. Thus what is presented here is a counter to any simplistic reductionist view of how the brain works.

Some of these issues are very controversial. In the space available, I cannot possibly respond to all the counterviews to what I present, of which there are many, and will not attempt to do so. Furthermore, the literature is vast and it would not be

G. Ellis, *How Can Physics Underlie the Mind?*, The Frontiers Collection,
DOI 10.1007/978-3-662-49809-5_7

practical to survey it all. Rather I will aim to give sufficient references to literature that I regard as forming a golden thread supporting my case. They have the feeling of being right, and are supported by much data. Taken together, they fit into a coherent view that takes seriously what seem to me to be deep approaches to the various topics discussed, and are consonant with what has been presented in earlier chapters in this book. This thread involves the ideas of the embodied mind [82, 100, 160, 161], the realisation that much mental processing is unconscious [147, 224], the top-down nature of the way senses such as vision work [99, 201], the importance of emotions as well as rationality [50, 116], and the significance of the social mind [66, 71], which have led to language being a key feature of humanity [54, 236]. For a single book that gives a profound overview of brain structure and function that supports what I propose here, see Erik Kandel's magnificent book *The Age of Insight: The Quest to Understand the Unconscious in Art, Mind, and Brain, from Vienna 1900 to the Present* [147] (he is a Nobel prizewinner for his work on memory, and the first author of one of the major standard works on neuroscience [148]), or his excellent smaller book *In Search of Memory: The Emergence of a New Science of Mind* [146].

The sections in this chapter are as follows:

- Section 7.1. Introduction.
- Section 7.2. Basics of the brain.
- Section 7.3. Top-down processes.
- Section 7.4. Purpose and meaning as the key drivers.
- Section 7.5. Symbolism and effectiveness of thought.
- Section 7.6. The effects of Platonic entities.
- Section 7.7. The complex whole.

Perhaps the most controversial theme I suggest is the relation of the mind to Platonic possibility spaces, such as that for mathematics, which it is able to comprehend via its neural net structure (Sect. 7.7). My views in this regard are strongly supported by Paul Churchland's profound book *Plato's Camera: How the Physical Brain Captures a Landscape of Abstract Universals* [43]. Even if details of his argument are debatable [39], something like this must be the case, because we do indeed comprehend timeless and eternal mathematical truths [40, 195].

7.1 Introduction

In this section, I concentrate on a few key threads that set the framework for the thesis of the book. What I present is consonant with mainstream views of the brain as set out in standard texts on neuroscience [15, 115, 148, 154], psychology [114], physiology [203], and regarding the evolution of the mind [71, 100], as well as popular books on these themes by major figures in these fields [99, 147]. I will briefly summarise these threads in this introductory section, and then develop them further in the rest of the chapter. They are as follows:

- Dynamical systems vs plasticity, learning (Sect. 7.1.1).
- Modular hierarchical structure: neural nets (Sect. 7.1.2).
- Basic functioning: rationality, intuition, emotion in a social context (Sect. 7.1.3).
- Bottom-up and top-down effects (Sect. 7.1.4).
- The effectiveness of thoughts: symbolism and language (Sect. 7.1.5).
- The key role of purpose and meaning (Sect. 7.1.6).
- The role of Platonic spaces (Sect. 7.1.7).
- An integrational view: mental powers and free will (Sect. 7.1.8).

7.1.1 Dynamical Systems Versus Plasticity and Learning

It is crucial to the nature of the brain that it is able to learn: it is dynamically adapted to the physical, ecological, and social environment in which an individual is situated. This adaptive nature is the source both of our common basic human abilities through the way it affected evolution over geological timescales, as well as of our individual abilities based on specific responses whereby learning takes place in our personal lives as we grow to maturity, and then as we continue to learn on a day by day basis. Thus adaptation occurs on evolutionary, developmental, and functional timescales. This crucial adaptive behaviour at the macro-level is based on brain plasticity at the micro-level. One might try to develop models characterising the way brain mechanisms operate at a fundamental level, based on statistical physics, energy principles, dynamical systems theory, or Hamiltonian dynamics. These certainly are all an important part of the story. However, Hamiltonian systems (the fundamental underlying physics) and dynamical systems as usually defined are not adaptive: that is, any attractors, saddle points, etc., are fixed by the structure of the dynamical system at the outset, and so they cannot learn. Such dynamics can characterise the way neural circuits operate once they have a form that is adapted to the environment, but they cannot explain how that adaptation takes place in the first place. One needs something more.

It appears that the main underlying mechanism is adaptive selection of synaptic connections [41, 72, 83]. The initial set of relatively non-specific connections are refined to produce a precise pattern of connectivity, whose weights are then adjusted through learning processes. Non-local neuronal systems diffuse neuromodulators (noradrenaline, dopamine, serotonin, glutamate, etc.) from the limbic system to the cortex via non-specific axonic projections, and these modulate neural connections [72]. Experience produces sustained changes in the effectiveness of neural connections by altering gene expression (see [143, pp. 46–47] and [144, p. 93]). There may also be shorter term coalitions of neurons that fire together in an oscillatory way for some purpose, being able to adapt these temporary networks to the situation at hand [35].

Table 7.1 The basic hierarchy for the brain

	Structure/entity	Function/interaction
Level 9	Society: culture	politics, economics, social interaction
Level 8	Individual: consciousness	Thoughts, emotions, intentions
Level 7	Brain: cortex, limbic system	Integration, control
Level 6	Assemblies of neurons: neural nets	Pattern recognition, prediction
Level 5	Neurons	Signal processing, computing
Level 4	Axons, dendrites	Nerve impulses
Level 3	Biochemical structures	Genetic and cell functions
Level 2	Molecules	Bonding, energy interactions
Level 1	Atoms	Quantum physics, electrical interactions

From Scott [214]

This theme of plasticity is developed in Sects. 7.2 and 7.3. Any brain model which omits it is incomplete. Dynamical systems alone cannot comprehend brain function.

7.1.2 Modular Hierarchical Structure, Neural Nets

The brain is made of interconnected neurons that make up a modular hierarchical structure (see [4, 154] and [95]) with emergent properties that depend on this layered structure [136]. Each level has modules composed of items from the level below, e.g., brain domains (brainstem, limbic system, cortex), neural nets, neurons, axons and dendrites, voltage-gated ion channels and molecule-gated ion channels, biomolecules, and atoms. Such structures are the only way to create complexity out of simple components [220].

There are interlocking networks with multiple hierarchies: the overarching ones being an implementation hierarchy and a logical hierarchy (compare with the case of digital computers discussed in Sect. 2.3), with structure and function as in Table 7.1.

Information flows from a neuron's dendrites to the nucleus to axons to synapses and on to other neurons, through motions of electrons that together comprise spike trains of action potentials [146]. The outcome depends on the specific pattern of connections between neurons. These structural relations form a network that has to be specified in addition to the properties of the neurons in order to determine how they process information. Network motifs [2] process information in specific ways that filter information and give simple control circuits that can be assembled together to form functional networks leading to complex behaviour [115, 129, 218]. This is the key to how networks work, rather than the statistics of connections. Different networks of connections will give different outputs. Personality and memory are determined by details of this network structure in each individual. This is developed in Sect. 7.2.1.

7.1.3 Rationality, Intuition, and Emotion in a Social Context

The basic functioning of the brain is to balance rationality, intuition, emotion, and meaning (Fig. 7.6), all the time learning from its interaction with the physical, ecological, and social environment in an adaptive way. Adaptive emergence of the brain is both diachronic (taking place over historical time) and synchronic (taking place continually on an ongoing basis as the brain functions in its local context). All this occurs in the context of the society in which we live (Fig. 7.16), and indeed individual minds cannot be understood on their own: their nature and existence is a result of interaction with society [66].

To save computational resources, many functions are automated on the basis of previous experience so that we do not have to pay conscious attention to them. This underlies what we call intuition. All the time the brain is engaged in unconscious inferences [224] and making predictions as to what to expect [129], which thereby shape our perceptions of the world [99]. This process of perception creates the illusion that we have direct contact with objects in the physical world and that our own mental world is isolated and private [99, p. 17]:

> Through these two illusions we experience ourselves as agents acting independently on the world. But at the same time we can share our experience of the world. Over the millennia this ability to share experience has created culture that has in its turn modified the functioning of the human brain.

Rationality is not separate from emotion. Rather emotion plays a key role in shaping rationality [49, 50], in particular focusing attention on key items of concern to which attention should be paid. Indeed they cannot really be regarded as separate from each other [191]. Primary (genetically determined) emotional systems play a key role in function and development, and hence were crucial in evolutionary history (Sect. 7.2.5).

Attention can be directed to relevant items [144] and choices made as to what actions should be taken in order to achieve chosen goals [114], resulting in appropriate muscle movements directed by the brain in a top-down way. This is developed in Sect. 7.2.2.

7.1.4 Bottom-Up and Top-Down Effects Both Occur

Top-down causation [79] is crucial to brain function and its relation to the body and society. It takes place among other things in the following:

- From the motor cortex down to peripheral muscles.
- In terms of prediction and expectation that shape what we see and experience.
- In terms of controlling attention and planning actions.
- In terms of choosing goals in the context of aesthetic preferences and ethical beliefs.

- From the brain to the immune system and to bodily welfare, as evidenced for example by the effectiveness of placebos.
- In terms of social influences on the brain moderated by teaching, role models, and ongoing social experiences.
- In terms of the evolutionary adaptation of the brain to the developing social context on evolutionary timescales, which led in particular to the development of language.

In brief, developmental processes lead to the emergence of the higher levels of structure and function, which then lead to contextual effects partly shaping what happens at the lower levels and thereby in turn crucially affecting those developmental processes. Through this process, society affects details of neural connections related *inter alia* to language, social customs, roles, ethics, and technology as well as geography and history. This is developed in Sect. 7.3. It relates to the free will debate (Sect. 7.7.4).

7.1.5 The Effectiveness of Thoughts: Symbolism and Language

Humankind is famously the symbolic species [54], with the development of language having been a crucial feature in our evolutionary history because it is what led to the development of a social brain [70, 71]. This is what enabled us to surpass all the other primates and conquer the world.

This development of symbolism made it possible to name objects, actions, qualities, and qualifiers referring to generic and specific entities and events. Being able to label patterns of symbols then led to the recursion that gives symbolism its full power, in particular enabling offline contemplation of actions and their consequences, and hence the rise of planning and technology. Like digital computers, the brain is thus a prime example of a physical mechanism allowing non-physical entities to have causal powers.

Our ideas and plans, driven by emotions and expectations, change the world around us today, and in the past led to the rise of civilisation [33]. This is developed in Sect. 7.5.

7.1.6 The Key Role of Purpose and Meaning

Meaning and values guide this whole process of decision-making and action [103, pp. 128–157]. Our system of values and ethics shapes what we regard as desirable or undesirable actions, and what we see as meaningful or not. The brain is searching all the time for meaning [88], and our concept of what is meaningful changes our life choices. We will act differently if we aim to enrich ourselves or to dedicate ourselves to uplifting the poor, if we seek power for ourselves, or to fight for freedom and justice for all [216], if we devote our lives to sport or to art, or to leisure and recreation. These choices at the top level of the hierarchy of goals shape all the subsidiary goals we choose in order to fulfil these higher level goals, and then lead to changed outcomes in the real world.

These value choices (whether made explicitly or implicitly) are the top-level decisions that then, in a top-down way, change all else that happens in our lives, including what the cells and genes in our bodies do [143, 144], as well as the society in which we live [86]. This is developed in Sect. 7.4.

7.1.7 The Relationship with Platonic Spaces

The mind can discover unchanging eternal relationships and possibilities that were always there and exist independently of the human mind: that is, they can reasonably be called Platonic entities. However, they make a real difference in the world.

One of them is the nature of possible algorithms that can be realised in computer programs (Sect. 2.7.5), which shape important uses of computers [168]. Another is mathematical equations or theorems describing relationships between mathematical entities, such as $e^{i\pi} + 1 = 0$, Pythagoras' theorem, or the fact that $\sqrt{2}$ is irrational.

These are timeless and eternal mathematical truths [40, 194], comprehended by the mind through the nature of the pattern-recognition abilities of neural nets as explained by Churchland [43]. They then change the world through their applications in science, engineering, and commerce [229]. This is developed in Sect. 7.6.

7.1.8 Mental Powers and Free Will

A summary concludes the chapter (Sect. 7.7) and refers in particular to the issues of mental powers and free will. The view will be that they both have to exist, otherwise the very process of writing and reading this book make no sense (as remarked in Sect. 1.7.2). The existence of top-down causation undermines the foundations of any strong reductionist view that is taken to preclude the existence of meaningful mental action. We do indeed have mental powers that enable our purposes to have physical outcomes in the real world: the existence *inter alia* of spectacles, aircraft, computers, and cellphones is sufficient to prove this. Physics per se cannot lead to these results [76]. The theme is developed in Sect. 7.7.4.

7.2 Basics of the Brain

The mind is inherently embodied [14], with structure and function intimately linked in a dual way, as in all biology (see Kandel [147, pp. 226–237]). Both structure and function have a modular hierarchical nature, so that in each case complex outcomes can emerge from simple parts. Thus new kinds of organisation occur at higher levels, based on complex combinations of lower level elements; this allows new kinds of behaviour and function to arise at higher levels, quite different from what can occur at lower levels. I look in turn at the following:

- Brain anatomy (Sect. 7.2.1).
- Basic brain function (Sect. 7.2.2).
- Large scale function (Sect. 7.2.3).
- Environmental and genetic influences: brain plasticity (Sect. 7.2.4).
- The origin of humanity: the social mind and language (Sect. 7.2.5).

7.2.1 Brain Anatomy

The structure of the brain is described in [103, pp. 17–37], [224, pp. 8–19], and [258, pp. 37–74], and major texts such as [4, 148, 186, 218]. It has macro, micro, and meso aspects.

7.2.1.1 Macrostructure

As a first approximation, on a **macro scale** the brain consisting of three main parts, literally stacked one on top of the other:

- The *reptilian brain*, the seat of instinctive behaviour, consisting of the brain stem, cerebellum, and basal ganglia.
- The *paleomammalian brain*, or *limbic system*, the main seat of emotions, containing a collection of structures like the telencephalon, diencephalon, and mesencephalon, and including the hippocampal formation, hypothalamus, amygdala, anterior cingulate gyrus, septal nucleus, nucleus accumbens, ventral tegmental area, some thalamic nuclei, and other nearby areas (there is some dispute about precisely which ones to include).
- The *neomammalian brain*, the seat of intellect and intuition, consisting of two hemispheres linked together by the corpus callosum, each hemisphere being made up by a number of lobes, the frontal lobe, parietal lobe, occipital lobe, and temporal lobe (see Fig. 7.1).

These together form the triune brain. At a rough approximation they represent an evolutionary sequence of brain developments. The central nervous system (CNS) consists of the above plus the following:

- Our *sensory organs* (eyes, ears, tongue, nose, balance organs, and touch), linked to both the limbic system and the cortex.
- The *spinal cord*, linking the brain to muscles and nerves in the periphery.
- The *enteric nervous system* or *second brain*, consisting of nerve cells in the gut that act together independently of the main brain [105].

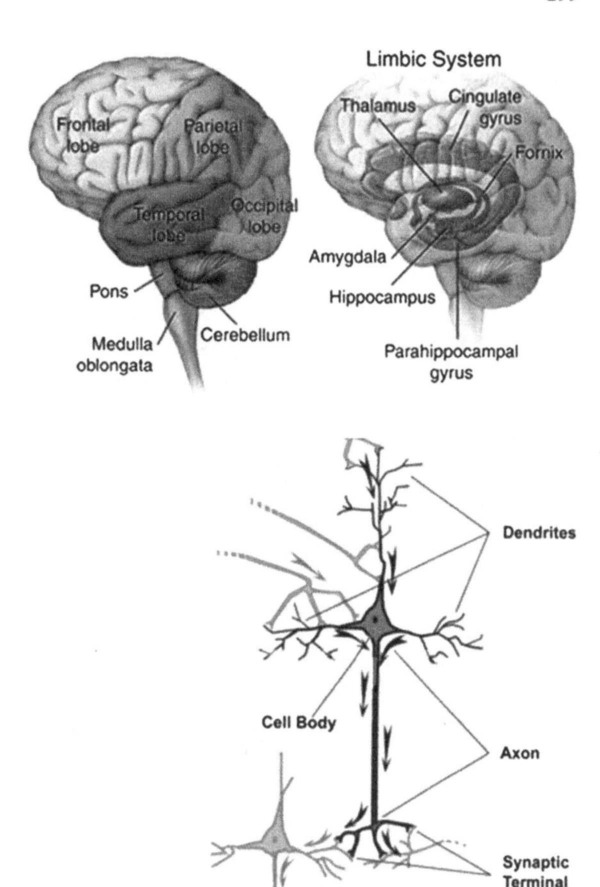

Fig. 7.1 The major brain domains. The cortical lobes (*left*) above the pons, cerebellum, and medulla oblongata, and some of the limbic structures (*right*). Credit: Courtesy of Alzheimer's Disease Research, a Bright Focus Foundation program http://www.brightfocus.org/alzheimers

Fig. 7.2 Neurons have dendrites. a nucleus, and axons. They are connected to other neurons via synaptic terminals. Credit: Wikimedia

7.2.1.2 Microstructure

On a micro scale, the brain consists of neurons, made up of dendrites, nuclei, and axons (Fig. 7.2), linked together by synapses to form neural networks (Fig. 7.3). An electric signal is carried down many dendrites to the nucleus, where some form of computation takes place that generates an output signal when input signals collectively exceed a threshold. This output signal is then carried away from the nucleus along numerous axons by spikes in the electrical action potential [146].

Synapses can be either electrical synapses, where the signal is transmitted from one neuron to the next by a direct electrical connection, or chemical synapses, where the electrical signal gets converted to a chemical signal that gets conveyed by diffusion of neurotransmitters across the synaptic gap from one neuron to the next. Neuromodulators alter the synaptic strengths [224, pp. 34–41]. Taken together, the interconnected neurons are joined together in repeating patterns to form higher level network structures [125].

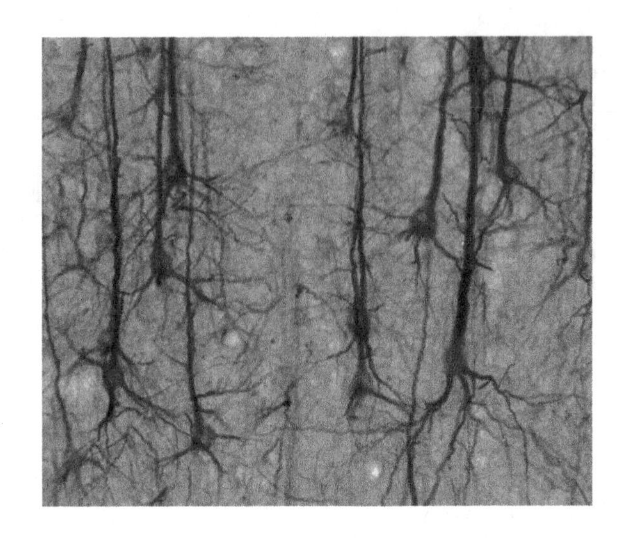

Fig. 7.3 Neurons are joined together across synapses to form networks. *Source* Wikimedia [By UC Regents Davis campus—http://brainmaps.org]

7.2.1.3 Mesostructure

Neural networks process information in specific ways, often via network motifs [2], i.e., recurring patterns of connections such as the feedforward loop motif (positive or negative) that filters information and gives simple control circuits that can be assembled together to form functional networks leading to complex behaviour [11, 115], e.g., in the worm (*C. elegans*) and the fish C-start circuit [2]. When linked together, these brain microcircuits form higher level structures.

On a mesoscale, the networks of neurons in the neocortex are arranged in six main horizontal layers (Fig. 7.4), with interconnections being mainly local up and down connections in vertical columns that run perpendicular to the layers [129, pp. 138–140]. These form neural nets that can perform computations such as pattern recognition and generalisation [4, 68, 218], and in particular naming of patterns [129]. Other regions differ: the hippocampus has three such cellular layers, but they still represent local interactions in the brain, largely organised in a columnar fashion. They can be represented quite well as layered neural networks, modelled by artificial neural nets [21].

However, there are also non-local reciprocal connections between distinct brain areas and between the brain and the sensory organs. In particular, there are on the one hand recurrent connections between different cortical regions [74], and on the other diffuse projections from the limbic system to the cortex known as monoamine systems or ascending systems [72, 154]. From their nuclei of origin, they send axons up and down the nervous system in a diffuse spreading pattern. The ascending systems conveying neuromodulators such as dopamine and serotonin from the limbic system to the cortex, as well as aspartate. The effect of these systems projecting profusely is that each associated neuromodulator (for example, norepinephrine and dopamine) affects large populations of neurons, allowing non-local interactions to occur in the

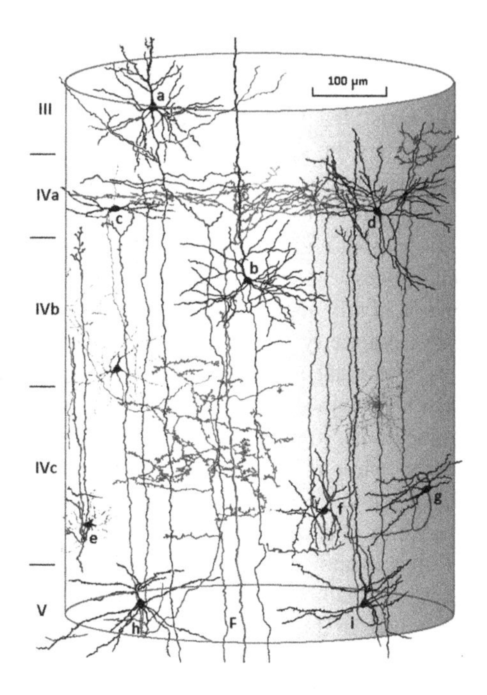

Fig. 7.4 Neurons form layers in the cortex. The local connections here are to be contrasted with the diffuse connections in Fig. 7.8. Credit: Wikimedia commons (Mavavf)

brain. Their release affects the probability that neurons in the neighbourhood of these axons will fire after receiving glutamatergic input, thus they are an important mechanism effecting neural plasticity [186, pp. 271–288]. These systems bias neuronal responses affecting both learning and memory by guiding neuronal group selection, and for this reason that they are sometimes termed value systems [72, 74].

Other classes of non-local effects are due to nitric oxide (NO), which is a freely diffusing neurotransmitter [186, pp. 277–278], and immune system molecules which are also neuromodulators [225–227]. These forms of modulation allow a kind of plasticity in the network, in which the intrinsic properties of units are changing as the network operates.

Putting this together, one needs hierarchical models of the brain relating these different levels, such as those by Karl Friston [95]. These models comprise hidden layers of state-space or dynamic causal models, arranged so that the output of one provides input to another. The hierarchical structure of these models shapes the interactions. Dynamical priors are mediated by the equations of generalised motion, and structural priors by the hierarchical form, under which states in higher levels provide constraints on the level below. This is a mechanism by which top-down causation takes place between the different levels.

7.2.2 *Basic Brain Function*

At a broad level, the brain can be thought of as an information processing system: data enters the system and is processed, and an output is generated [15, 173, 224]. However, brains are more than just input–output devices [115, p. 3]: "they are complex organ systems capable of arousal, perception, context appropriate reactions, anticipation, and sophisticated discriminations." Contextual knowledge modulates this process, and memories are stored for future use. Even in the honey bee, one finds a highly sophisticated ability to make abstract discriminations, e.g., to recognize asymmetry *per se* [115, p. 3]. In humans, the brain stem, limbic system, and cortex interact in bottom-up and top-down ways [224]. In particular [224]: "The cortex is connected to the midbrain and brain stem centres, and by way of these connections, higher order cognition can turn on and turn off emotion and arousal centres while also guiding the selection of behaviours to be expressed." This is top-down action. Mental processing is mostly unconscious [143, p. 70], but is crucially guided by consciously formed abstract concepts (such as concepts of generic classes of objects and actions) and models (such as theories and mental maps), using language that is largely metaphorically based [160, 161], resulting in action plans and decisions. In this section, I shall look in turn at the following:

- Micro-level brain function.
- Macro-level brain function.
- The senses: vision.
- Patterns and classification, filtering and prediction.
- Conscious and unconscious.
- Symbolic representation.
- Emotional systems.

7.2.2.1 Micro-Level Brain Function

The complexity of brain function and structure boils down to using, making, and modifying neuronal connections (see [201, pp. 225–229] and [147, 148, 186]).

Neurons. These are connected to many other neurons via dendrites (on the input side) and axons (on the output side). They function like transistors in that they are context-dependent gates that send signals to other neurons via spike trains of action potentials, that is, spikes in electric potential enabled by voltage-gated ion channels that control the flow of potassium and sodium ions into and out of the axons [186, pp. 26–132]. Through these signals they can inhibit or excite neighbouring neurons at synapses [186, pp. 156–269]. While this takes place, energy is used, so oxygen and sugars must be continuously supplied to every neuron and heat and waste removed (the temperature must be tightly controlled in order that neuron action is reliable). They can be connected together to perform logical operations, in analogy with the way this can be done for transistors in a computer [170].

Networks. The interconnections of neurons form hierarchically structured networks. They are extremely complex, but their function can be understood in terms of repeating themes at the micro-level (network motifs) that perform basic logical operations, layered structures (neural nets) at the meso-level that are able to carry out pattern recognition, naming, and generalisation, and at the macro-level, recurrent connections between different brain areas that transfer information between them and may temporarily bind them in oscillatory circuits.

Network Motifs. These carry out simple information processing functions, e.g., negative autoregulation can become bistable and lock the system into a specific state, thus acting as a kind of memory [2, p. 37]. The feedforward motif can act as a delay element allowing persistence detection [2, p. 5], or it can act as a pulse generator [2, p. 58]. Single input network modules can generate temporal expression programs [2, pp. 77–81], as can multi-output feedforward loops which can regulate simple behaviour [2, pp. 83–87]. Network motifs occur in signal transduction networks in cells, which are equivalent to multilayer perceptrons [2, pp. 104–115] that can perform detailed computations: they have activation thresholds, are able to perform discrimination and make generalisation, and have graceful degradation. Furthermore, they do not grind to a halt if individual elements are damaged [2, p. 113]. This all happens *within* neurons. Network motifs also occur in neuronal networks when neurons are connected together, for example in *C. elegans* [2, pp. 118–127]. Two-input feedforward loops can generate AND or OR gates [2, p. 123] and so provide the basis for logical computations, just as simple combinations of transistors provide the basis for logical computations in digital computers [170].

In neocortical circuits there are similarities in circuit organization across areas and species, suggesting a common strategy to process diverse types of information, including sensation from diverse modalities, motor control, and higher cognitive processes [125]. There is a basic circuit pattern that appears to be repeated across neocortical areas [218, pp. 7–21], with area- and species-specific modifications adapting individual neocortical regions to the type of information each must process. Cells receive input from higher order cortex and inhibit other interneuron classes, with the sign of modulation (excitatory or inhibitory) depending on top-down cortical input.

Layered Neural Nets. In the cortex, neurons are broadly speaking structured as layered neural nets (see Fig. 7.5). These have an input layer, one or more hidden layers, and an output layer, with neurons in each layer connected to those in the next by links with adjustable strengths. Given an input pattern, it will select one of the output nodes as its response to that input. Such a network can be trained to recognise patterns, for example the letters of the alphabet, by repeatedly showing the network the input and desired output, and adjusting the link strengths in an adaptive way so that the output eventually corresponds to the input. Thus neural nets are capable of pattern recognition [21], for example, face recognition or recognising signatures. They can then generalise by recognising similar rather than identical input patterns. This is a form of memory, implemented via adaptively selecting the connection strengths [146, 147]. While they can then very successfully recognize specific sets of input patterns such as the letters A to Z, there is no algorithmic process of pattern

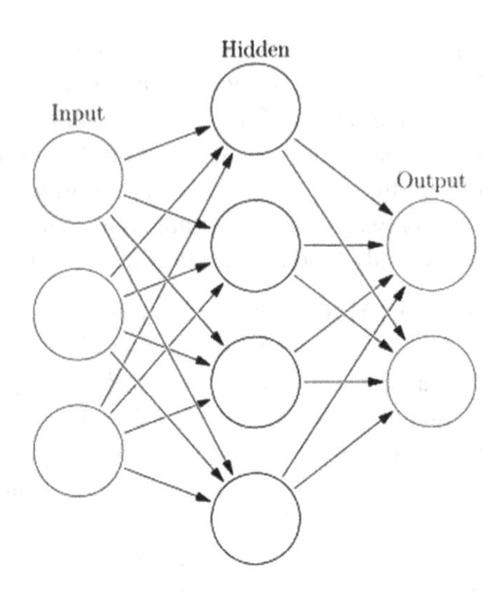

Fig. 7.5 A simple neural net with an input layer, one hidden layer, and an output layer. Credit: Wikimedia commons (Glosser.ca)

recognition involved, and a single network can be trained to recognise all the letters of the alphabet in a holistic way. This is a property of the network as a whole. There is no connection strength of a specific neuron in the hidden layers that for example encodes the letter A. The ability to recognise each letter is embodied in the set of hidden connection strengths as a whole. This also means the network is robust: if one or more nodes are removed, it will still perform in an adequate way.

This way that neural nets work represents how much of the cortex functions. It is a key feature that they are not intrinsically rule-based mechanisms, but recognise patterns [43, 129]. While they can be trained to work in a rule-based way, this is not their natural mode of operation. They illustrate clearly how plasticity at the network level is based on synaptic plasticity at the micro-level [186, pp. 228–242], affected by neurotransmitters [186, pp. 272–288].

Emergence of Higher Causal Levels. Connecting meso-structures together to create macro-domains, and connecting the various macro-domains together via recurrent (vertical and horizontal) connections [74], one obtains structures capable of higher level cognitive work. Hoel at al. [131] characterise how this happens as follows:

> We use a measure [effective information (EI)] that depends on both the effectiveness of a system's mechanisms and the size of its state space: EI is higher the more the mechanisms constrain the system's possible past and future states. [...] we show that for certain causal architectures EI can peak at a macro-level in space and/or time. This happens when coarse-grained macro mechanisms are more effective (more deterministic and/or less degenerate) than the underlying micro mechanisms, to an extent that overcomes the smaller state space. Thus, although the macro-level supervenes upon the micro, it can supersede it causally, leading to genuine causal emergence—the gain in EI when moving from a micro- to a macro-level of analysis.

That is, as emphasized in this book, the core of complex behaviour lies in setting suitable constraints on the context in which lower level processes operate so that macro processes control what happens, something which does not in any way undermine the underlying physics (rather it channels it).

Emergence and the Underlying Physics. Multiple layers of emergent behaviour occur on the basis of the underlying physics, but are not deducible from that physics. Here are just two examples: Alwyn Scott [214] emphasizes that the Hodgkin–Huxley equations governing the propagation of action potentials down axons can be traced to the underlying physics, but are not themselves laws of physics: they are new laws that emerge at the hierarchical level of the axon to govern the dynamics of nerve impulses, because of the neuron structure [214, p. 52]:

> One cannot derive these laws from physics and chemistry because they depend upon the detailed organisation of the intrinsic proteins that mediate sodium and potassium current across the membrane and upon the geometric structure of the nerve fibres.

He emphasizes the difference between these equations and the underlying Schrödinger equation: the latter is based on energy conservation, while the former do not conserve anything. The latter has no arrow of time, but the former has a past, present, and future [214, p. 52]: "they push forward in one direction with respect to time, just like our bodies and our thoughts". Furthermore, there is no parameter in the Hodgkin–Huxley equation that could be predicted or derived from Schrödinger's theory [214, p. 53]. This is the hallmark of genuine emergence, and underlies what happens in neural networks.

Secondly, the logical operations of the brain are not determined by either the laws of electromagnetism or the Hodgkin–Huxley equations *per se*, even in conjunction with the laws of synaptic behaviour. They are determined by the way the neurons are joined together in specific types of networks that can perform logical operations due to their structure, for example, their network motifs [2] and microcircuits [218], or layered neural network structure [43, 129]. This is like the way digital computer circuits embody logical principles not implied by their parts on their own [170]. Higher level logical operations can be generated by appropriate combinations of lower level operations based on the lower level physical components, and underlie the macro brain functions we consider next. It is the specific structuring of interconnections of these components, together with their individual logical behaviour, that is the interface between logical function and the underlying physics. This is based on the ability of neural networks to recognise patterns and generalize [167]:

> The ability to predict future states of the environment is a central pillar of intelligence. At its core, effective prediction requires an internal model of the world and an understanding of the rules by which the world changes.

Internal models developed by deep neural networks have this capacity.

Where do these network structures come from? Through adaptive selection on evolutionary, developmental, and functional time scales in the local physical, ecological, and social context, each playing a key role in human brain development. The result is networks that have the same function but differ in internal representations, because the needed learning is a top-down process to the level of connections, selecting any member of an equivalence class that works [164], hence with different representations occurring at the lower level for the same higher level function (Sect. 3.5). The emergence of the brain is possible only because of all these forms of top-down causation, as discussed below.

7.2.2.2 Macro-Level Brain Function

Given the emergence of higher level structures, what are the key macro-level functions that emerge? Data comes in from the viscera and via the senses [224]. On this basis, the brain has the following capacities:

- An ability to perceive complex objects and events and patterns [66, p. 124], classify them, and predict what might happen in the future on this basis.
- A capacity for delayed response, which shows that the animal can carry around an image or idea in its head, independently of the environment [66, p. 125] implying at least short term memory storage. Thus it has an independent mental model of the world that transcends its immediate environment [66, p. 123].
- Long term memory, which is key to learning [87, pp. 29–40, 71] and social cognition [99]. Semantic memory is a network of associations and concepts that underlies our basic knowledge of the world [224, pp. 150–156], while procedural memory is a kind of bodily memory for motor skills [224, pp. 156–160]. Episodic memory involves the re-experiencing of past events [224, pp. 160–167].
- Adaptive flexibility, in particular adaptability to environmental novelty [66, pp. 126–127] and the ability to solve problems and adapt to new situations [66, pp. 124–125]. This is facilitated by imitation and play [103, pp. 160–165].
- Consciousness, an egocenter that ties all these maps together and provides a unified perception of self-hood [66, pp. 134–136] through its grounding in physical embodiment. This involves selectivity of attention [66, pp. 127–128] and the ability to focus or concentrate resources [66, pp. 130–131], and provides the basis for a suite of domain general skills: self-monitoring, divided attention, self-reminding, autocuing, self-recognition, rehearsal and review, whole-body imitation, mind-reading, pedagogy, gesture, symbolic invention, and complex skill hierarchies [66, pp. 139–148]. Together these make up self-awareness [103, pp. 189–194].
- Social intelligence, the ability to cultivate and remember individual relationships within a working social group [66, p. 129], and mind-reading or having a theory of mind (see [66, pp. 129–130] and [103, pp. 48–54]), leading to the 'social brain' [71].
- The brain hides most of what is going on and presents us with a picture of our present situation that makes sense to us: a huge amount goes on unconsciously [99]. Routine activities are automatic and do not require extra resources or attention [66, p. 132].

The brain needs sleep and rest to recover and reset after prolonged operation, in order to maintain all these functions. Sleep is a vital part of brain function.

7.2.2.3 The Senses: Vision

All the senses operate in essentially the same way, filtering input all the time, neglecting vast amounts of information (the taken-for-granted background of life), and presenting us with a constructed image of the way things are that does not correspond directly to the incoming sense data [99, pp. 21–60, 111–138]. This is particularly clear in the structure and function of the visual system [147, pp. 233–255, 260–303].

The structure of the visual system [171, pp. 105–119] underlies vision, but the key issue is the inverse problem discovered by Berkeley: the information in the retina is a 2D projection of the 3D real world, so how is it that we perceive positions and motions of objects in a 3D world [201, p. 120]? Helmholz argued that vision has to depend on learning from experience in addition to the information supplied by neural connections to the brain. This is confirmed by detailed study of visual illusions, e.g., to do with brightness and colour contrasts [201, pp. 124–142, 146–159, 161–169]:

> The colors we see are not the result of the spectra in retinal images *per se*, but result from linking retinal spectra to real world objects and conditions discovered empirically through the consequences of behaviour.

The same applies to perceiving motion [201, pp. 201–217]. This means that it is likely that all perceptions are illusory constructions produced by the brain to achieve biological success in the face of the inverse problem (see [99] and [201, p. 121]). I return to this in Sect. 7.3.3 below.

7.2.2.4 Patterns and Classification, Filtering and Prediction

Animals learn to distinguish events that occur regularly together from those that are only randomly associated. Thus the brain has evolved a mechanism that 'makes sense' out of events in the environment by assigning a predictive function to some events, based on past experience and learned rules of behaviour [144, pp. 75–78]. This predictive capacity is built into the continually adapting connections in neural networks in the brain [129]. Each region of the cortex has a repertoire of sequences it knows, and it has a name for each sequence it knows [129, p. 129]. And the known must be distinguished from what is new [115, p. 105]:

> The ability to recognize novelty is one of the cardinal features of animal brains [...] apart from perception itself, novelty detection requires that perceived objects be categorised so that similar items are not perceived as novel. This ability to discriminate also requires memory of previously seen objects so that they may be compared to current ones.

Prediction takes place all the time, based on comparison with known situations and quantities. This is the memory-predictive framework of intelligence: the human cortex is constantly predicting what we will see, hear, and feel, mostly in ways we are unconscious of [129, p. 104]. Learning takes place all the time from experience by Hebbian processes [129, pp. 164–166] and by neuronal group selection [72], discussed below (Sect. 7.2.4.2).

Making Sense of It. The superficial layer neurons within local patches of cortex and within areas cooperate to explore possible interpretations of different cortical input and cooperatively select an interpretation consistent with their various cortical and subcortical inputs [68]. And a key element here is reading the intentions of others [99, pp. 85–110, 139–159], which is crucial in the social environment and underlies the possibility of the social brain for which there is archaeological [71] and psychiatric [228] evidence. Thus theory of mind is key to the function of the *social brain* (see [100, p. 22], [66, pp. 59–63], and [99]):

> We can read not only our own minds but the minds of others. We can do this effortlessly, without direct training, because of a specialized form of social intelligence that is more highly evolved than in other primates. As something perfected in the evolutionary arena, it is a survival skill and we must have it to survive in human society.

This leads to the *intentionality bias*: the constant search for meaning/intentions in the actions of others [206], as demonstrated by brain imaging studies of the sequence of brain events when harmful or beneficial actions take place. Different animals can understand different levels of intentionality [100, pp. 175, 318]. Level 4 level of intentionality is "John thought that Mary was worried about what Jake would think of Ethel's idea". Humans can handle about 5 or 6 such levels.

7.2.2.5 Conscious and Unconscious

Most of what happens is subconscious. All the time the brain is engaged in unconscious inferences (see [99, p. 17], [144, pp. 70–72], and [224, pp. 79–86]). We automatise so that we only have to pay attention to what is novel [66, pp. 58, 90]. Conscious processes are largely responsible for setting up the automatized cognitive routines of the human mind, except for certain basic built-in reflexes and instinctive responses. However, a small change in temporal sequence, e.g., something unexpected, changes an instance of psychic determinism from unconscious to conscious [144, pp. 77–78].

Consciousness is our awareness of our state of being [133, 258]. The most basic function of consciousness is to monitor the state of homeostatic systems and report whether they are contented or not (this is bodily self-monitoring), but its more advanced function is to bring us into connection with the current state of the world around us [224, p. 91] and reflect on that state. It is the generator of the rich texture of perceptual qualities (sights, sounds, smells, etc.) that we are able to experience and it thereby provides an awareness of what is happening around us, grounded in a background of self-awareness (see [133], [224, pp. 93–94], [258]). One can further move from perception (simple or primary consciousness) to thinking about present and past perceptions (secondary or reflexive consciousness), which spreads consciousness over time and so involves memory, so leading to the development of the autobiographical self [224, pp. 95–98].

Our brain effortlessly creates a perception of the physical world [99, pp. 111–138] that is contextually driven: we have prior beliefs of what objects should be where, and on this basis predict what signals our eyes and ears should be receiving. These predictions are compared with the actual signals and our model of the world is updated on that basis, in a continual perceptual loop, which also uses information derived from our actions [99, pp. 126–127, 130–131]. The prior knowledge used in this Bayesian process is mainly acquired through previous experience. Our brain's model of the world is constantly updated on the basis of sensory signals that we compare with predictions and thereby resolve errors in our model [99, pp. 132–137].

Action and Goal-Setting. A key function of consciousness is setting goals [247]. We act on the basis of hierarchically structured sets of goals based on our understanding of the situation and our values. This is the way we influence the world. Goal-directedness and evaluation are present in even the most rudimentary biological activity [127]. The distinctiveness of intelligent action lies in the ability of the organism to detach itself from the immediate biological and environmental stimuli in order to evaluate the options available for attaining these goals. Such evaluation depends on hierarchical structuring of cognitive processes [180, p. 12]. Human goal-seeking has two distinctive characteristics: explicit use of symbolic systems in making goal choices, and meta-analysis: thinking about thinking. This involves an adaptive choice of a hierarchy of goals. We learn to do this at a young age [104]. This is the topic of Sect. 7.3.5.

7.2.2.6 Symbolic Representation

Through its neural network structures, the brain can form a classification of patterns [43], and group together patterns that are part of the same higher level object, in such a way that they can then be named [129, p. 165]. Systematic classification of objects, actions, and ideas requires a language or other symbolic representations by which we can refer to them, and this symbolic representation, to a large degree based on metaphor, is our key distinctive human mental ability, on which intelligence builds (see [54] and [129, pp. 126–130]):

> **Key Ability**. Firstly, to be able to recognize, name, and classify recurrent patterns in the environment: enduring ones (objects), repeating ones (actions), and relations between them (causal patterns). Secondly, to be able to do this recursively, that is, giving names to patterns of names and storing named sequences of sequences. This leads to our unique symbolic ability.

By the process of abstraction, that is, collapsing predictable sequences into named objects at each level in our logical hierarchy, we achieve more and more stability the higher we go. This creates invariant representations [129, p. 130], and happens in a modular way: a hierarchy of nested sequences allows sharing and reuse of lower label objects in new contexts [129, p. 131]. It is this recursive property that allows higher order thinking (see Sect. 7.3.4).

7.2.2.7 Emotional Systems

Conscious awareness is grounded in emotional awareness, which tell us how we feel about things and motivates our actions [224, pp. 105–138]. We share with other mammals the basic emotion command systems and the feelings that correspond to them (see [192] and [224, pp. 113–133]). These are our genetic heritage. Secondary emotions are socially determined in response to our daily experiences (Sect. 7.2.4).

The associations that lead to emotions are unconscious, but the output of the emotional systems can be inhibited by the frontal lobe [224, pp. 136–137]. The executive system allows delays in decisions in the interests of thinking, which can be regarded as imaginary acting whereby the outcome of a potential action is evaluated [224, p. 281]. This is achieved by running the envisaged action program while motor output is inhibited [224, p. 282] and using language as a powerful tool of self-regulation through inner speech [224, p. 283]. One should note here that in contrast to other cognitive processes, emotions cannot, by definition, be unconscious, as already pointed out by Freud [223].

7.2.3 Large Scale Function

It is common to think of the brain as primarily a machine for logical thinking: determining in a rational way the best options of action in a given context. Here 'best' means that it maximises our welfare. This is the economists' view of humans as rational agents. However, things are rather more complex than that, quite apart from the issue of handling judgement under uncertainty by various heuristics that lead to biases [7, 141], suggesting that prospect theory is more appropriate than utility theory [142]. A plausible diagram of the large scale interactions involved is given in Fig. 7.6. The following are the important issues I discuss here:

- Interaction of rationality, faith, and hope.
- Effects of primary and secondary emotions.
- Effects of values and meaning.

Inputs from society (the variable and mutable social world) are considered in Sect. 7.2.4, and inputs from Platonic entities (eternal and unchanging entities such as mathematics) are discussed in Sect. 7.6.

7.2.3.1 Rationality, Faith, and Hope

Firstly, in order to live our lives, we need faith and hope, because we always have inadequate information for making any real decision. Faith is to do with understanding what is there, hope with the nature of the outcomes. When we make important decisions like whom to marry, whether to take a new job, or whether to move to a new place, we never have enough data to be certain of the situation or the outcome. We can keep gathering evidence as long as we like, but we will never be truly sure as

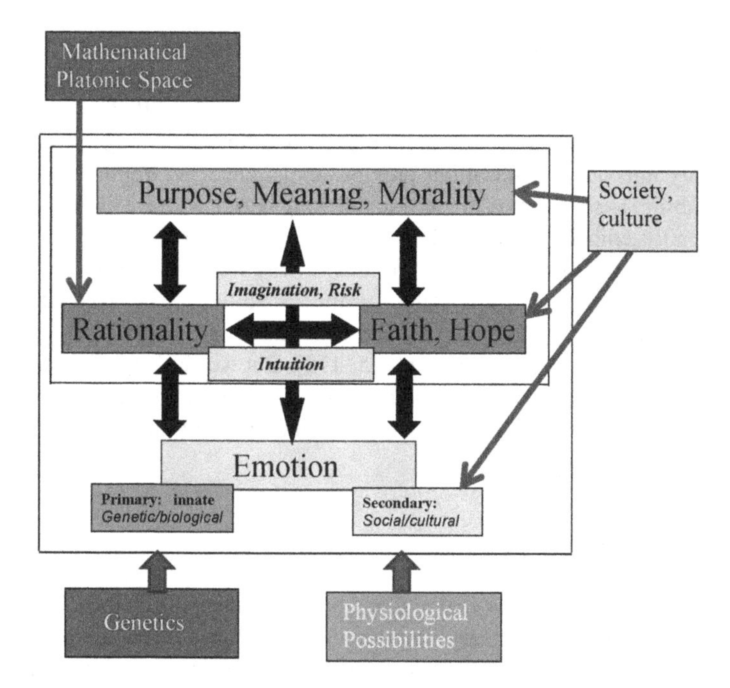

Fig. 7.6 The major aspects of brain function. Rationality underlies economic choices and everyday decisions, but is completed by faith and hope, because we never have certainty about outcomes in any complex situation. Rationality operates on the basis of imagination of alternatives and attitudes to risk, e.g., how much data do I need before making a decision?) Intuition is a form of fast-track judgement, enabling fast deployment of learnt skills. Primary genetically determined emotional systems guide rationality as to what is liable to be dangerous, while socially determined secondary emotional systems help negotiate the social world. The balance between rationality and emotion takes place within the higher level choices of values and goals determined by purpose, meaning, and morality. These act down to determine acceptable and unacceptable paths of action. Platonic unchanging relationships such as mathematical truths can be learnt by the brain and enable better informed decision-making. All this takes place within the constraints of the underlying physiological possibilities on the one hand, and genetic inheritance on the other. The latter includes instinct (not shown) which operates automatically

to how many people will buy our product, what the weather will be like, how people will treat us, and so on. Thus our choices in the end have to be concluded on the basis of partial information and are necessarily to a considerable degree based on faith and hope: faith about how things will be, hope and trust that it will work out all right. This is true even in science. When my scientific colleagues set up research projects to look at string theory or particle physics, they do so in the belief that they will be able to obtain useful results when their grant applications have been funded. They do not know for sure that they will succeed in their endeavours. They believe that their colleagues will act honestly. So embedded in the very foundations even of science, there is a human structure of hope and trust [47].

Together with our attitudes to risk, perceptions of how things are now and will be in the future are crucial in making real-world decisions. Do we tend to see things in a threatening or optimistic way? Are we willing to act on the basis of little evidence, or do we demand very detailed analysis before proceeding? This sets the balance we make between rationality on the one hand and faith and hope on the other. Helping make decisions are intuition and imagination [103, pp. 187–189]. The intuition of a doctor, a motor car mechanic, a football player, or a financial analyst, for example, is the deeply imbedded result of our previous experience and training [182]. It is a fast-track ability to see the guts of the situation long before we have had time to figure it out rationally, embodying in rapid-fire form the results of previous experience and rational understanding [140]. We can rationalise why we came to such a conclusion after the event. Imagination helps us think of the possibilities to be taken into account in making our rational choices and to envisage what might occur, setting the stage for our analysis of options and choices [25, 178]. But we can never imagine all the options: the completely unexpected often occurs and undermines the best laid plans of mice and men, and even the widest lateral thinking only uncovers some of the possibilities.

7.2.3.2 Emotions

Secondly, our emotions are a major factor in real decision-making [103, pp. 66–68]: both the hard-wired primary emotions that are our genetic inheritance from our animal forebears, and the socially determined secondary emotions that are our cultural inheritance from society. As explained so well by Antonio Damasio [49, 50] no decisions are made purely as a result of rational choice. The first factor effecting what we tend to do is the emotional tag attached to each experience, memory, and future plan. For example, the hoped-for joy of successful achievement underlies most work in science. Without it, science would not exist. In a full human life, emotional attachment is one of the most important driving factors, determining how we deploy our rationality. The importance of emotions derives from the fact that the primary emotions have evolved over many millions of years to give us immediate guidance about what is good for our survival in a hostile environment. They then guide the development of secondary emotions, which tell us what is good for us in terms of fitting into society, and intellect [147, pp. 325–327, 349–361]. However, there is also top-down control of emotional information [147, pp. 366–377, 434–436] and elicitation of emotional response: the basomedial amygdala mediates top-down control of anxiety and fear, for example [1]. The functionally distinct populations of neurons associated with positive and negative associations is described by Namburi et al. [185].

As explained by Solms, Freud emphasized that there is no such thing as unconscious affect, and emotions in higher animals also include an instinctual component, that is, they include innate mental organizations (Solms [223]):

> They do not consist solely in the affective expressions of current drive oscillations, which give emotion its 'keynote', but also in a number of specific varieties of pleasure and unpleasure which are built into the brain by virtue of their contribution to survival and reproductive success. Examples of these are sexual lust, anxious fear, rageful aggression, affectionate attachment and joyous play [192, 193]. These instinctual dispositions—known as 'basic emotions'—are 'hard wired' tools for living, giving rise to stereotypical behaviours designed

> to contend automatically with predictable situations of universal biological significance. On top of these mechanisms, higher in the brain, are added the further complexities introduced by individualised learning.

Thus processing of and response to incoming sensory data are mediated by memory, and guided by affect [87, pp. 41–49]. It is probable that affective systems played a key role in evolutionary development, because they are very important in behavioural terms. This is developed further in Sect. 7.2.4.

7.2.3.3 Values

Thirdly, we need values to guide our rational decisions; ethics, aesthetics, and meaning are crucial to deciding what kind of life we will live. They are the highest level in our goals hierarchy, shaping all the other goal decisions by setting the direction and purpose that underlies them: they define the *telos* (purpose) which guides our life. They do not directly determine what the lower level decisions will be, but set the framework within which choices involving conflicting criteria will be made and guide the kinds of decisions which will be made, particularly by setting constraints on what is acceptable behaviour and guiding as to what is desirable. Haidt and Kesebir give a useful introduction to how views of morality have evolved in recent decades [123]:

> The 'new synthesis' in moral psychology has shifted attention away from reasoning (and its development) and onto emotions, intuitions, and social factors (which are more at home in social psychology than in developmental psychology).

From the viewpoint of this book, emotional intuitions are necessary to moral decision-making [123], but do not fully encompass them, for rational reflection and self-searching is a key element of higher level morality [180]. Indeed this is all done in the context of overall meaning and purpose (*telos*), for the mind searches all the time for meaning [88], both in metaphysical terms and in terms of the social life we live [86]. These highest level understandings, and the associated emotions, drive all else.

Our minds act, as it were, as an arbiter between three tendencies guiding our actions:

- First, what rationality suggests is the best course of action: the cold calculus of more and less, the economically most beneficial choice.
- Second, what emotion sways us to do: the way that feels best, what we would like to do.
- Third, what our values tell us we ought to do: the ethically best option, the right thing to do.

It is our personal responsibility to choose between them on the basis of our best wisdom and integrity, making the best choice we can between these usually conflicting calls, informed by the limited data available, and in the face of the pressures from society on the one hand (which we must understand as best we can) and from our inherited tendencies on the other. Our ability to choose is a crucial human capacity.

The Search for Meaning. Values are closely related to the search for meaning and understanding, both in relation to other minds and intentions, and in relation to ones' own self, giving purposes to ones actions and shaping the goals one strives for [180, p. 11]:

> Meaning is fixed by action in a social world. Morally responsible action is enabled both by rationality and sophisticated symbolic language and first appears in the human species when it becomes possible to direct higher order evaluative processes towards one's own cognitive and lower order evaluative processes, influenced by the environmental scaffolding of moral language.

The search for meaning is not an optional extra: as made so clear by Viktor Frankl, it is a basic need, and a core driver of mental life [88]. I return to this in Sect. 7.4.

7.2.4 Environmental and Genetic Influences: Brain Plasticity

A key question, of course, is how the brain gets to be what it is. This has intertwined evolutionary (timescales of hundreds of millions of years) and developmental (timescales of decades) aspects. But the primary point is that the brain is not developed in a predetermined way through genetic influences: rather it adapts to the environment in which it finds itself. Brain plasticity at the micro-level allows adaptation at the macro-level. This development is guided by experience, evaluated on the basis of the primary emotional systems. Here I consider in turn the following items:

- Developmental systems.
- Nature and nurture: emotion and rationality.
- The key primary emotional systems.
- Effective modules.
- Further inbuilt systems?

7.2.4.1 Developmental Systems

Griffiths and Stolz express this developmental view nicely as follows [117]:

> The 'developmental systems' perspective in biology is intended to replace the idea of a genetic program. This new perspective is strongly convergent with recent work in psychology on situated/embodied cognition and on the role of external 'scaffolding' in cognitive development. Cognitive processes, including those which can be explained in evolutionary terms, are not 'inherited' or produced in accordance with an inherited program. Instead, they are constructed in each generation through the interaction of a range of developmental resources. The attractors which emerge during development and explain robust and/or widespread outcomes are themselves constructed during the process. At no stage is there an explanatory stopping point where some resources control or program the rest of the developmental cascade. [...] we suggest that what is distinctive about human development is its degree of reliance on external scaffolding determined top-down by interaction with the environment.

Because the brain is not isolated it creates culture [99, pp. 163–183], which then modifies the functioning of the brain [231]:

> Environments, particularly in the form of developmental environments, do not just select for variation, they also create new variation by influencing development through the reliable transmission of non-genetic but heritable information. This approach stresses particularly views of embodied, embedded, enacted and extended cognition, and their relationship to those aspects of extended inheritance that lie between genetic and cultural inheritance, the still gray area of epigenetic and behavioral inheritance systems that play a role in parental effect. These are the processes that can be regarded as transgenerational developmental plasticity and that I think can most fruitfully contribute to, and be investigated by, developmental psychology.

Thus, as emphasized also by Merlin Donald [65, 66], human beings are deeply integrated organisms embedded in and transformed by their genetic, epigenetic (molecular and cellular), behavioral, ecological, socio-cultural and cognitive–symbolic legacies.

7.2.4.2 Nature and Nurture: Emotion and Rationality

So which cognitive capacities are innate and which are culturally developed? The instinctive and sensory systems are hardwired, as are the subcortical structures such as the amygdala and other nuclei in the limbic system. What about the cortex? Neurons in the cortex are arranged in a basic structure of folds, layers, and columns. This is written into our genes, and enables the basic capacities of the cortex discussed above, namely, pattern recognition, prediction, and generalisation. Thus these are inherited. But it is not remotely plausible that detailed cortical wiring—the specific synaptic connectivity of many billions of neurons—is genetically determined, partly because there is simply not enough genetic information, and partly because there is no plausible developmental mechanism that could do this. In any case, it does not fit with the developmental view just discussed. Most of our thoughts are shaped by society [18].

The key clue is given by Damasio [49, 50], who points out that emotion is essential for rationality through its role in guiding thought and attention. It is biologically adaptive to have the ability to respond defensively to danger signals before the real danger is present [144, pp. 77–78]. These capacities are provided by the primary affective systems that have been investigated in depth by Panksepp [192, 193]. They are based on the subcortical brain structures that have been called the limbic system and associated diffuse connections (the 'ascending systems', see Fig. 7.7). These are indeed genetically inherited, so they must play an important role in brain function, and require an evolutionary explanation.

Psychological Function. Emotions play a key role not only through immediate effects on cognition, but also in development through dual coding: each sensation as it is registered by the child also gives rise to affect or emotions [116, p. 56], which are stored as part of the memory. This dual coding of experience is the key to understanding how emotions organize intellectual abilities and indeed create the sense of self. For example, as explained by Greenspan and Shanker, a baby first learns causality not through pulling a string to a bell but through the exchange of emotional

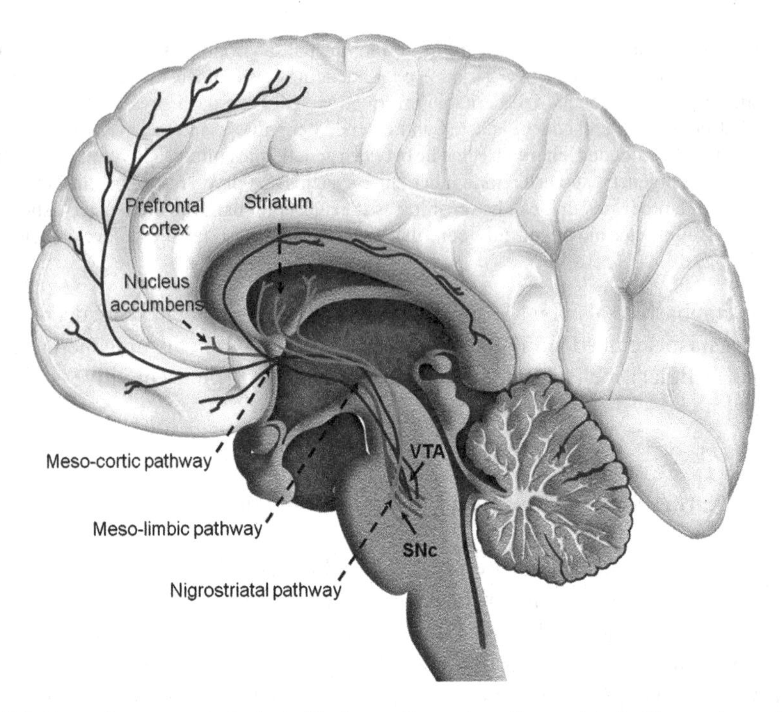

Fig. 7.7 Reward structures in the brain (the 'ascending systems'), associated with neurotransmitters such as serotonin. They project diffusely rather than connecting specific neurons. Note the contrast with Fig. 7.5. Credit: Wikimedia commons. From Arias-Carrión et al. [6]

signals (I smile, you smile back) [116, p. 51]. This is based on the attachment system identified by Panksepp, which results in separation anxiety and distress when infants are separated from their parents.

Physiology and Development. Part of developmental plasticity takes place through Hebbian processes: neurons that fire together wire together [115]. But that process does not distinguish between what is good and what is bad. It has no adaptive capacity to respond positively to what is good for survival and negatively to what is not.

That capacity is provided by the process identified by Gerald Edelman as neural Darwinism [72, 73] whereby neuromodulators such as serotonin and dopamine project diffusely from the limbic system to the cortex and then can selectively strengthen or inhibit the pattern of synapses that are active at that time. They originate in the structures recognised by Panksepp as the physiological basis of the primary affective systems. This is therefore a powerful mechanism for shaping neural plasticity to support survival needs as recognised by these primary affective systems. The psychological correlates of these systems are emotions such as fear and joy, which are powerful behavioural motivators. There has to be a functional and evolutionary explanation for the existence of these systems (Fig. 7.7), and that explanation is their role in shaping plasticity. This role, as set out here, must have been so important that they were selected for, in evolutionary terms.

In this way, combining combining Panksepp's affective neuroscience and Edelman's neuronal group selection, one arrives at the idea of affective neural group selection [81], which identifies the functional purpose of the ascending systems as both immediate behavioural guidance and longer term neural adaptation. From a neural net viewpoint, one arrives at salience affected neural networks (SANN) [202]. From a psychological viewpoint this makes clear how emotional drives underlie both behaviour and brain development.

7.2.4.3 The Key Primary Emotional Systems

Given this understanding, it is important to identify the genetically determined primary emotional systems, an area of considerable disagreement. Panksepp [192, 193] identifies three categories of affective systems: primary (raw instinctual–affective), secondary (unconscious learning and memory related processing), and tertiary (higher cognitive manifestations) levels.[1] He identifies seven blue ribbon instinctual emotions, in addition to homeostatic affects and the sensory affects. Toronchuk and Ellis [244] propose two others, leading to nine basic instinctual emotional systems as set out in Fig. 7.8. As each of them is a genetically inbuilt module, each is of evolutionary importance, and this has to be because it is behaviourally important. The justification for each of them is given in [193, 244]. Here I will just identify four of them, operating in dual pairs.

- **The Drive for Meaning**. The need to understand as emphasized by George Kelly [150] and Jeff Hawkins [129], and so the drive to find meaning [88]. The corresponding affective system is labelled the SEEKING system by Panksepp [192, 224]. It has dual components: a wanting system that searches for what is meaningful or wanted, and a liking system, the latter turning off the former when satisfaction is achieved.

Operating alongside this is:

- **The Play System**. In animals, this is mainly, but not exclusively, rough and tumble play [192], which is crucial in developing skills, while in humans it extends to the imaginative play that is the key to creativity and learning [114, 244].

Then there are two systems that operate in tandem to encourage and enable communal living. They are firstly:

- **The Attachment System**. The need for affiliation-belonging [209] (starting with the mother–child bond, and then extending to peers) which underlies the existence of societies. The component of the environment thought to be most important for humans, and in fact for all mammals, is the infant's major caretaker, usually its mother [144, pp. 80–84], so this starts with the need to bond to the mother [116],

[1] For a brief summary, see http://emotionresearcher.com/the-emotional-brain/panksepp/.

TABLE 1: Evolutionary needs, and the emotional systems that have evolved to meet them. E1 is a generalised system providing incentive for the others and this dependence is noted only once.

EVOLUTIONARY NEEDS MET	*PRIMARY EMOTIONAL SYSTEM*	*Works With:*	*FUNCTIONS*
INDIVIDUAL NEEDS			
Basic Functioning	**E1: SEEKING system**	**E2-9**	Situation evaluation, incentive salience, hedonic appraisal, facilitates learning
Basic Survival	**E2: DISGUST system** (repulsion)		Avoiding harmful foods, substances, environments
	E3: RAGE system	**E4,E9**	Defence: protection of organism, resources, and con-specifics, limiting of restraint on movement
	E4: FEAR System	**E3,E9**	Defence: flight, limiting of tissue damage
SOCIAL NEEDS			
Reproduction	**E5: LUST** system (sexual desire, satiation)	**E6,E7**	Ensuring procreation, enhancement of bonding
Group cohesion: Bonding & Development	**E6: PANIC/attachment** (affiliation, separation distress)	**E5,E7**	Protection of vulnerable individuals; creates bonding through need for others
	E7: CARE system	**E5,E6**	Caring for others, particularly offspring
	E8: PLAY system	**E6,E7**	Bonding with con-specifics, development of basic adaptive and social skills, creativity
Group function: Regulating conflict	**E9: POWER/dominance** system (rank, status, submission)	**E3,E4,E5**	Limiting aggression in social groups: allocating resources, esp. sexual ones.

Fig. 7.8 The basic genetically determined affective systems (extension of Panksepp's list [192] by Toronchuk and Ellis [244]). Each system has associated emotional feelings. E1–E5 are related to individual survival needs, while E5–E9 are related to group survival needs, underlying the social brain and so playing a key role in evolutionary history [71]. E1 is the basic system, and the others piggy-back off it. These systems are shared with all the higher animals [192, 193], but the PLAY system E8 extends to imaginative play, crucial in learning and imagination. The disgust system E2 [207], related to gut feelings and the second brain [105], may be the oldest system, but gets coopted to respond to social events in society

associated with panic on separation. However, it then develops to a strong need to belong to some kind of community [228], and then is the basis of the social brain [71], and so underlies the formation of social groups.

Secondly, once societies form, there will be all sorts of tensions in them related to resource allocation that need to be regulated. This is the function of:

- **The Dominance/Ranking System**. Associated with territory in the broadest sense, this allocates positions and resources in society (see [12], [103, p. 73], [228, 249]), as characterised by the pecking order and the alpha male, and thereby reduces conflict. This is again a dual system: the competitive drive that seeks for territory and rank (Ardrey [5]), and the obedience system that accepts the outcome of that competition, hence defusing the conflict.

Together these two systems give the emotional underpinning for forming stable social groups. This formation is of key importance in our evolutionary history [71].

In Summary. The evolutionary explanation for the existence of affective system structures [192] (such as the ascending systems [154]) and associated neurotransmitters (such as dopamine and noradrenaline) is that they enable the primary emotional system to guide the development of the cortex as the individual interacts with the physical and social world around. From the day we are born they provide protection from potential dangers and encourage us to learn and to form communities with others; and they shape brain plasticity in response to our experiences.

7.2.4.4 The Overall View

Overall the view is as set out by Merlin Donald[2]:

> Our distinctively human consciousness seems to be contingent on four things: an expanded executive brain system, extreme cerebral plasticity, a greatly expanded working memory capacity, and especially a process of brain–culture symbiosis that I have labelled 'deep enculturation'. Constructivism [is] an approach to human cognition that originated in French philosophy, with Condillac. Constructivism holds that the mind self-assembles, according to the dictates of experience, guided by a set of innate propensities, which correspond roughly to the basic components of conscious capacity.

I have here identified those innate propensities as firstly basic memory, pattern recognition, generalisation, and prediction abilities, and secondly the primary emotional systems indicated in Fig. 7.8. Dunbar [71] emphasizes how it was essential for primates to live in groups in order to have a good chance to survive, so the emotional systems that lead to group formation and stabilisation (the attachment and dominance/ranking systems) were crucial for evolutionary development. They are also central to psychic well-being [228].

---

[2] http://newlearningonline.com/new-learning/chapter-6/donald-on-the-evolution-of-human-consciousness.

7.2.5 The Origin of Humanity: The Social Mind and Language

A key issue is how humankind emerged form the higher primates: what specific features of structure and function, and specifically brain and mind, separate us and give us our huge advantage over all the rest of the animal kingdom? What kinds of evolutionary pressures led to this extraordinary emergence, which took place very rapidly once it was under way?

7.2.5.1 The Social Context

Minds cannot be understood on their own: we have a distributed cognition that is not contained solely within the head of the individual (see [65] and [100, p. 14]). The human mind is unlike any other on this planet, not because of its biology, which is not qualitatively unique, but because of its ability to generate and assimilate culture, which provides us with symbolic tools such as language that then shape the way we see reality [18].

7.2.5.2 Evolution of Humanity

Social behaviour has biological origins: in order to survive and prosper, we had to become social [103, pp. 82–112]. This also then shapes evolution [71, 100, 204]. While there is considerable debate about specifics, a broad overview is emerging that makes sense in the light of the viewpoint put forward in this book.

Environmental changes placed pressures on roving bands of hominids making it very important that they develop social cohesion so that they could face these challenges together in order to survive [71]. This led to the emergence of the social mind [66] through a series of developments whereby full-blown symbolic systems (languages) developed from simpler modes of communication [236], and enabled the sharing of intentions and ideas between individuals [242], offline development of conceptual thought [20], and expansion of the mind to include storage systems and thinking aids in the environment [236]. Physical changes such as development of the vocal tract and control of breathing enabled language development. All of this enabled large social groups to arise and develop social interactions implementing the technology which was the foundation of the rise of civilisation [33].

7.2.5.3 The Social Brain Hypothesis

There have been various proposals as to how this happened, largely in agreement with each other, based on the importance of this group functioning:

- The *social brain hypothesis* (Dunbar [71, pp. 59–68]) says that the prime mover in primate brain evolution is the evolution of more complex forms of sociality, with increasing brain size allowing ever greater behavioural complexity, and group size itself being an emergent property. Time budgets are a key limitation: they must allow for rest, foraging, movement, and socialisation activities that bond the group together [71, pp. 84–93]. They must also allow for learning activities, not mentioned by Dunbar, which is the key function of play. Inventions such as cooking (reducing foraging times) and music and story-telling (allowing bonding of larger groups) allowed a balancing of the time budget as band size increased. Language was needed to facilitate exchange of information and make agreements, and for social bonding [71, pp. 263–275].
- Gamble [100] expresses the social brain hypothesis as follows: large brain size in primates and hominims is selected for by social rather than simply ecological factors concerned with getting food, i.e., our social lives drove our remarkable encephalization [100, pp. 15, 68–73, 319]. Climate change was a factor driving the need for adaptation [100, pp. 33–68, 76–84], leading to mobility together with group fission and fusion. Technology makes greater mobility possible and develops hierarchically organised cognition associated with an imaginative capacity and recursive analogical reasoning [100, p. 174]. Dispersals enabled settlement in ecologically favourable locations. Overall [100, pp. 317–318]:

  > An essential element of a new framework for deep human history is the concept of the relational, rather than rational, mind. Hominínims will never make sense to us if we only deal with them as rational creatures interested in food and reproduction and not much else. [...] We need to replace the isolated, rational mind with one that recognizes that cognition is distributed and the mind extended.

- Bickerton [20] sees the key feature in development of exceptional mental competences to be a climate-change induced switch from individual to cooperative scavenging. The resultant communication needs in a group led to brain reorganisation allowing the hierarchical organisation of strings of words and so the rise of language, enabling 'offline' thinking.
- Tomasello [242] similarly argues that a change in the mode of thinking was evoked by the development of cooperative foraging. The development of joint cooperative action demanded recursive 'mind reading' to establish common goals and assumptions, together with a reciprocity of perspectives on the joint enterprise. A ratchet effect worked to allow cumulative cultural and technological evolution [239]. The further key point was the establishment of sanctioned group norms underlying increasing cultural complexity. Norms brought social institutions and so a conventionalised grammar.
- Stringer [236] sees the development of mind-reading ability as regards both prey and fellow hunters as the key to the human advantage over other species [236, pp. 105–112], which is of course a form of the social brain hypothesis. He supports Darwin's view that, apart from language, the art of making fire was probably the greatest discovery ever made by our prehistoric ancestors: once cooking became central to human life, it influenced our evolution as regards digestion and

gut, teeth and jaw, and muscles for mastication [236, pp. 139–141]. Division of labour obviously needed social agreement, and allowed the rise of technology [236, pp. 141–146]. The rise of increasingly complex societies led to a need to communicate increasingly intricate and subtle messages, and the development of language [236, pp. 213–214]. The unifying effects of shared beliefs amid increasing social complexity would have provided the glue that bound people together, encouraged self-restraint, and put group needs ahead of individual ones [236, p. 223], that is, it would have led to the development of social norms.

- Greenspan and Shankar emphasize that the social brain is an emotional brain [116, p. 9]:

> The growth of complex cultures and societies and human survival itself depends on the capacities for intimacy, empathy, reflective thinking and a shared sense of humanity and reality. These are derived from the same formative emotional experiences that lead to symbol formation.

The point then is that, although there is a degree of divergence in the detail, in each case the evolutionary process is driven by the way the developing human line adapted to changing environmental conditions and thereby changed not just physical features (e.g., developing bipedalism), but also the nature of cognition, i.e., what we can do inside our minds [71], specifically developing the ability to handle complex social relations. But these can only exist if groups exist and selection acts as regards group properties. These pressures in turn select individuals with capacities that enable the desired group properties [78]. The adaptive process that made our brains what they are was thus a multilevel top-down process [103, pp. 80–82], as discussed in Sect. 4.3.7.

7.2.5.4 The Issue of Cheating

A counterview to this is that, while group selection is possible, it rarely happens because of the cheating problem studied in depth by evolutionary theorists.[3] However, as regards hominids, the view presented here is supported by the data on human evolution that supports the social brain hypothesis [71, p. 59]. In that case selection took place for those features that led to the formation of social groups, and this did indeed take place in our evolutionary history, as a matter of historical fact. Thus the causal chain is from the evolutionary advantages of group living [71, p. 85] down to the features of human nature that support group living (the primary emotional systems discussed above), and hence down to those genes that support that nature. The cheating issue did not prevent this multilevel selection taking place.

How then is cheating dealt with? It is probably the case that, in developing social groups, cheating is largely dealt with by norms enforced by social structures [17, 123, 166]. The individual is not free to get away with cheating at will, because strong social structures develop culturally to deal with it. We become expert cheat detectors, and shame those who indulge in it. This is where the secondary (social) emotions play

---

[3] I thank Michael Ruse for discussions on this topic.

a key role in our behaviour and hence in our evolutionary history. In any case, the historical record shows that the behaviour of cheats did not prevent the formation of social groups [71], and that fact had a profound effect on our evolutionary history.

7.2.5.5 Effective Modules

Given this understanding, we can look at the nature and nurture issue in regard to the alleged existence of a variety of cognitive modules, such as innate language modules proposed by Chomsky and Pinker and innate folk physics modules proposed by David Geary. The position taken here is that there are no genetically fixed cognitive modules whatever, and in particular no innate language modules that control grammatical structure. Griffiths and Stolz state the issue of innate modules as follows [117]:

> What individuals inherit from their ancestors is not a mind, but the ability to develop a mind. The fertilised egg contains neither a 'language acquisition device' nor a knowledge of the basic tenets of folk psychology. These features come into existence as the mind grows. A serious examination of the biological processes underlying such easy terms as 'innateness', 'maturation', and 'normal development' reintroduces the very themes that are usually taken to be excluded by an evolutionary approach to the mind—the critical role of culture in psychological development and the existence of a plethora of alternative outcomes for the developing mind.

This is the position taken here. Thus what I suggest is that we do not have

$$\texttt{evolution} \rightarrow \texttt{folk modules},$$
$$\texttt{folk modules} \rightarrow \texttt{learning effects},$$

as some propose. Rather the genetically fixed affective systems studied by Panksepp and associated ascending systems shape the way the cognitive system learns as it interacts with its environment:

$$\texttt{evolution} \rightarrow \texttt{emotional systems},$$
$$\texttt{emotional systems} + \texttt{experience} \rightarrow \texttt{effective folk psychology behavior},$$
$$\texttt{effective folk psychology behavior} \rightarrow \texttt{learning effects}.$$

Where then do the language regularities come from that Chomsky recognised and categorised as being due to a deep grammar module? On this view they are due to essential syntactic limitations on any language whatever in order that it be an adequate symbolic system for describing the world around. They are due to fundamental semiotic constraints on any symbolic representation of our experiences and environment, as explained in detail by Deacon [55]. How then is language learned? Through ongoing experience of the use of language in meaningful contexts [241], developing an embodied construction grammar [16, 82], particularly via mother–child bonding and the child's search for meaning in this developing relationship [116]. This process is beautifully described in *Chloe's Story* by Carole Bloch [23]. Is there a poverty of stimulus, as claimed by Chomsky and others? Certainly not, if we take the intense mother–child interaction into account [23, 116]. Empirical evidence in this regard is given in [200], but the argument here is based in the end on the view that there is no plausible biological mechanism that could either create such modules, or cause the detailed cortical wiring that would be required to realise them.[4]

This is an essentially top-down process, rather than a bottom-up process controlled in a genetic way: the interaction with the cultural environment shapes detailed cortical connections with the primary emotions acting as modulating factors. The behavior that gets built in will be suitably tuned to the culture in which the individual lives, because it is created through interaction with that culture. This experiential shaping of these systems to fit the local environment in a meaningful way is an aspect of the crucial feature of brain plasticity [66].

7.3 Top-Down Processes

The above sections have shown many ways in which brain functioning proceeds as a result of a combination of bottom-up and top-down influences. Elder–Vass puts it this way [75, p. 59]:

> Downward causation works in just the same way as any other type of causation. The causal mechanism depends ultimately on the presence of the levels of organisation represented by the causing entity. The operation of the higher level causal effect will depend on the causal effects of the parts, but it is only when they are organised in the form of the 'whole' causing entity that they have this effect.

Murphy and Brown write as follows:

> Our account of downward causation involves selection or constraint of lower level processes on the basis of how thoses lower level processes or entities fit into a broader (higher level) causal system [180, p. 12]. The essence of downward causation is selective activation of the causal powers of the entities constituents, which takes place via context sensitive constraints exercised by the whole of a complex adaptive system on its components [180, pp. 43, 87–90]. The binding of components into a dynamic system limits the degree of freedom of the components but the system as a whole gains a broader causal repertoire [180, p. 89]. Complex adaptive systems can be goal-directed, making them semi-autonomous from lower level control, and are capable of selecting the stimuli in the environment to which they will respond, making them semi-autonomous from environmental control as well [180, pp. 90, 95].

Evolution by natural selection is a classic example [180, pp. 57–58]. In each case, the system will be imbedded in a larger system, and understanding it fully will require understanding that larger system [180, pp. 196–203]. For example, a thermostat system keeping a plant at 65 °C can be understood as a purely mechanical system, but understanding why it is set to this temperature requires knowledge of biology and intentions. There are always multiple levels of cause in action. The bottom level

[4]Claims of their existence based on Gold's proof and infinite recursion [139] have nothing to do with natural language, because infinity has nothing to do with any realisable language [80].

physics is only one of them [79, 189]. The distinction between top-down and bottom-up effects plays a central role in experimental psychology [217]. Here I shall look in turn at the following items:

- The different kinds (Sect. 7.3.1).
- Memory, learning, and deleting (Sect. 7.3.2).
- Vision (Sect. 7.3.3).
- Language and reading (Sect. 7.3.4).
- Goal-directed behaviour and attention (Sect. 7.3.5).
- Health (Sect. 7.3.6).
- Social neuroscience (Sect. 7.3.7).

7.3.1 The Different Kinds

All the different kinds of top-down causation identified in this volume (as discussed in Chap. 4) occur in the brain:

Dynamical Systems Aspects (TD1). Micro and macro dynamical systems abound in the brain. They include propagation of action potentials from one neuron to another, channelled by axon structures [148], pattern recognition by neural networks with weights that have been determined by previous experience [21], and reflex action [180, pp. 111, 114]. In each case micro-level interactions are channelled by higher level structures.

Cybernetic (Goal-Directed) Aspects (TD2). These occur at micro- and macro-levels [180, p. 106]. At the micro-level there are hundreds of feedback control loops in cells [203], at the physiology level feedback control loops preserve homeostasis [188, 203], and at the macro-level tasks such as driving an automobile involve continual feedback control. In each case outcomes are determined by a set of predetermined or chosen goals.

Learning and Plasticity (TD3). The brain is built to learn so that it can adapt to and respond to changing circumstances [114], and this adaptivity at the macro-level is based on underlying plasticity at the micro-level (see [147] and [180, pp. 107, 114–117]). Learning and adapting is the key to ecological and social advantage. In each case outcomes are shaped by selection criteria guiding the adaptive process, as well as the context in which the selection process takes place.

Adapting Feedback Responses (TD4). A key aspect of operant conditioning (Pavlovian responses) is that they change if the context changes, i.e., the embodied goals are adaptively selected according to the environment [114]. The outcome depends on the criteria used to select the goals, as well as the context in which this takes place. The neural basis for this is described by Donoso et al. [67]

Adapting the Mode of Adaptation (TD5). The essence of intelligence is the ability to adaptively select the criteria guiding adaptive selection of actions, which requires the capacity to think offline, facilitated by symbolic representation [180, pp. 120–127]. This is a key advantage provided by language. The outcome depends on the higher level criteria used to select the lower level selection criteria, as well as the context in which this takes place.

Crucial Contextual Effects. Through these various mechanisms, the individual mind and culture coevolve, with each affecting the other in crucial ways. Overall the view is as set out by Merlin Donald[5]:

> [O]ur conscious capacity provides the biological basis for the generation of culture, including symbolic thought and language. Conversely, culture also provides the only explanatory mechanism that can unlock the distinctive nature of modern human awareness. [...] the specifics of our modern cognitive structure are not built in. Our brains coevolved with culture and are specifically adapted for living in culture—that is, for assimilating the algorithms and knowledge networks of culture. [...] Cultural mind-sharing is our unique trait. Human culture started with an archaic, purely non-linguistic adaptation, and we never had to evolve an innate brain device for language *per se* or for many other of our unique talents, such as mathematics, athletics, music, and literacy. On the contrary, these capacities emerged as by-products of our brain's evolving symbiosis with mindsharing cultures. Language emerges only at the group level and is a cultural product, distributed across many minds.

In the rest of this section, I look at various specific aspects of top-down causation in the brain: memory, learning, and deleting (Sect. 7.3.2), vision (Sect. 7.3.3), language and reading (Sect. 7.3.4), goal-directed behaviour and attention (Sect. 7.3.5), health (Sect. 7.3.6), and the relation to society, i.e., social neuroscience (Sect. 7.3.7). I conclude by commenting on how this relates to the physical substrate (Sect. 7.3.8).

7.3.2 *Memory, Learning, and Deleting*

Personality and behaviour is crucially based on learning and memory. Learning works by starting with a set of methods or hypotheses, testing them, and keeping those that work while discarding those that fail. It is a classic example of adaptive selection, which together with creativity is the key element allowing useful new ideas and information to come into being. Adaptive selection is also key to memory, when we keep important memories and information and discard that which is no longer useful.

The basic idea of adaptive selection is shown in Fig. 7.9. There is an ensemble of incoming stuff (senses, thoughts, theories) and whatever meets a certain selection criterion is kept, whatever does not is discarded. In this way a more ordered outcome (thoughts, theories, memories) is created from a more disordered set of input states. The incoming ensemble may be random, the result of experience, or the result of imagination. Adaptive selection of what is meaningful enables a selected set of entities to become causally effective as their competitors are discarded. It is the basis of all scientific discovery and learning as it shapes our understanding. It is a form of top-down causation because it is guided by higher-level selection criteria that crucially determine the outcome [77].

---

[5] http://newlearningonline.com/new-learning/chapter-6/donald-on-the-evolution-of-human-consciousness.

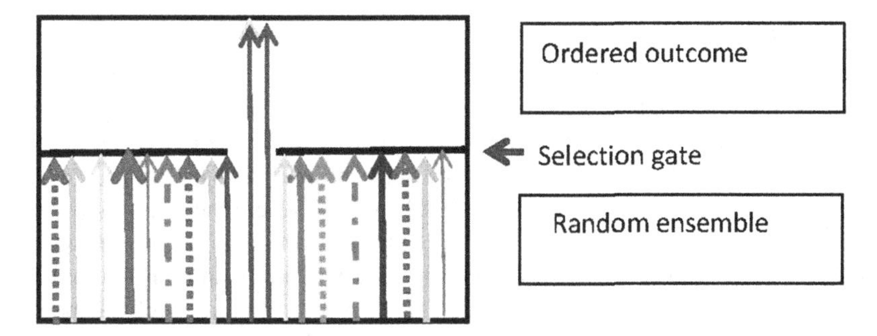

Fig. 7.9 The basic selection process. A selection gate creates order out of disorder by letting through only entities in a random initial ensemble that satisfy some selection criterion. This is an irreversible process: the initial state cannot be determined from the final state (the needed data is missing). Selection processes are important in the brain [84]

Memory. There are three memory systems: procedural or implicit memory [147, pp. 308–309, 311, 313], declarative or conscious memory [144, pp. 70–71], and working memory [144, pp. 86–87]. The last is short term memory, the first two are long term memory [114]. In each case memory is formed by storing traces of experiences and thoughts, perhaps by altering synaptic connection strengths, but perhaps in other ways. There are two points to note:

Categorisation. The brain processes senses by categorising them in a hierarchical way. Categorisation is essential: without it raw data would be meaningless. In order to be useful memories must be classified and indexed so that they can be recalled in relevant contexts. On the one hand the emotional tag attached will be stored, assigning it an emotional significance and sign (good or bad). This tells us the importance of the memory. On the other hand, conceptually speaking, it will be linked to other related memories in a hierarchical classificatory system (explicit or implicit). The brain could not establish categories without memory, and conversely cannot usefully store memory without categorisation, which is a process of selection: specific items are accepted as belonging to one category and discarded from others.

Selection. More than that, the memory system must decide what memories are worth saving in long term memory and what should be discarded. With finite memory resources available, we cannot afford to let our brain get swamped with unnecessary irrelevant details of what has happened. But we still need to retain what is important. So in all memory systems, forgetting is crucial in order to retain what is meaningful. In computer systems this is formalised as a process of garbage collection and deletion. A key point is that this is where irreversibility takes place in computation: in a crucial way, it is where the arrow of time comes in and entropy is generated. An everyday example is the deletion of files and emails.

Thus overall we need a process for forgetting what is not needed, remembering what is meaningful, and organising and sorting what is retained. This top-down processing of information in relation to selection criteria related to meaning is basic to memory use in selecting meaningful action paths (Kandel [147, pp. 304–321]). Selection processes are an important aspect of brain function [84].

7.3.3 *Vision*

Vision is the most studied of all the senses. An obvious bottom-up view of how it works is that photons arrive at the retina and get detected by retinal receptors (rods or cones), whence the resulting signals are sent up the optic nerve to the thalamus and visual cortices in the occipital lobe [201, p. 106], where they get assembled to form the visual picture we see. But this is not how it works. Rather we see what we expect to see, on the basis of our immediate past experience and our interpretation of the context in which we are situated. That is, vision is predictively shaped in a top-down way on the basis of our expectations [99, 108, 201]. As stated by Kandel, we do this in a gestalt way [147, p. 199]:

> Overall in perceiving an object, a scene, a person, or a face we respond to the whole rather than the individual parts. We do this because the parts affect one another in such a way that the whole ends up being much more meaningful than the sum of its parts.

Indeed this is inevitable. Because our senses receive only a projection onto the retina of the 3D outside world, what we see has many interpretations and we must construct the image we see from this limited information [147, pp. 200–202]. Thus vision is based on a process of guessing and hypothesis testing in the brain, on the basis of past experience.

Kandel explains that we do this by an automatic contextual process of hypothesis testing (unconscious inference): before we perceive an object, our brain has to infer what that object might be, based on information from the senses [147, p. 203]. This idea was developed by Gombrich in relation to visual illusions [147, pp. 208–212]: once we have formulated a successful hypothesis about the image it not only explains the visual data but excludes alternatives. The viewer fills in, through top-down processing, any missing needed information [147, p. 314]. Frith confirms this view when he states that our perception of objects in the world is not immediate. Before we can perceive an object, the brain has to infer what the object might be on the basis of the information reaching the senses [99, pp. 41–44]. Why should we believe this paradoxical view of how vision works? Firstly, because of visual illusions, and secondly because of evidence from neuroscience studies.

7.3.3.1 Illusions

The primary illusion is that we do not see the blind spot where there are no optical receptors (see [156, pp. 53–54] and [99, p. 135]). Our mind fills in what it cannot see because no signals are coming in from that part of our field of view. Furthermore, only the central part of our field of vision can be seen in detail and in colour, while the rest remains blurred and in grey and white [99, pp. 41–42].

However, there are many other illusions that prove this view of 'unconscious inference' to be correct (see [99, 201] for many examples). I give just two here so that you can experience it for yourself. First, any normal human mind sees two triangles in Fig. 7.10. However, there are no triangles there: just three angle segments

Fig. 7.10 Kanizsa's triangle illusion. Credit: Wikimedia Commons (Jaap Pol)

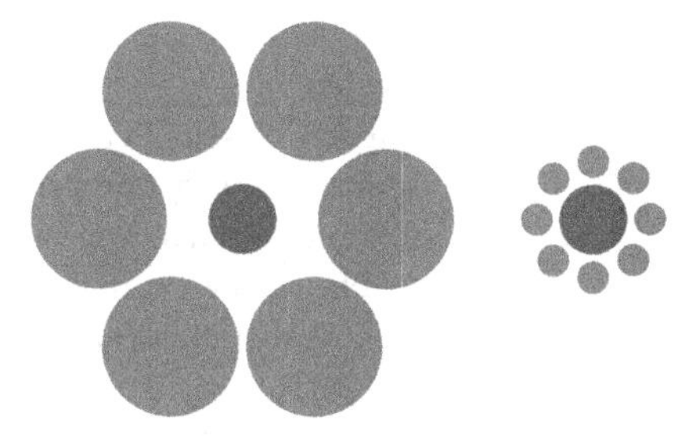

Fig. 7.11 Size comparison illusion (the moon illusion). The inner circles are the same size. Credit: Wikimedia Commons

and three partially filled discs. You can break the illusion by covering up the rest of the figure so as to see just one of the angles or one of the disc segments. The mind fills in what is not there, and the point is that it's not that you *think* you see two triangles: that *is* what you see! This experimentally demonstrates how perception works by model fitting.

Second, in Fig. 7.11, it appears that the two inner circles are different sizes, but they are the same (measure them!). This misleading illusion is an environmental effect due to the proximity of the outer circles. It is a consequence of visual methods of judging sizes and distances in natural habitats.

7.3.3.2 Neuroscience of Top-Down Visual Processes

Some papers discussing top-down influences on visual processing are as follows:

- Gilbert and Li [107]. Re-entrant or feedback pathways between cortical areas carry rich and varied information about behavioural context, including attention, expectation, perceptual tasks, working memory and motor commands. Neurons receiving such inputs effectively function as adaptive processors that are able to assume different functional states according to the task being executed. Recent data suggest that the selection of particular inputs, representing different components of an association field, enable neurons to take on different functional roles. Their review discusses the various top-down influences exerted on the visual cortical pathways and highlight the dynamic nature of the receptive field, which allows neurons to carry information that is relevant to the current perceptual demands.

- Peterson and Cacciamani [197]. Object perception is a dynamical, integrated process in which (a) high-level memory representations are accessed before objects are perceived, (b) potential objects compete for perception and only the winners are perceived, and (c) there is no clear dividing line between perception and memory. There is accumulating evidence favoring a dynamical model involving feedback as well as feedforward processing and interactions between high and low levels of the visual hierarchy.
- Peterson [196]. Top-down effects are effects that originate in high levels of the hierarchy of visual processes and exert an influence at lower levels. Examples of visual information thought to reside at, or arise from, high levels in the hierarchy are: the perceiver's intentions, expectations, attentional goals, and memories established on the basis of past experience.
- Nakabayashi and Liu [183]. According to a classical view, young children process faces in a piecemeal fashion before adult-like holistic processing starts to emerge at the age of around 10 years. This is known as the encoding switch hypothesis. These authors conclude, quite contrary to the classical view, that holistic processing is not only present in early child development, but could even precede the development of part-based processing.
- Kuo et al. [158]. Retrospective cues trigger activity in a large-scale network implicated in attentional control and lead to retinotopically specific modulation of activity in early visual areas V1–V4. Shifts of attention during visual STM (VSTM) maintenance are associated with changes in functional connectivity between pFC and retinotopic regions within V4. These findings provide new insights into top-down control mechanisms that modulate VSTM representations for flexible and goal-directed maintenance of the most relevant memoranda.
- Stokes et al. [230]. Long-term memory prepares neural activity for perception.
- Layher et al. [162]. The categorization of real world objects is often reflected in the similarity of their visual appearances. Such categories of objects do not necessarily form disjunct sets of objects, neither semantically nor visually. The relationship between categories can often be described in terms of a hierarchical structure. For instance, tigers and leopards build two separate mammalian categories, both of which are subcategories of the category *Felidae*. These authors propose a recurrent computational network architecture for the unsupervised learning of categorial and subcategorial visual input representations. During learning, the connection strengths of bottom-up weights from input to higher-level category representations are adapted according to the input activity distribution. In a similar manner, top-down weights learn to encode the characteristics of a specific stimulus category. Feedforward and feedback learning in combination realize an associative memory mechanism, enabling the selective top-down propagation of a category's feedback weight distribution. They suggest that the difference between the expected input encoded in the projective field of a category node and the current input pattern controls the amplification of feedforward-driven representations. Large enough differences trigger the recruitment of new representational resources

and the establishment of additional (sub-)category representations. The proposed combination of an associative memory with a modulatory feedback integration successfully establishes category and subcategory representations.

- Bressler et al. [31]. Information about an impending stimulus facilitates its subsequent identification and ensuing behavioral responses. This facilitation is thought to be mediated by top-down control signals from the frontal and parietal cortex that modulate sensory cortical activity. Here the authors show, using Granger causality measures on blood oxygen level-dependent time series, that frontal eye field (FEF) and intraparietal sulcus (IPS) activity predicts visual occipital activity before an expected visual stimulus. Top-down levels of Granger causality from FEF and IPS to visual occipital cortex were significantly greater than both bottom-up and mean cortex-wide levels in all individual subjects and the group. These results suggest that FEF and IPS modulate visual occipital cortex, and FEF modulates IPS, in relation to visual attention.
- Sigman et al. [219]. Top-down reorganization of activity in the visual pathway after learning a shape identification task.

7.3.4 Language and Reading

Top-down causation is a key feature in language production and understanding: holding context in mind is crucial to understanding speech and in reading [110–112, 221].

7.3.4.1 Language

This is a contextual form of communication where reference is continually made to the social, historical, political, and environmental context. Indeed the way meaning is embodied in the hierarchical structure of language is context-dependent all the way down: the individual units at each level (sentences, phrases, words, phonemes) only attain their meaning and function in the larger context of the whole meaningful situation [111, 157], and this is embodied in words and phrases such as 'they', 'then', 'their', 'there', and 'at that time'. Words are often omitted because they are assumed as implied by the context, and allusions are made to social and political issues that are taken for granted as commonly known [66]. Listening and reading are active rather than passive processes: the reader brings her knowledge to bear on the text that is presented in order to interpret it (among other things because much of language is metaphorical [159]).

7.3.4.2 Reading

Reading is the same as vision: the words convey meaning that is hinted at and realised by interpretation and guessing at what is meant, on the basis of the understood context, just as happens in listening. Words hint at feelings and understandings that can only partially be expressed in words, we learn to fill in the rest. It does not work by reading phonemes and assembling them to understand words:

> **Basic Principle.** As in the case of vision, an ongoing holistic process takes place whereby the cortex predicts what should be seen, fills in missing data, and interprets what is seen on the basis of expectations in the current context.

Contextual Meaning. The lower levels do not make sense without the higher level context, so text is read in a top-down manner. Just as in the case of vision, reading is a predictive activity that has the nature of a psycholinguistic guessing game [110, 111, 221]. The higher levels set the context for understanding lower level entities and can even select their pronunciation and meaning. Here are some examples:

The grammatical nature of a word can depend on context: is 'flies' a noun or verb?

- Time flies like an arrow.
- Fruit flies like a banana.

You deduce what the word 'plane' means from the sentence as a whole:

- The horses ran across the plane.
- The plane landed rather fast.
- I used the plane to smooth the wood.

English is only weakly a phonics-based language, as emphasized by Steven Strauss [232]. We cannot reliably read in a bottom-up manner, because not only the meaning, but even the pronunciation depends on context:

- How should we pronounce 'Wound'? Her wound hurt. She wound the clock.
- How should we pronounce 'ough'? A rough-coated, dough-faced, thoughtful ploughman strode through the streets of Scarborough; after falling into a slough, he coughed and hiccoughed.

Thus language functions in a coherent contextual way, with ambiguity and redundance as central features [94].

7.3.4.3 Evidence

Firstly, the above is evidence. But can we provide more detailed systematic studies that support this view? Detailed studies of the reading process [85, 233] prove the above is the way reading works in a meaningful context.[6] This has been done in three ways:

---

[6] Reading studies in meaningless contexts, such as those by Dehaene [60], clarify partial aspects of the reading process, but simply fail to get at the core of what is going on in real reading. Similarly reading tests based on meaningless phonemes fail to relate to genuine reading ability.

Eye Movement Tracking Studies. These show that we do not read the words or phonemes on the page consecutively, but skip words and jump back and forth between them [85]. Thus we do not assemble the thoughts in a strict bottom-up way on the basis of the consecutive syllables printed on the page.

Miscue Analysis. This shows that we often substitute words of essentially the same meaning for what is written on the page [85], proving that reading is driven holistically by meaning rather than bottom-up by the details of what is on the page.

Reading Miswritten Text. One can experience how the mind fills in missing text in a meaningful way by reading the following examples: Y u cn re d this evn tho gh it is not phonem cally correct, and this thuogh letters are wron g, and this though words missing. The fact that you could read the above proves that the mind continually guesses and fills in, all the time searching for meaning. This process is driven top-down, otherwise it could not work. Here is another well known example:

7H15 M3554G3 53RV35 7O PR0V3
HOW 0UR M1ND5 C4N D0 4M4Z1NG 7H1NG5!
1MPR3551V3 7H1NG5!
1N 7H3 B3G1NN1NG 17 WA5 H4RD
BU7 NOW, 0N 7H15 LIN3 YOUR M1ND 1S R34D1NG 17 4U70M471C4LLY
W17H 0U7 3V3N 7H1NK1NG 4B0U7 17.

We can make sense in each case because of the contextual way we read.

fMRI Studies. There are very limited fMRI studies that look at natural reading, as discussed in [42] and references therein. This study showed that natural reading versus pseudo-reading showed different patterns of brain activation: normal reading produced activation in a well-established language network that included superior temporal gyrus/sulcus, middle temporal gyrus(MTG), angular gyrus(AG), inferior frontal gyrus, and middle frontal gyrus, whereas pseudo-reading produced activation in an attentional network that included anterior/posterior cingulate and parietal cortex. This is to be expected: pseudo-reading is not a meaningful activity, so the brain centres that are searching for meaning and trying to predict what will come next are stymied and do not know what to do.

7.3.4.4 The Holistic View

This is the same as all perception [147] (see the previous section on vision). Sohoglu et al. express it this way [222]:

> A striking feature of human perception is that our subjective experience depends not only on sensory information from the environment but also on our prior knowledge or expectations. [...] We used concurrent EEG and MEG recordings to determine how sensory information and prior knowledge are integrated in the brain during speech perception. We manipulated listeners' prior knowledge of speech content by presenting matching, mismatching, or neutral written text before a degraded (noise-vocoded) spoken word. When speech conformed to prior knowledge, subjective perceptual clarity was enhanced. This enhancement in clarity

> was associated with a spatiotemporal profile of brain activity uniquely consistent with a feedback process: activity in the inferior frontal gyrus was modulated by prior knowledge before activity in lower-level sensory regions of the superior temporal gyrus. [...] The data are best explained within the framework of predictive coding in which sensory activity is compared with top-down predictions and only unexplained activity propagated through the cortical hierarchy.

This predictive kind of process applies in particular to reading.

7.3.5 Goal-Directed Behaviour and Attention

We plan what to do and then act according to those plans. The planning is a macro-level psychological activity, based on our feelings on the one hand and our rationality on the other: is this a good investment? Will I feel better if I move to France? Is it safe to go to the football game? We make a decision on the basis of our hierarchy of goals, and then that decision results in our arms or legs or tongue moving because electrons in motor neurons cause protein filaments made of actin and myosin to slide past one another, producing a muscle contraction in accordance with our action plan.

7.3.5.1 Rational Decisions

Our rational decisions act down from the psychological level to the level of electrons and then up through muscles to our limbs and bodies [203]. Animals including humans engage in goal-directed behavior flexibly in response to events and their background, which is a form of contextual behavior [163]. In order to make decisions we focus our attention on the issue at hand [103, pp. 285–286], consider it in terms of options and outcomes, and reach a conclusion which we then pass on to the motor cortex to initiate action. This results in a coordinated motion of electrons in our muscles, i.e., a top-down effect from our plans and intentions, realised through coordinated patters of neuronal excitations in the cortex, to the underlying microphysics. This is in analogy with what happens in digital computers (Chap. 2), where the logic of the high level applications program ultimately determines which set of electrons flow through which gates at the micro-level. How this happens is described by Deco and Rolls [59]:

> Cognitive behaviour requires complex context-dependent processing of information that emerges from the links between attentional perceptual processes, working memory and reward-based evaluation of the performed actions. We describe a computational neuroscience theoretical framework which shows how an attentional state held in a short term memory in the prefrontal cortex can by top-down processing influence ventral and dorsal stream cortical areas using biased competition to account for many aspects of visual attention. We also show how within the prefrontal cortex an attentional bias can influence the mapping of sensory inputs to motor outputs, and thus play an important role in decision-making. [...] The models also directly address how bottom-up and top-down processes interact in visual cognition, and show how some apparently serial processes reflect the operation of interacting parallel distributed systems.

According to Lee and Lee, the rhinal cortical regions and the hippocampal formation play key roles in mnemonically categorizing and recognizing contextual representations and the associated items [163]. In addition, it appears that the fronto-striatal cortical loops in connection with the contextual information processing areas critically control the flexible deployment of adaptive action sets and motor responses for maximizing goals. Goal-directed control is exerted by cortical units that are regulated by both top-down feedback and oscillatory coherence [152]. Top-down modulation mediated by the prefrontal cortex is a causal link between early attentional processes and subsequent memory performance [257]. Memory retrieval in the hippocampus is thought to be influenced by top-down inputs from the prefrontal cortex (PFC) through neurons that project from the dorsal anterior cingulate cortex (dACC) to the CA1 and CA3 regions of the hippocampus [29].

7.3.5.2 Attention

This all depends on the focus of attention, and attentional prioritization modulates sensory processing by a top-down signal for attentional allocation [69, 187]. Top-down modulation bridges selective attention and working memory [102]. According to Deco and Rolls [58], recent neurophysiological experiments support a 'biased competition hypothesis' of the neural basis of attention. According to this hypothesis, attention appears as a sometimes non-linear property that results from a top-down biassing effect that influences the competitive and cooperative interactions that work both within cortical areas and between cortical areas. They describe a detailed dynamical analysis of the synaptic and neuronal spiking mechanisms underlying biased competition.

7.3.5.3 Integration

Finally, it is important to note that the different senses are integrated in this attentional and decision-making process. The prefrontal cortex (PFC), specifically BA44, may function as the essential region for hierarchical processing across the domains [136]: the hierarchical organization of cognitive controls within the PFC forms a cascade of top-down hierarchical processes operating along a posterior-to-anterior axis of the lateral PFC including BA44 within the network [136]. This takes place in an adaptive contextual way, as described by Gruber and McDonald [119]:

> In interactions among multiple brain systems controlling motivated behavior, multiple brain systems acquire information in parallel and either cooperate or compete for behavioural control The hippocampus provides contextual specificity to the emotional system, and provides an information-rich input to the goal-directed system for navigation and discriminations involving ambiguous contexts, complex sensory configurations, or temporal ordering.

Thus context, emotion, and the strategic pursuit of goals determine the choice of action. Top-down action from the cortex to the periphery then realizes the outcomes [14, 203].[7]

7.3.6 *Health*

Because the mind is an embodied mind, there are both bottom-up and top-down influences between it and our physical state of health. In this section, I look in turn at:

- Placebo effects.
- The immune system.
- Stress and anxiety.
- Hierarchy and health.
- Loneliness or exclusion.
- Meditation.
- Psychotherapy.

7.3.6.1 The Placebo Effect

The placebo effect occurs when a non-active pill is perceived by the patient to be a source of healing and effects an improvement in her physical well-being that cannot be associated with bottom-up causation because there is no physiologically active ingredient present. It is very well documented that this occurs [132, 134], and indeed in testing medications one must test, not against the zero effect of the inactive control medicine, but against the placebo effect level [254], as otherwise one will show a positive result that is caused by the placebo effect [155]. This efficacy is conclusive proof of top-down effects from the mind to the physical welfare of the patient.

Humphrey and Skoyles attribute this efficacy to what they call a 'health governor' that has evolved to perform an analysis of the health situation at the moment and to predict what physiological actions should be taken to maximise satisfaction in the short-term while minimizing long-term risks [134]: "Crucially, he needs to be able to make an informed guess about future needs and opportunities, so that he can budget accordingly." This governor is influenced by the subject's optimism at whatever level it is experienced, which may be a mediating factor in most if not all the health governor's decisions. And when placebos are used, they work because patients expect them to work: they induced the health governor, with an improved forecast, to release selfcure [134]. This predictive function is obviously analogous to the predictive nature of sensory systems: it is a basic cortical function [129].

[7] This is just a selection of papers supporting this view: there are many others coming out all the time. The details may differ, but the overall picture is consistent across them.

Whether or not this is the explanation, something like this must be the case, as is confirmed by brain imaging studies by Wager et al. examining top-down effects from expectations and beliefs to brain activation and hence to levels of pain experienced [253]:

> The experience of pain arises from both physiological and psychological factors, including one's beliefs and expectations. Thus, placebo treatments that have no intrinsic pharmacological effects may produce analgesia by altering expectations. However, controversy exists regarding whether placebos alter sensory pain transmission, pain affect, or simply produce compliance with the suggestions of investigators. In two functional magnetic resonance imaging (fMRI) experiments, we found that placebo analgesia was related to decreased brain activity in pain-sensitive brain regions, including the thalamus, insula, and anterior cingulate cortex, and was associated with increased activity during anticipation of pain in the prefrontal cortex, providing evidence that placebos alter the experience of pain.

This effect is in addition to that due to distraction, as shown by Buhle et al. [34]:

> In this study, the authors examined whether two common forms of pain control—placebos and distraction—work through independent or shared processes. Researchers placed a cream on participants' arms and told them it was either a nonanalgesic control cream or a strong analgesic cream. Researchers then induced mild and moderate pain in the participants. During pain induction, participants alternated between performing a working memory task and fixating on a cross. The pain-reducing effects of the placebo and the working memory task were additive, suggesting that placebos do not work by redirecting a person's attention and that using a combination of the two techniques to reduce pain increases the benefit.

The functional anatomy leading to the effect is described by Mayberg et al. [174]:

> Administration of a placebo can result in a clinical response indistinguishable from that seen with active antidepressant treatment. [...] The common pattern of cortical glucose metabolism increases and limbic–paralimbic metabolism decreases in placebo and fluoxetine responders suggests that facilitation of these changes may be necessary for depression remission, regardless of treatment modality. Clinical improvement in the group receiving placebo as part of an inpatient study is consistent with the well-recognized effect that altering the therapeutic environment may significantly contribute to reducing clinical symptoms. The additional subcortical and limbic metabolism decreases seen uniquely in fluoxetine responders may convey additional advantage in maintaining long-term clinical response and in relapse prevention.

This suggests that placebo effects are short term rather than long term, but confirms that they do indeed exist as a top-down effect from the therapeutic environment. Overall [254]:

> For psychological disorders, particularly depression, it has been shown that pill placebos are nearly as effective as active medications, whereas psychotherapies are more effective than psychological placebos. However, [...] when properly designed, psychological placebos are as effective as accepted psychotherapies.

7.3.6.2 Immune System

The adaptive control system is based on immune molecules that are also neurotransmitters, and hence enable two-way interaction between the immune system and the brain, as discussed by Esther Sternberg [225, 226]. Indeed top-down neuronal control of immunity has been found to exist even in nematode worms [132]. In the case of humans [227]:

> The central nervous system (CNS) regulates innate immune responses through hormonal and neuronal routes. The neuroendocrine stress response and the sympathetic and parasympathetic nervous systems generally inhibit innate immune responses at systemic and regional levels, whereas the peripheral nervous system tends to amplify local innate immune responses. These systems work together to first activate and amplify local inflammatory responses that contain or eliminate invading pathogens, and subsequently to terminate inflammation and restore host homeostasis. [...] the CNS can be considered as integral to acute-phase inflammatory responses to pathogens as the innate immune system.

Thus this is a direct route whereby the mind can influence bodily health.

7.3.6.3 Stress and Anxiety

It is well known that stress and anxiety cause physical symptoms [46, pp. 239–261]. For example, up to 20 % of cardiac surgery patients do not show improvements in health-related quality of life after surgery, despite apparently successful surgical procedures, because of exposure to high stress in the cardiovascular intensive care unit [213]. Because it originates in the mind and is related to lifestyle, it is a top-down phenomenon which McEwen describes as follows [175]:

> Stress begins in the brain and affects the brain, as well as the rest of the body. Acute stress responses promote adaptation and survival via responses of neural, cardiovascular, autonomic, immune and metabolic systems. Chronic stress can promote and exacerbate pathophysiology through the same systems that are dysregulated. The burden of chronic stress and accompanying changes in personal behaviors (smoking, eating too much, drinking, poor quality sleep; otherwise referred to as "lifestyle") is called allostatic overload. Brain regions such as hippocampus, prefrontal cortex and amygdala respond to acute and chronic stress and show changes in morphology and chemistry that are largely reversible if the chronic stress lasts for weeks. However, it is not clear whether prolonged stress for many months or years may have irreversible effects on the brain. [...There is] top-down regulation of cognitive, autonomic and neuroendocrine function. This concept leads to a different way of regarding more holistic manipulations, such as physical activity and social support as an important complement to pharmaceutical therapy in treatment of the common phenomenon of being 'stressed out'. Policies of government and the private sector play an important role in this top-down view of minimizing the burden of chronic stress and related lifestyle (i.e., allostatic overload).

Two particular mechanisms leading to stress, characterised by Stevens and Price [228], have been mentioned above, related to the primordial emotional systems: namely malfunctions of the hierarchy/status system and the attachment/belonging system.

7.3.6.4 Hierarchy and Health

Social status and health in humans and other animals plays an important role in social life (see [103, p. 73], [210, 211]) and is associated with a primordial emotional system [244]. Malfunctioning of this system is a major cause of psychiatric illness [228]. Sapolsky expresses it this way [211]:

> Dominance hierarchies exist in numerous social species, and rank in such hierarchies can dramatically influence the quality of an individual's life. Rank can dramatically influence also the health of an individual, particularly with respect to stress-related disease. Socioeconomic status (SES) is the nearest human approximation to social rank and SES dramatically influences health.

This is a top-down influence from social perceptions to bodily health.

7.3.6.5 Loneliness or Exclusion

The deep need for attachment [116] is also based on a primordial emotional system [192] whose malfunction is again a major cause of psychiatric illness [228]: loneliness can be hazardous to your health [177]. It is felt as exclusion or loneliness. The point in both cases is that it is a mental distress that drives the physical distress. The mind is the source of a problem that is then manifested in the body. By contrast, happiness has a positive causal effect on longevity and physiological health [93].

7.3.6.6 Meditation

Given that stress causes ill health and meditation can reduce stress, one might suppose that meditation would improve health, and a variety of studies suggest that this might indeed be so. As an example, Davidson and McEwen, writing about the social influences on neuroplasticity, and stress and interventions to promote well-being, state [52]:

> Experiential factors shape the neural circuits underlying social and emotional behavior from the prenatal period to the end of life. Although the precise mechanisms of plasticity are still not fully understood, moderate to severe stress appears to increase the growth of several sectors of the amygdala, whereas the effects in the hippocampus and prefrontal cortex tend to be opposite. Structural and functional changes in the brain have been observed with cognitive therapy and certain forms of meditation and lead to the suggestion that well-being and other prosocial characteristics might be enhanced through training.

Evidence suggests this maybe so. For example [118, 252]:

> Mindfulness-based stress reduction (MBSR) is a structured group program that employs mindfulness meditation to alleviate suffering associated with physical, psychosomatic, and psychiatric disorders. The program, nonreligious and nonesoteric, is based upon a systematic procedure to develop enhanced awareness of moment-to-moment experience of perceptible mental processes. The approach assumes that greater awareness will provide more veridical

> perception, reduce negative affect, and improve vitality and coping [...] A meta-analysis suggests that MBSR may help a broad range of individuals to cope with their clinical and nonclinical problems.

Similarly, [184] give evidence suggesting that meditation reduces pain-related neural activity in the anterior cingulate cortex, insula, secondary somato sensory cortex, and thalamus. Again, top-down effects from the brain can possibly improve physical health. This is supported by a study of the effectiveness of a brief meditation and mindfulness intervention for people with diabetes and coronary heart disease [153].

7.3.6.7 Psychotherapy

A structured way of promoting mental health in a top-down way is psychotherapy ('the talking cure'). Eric Kandel writes about how this can work, consistently with our current knowledge of neuroscience, as follows [143, 144]:

- All mental processes derive from operations of the brain.
- Genes determine neuronal functioning.
- Social and developmental factors contribute importantly to the variance in mental illness. These factors express themselves in altered gene expression. Nurture is ultimately expressed as nature.
- Altered gene expression induced by learning gives rise to changed patterns of neuronal connections, which give rise to different forms of thinking and behaviour.
- Psychotherapy produces changes in long-term behaviour by learning which produces changes in gene expression, and hence changes in neuronal interconnection.

This is of course a controversial area, with many competing forms of psychotherapy and various claims as to their success. Without taking a stand on this, Kandel makes clear that there is a physical mechanism whereby this can work in a top-down way.

Overall, a holistic view of medicine, dealing with physical trauma and psychological stress such as anxiety in a way that takes meaning and purpose seriously, is likely to be the most productive. This kind of integration of a humanistic view and a scientific perspective is possible. For example, it is the basis of treatment in the systematic anthroposophic approach to treatment via art therapy.[8]

7.3.7 Social Neuroscience

This all takes place in the context of society, which exerts a major top-down effect on individual human minds. An individual mind cannot thrive: we are designed to develop attachments [46, 116, 138] and to live in communities [64–66] based

---

[8]This is made clear for example in the lecture Ringvorlesung Topos Kunsttherapie 03, November 2015: "Anthroposophische Kunsttherapie: Beispiele aus der Praxis" Margaret Ellis, Alanus Hochschule. See also [101] for a holistic view.

on language and culture [17], and this is reflected in our brain. There is a large literature on social neuroscience [36, 37, 124], the neuroscience of empathy [56], and evolutionary cognitive neuroscience [198]. This is necessarily a multilevel affair: Cacioppo and Decety explain [37]:

> Social species are so characterized because they form organizations that extend beyond the individual. The goal of social neuroscience is to investigate the biological mechanisms that underlie these social structures, processes, and behavior and the influences between social and neural structures and processes. Such an endeavor is challenging because it necessitates the integration of multiple levels.

The main sections in the major text *Foundations in Social Neuroscience* [36] are entitled as follows: (A) multilevel integrative analysis of social behaviour, (B) social cognition and the brain, (C) social neuroscience of motivation, emotions, and attitudes, (D) biology of social relationships and interpersonal processes, and (E) social influences in biology and health. I cannot discuss here all the issues raised in the 83 chapters of that book, some of which have been dealt with in other sections of this chapter. Rather I shall look here briefly at the following:

- Culture.
- The social mind and social cognition.
- Environmental effects on the brain.

7.3.7.1 Culture: The Taken-for-Granted Reality

We live in a cultural ambience that we take for granted because it is our experience that this is the way things are, and that experience shapes the way we think and act. Indeed, as emphasized in Berger and Luckman's book *The Social Construction of Reality* [18], it determines what we perceive as reality. It does so particularly by incorporating particular conceptual schemas, roles, and frames that structure social life [166]. It is very difficult for us to step outside this framework of thought into which we are born [17]. Thus it represents a major top-down effect from culture (a non-physical phenomenon) to the neural connections that determine what we understand and believe. This will affect decision-making [248].

Tomasello expresses it this way [240]:

> Human beings are biologically adapted for culture in ways that other primates are not, as evidenced most clearly by the fact that only human cultural traditions accumulate modifications over historical time (the ratchet effect). The key adaptation is one that enables individuals to understand other individuals as intentional agents like the self. This species-unique form of social cognition emerges in human ontogeny at approximately 1 year of age, as infants begin to engage with other persons in various kinds of joint attentional activities involving gaze following, social referencing, and gestural communication. Young children's joint attentional skills then engender some uniquely powerful forms of cultural learning, enabling the acquisition of language, discourse skills, tool-use practices, and other conventional activities. These novel forms of cultural learning allow human beings to, in effect, pool their cognitive resources both contemporaneously and over historical time in ways that are unique in the animal kingdom.

Language. One of the most important ways this happens is in terms of language, which is the essential tool enabling us to have a culture (see Sect. 7.5.2) [54, 215]. Society shapes neural connections in the individual, in particular through our learning a specific language and associated symbolism [116, 241]. This develops through our ability to conceptualise a hierarchically structured recursive phrase structure which emerged in relation to a mutation leading to the FOXP2 gene [45]. The exact way this happened is not fully clear, but what is clear is that it was a top-down process driven by the selective advantages of group living, where intentions can be communicated between minds [71].

Mechanisms. As explained above, this development will be driven by the interplay between emotional systems and intellect. Freeman et al. explain how this happens in terms of brain structure by examining the relation of social mores to the dominance system [91]:

> It has long been understood that culture shapes individuals' behavior, but how this is accomplished in the human brain has remained largely unknown. To examine this, we made use of a well-established cross-cultural difference in behavior: American culture tends to reinforce dominant behavior whereas, conversely, Japanese culture tends to reinforce subordinate behavior. [...] activity in the right caudate and medial prefrontal cortex (mPFC) correlated with behavioral tendencies towards dominance versus subordination, such that stronger responses in the caudate and mPFC to dominant stimuli were associated with more dominant behavior and stronger responses in the caudate and mPFC to subordinate stimuli were associated with more subordinate behavior. The findings provide a first demonstration that culture can flexibly shape functional activity in the mesolimbic reward system, which in turn may guide behavior.

Emergence of Culture. David Sloan Wilson remarks [256] that the transition from bottom-up to top-down dominated causation in the relation of mind to the society in which it is imbedded is a major evolutionary transition in the historical development of humanity, resulting in the emergence of the social order as a higher level entity in its own right, and a consequent change in the nature of the evolutionary processes at work. This accords with the view of Dunbar and the social brain hypothesis [70, 71].

7.3.7.2 The Social Mind and Social Cognition

The social mind [70, 71] is based on the social brain [46, pp. 50–65] and the attachment system (see [46, pp. 176–236], [116, 192]). Cardoso et al. explain the underlying mechanisms as follows [38]:

> Group-living animals must adjust the expression of their social behaviour to changes in their social environment and to transitions between life-history stages, and this social plasticity can be seen as an adaptive trait that can be under positive selection when changes in the environment outpace the rate of genetic evolutionary change. Here, we propose a conceptual framework for understanding the neuromolecular mechanisms of social plasticity. According to this framework, social plasticity is achieved by rewiring or by biochemically switching nodes of a neural network underlying social behaviour in response to perceived

> social information. Therefore, at the molecular level, it depends on the social regulation of gene expression, so that different genomic and epigenetic states of this brain network correspond to different behavioural states, and the switches between states are orchestrated by signalling pathways that interface the social environment and the genotype.

This is of course essentially the same mechanism as identified by Kandel in the case of psychotherapy. Blumstein et al. concur [24]:

> Social interactions among conspecifics are a fundamental and adaptively significant component of the biology of numerous species. Such interactions give rise to group living as well as many of the complex forms of cooperation and conflict that occur within animal groups. Although previous conceptual models have focused on the ecological causes and fitness consequences of variation in social interactions, recent developments in endocrinology, neuroscience, and molecular genetics offer exciting opportunities to develop more integrated research programs that will facilitate new insights into the physiological causes and consequences of social variation. Here, we propose an integrative framework of social behavior that emphasizes relationships between ultimate-level function and proximate-level mechanism, thereby providing a foundation for exploring the full diversity of factors that underlie variation in social interactions, and ultimately sociality.

Two things are crucial here: this is a two-way multiple level interaction, and in the end it is driven by ultimate level function, that is, purpose, which drives human lives. Key components of these interactions are theory of mind, empathy, and mirror neurons.

Theory of Mind. A network of structures in the brain is dedicated to social cognition and theory of mind [147, pp. 410–420]. We engage in mind-reading all the time [99]. This involves:

- a face recognition system,
- recognizing another person's presence and emotions,
- interpreting actions and intentions through analysis of biological motion,
- imitating the actions of others by means of mirror neurons,
- developing a theory of mind which identifies what others are thinking and so are likely to do.

An example of the mechanisms at work is that social interaction modifies neural response to gaze shifts because the direction of gaze can reveal intentions and help to predict future actions [32]. The medial frontal cortex is finely tuned for social cognitive processes, such as self-reflection, person perception, and making inferences about others' thoughts [3].

Empathy and Mirror Neurons. The social brain is crucially based on mechanisms that generate empathy [56, 57], which are closely related to the attachment system and tend to dampen conflict. Empathy is largely effected through mirror neurons [224, pp. 282–283].

Cheating and Evolution. Developing these structures enabling social interactions and group formation has been a key part in human evolutionary development [71]. It is notable that the evolutionary psychology literature focuses strongly on problems arising for social group formation because of the issue of cheating; this is even used as an argument against multi-level evolutionary processes. However, this issue is not mentioned in the social neuroscience literature [36, 46, 124] or the literature on the social brain [71, 100]. The focus instead in these writings is on those neural systems that make us want to live in groups, for which there is a great deal of evidence. The cheating issue does not appear to be a real problem in social neuroscience terms, or in its outcomes in terms of effects on evolutionary history [71].

7.3.7.3 Environmental Effects on Mind and Brain

All the above sections have made clear that, as the mind is continually adapting to its environment, there are strong top-down effects from that environment to brain micro-structure. There are many examples. Apart from language, the classic case is London taxi drivers, who, because of their work, have a map of London in their brain [169]:

> Gray matter differences could result from using and updating spatial representations, but they might instead be influenced by factors such as self-motion, driving experience, and stress. We examined the contribution of these factors by comparing London taxi drivers with London bus drivers, who were matched for driving experience and levels of stress, but differed in that they follow a constrained set of routes. We found that compared with bus drivers, taxi drivers had greater gray matter volume in mid-posterior hippocampi and less volume in anterior hippocampi. Furthermore, years of navigation experience correlated with hippocampal gray matter volume only in taxi drivers, with right posterior gray matter volume increasing and anterior volume decreasing with more navigation experience. This suggests that spatial knowledge, and not stress, driving, or self-motion, is associated with the pattern of hippocampal gray matter volume in taxi drivers.

This is top-down in two separate ways. Firstly, the fact that they work as taxi drivers rather than bus drivers (social occupation) affects their brain structure: different hippocampal areas develop differently. Secondly, the details of the geography of London (the relation between Piccadilly Circus and Trafalgar Square, say) is an abstract geometric relation that is learnt by their training process, and so is embodied in details of synaptic connections [250].

Two more examples are as follows: there are more hippocampal neurons in adult mice living in an enriched environment as compared with littermates housed in standard cages [151]; and childhood socioeconomic status (SES) is associated with cognitive achievement throughout life because of differences in neural processing [120].

The overall situation is expressed by Merlin Donald as follows [66, p. 153]:

> In humans, even after our expanded brain is factored in, something remains that cannot be accounted for by innate properties. That additional element is enculturation. The specific form of the modern mind has been determined largely by culture.

7.3.8 The Physical Substrate

This section has given many examples of top-down causation in the brain: one cannot understand the brain without taking it into account. But how does it happen, given the

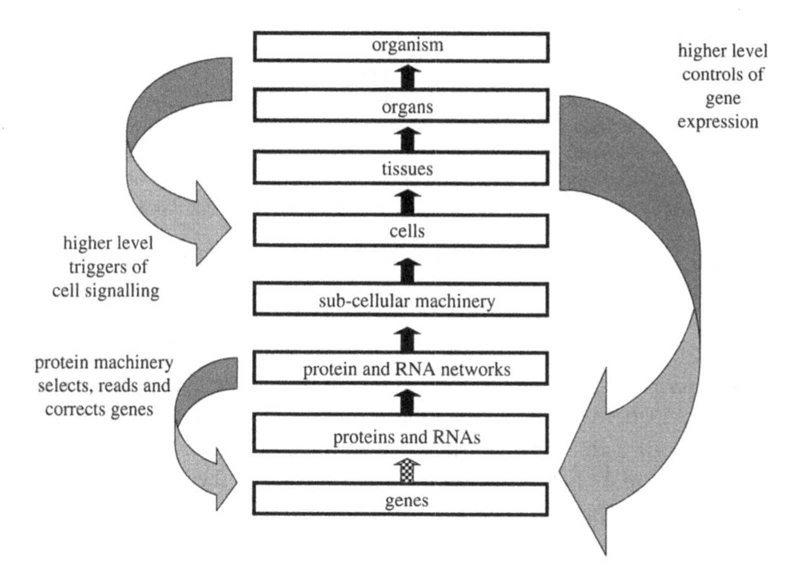

Fig. 7.12 Top-down effects from organs to genes apply to the brain in particular. From [189], courtesy of Denis Noble

causal completeness of physics at the bottom? How is there freedom at the bottom for this to occur? This issue has been examined in depth in the more general context in Chap. 5, so here I will just comment on the issue as regards the brain in particular, considering in turn:

- Cells and genes.
- Contextual effects.
- Adaptive selection.

7.3.8.1 Cells and Genes

In terms of the relation to the underlying cellular machinery, genes, and proteins, the top-down causal effects are as shown in Dennis Noble's diagram (Fig. 7.12). This is explained in his article [189]. The organism level acts down to the level of cells, the organ level acts down to the level of genes, and the protein machinery acts down to the level of genes [106]. This has already been touched on in Sect. 7.3.6 as regards psychotherapy and Sect. 7.3.7 as regards social plasticity.

7.3.8.2 Contextual Effects

In terms of neural structures, Bressler and McIntosh explain the role of neural context in large-scale neurocognitive network operations as follows [30]:

> This chapter examines the role of a critical aspect of brain function, which we call neural context, in the selective functioning of interacting neural systems in cognition. We define neural context as the local processing environment of a given neural element that is created by modulatory influences from other neural elements. Neural context allows the response properties of one element in a network to be profoundly affected by the status of other neural elements in that network. As a result of neural context, the relevance of a given neural element for cognitive function typically depends on the status of other interacting elements.

At the neural level this works by recognizing a sequence of patterns in a cortical region which will predict its next input and tell the region below what to expect (Hawkins [129, p. 135]). Information flows up and down the layers of the cortex [129, pp. 142–143], which has converging patterns going up the cortical hierarchy, diverging patterns going down the cortical hierarchy, and delayed feedback through the thalamus [129, p. 147]. This enables the brain to classify inputs, to learn sequences of patterns, to form a constant pattern, or name, for a sequence, and to make specific predictions [129, pp. 147–156]. Observed patterns flow up the hierarchy and predictions flow down the hierarchy [129, pp. 156, 159], enabled by reciprocal connections [129, pp. 161–164]. Detailed studies of the cortical circuits that make logical operations possible [4, 218] are analogous to studies of the circuits that make logical operations possible in a computer [170]. This all takes place by constraints on the flow of electrons between synapses due to the detailed neuronal structuring.

7.3.8.3 Adaptive Selection

Additionally, adaptive selection at each level (underlying plasticity at each level) is crucial. Kerr et al. express this as follows [152]:

> The brain is able to flexibly select behaviors that adapt to both its environment and its present goals. This cognitive control is understood to occur within the hierarchy of the cortex and relies strongly on the prefrontal and premotor cortices, which sit at the top of this hierarchy. Pyramidal neurons, the principal neurons in the cortex, have been observed to exhibit much stronger responses when they receive inputs at their soma/basal dendrites that are coincident with inputs at their apical dendrites. This corresponds to inputs from both lower-order regions (feedforward) and higher-order regions (feedback), respectively. In addition to this, coherence between oscillations, such as gamma oscillations, in different neuronal groups has been proposed to modulate and route communication in the brain. [...] We demonstrate that more sophisticated and flexible top-down control is possible when the gain of units is modulated by not only top-down feedback but by coherence between the activities of the oscillating units. With these types of units, it is possible to not only add units to, or remove units from, a higher-level unit's logic operation using top-down feedback, but also to modify the type of role that a unit plays in the operation. Finally, we explore how different network properties affect top-down control and processing in large networks.

In Conclusion. Interlevel causation, both bottom-up and top-down, is key to brain function. Evolution has selected for it to occur. The underlying physics is channelled and constrained to enable this to happen.

7.4 Purpose and Meaning as the Key Driving Forces

Given that there is top-down causation, what is the top level? It is not clear that there is a fixed top level in physical terms: there may rather be transient oscillatory bindings of neuronal groups that act as the top for a while [35]. However, in psychological terms there is indeed a top level: it is the level of meaning or *telos*, associated with ethics and aesthetics (Sect. 7.2.3.3 and Fig. 7.6).

I cannot do proper justice to this theme here, as it is the subject of all literature devoted to human wisdom through the centuries. Rather I will just make a few remarks about the following:

- Goals and purposes.
- The human search for meaning.
- Purpose, ethics, and understanding.
- The meaningful social context.

7.4.1 Goals and Purpose

The Goal Hierarchy. We have a hierarchy of goals that guide what we do as we make our shorter term and longer term decisions about life. These goals are chosen in order to fulfil some kind of purpose, ultimately related to meaning (see Sect. 7.4.2). At each stage we have a hierarchical goal structure that interacts with our immediate and longer term situation, e.g.,

- I wish to help the sick, because there are so many who need help.
- I have no medical training. I therefore need a medical education.
- I'll need to find a good college so I can apply.
- I'll ask friends to find out what is a good medical college, so I can choose one that is suitable.
- I'll need to get finance so I can afford it, so I'll have to fund a bursary.

And so the list continues. Each goal is dependent on the one above and on the social context. Eventually, this chain results in actions such as phoning a friend to ask their advice. The reference [122] gives a nice set of examples of such goal chains, such as a chimpanzee using a probe to feed on termites to satisfy the basic need of 'hunger', and cases relevant to the human archaeological record, for example, the manufacture and use of a bow-and-arrow tool set.

Goal Choice. How are goals selected? We test them and see how they work out, and if they seem right. That is, they are adaptively selected (which is TD4). But that selection takes place in terms of some kind of adaptive selection criteria. Where do these come from? They, too, are adaptively selected by trial and error, so we have adaptive selection of selective criteria (which is TD5). The neural basis of this process is elicited by Friston et al. [98], who show that if value-dependent modulation is extended to the inputs of neural value systems themselves, initially neutral cues can acquire value.

The Highest Level. Obviously this process faces a problem of infinite recursion: where does the next higher level of selection criteria come from? At some point we have to draw a line and say, this is where I stand: these are my founding principles, this is the purpose in my life. That is where one makes value choices based on one's view of meaning. This may be to help others, to serve God, to save the Earth, to create great art, to maximise one's own welfare, to understand how things function, or just to survive. This then leads to career choices: to be a politician, a business person, a scientist, an artist, a welfare worker, an environmentalist, a ballet dancer, and so on. These choices relate to how one sees meaning in one's life and how one sees the good life [90]. These are abstract ideas that shape what one does, and thereby act down to muscles, neurons, and genes, and on to electrons and atoms as we try to fulfil these goals and purposes. They are crucial to society [86].

7.4.2 The Human Search for Meaning

Humans are built to search for understanding all the time: not just to predict, important as this is [150], but in order to make sense of things: to find meaning, as set forth profoundly in Viktor Frankel's book *Man's Search for Meaning* [88] and Saint-Exupery's book *Flight to Arras* [53] in the case of individuals, and in Charles Taylor's book *Sources of the Self: The Making of Modern Identity* [237], in terms of the way Western society has developed its sense of meaning over centuries through a strenuous process of struggle.

Frankel concludes that the meaning of life is found in every moment of living, and that life never ceases to have meaning, even in suffering and death. Frankl concludes from his experience that a prisoner's psychological reactions are not solely the result of the conditions of his life, but also of the freedom of choice he always has even in severe suffering. He records his feelings in the desperate conditions he was subjected to as a prisoner in Auschwitz [88]:

> For the first time in my life I saw the truth as it is set into song by so many poets, proclaimed as the final wisdom by so many thinkers. The truth? That love is the ultimate and the highest goal to which man can aspire. Then I grasped the meaning of the greatest secret that human poetry and human thought and belief have to impart: the salvation of man is through love and in love. I understood how a man who has nothing left in this world still may know bliss, be it only for a brief moment, in the contemplation of his beloved. In a position of utter desolation, when man cannot express himself in positive action, when his only achievement may consist in enduring his sufferings in the right way—an honourable way—in such a position man can, through loving contemplation of the image he carries of his beloved, achieve fulfilment.

Saint-Exupery meditates on why he flies his aircraft and faces death in the face of the unstoppable German onslaught on France in 1941 [53]:

> When, in the Sahara, the Arabs would surge up in the night round our campfires and warn us of a coming danger, the desert would spring to life for us and take on meaning. Those messengers had lent it distance. Music does something like this. The humble odor of an old cupboard does when it awakens and brings memories to life. Pathos is the sense of distance.

> But I know that nothing which truly concerns man is calculable, weighable, measurable. True distance is not the concern of the eye; it is granted only to the spirit. Its value is the value of language, for it is language which binds things together. And now it seems to me that I begin to see what a civilization is. A civilization is a heritage of beliefs, customs, and knowledge slowly accumulated in the course of centuries, elements difficult at times to justify by logic, but justifying themselves as paths when they lead somewhere, since they open up for man his inner distance.

The search for meaning and understanding is with us all the time, as an ongoing interaction between intellect and emotion, guiding what actions we undertake.

7.4.2.1 Stories and Literature

We often understand meaning in terms of stories, which play a key role in individual and communal life. Daniel Taylor writes as follows [238]:

> You are the product of all the stories you have heard and lived—and of many that you have never heard. They have shaped how you see yourself, the world, and your place in it. Your first great storytellers were home, school, popular culture, and, perhaps, church. Knowing and embracing healthy stories are crucial to living rightly and well.

Jonathan Gottschalk's book *The Storytelling Animal* [113] claims that stories help us navigate life's complex social problems, just as flight simulators prepare pilots for difficult situations. Storytelling has evolved, like other behaviour, to ensure our survival. He presents this view in the light of current research in neuroscience, psychology, and evolutionary biology.

It is not only the case that we shape stories; stories also shape us. This is why the great literature of the world, Dante, Shakespeare, Dickens, Hugo, Dostoyevsky, Tolstoy, and so on, has such a hold on the human mind. It presents us with the kinds of dilemmas we face and the moral choices that shape outcomes in the world [27]. And of course the above stories by Saint-Exupéry and Frankl are examples of just such writings. This importance of stories is apparent also in social history and politics: stories about the past crucially influence the present (think of the conflicts in Syria and Palestine, for example).

7.4.3 Purpose, Ethics, and Understanding

Setting purpose involves an ethical stance, whether it be to do all one can to help others (the good life is selflessness), to care for the world (the good life is caring for the plants and animals and birds), or to maximise one's own wealth (the good life is to be rich), or to make great art (the good life is creating beauty), or whatever. Whatever this choice is constrains what other goals will be set and what choices will be made. It is a key element in social life [86].

7.4.3.1 Intention

Intention is key to morality, so morality cannot be encompassed by biological approaches that deal only with sociability and cooperation, because no one knows how to observe an animal's intentions [138, p. 169]. Thus the biologists' definition of altruism, defined only in terms of the consequences of an agent's behaviour on another, misses the key human attribute of intention and rational reflection. It cannot deal with real morality, which takes these into account [138, pp. 157–193]. Rosati expresses this as follows [205]:

> In our everyday lives, we confront a host of moral issues. Once we have deliberated and formed judgments about what is right or wrong, good or bad, these judgments tend to have a marked hold on us. Although in the end, we do not always behave as we think we ought, our moral judgments typically motivate us, at least to some degree, to act in accordance with them.

Meyer and Braga state this as follows [176]:

> Most recent developments in the study of social dilemmas give an increasing amount of attention to cognition, belief systems, valuations, and language. However, developments in this field operate almost entirely under epistemological assumptions which only recognize the instrumental form of rationality and deny that 'value judgments' or 'moral questions' have cognitive content. This standpoint erodes the moral aspect of the choice situation and obstructs acknowledgement of the links connecting cognition, inner growth, and moral reasoning, and the significance of such links in reaching cooperative solutions to many social dilemmas.

This is related to Maslow's concept of actualisation: the need we have to fulfil our human potential when we have fulfilled all our physical needs [90, pp. 28–42].

7.4.3.2 Moral Responsibility

The moral view is that virtue development is a key to a whole human life [90]. It is based on understandings of meaning, values, and morality. Morally responsible action depends on the ability to evaluate ones own reasons for acting in the light of a concept of the good. The cognitive prerequisites of such action include a sense of self, the ability to predict and represent the future, and higher order symbolic language [180, p. 13].

Social Meaning. The choices we make in this regard, such as respect for the life, integrity, and well-being, even flourishing of others, is important in our own lives, but equally has shaped society [86] and history, and has even shaped modern identity, as explored by Charles Taylor in his masterful book *Sources of the Self: The Making of Modern Identity* [237]. Key are our understandings of what makes a good life, as mentioned above, and issues to do with dignity [237, pp. 15–16] and virtue [90].

7.4.4 The Meaningful Social Context

This section has been emphasizing the crucial role of the levels of ethics and meaning from the standpoint of high level brain function as shown in Fig. 7.6, where meaning, aesthetics, and ethics are causally effective in shaping what happens at lower levels. They are in turn modulated by the society we live in: they cannot be understood in isolation [65]. They act down to the level of neurons and shape what happens at that level, which then, through our muscles, reaches out and shapes the world and atoms in it. And this interaction is based on our abstract understandings of social and individual events. David Deutsch has a classic comment on the topic in his book *The Fabric of Reality* [63, p. 26]:

> Consider one particular copper atom at the tip of the nose of the statue of Sir Winston Churchill that stands in Parliament Square in London. Let me try to explain why that copper atom is there. It is because Churchill served as prime minister in the House of Commons nearby; and because his ideas and leadership contributed to the Allied victory in the Second World War; and because it is customary to honor such people by putting up statues of them; and because bronze, a traditional material for such statues, contains copper, and so on. Thus we explain a low-level physical observation (the presence of a copper atom at a particular location) through extremely high-level theories about emergent phenomena such as ideas, leadership, war and tradition. There is no reason why there should exist, even in principle, any lower-level explanation of the presence of that copper atom than the one I have just given.

Another example is particle collisions at the Large Hadron Collider (LHC) at CERN. These micro events are the result of the top-down effect of abstract thoughts in the minds of experimenters, who wished to establish whether or not the Higgs boson actually exists, on the particle physics level. Without these abstract thoughts, related to high level physics theories, there would be no such collisions. And they would not have taken place without the extraordinary social organisation that led to the existence of the LHC at CERN. This is all a monument to the causal effectiveness of the individual and social human mind.

7.5 Symbolism and Effectiveness of Thought

Abstract reasoning is the basis of the effective power of thought and the causal power of social constructions, enabled by symbolism such as language and mathematics. But how can this causal power of abstract thoughts be realised through the actions of the physical brain? As in the case of digital computers (Chap. 2), the modes of action at the higher and lower levels are quite different. Brown and Murphy express it this way [180, p. 13]:

- How is it that a series of mental/neural events come to conform to rational (as opposed to merely causal) patterns? One wants evolutionary as well as functional explanations.

- And what difference does the possession of mental capacities make to the causal efficacy of an organisms interaction with its environment? Why does it lead to major evolutionary advantage?

The key point is this:

> **Logical Functioning**. The neural network structure of the brain, beautifully adapted for pattern recognition and prediction [129], can also, via language and symbolism, engage in rule-based logical reasoning. This enables the understanding and generalization of abstract patterns of causation, including their description in logical symbolism or mathematical notation.

This gives evolutionary advantage because this understanding enables very effective forward planning on the one hand, and cumulative technological development on the other, and both of these are major advantages in terms of survival [33]. Prediction is not just one of the things our brain does. It is the primary function of the neocortex [129, pp. 89–97]. Symbolic manipulation and quantitative analysis allows it to be done much more effectively. This has various aspects. I look here in turn at:

- Logical functions.
- Naming, symbolism, and language.
- Effectiveness of thought.
- Thoughts and neural networks.
- The causal power of social constructions.

7.5.1 *Logical Functions*

How can it be that mental processes obey the dictates of reason rather than merely being dictated by the laws of microbiology or physics? According to Brown and Murphy [180, pp. 229–230]:

> How does *modus ponens* get its grip on the causal transitions between brain events? The reason is that the technological structure on which our higher causal processes depend is designed so that its causal processes realize rational transitions.

In computers, the software drives the hardware and the physics is controlled by the logic of the algorithms, because of the design of the connections between the gates at the micro-level [170]. In the cortex, this is not the case, at least initially: the layers in the cortex, based on recurrent circuits [68], are designed for generic pattern recognition and prediction. The patterns of logical argumentation have to be learnt. This must then lead to corresponding appropriate detailed synaptic connectivity, analogous to the connectivity in a computer.

7.5.1.1 Rules of Logic

The basic logical operations of AND, OR, NOT, and so on, need to be implemented at the micro-level, plus rules for combining them using the basic logical identities to give ways of rearranging them (this is the process of algebraic manipulation) [170, pp. 53–72]. Thus one has rules of structure (grammar) and rules of manipulation (equivalent reformulations), building up higher order hierarchically designed structures [170, pp. 122–125], including, crucially, branching structures (IF/THEN) via truth tables, as in control circuits [170, pp. 133–139]. This then allows recursion: imbedding of the same logical pattern hierarchically within itself, which is the key to complex logical analysis.

Once learned, there is an isomorphism between the desired logical operations and their physical realisation: and that isomorphism came about through adaptive selection of the brain's neural circuits to the desired logical operations. That is, once trained, a neural net recognizes patterns of allowed changes and can carry them out in sequence so as to logically deduce outcomes of premises. Hence rule-based systems can be supported by neural nets, and can be used to analyse consequences of thoughts and actions.

Note that, while the brain can indeed be trained to carry out these logical operations, they are not its natural mode of operation, which is pattern recognition and prediction. That is why logical rules, which are abstract patterns independent of the human mind (see Sect. 7.6), have to be discovered and learnt.

7.5.1.2 The Essential Downward Link

How is this possible? How do we transcend bottom-up mechanisms arising through the nature of physics alone, allowing reasoning through brain processes? Murphy and Brown suggest that we need to understand the mental as [180, p. 11]:

> [...] pertaining to a higher level dynamical system that is the brain in the body involved in interaction with the world, both physical and social. Thus we shall argue for the essential action relevance of perception and memory; mind is paradigmatically manifested in informed engagement in action-feedback-evaluation-action loops in the environment [...] we take rationality to depend basically on downward causation form the environment reshaping neural networks, and further, on the hierarchical ordering of cognitive processes such that higher level evaluation selects among lower level processes in accord with demands of the environment, both physical and social.

This is of course a statement about the power of adaptive selection. Tse [247] fleshes this out in terms of neural and synaptic processes. It is possible in causal terms because an effective information analysis shows that, for certain causal architectures, coarse-grained macro mechanisms are more effective than the underlying micro mechanisms, so that, although the macro-level supervenes on the micro-level, it can supersede it causally, leading to genuine causal emergence [131]. It is through top-down activation of perceptual symbols that 'controlled' conscious pathways can interact with what have been regarded as the unconscious action pathways, such as those of encapsulated affective/incentive response systems [179]. A functional architecture in the left mid-superior temporal cortex (lmSTC), which in key respects resembles that of a classical computer, may play a critical role in enabling humans to flexibly generate complex thoughts [89].

But in order to enable complex logical operations, we need a system of symbols to be manipulated, that is, we need a suitable language.

7.5.2 Naming, Symbolism and Language

Our unique human ability is to be able to use symbolic systems [54], that is, systems of referencing to things, events, and actions that enable us to make coherent symbolic models of the physical, ecological, and social world around us that represent it reasonably well. This labelling system forms a language with an agreed grammatical form, and this enables us to formulate and share abstract concepts and ideas.This section looks at language. The other great symbolic system is mathematics, which I comment on in Sect. 7.6.1.

Language is a symbolic system [19, 54] with a semiotic function [246]: its purpose is to convey meaning, emotions, facts, and concepts in a social context through systematic use of symbols. It represents the world of objects, actions, feelings, and qualities, as well as relationships, ideas, and theories. This representational function involves naming, indexing, and using metaphor [160, 246]. Facts represented are both contingent (historical, geographical, and other specific features of the world and of narratives) and generic (universal patterns characterising the way it all works in general). The relation between these two features (concrete/specific and abstract/generic) is a key aspect of thought and of language, involving development of classes of entities and classification of specific instances.

While language has an abstract character, it is embodied via an equivalence class of physical representations. In particular it has spoken and written forms. Its existence enables the cumulative buildup of understandings and ideas in individuals and in society, and (through technological means) their long-distance sharing in geographic terms as well as their storage and preservation in various media, enabling their propagation over long times. This storage of ideas in external media (ranging from diaries and memos to encyclopedias and the internet) greatly facilitates our mental powers and underlies an exponential growth of knowledge [44]. Language attains its social power by enabling mediation of social interactions on a small scale, and by enabling the utility of mass media and communication systems on a large scale. These in turn enable widespread dissemination of facts, ideas, and meanings, extending the cognitive web from local communication to a global system. Minds cannot therefore be understood on their own [65]: they are part of a society that is in turn part of a global intercommunication network. It has the following aspects:

- Naming
- Symbolism and language structure.
- Modular hierarchical structure.
- Equivalence class of representations and embodiment.
- Iconicity.
- Metaphor.
- Language function.

7.5.2.1 Naming

Language relies on the ability to use words to refer to objects, actions, and qualities in a way that can be recalled, i.e., it relies on naming things in an indexical way [180, p. 191]. This naming takes place via neuronal structures in the cortex (see Hawkins [129, pp. 136–167, 147–156]).

The cortex has a hierarchical design with a six-layered form [129, p. 107] plus feedback connections, so information can flow both upward and from higher to lower regions [129, p. 113]. The job of any cortical region is to find out how its inputs are related, to memorize the sequence of correlations between them, and to use this memory to predict how the inputs will behave in the future [129, pp. 123–124, 153–156]. Thereby the hierarchical structure of the cortex stores a model of the hierarchical structure of the real world [129, p. 125]. A predictable sequence of events gets identified with a 'name', i.e., a constant pattern of cell firing [129, pp. 126, 129]. A column is the basic unit of prediction, synaptic connections to other parts of the brain giving it the context it needs to deal with ambiguity [129, pp. 141, 153–156]. This structure thus enables recognition and naming of abstract patterns (see Hawkins [129, p. 57] and Churchland [43]), which then provides the foundation of language.

7.5.2.2 Symbolism and Language Structure

The major function of language is its labelling of specific and generic objects and instances, as well as abstract entities, through use of indices and symbols. It is based on iconicity, which enables it to refer beyond what is immediately present and hence is disjunct from physical referents. It has the following features [246]:

- **Arbitrariness**. The absence of any necessary connection between the form of a word and its meaning.
- **Stimulus Freedom**. Our ability to say anything at all in any situation, so enabling discourse that is freed from the immediate situation and stimuli. We can think about what we do.
- **Displacement**. The ability to speak about things other than here and now: the future, the past, the possible, even the impossible.
- **Open-Endedness**. The ability of language to say new things, virtually without limit.
- **Redundancy**. The full message is entailed by part of the given text/message, hence one can determine the full message by partial information (if the context is known).
- **Recursion**. A key feature in the development of language is the emergence of recursion [55, 246]: the occurrence in a sentence of a syntactic category containing within it a smaller version of the same category.

7.5.2.3 Modular Hierarchical Structure

The key functional element in representing complex aspects of reality is a modular hierarchical structure [26] with a class structure involving inheritance and modularity, where larger elements are built up from smaller ones. This is also called *duality of patterning*: a type of structure, encoded in a grammar with a particular syntactical structure, in which a small number of meaningless units are combined to produce a large number of meaningful (semantic) units [245]. This introduces the crucial feature of discreteness of structural units (the quantum principle), which units can then be repeatedly combined as recognised and named higher level units, giving hierarchy and allowing arbitrary complexity of combinations to be built up through recursion. The combinatorial structure of language is based on this principle [245].

This modular hierarchical structure of language enables its completely flexible representational and social function. This is a very general way of handling complexity: break up a complex task into simpler tasks that are completed first [26, 220]. This structure is bound by strict semiotic requirements [55], leading to a necessary set of implicit rules, but with a great variety of possible realisations (different languages/dialects).

Because of recursion (see [55, 246], and [245, p. 288]):

> The recognition of a suitable set of syntactic categories allows us to analyse all the sentences of a language as being built up, by means of a fairly small set of rules allowing recursion, from just these few categories.

This is based on chunking and labelling, that is, naming patterns of words so that they can be treated as a single unit. Thus a crucial aspect of language is identifying as a single unit and naming compound experiences or concepts, thereby allowing hierarchical structuring and building up patterns of patterns [245, p. 244]:

> Recursion is pervasive in the grammars of the languages of the world, and its presence is the chief reason we are able to produce a limitless variety of sentences of unbounded length just by combining the same few building blocks.

7.5.2.4 Equivalence Class of Representations and Embodiment

The same thought can be represented in many different ways in the same language, and of course in a much greater variety of ways in other languages. This is the arbitrariness of iconic representation. The meaning is invariant, but the representation is flexible.

Physical realisation of language can be neural (in an individual's brain), spoken (sound), written (visual), electronic (digital), or in visually transmitted sign patterns (sign languages). The same representational patterns, carrying the same semantic meaning, are embodied in these different representations. They are all enabled by the physical structure of the brain, which is hierarchically structured so as to enable an interplay of sensory interpretation and prediction, based on pattern recognition, classification, memory, and extrapolation [129]. Meaning is embodied in an equivalence class of such surface representations: it is independent of whether language is spoken, written, or signed, and of language family, dialect/pronunciation, and font. A profound ability of the mind, underlying the flexibility of language usage, is to recognize them all as functionally equivalent. The concepts represented are recognized as entities that exist in their own right, which can be labelled and represented in many different ways.

Thus thoughts are abstract entities with multiple possible instantiations. The thought itself is not the same as any particular one of them. It is essentially the equivalence class of instantiations that mean the same thing. There is a complex relation between thoughts and the truths they express [92].

7.5.2.5 Iconicity

The features described above enable language to give an excellent representation of arbitrary objects, situations, events, and ideas, on the basis of combinations of an arbitrarily chosen set of atomic symbols together with suitable combination rules. This flexibility of representation, provided a certain set of basic semiotic constraints are fulfilled [55], is remarkable. The resulting sentences can be printed on paper, written by hand, spoken, or stored in electronic media. These physical instantiations represent our understandings, theories, fears, emotions, and hopes in symbolic form. They are of course the result of specific neural connectivities and excitations which are shaped by our experience and intentions.

7.5.2.6 Metaphor

A key feature of the way language functions is the use of metaphor [103, pp. 54–66], often based on action experiences. This plays a major role in cognition and meaning-making [160]. This takes place in the context of conceptual schemas and cultural frames [82], which are the context of our understandings.

7.5.2.7 Language Function

Language enables the sharing of ideas and emotions and plans, so making social life possible. It allows us to live in a virtual world, enabling us to imagine other worlds that might exist, and share them via parables and stories [103, pp. 54–66]. Sharing stories with others enhances group bonding. This would have played a fundamental role in human evolution by underpinning group formation [71, p. 19]. It enables us to systematically propose and examine ideas and theories that change the way we understand and live. It is contextual all the way (Sect. 7.3.4): we use the context of known sequences to resolve ambiguity [112] and of understood context to understand conversations and texts.

Symbolic language is a prerequisite for both reasoning and morally responsible action [180, p. 13]. It is what enables us to be fully human [54, 66, 71].

7.5.3 *Effectiveness of Thought*

Language enables us to think about things in a systematic way, on the one hand making theories of how things work in general and testing these theories (the scientific process), and on the other hand, making specific action plans on the basis of these theories.

7.5.3.1 Action Plans

Making plans involves considering alternatives, assessing our best plans in terms of some criteria of choice, and making plans to achieve our goals. Mathematics and computational simulation allow us to do this in a quantitative way, for example, in designing a bridge, a dam, or an airliner. We can then implement these plans, and they result in major or minor changes in the world around us: the existence of aircraft and computers, of paper and paper clips, of dams and bridges and cities. Because most of the things we create are hierarchically structured, the plans, too, will be hierarchically structured.

These plans and theories are all consequences of our thought processes. They are the result of top-down effects from abstract ideas to neuronal excitations and into the world, down to the level of atoms. We are surrounded by proof of the efficacy of mental thought (see Fig. 7.13). This effectiveness is based specifically on models and theories.

Models. These are the core of understanding, ranging from metaphors to maps to spreadsheets to complex theories such as general relativity theory to massive computer simulations like those used in weather prediction and computer-aided design of aircraft. They extend to social and psychological theories ('folk psychology') which aid understanding of interpersonal relations, politics, and economics, with varying degrees of success. They underlie our behaviour, because we use them for predictions of what might happen, and we use them to plan manufacture of objects such as an aircraft or computer.

Their success is based on the hierarchical symbolic structures discussed above, using symbolic manipulation rules on stored data and patterns related to the real world by correspondence rules that are verified by observation and experiment. The models can be verbal, geometric (such as maps), symbolic, or mathematical. In each case they can be held in the mind as mental models, or spoken about, written down on paper, or stored in a computer in electronic form. These various representations all represent the same hierarchically structured abstract model of some aspect of the world (see Fig. 7.14).

Fig. 7.13 Proof of the power of thought. An Airbus (the physical object). Credit: Wikimedia Commons (Julian Herzog)

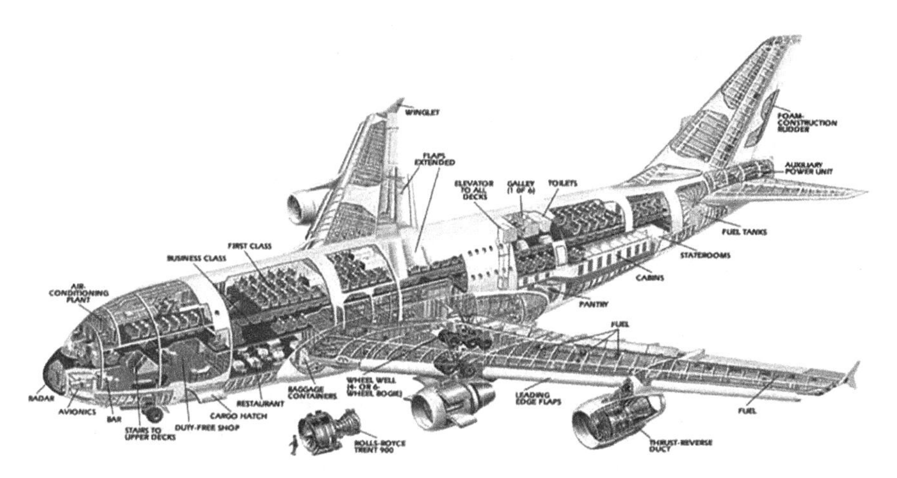

Fig. 7.14 Conceptual model of an Airbus (the abstract plan). Credit: From avionale.com, with permission

7.5.3.2 Abstract Theories

General theories underlie successful models, and in particular physics and chemistry underlie engineering. I will give just one example: the theory of electromagnetism.

Maxwell's Theory of Electromagnetism. This theory is not the same as any single person's brain state. It is an abstract entity, that can be represented in many ways and formalisms: by 3D partial differential equations, by 4D covariant equations, by

Hamiltonian variational principles, by differential equations for spinors, for example. These equations can be thought about, printed on paper, presented on a computer screen, embodied in a computer program.

The theory is not the same as any of these representations: it is the equivalence class of all of them. Because this abstract theory (a mental construct) describes the physical phenomena of electricity and magnetism in the real world very well, it can be used to predict how physical systems will behave. It has been shown to be a valid description of physical reality, within its domain of application,[9] by experiments confirming these predictions. Consequently, it can be used to design electric motors and generators, magnets and relays, circuits and transmission lines, and so on. Crucially, these mathematical equations led to the realisation that light is an electromagnetic wave. This implies that such waves can exist at other wavelengths, carrying information from one place to another. This led to the development of radio, TV, radar, cell phones, and so on, which have played a major role in technology and social life, based on manipulation of physical materials composed of atoms and electrons.

This abstract theory has therefore altered physical configurations in the real world in a major way, and hence is causally effective. It is an irreducible higher level causal factor (it cannot be derived by coarse-graining any lower level variables). It demonstrates how non-physical entities, underlying the development of technology, can have enormous causal power [33]. This is enabled by the power of equations we have articulated in our minds, that are then used in engineering design. This is just one example: for many others, see Ian Stewart's book *Seventeen Equations that Changed the World* [229].

7.5.4 Thoughts and Neural Networks

What is the difference between the thoughts discussed here and the neural networks that correlate with them? This is not a reductionist question. What is the difference between the thoughts and the neural processes that correlate with them at the same hierarchical level?[10] The view I take is that this is similar to the relation between software and hardware in computers, discussed in Chap. 2: the neural processes are the physical correlates of the abstract thoughts that are realised through them. This does not mean I think brains can be considered as digital computers in all ways, but that there are some similarities that are illuminating. This is considered by Marr and Poggio, who develop the following theme [172]:

> The CNS needs to be understood at four nearly independent levels of description: (1) that at which the nature of a computation is expressed; (2) that at which the algorithms that implement a computation are characterised; (3) that at which an algorithm is committed to a particular mechanism; and (4) that at which a mechanism is realised in hardware. In general, the nature of a computation is determined by the problem to be solved, the mechanisms that

[9] It is a classical theory, i.e., it does not include quantum effects.

[10] I thank Mark Solms for raising this issue.

are used depend on the available hardware, and the particular algorithms chosen depend upon the problem and the available mechanisms. Examples are given of theories at each level.

There is a modular hierarchically structured physical system—the brain—which corresponds to computer hardware (the implementation hierarchy) and which enables mental processes to be processed in terms of their own logic (the logical hierarchy). Thoughts correspond to and are enabled by excitations of the brain (their neural correlates), but are not themselves physical entities, just as a computer program is not a physical entity. The thought is an abstract process with its own logic (think of a mathematician giving a proof of Pythagoras' theorem), which is enabled by the pattern recognition, prediction, and generalisation processes enabled by the underlying neural networks (see Hawkins [129] and Churchland [43]). But in a computer, what happens at the physical level is determined by the algorithm at work (an abstract entity), realised through electron flows through networks of wires connecting gates. Similarly, thoughts are abstract entities realised by electron flows through networks of axons and dendrites joined by synapses.

When they have been internalised, the basic logic of the thoughts may be represented by specific adjusted weights in the neural networks, so that recall is automatised (this is the basis of intuition). This process of automatisation is similar to the way that, in computers, software can be developed in a test system and then burnt into a chip so that it is thereafter hardwired. However, the biological process is of course far more flexible (and digital computers are becoming similarly flexible through the use of memresistors).

7.5.5 Causal Power of Social Constructions

Social structures are inventions of the mind that shape society. They get embodied in material form to some degree through buildings, uniforms, logos, texts of various kinds, and so on, but they are themselves abstract entities. The causal power of social structures [75] is enabled by top-down causation from those structures to the mind. In this section, I look in turn at:

- Language.
- Roles and norms.
- Games.
- Institutions and organisations.
- Money.

7.5.5.1 Language

The generic nature of language has been considered above, but any specific society has specific languages that shape social interactions in that society. They are in effect social agreements reached over time by our ancestors that are passed down to us. They are embodied in present day social institutions, in particular being taught in schools. It is our shared language that enables social institutions to exist.

7.5.5.2 Roles and Norms

Berger [17] and Longres [166] emphasize the importance of roles and norms in society, as do Haidt and Kesebir [123].

Roles. These are part of the social system to which we belong and they somehow exist separately from ourselves [166, pp. 42–47]. They lead to role behaviour [17, pp. 94–99], as the role provides patterns according to which the individual is expected to act in particular situations: the role of the teacher, the pupil, the doctor, the patient, the policeman, and so on. These expectations then to a large degree shape our responses. They relate to status functions, which have 'deontic powers', as explored by Searle [215].

Again this is a form of top-down constraint from society to the individual: roles are abstract aspects of the social order that get imbued into individual brains by socialisation processes [17], until they become part of the perceived reality [18].

Norms. Social control is essential to the existence of society [17, p. 68], and a powerful way this occurs is via norms (see [166, p. 35] and [75, pp. 116, 117]):

> Norms are regularized practices encouraged by dispositions or beliefs about appropriate ways of behaving that are shared by a group of people.

They lead to regularities in society, by influencing individual behaviour (see [17] and [75, pp. 122–137]). The commonest strategy in the literature is to ascribe the causal role to norms themselves, rather than the agents who enforce the norms when necessary [75, p. 117]:

> The most important attribute of a social system is the social norms which hold it together. Norms consist of all agreements, formal or informal, explicit or implicit, which regulate and give order and purpose to a system, be it a primary or secondary group. Examples include goals and objectives, values and ideologies, traditions, lifestyles, and folkways or mores; dogmas, laws, policies, and procedures and rules; regulations, obligations, and duties. Social norms are experienced by individuals as expectations, the expectations of other people as well as the expectations that emerge from the self as a function of participation with other people.

This is what Berger calls morality customs and manners [17, p. 74].

Thus norms are abstract social agreements. They act top-down to constrain how individuals in the group function. Many of them are implicitly built into us by the social structures in which we live and affect our behaviour by shaping our world view [18].

7.5.5.3 Games

The rules of a game, e.g., chess, football, cricket, hockey, tennis, are social agreements that have a causal power over the players in the game. They play a strong role in social cohesion by allowing regulated competition. Each game has an arbitrary set of rules that are agreed on by participants (perhaps through their joining a social activity where these rules are already set through historical processes). This set of abstract rules then govern which moves are allowed and which are not. They therefore set the context within which the physical actions of the game take place. If the rules are changed, the game is different (some rules are better than others in terms of leading to interesting games and they will be adaptively selected over time). Thus they set top-down constraints on physical outcomes.

7.5.5.4 Institutions and Organisations

Social structures such as sports clubs, welfare organisations, firms and companies, banks, local and national governments, churches, schools and universities, and so on emerge in society through a variety of social interactions. They are then more than the members that make them up: eleven individuals is not the same as a football team, because each member now has a role assigned to them that shapes what they will do (a goalkeeper acts differently than a quarterback). Thus they are examples of relational emergence [75, pp. 66–68].

They are usually hierarchically structured, with multiple levels (see [14] and [75, pp. 48–53]). The organisation is shaped by abstract agreements that have causal power, e.g., constitutions, laws, contracts. These get made specific in terms of roles and contracts for personnel, and rules and operating procedures that determine what they actually do ('sorry, it's not in the book').

Thus institutions themselves have causal power because the institutional status functions and structures exist [75, 215]: the bank awards loans, the university awards degrees, the firm manufactures automobiles, the shop sells furniture, and so on, because they have the power to do so. The specific decisions are made by individuals or groups in the organisation, but they only gain their causal power because the organisation exists as a structured abstract entity: the president of the bank can be the president, and have corresponding executive powers, only because the bank exists as an organisation. Organisations have causal power over and above that of the individual people that make them up. These are all abstract ideas that have been institutionalised and made effective by social agreements and constitutive rules [215]. An organisation is not a physical thing, although it may have physical realisations such as buildings and equipment that belongs to them. They change what happens on the ground, and have this causal power because of the thoughts, embodied in language, that led to their coming into being.

7.5.5.5 Money

One of the most important social structures is the economic system and its facilitation through money. Money is not the same as the bank notes or coins that are its physical realisation: these are just physical instantiations of the idea of money, which abstract entity is what has the real causal power [75, p. 70], and can be realised in many ways, for example, through electronic transfer or promissory notes, as well as banknotes and coins. Money has immense causal power: it enables dams to be built, houses to be bought, people to be fed, and schools to be built, and a physicist cannot explain this causal power in terms of physics *per se*. Physics can explain the structure of the fibres in the bank note, but not why it can buy a meal. This causal power results from social agreements enabled by language and based on faith: a system of trust lies at the bottom of the economic enterprise. It is another outcome of the causal power of thought.

7.5.6 The Power of Emergent Levels

This section has made the case that, while other things are also going on (see Sect. 7.2.3), rational thought does indeed occur and have causal power. This is conclusively proved by the fact that we plan things such as buildings, cities, aircraft, automobiles, music festivals, sports competitions, and so on. And then in due course we are able to make them happen more or less according to plan. Thus for example, the plans for an Airbus (see Fig. 7.14) led to the existence of the physical entity (see Fig. 7.13).

In terms of the hierarchy of causation (Fig. 7.1), the process of designing the plane is same level action at the level of the mind/brain, enabled by the lower levels but not explained by them. There is no way you can explain the existence of the Airbus on the basis of physics, chemistry, molecular biology, genetics, cell biology, or neurology alone [76]. The rationale of designing and constructing an entity is required, and this consists of macro-level rational processes and actions taking into account physics, chemistry, design, manufacturing processes, economics, and so on. These each have their own rationality, developed by experts in each of these fields and taught to others.

Thus, as in the case of digital computers, the macro-levels have real causal power and are able to conscript the lower levels, viz., neurons, cells, genes, macromolecules electrons, ions, and so on, to enable their plans to be fulfilled. The lower levels carry out the work, but the higher levels decide what to do [10, 251]. If this were not so, the aircraft would not exist. If one denies the causal power of thought, one has to say it came into existence without a cause. The more sensible position is to recognise the power of thought, enabled by top-down action.

7.6 The Effects of Platonic Entities

The causal effectiveness of human thought discussed in the last section is based on exploring logical possibilities through a process of learning that is enabled by the underlying neural plasticity. It involves understanding abstract patterns that are not themselves physical phenomena, but are nevertheless part of the underlying nature of existence. They are timeless eternal relations that are true at all times and all places in the universe, independent of culture and even of the existence of the human mind. That is, they have something of the nature of Platonic entities. But they are able to be discovered by the human mind [43]. These relations can usefully be thought of as possibility spaces for both abstract relations and physical phenomena. They include possibility spaces for biology[11] and physics, but I will not focus on those here. Rather I will will just comment on the following:

- Mathematics (Sect. 7.6.1).
- Computational algorithms (Sect. 7.6.2).
- Accessing Platonic realms (Sect. 7.6.3).

I will call the corresponding possibility spaces Platonic spaces, to emphasize that they each represent an eternal unchanging set of possible relations.

7.6.1 Mathematical Relations

The existence of a Platonic world of mathematical objects is strongly argued by Penrose [195] and Connes [40], the point being that many mathematical entities and processes are discovered rather than invented. Integers, rational numbers, zero, irrational numbers, prime numbers, algebra, groups, integration, differentiation, Pythagoras' theorem, the fundamental theorem of calculus, and the Mandelbrot set (see Fig. 7.15) are classic examples.

These entities, operations, and theorems are not determined by physical experiment, but are rather arrived at by logical mathematical investigation. They have an abstract rather than embodied character. They are not made of any material substance, rather the same abstract quantity or method can be represented and embodied in many symbolic and physical ways, just as is true for any other thoughts. Although their representation is culturally dependent (octopuses would have a base 8 number system), they themselves (the abstract entities represented) are independent of culture. It is plausible that the same features will be discovered by intelligent beings in the Andromeda galaxy as here, once their mathematical understanding is advanced enough, which is why mathematics is advocated as the basis for interstellar communication. This Platonic world is being progressively discovered by humans, and represented by our mathematical theories. That representation is a cultural construct, but the underlying mathematical features they represent are not, they are timeless

[11] See A Wagner: http://aeon.co/magazine/philosophy/natures-library-of-platonic-forms/.

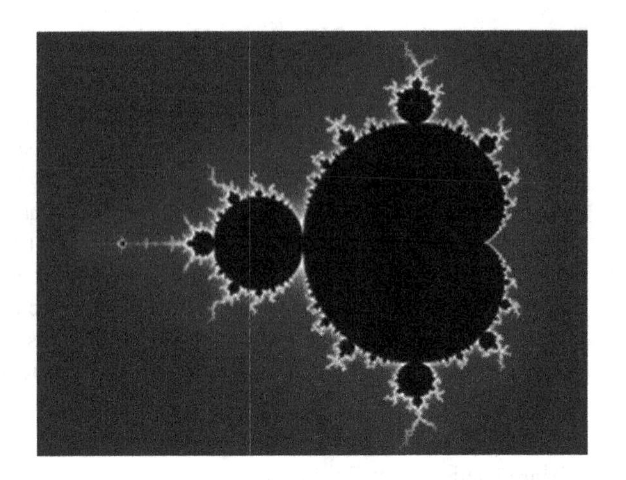

Fig. 7.15 Mandelbrot set. Credit: Wikimedia Commons (ArEb)

eternal realities whose existence does not in any way depend on the existence of human beings. Thus one must carefully distinguish the possibility space of mathematics Ω_{m} from its representation $\Omega_{\mathrm{m}}(S_i, t)$ in any specific society S_i at time t. The possibility space Ω_{m} is timeless and eternal. Its cultural representation $\Omega_{\mathrm{m}}(S_i, t)$, given by projection $\pi(S_i, t)$ from Ω_{m}:

$$\pi(S_i, t) : \Omega_{\mathrm{m}} \rightarrow \Omega_{\mathrm{m}}(S_i, t) , \tag{7.1}$$

varies across culture S_i and time t, and at any specific time t will represent only part of Ω_{m}. The process of mathematical discovery, i.e., the extension of $\Omega_{\mathrm{m}}(S_i, t)$ to cover ever more of Ω_{m} as time progresses, is the aim of the academic subject of mathematics. It does not in any way affect Ω_{m} itself.

An example is the value of the number π, which is (in decimal notation) approximately

$$\pi = 3.141\,592\,653\,589\,793\,238\,462\,643\,383\,279\,502\,884\,197\,169\,399\,375\,10\ . \tag{7.2}$$

Approximate values were known in Egypt and Babylon in the period 1900–1600 BC. Its value was known by Archimedes of Syracuse (287–212 BC) to lie between 3 1/7 and 3 10/71. Ever more accurate approximations to its actual value have been calculated since then by methods including polygon approximations and a variety of infinite series. The actual value, and the fact that its an irrational number, are eternal mathematical truths: they will be the same everywhere in the universe, and will be true whether intelligent beings like ourselves know about it or not. It plays an important role in mathematics (e.g., in complex analysis and Fourier analysis) and in probability theory, and underlies key properties of geometry that are important in physics and engineering. Of course, it can be expressed in other ways, for example in binary π is approximately

$$\pi = 11.00100100\,00111111\,01101010\,10001000\,10000101\,10100011\,00001000\,, \tag{7.3}$$

which is how it will be represented in digital computers. In hexadecimal its value is approximately

$$\pi = 3.243F6A8885A308D313198A2E037073\,. \tag{7.4}$$

The notational difference between (7.2) and (7.3) does not affect its intrinsic value. This is just like the way one can use different coordinate systems in general relativity theory to describe the same spacetime, or different sentences in different languages to express the same thought. The intrinsic meaning remains unchanged. Like physical laws, mathematical results are often unwillingly discovered, e.g., the irrationality of $\sqrt{2}$ and of π (hence they cannot be expressed in a finite decimal sequence). This is because mathematics often has a surprising nature (the existence of strange attractors for example or the deep unifications such as the development of analytic geometry and algebraic geometry and the Langlands program).

This abstract world is causally efficacious in two ways. First, it can be explored by the human mind and the resulting relationships shared and represented in many ways. One can, for example, print graphic versions of the Mandelbrot set in a book, resulting in a physical embodiment of this abstract pattern in the ink printed on the page (see Fig. 7.15). This is a projection from Ω_m to the atoms in the book. Second, the resulting relations can be used in commerce, physics, and engineering to analyse possibilities and so make changes in the world. For example, mathematics underlies decisions in commerce and construction projects in building and engineering that alter the world around us. It also underlies the way physics underpins engineering, e.g., the way Maxwell's equations underlie the telecommunications industry (Sect. 7.5.3). Thus mathematical relations make a real difference to what happens in the world. For other examples, see [229].

The causal variables here that enable this to happen, i.e., mathematical theories in the mind, or printed on paper, or represented in digital computers, are not coarse-grained lower level variables. They are high level relationships discovered and comprehended by the mind by analysis at the level of rational thought (Sect. 7.5.6). Their effectiveness is enabled by top-down action from mental contemplation of aspects of Ω_m to neurons in networks that recognise abstract platonic patterns, when we discover these relations, and down to motor neurons and electrons in muscles, when we use them to alter the world.

7.6.2 Computational Algorithms

The case of algorithms such as those used in digital computers is similar, as discussed in Sect. 2.7.5. Algorithms are step-by-step rules for obtaining a desired result. They are the basis for the enormous power of digital computers. What can be done by computers is characterized by a possibility space: the space of all possible computations Ω_c. This in turn is based on the set of all possible algorithms Ω_a, which includes the set of possible computer programs Ω_a(prog). The point is that there are only a limited number of possible ways of carrying out computations. For example, there are only a finite number of possible sorting algorithms, such as Insertion, Selection, Bubble, Shell, Merge, Heap, Quick, Quick3, which can be ranked according to efficacy in different circumstances.[12]

The existence of this set of sorting possibilities is once again a timeless, eternal, and unchanging fact. It has nothing to do with the existence of the human mind, even though the human mind is capable of discovering them. They are not mathematical equations: they are steps to obtain a desired outcome that can sometimes be used to solve equations. They are not based on the laws of physics (testing the validity of a sorting algorithm is not a physics experiment, it is a computer science exercise). They are based on logical possibilities characterised by timeless and unchanging possibility space Ω_a. They will be the same everywhere in the universe, at all times and all places. Computer scientists on other galaxies will discover the same set of possibilities, because they characterise all logical options that are possible.

Algorithms are causally effective and have made a great difference to the world, as discussed in MacCormack's book *9 Algorithms that Changed the Future: The Ingenious Ideas that Drive Today's Computers* [168]. For example, they underlie the computer-aided design of aircraft, the control of aircraft by autopilots, construction of automobiles by robots, bar-code recognition by laser readers, the functioning of GPS systems, and the functioning of cellphones, the internet, and search engines such as Google. Thus they underlie commerce, industry, engineering, and are of importance in much of individual and social life.

7.6.3 Accessing Platonic Realms

I have argued strongly for at least two types of abstract possibility spaces that have causal effects in the real world through the capacities of the human mind: those of mathematics (Sect. 7.6.1) and computer algorithms (Sect. 7.6.2). But philosophers have argued against the idea of Platonic spaces having such effects on the grounds that there is no way that an abstract space of possibilities could have any effect on the physical brain and consequently on the mind. If Platonic spaces exist, they will be causally ineffective.

This objection has been comprehensively answered by Paul Churchland in an important book entitled *Plato's Camera: How the Physical Brain Captures a Landscape of Abstract Universals* [43]. The brain operates by pattern recognition and prediction enabled by the overall pattern of neural connections in the cortex and the connection weights in these neural networks [43, 129], and possibly by synchronized

[12]See http://www.sorting-algorithms.com/.

patterns of oscillations between neurons [35]. When we engage in logical argumentation and mathematical thought, selection processes operating in neural networks develop in such a way as to recognise the abstract patterns of logic present in Platonic possibility spaces [43], which then become part of the causal processes in operation in the brain. This is best understood by using the activation-vector-space framework for the brain as an organ of thought, imagination, and reason explained in detail by Churchland [43].

Through this mechanism, a different form of necessity is causally effective in the brain than is embodied in physical laws: it is the necessity of abstract logic and mathematics, which underlies associated logical structures such as algorithms and digital computer programs. The patterns embodied in the abstract spaces of possibilities—a key part of necessity—are enabled to be causally effective through learning and exploration processes. In this way minds with real causal powers can apprehend Platonic realities and thereby bring new input that was not in the physical initial data, but was present in eternal and unchanging possibility spaces. Physics provides the necessary conditions for the existence of such higher level outcomes, but not the sufficient conditions to determine the resulting behaviour. Other causal factors: genetic, evolution, neural, social, are in action, but none of these deal with the core essence of the causal chain leading to the relevant logical outcomes. These are affected by relevant higher-level variables which are adapted to understanding mathematical issues or computer algorithms, and attain meaning and causal effectiveness at their own level.

Thus there is not only physical input in the universe: there is abstract input as well from these possibility spaces, which are knowable by the human mind, and represent crucial parts of necessity. These kinds of concepts and influences are causally effective but are not physical variables: they all lie outside the conceptual domain of physics, but they change things, e.g., by underlying the existence of cellphones and digital computers. The mind is able to be causally effective because it operates at the higher levels of the hierarchy (Fig. 7.4) in terms of the logic at that level, and can thereby understand and develop the implications of possibility spaces such as Ω_{m} and Ω_{a} discussed above.

7.7 The Complex Whole

Building on the previous chapters, this chapter has presented an integral view of the way the brain functions. This section will conclude the chapter by discussing the following aspects:

- A synthesis (Sect. 7.7.1).
- Genuine emergence (Sect. 7.7.2).
- Crick's callacy and the reality of higher levels (Sect. 7.7.3).
- Top-down action and the free-will debate (Sect. 7.7.4).
- Neuroscience and humanity (Sect. 7.7.5).

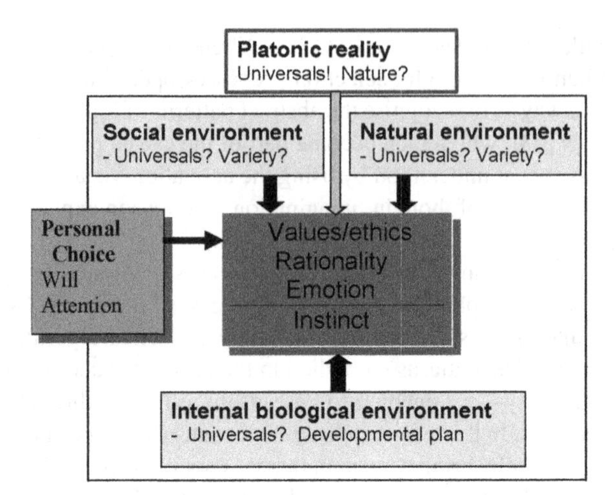

Fig. 7.16 Factors affecting the function of the brain. Psychological universals are based on universals in the social environment and in the natural environment, which act *top-down*, together with Platonic relations, which shape rational thought, while our inherited biological makeup (which underlies a universal human development plan), in turn based on physics and chemistry, acts *bottom-up* to enable us to carry out the choices we make within the constraints allowed by the underlying physics. They all interact with each other to produce the specifics of higher brain functioning via the processes of neuronal group selection, shaped by the primary emotions. However, variety in each environment as well as in the genetic inheritance leads to a variety of outcomes. Taken for granted is the environment of physical laws, which shapes the space of physical and biological possibilities

The view presented in this book is summarised in Fig. 7.6 (Sect. 7.2.3) and Fig. 7.16. Bottom-up effects occur from our cellular machinery, physiology, and inbuilt developmental plan, and top-down effects occur from the natural and social environments, as well as from Platonic spaces of possibilities, such as mathematics. Our personal choices are based on values, rationality, and emotions, and shape what we say and do.

7.7.1 A Synthesis

This chapter proposes an overview of brain functioning in context as follows:

- **Modular Hierarchical Structures**. The brain is comprised of modular hierarchical structures (the implementation hierarchy) with form adapted to enable modular hierarchically structured functions (the logical hierarchy), and development taking place through genetically enabled processes that respond to the environment adaptively.
- **Interaction Between Levels**. Bottom-up and top-down causation both take place, so one can only understanding the functioning of parts in the context of the whole.
- **Plasticity**. Parts and whole are complex adaptive systems adapting to their environments on evolutionary, developmental and functional time scales: hence all structures are shaped by functions that are adapted to physical and social context, but within the limits of what physiology and development will allow.

- **Basic Capacities**. These are sensations, pattern recognition, remembering, naming, based on neural networks and enabled by higher level contexts that feed down to enable ongoing prediction of future states that shape sensory perception.
- **Higher Level Capacities**. These include consciousness, comprehension, reasoning, choice, and action in reaction to perceptions, feelings, and context, enabled by language.
- **Subconscious and Automatised Processes**. These shape a lot of what happens, but can be overridden by consciousness when need be.
- **Integration**. Overall coordination is via drives and affective states enabling appropriate responses to environmental situations and including the drive to understand and find meaning.
- **Consciousness**. This is an emergent property of a dynamic core of neurons [243]. It is a higher level process enabled by the properties of the underlying neurons and genes, in turn enabled by the properties of the underlying molecules and physics, but it is not reducible to them, among other things because it is deeply meshed into ongoing interactions with the physical, ecological, social, and intellectual environment.

It is its complex organisational arrangement, arrived at by adaptation on evolutionary, developmental, and functional timescales that enables this all to happen. We have a social brain, facilitated by language, which alters consciousness and evolutionary history [71]. The mind must always be seen in context, as explained by Merlin Donald[13]:

> We have evolved a novel evolutionary strategy, which relies on off-loading crucial replicative information into our cultural memory systems. The algorithms that define the modern human mind may have been originally generated by collectivity of conscious brains living in culture, but these accumulated storehouses have now assumed a certain autonomy and have become an essential part of the mechanism by which we replicate, and continue to extend, the domains of our awareness. We have evolved into 'hybrid' minds, quite like any others, and the reason for this does not lie in our brains, which are unexceptional in their basic design. It lies in the fact that we have developed such a deep dependency on our collective storage systems, which hold the key to self-assembly. The ultimate irony of human existence is that we are supreme individualists, whose individualism depends almost entirely on culture for its realization. It came at the price of giving up the isolationism, or cognitive solipsism, of all other species and entering into a collectivity of the mind.

Adaptive selection is the key to learning and neural plasticity, with plasticity at the different levels supporting each other: higher level plasticity is based on lower level plasticity, which, via top-down causation, is shaped by higher level effects. Together this allows the effectiveness of consciousness [180, 247], whereby our plans and intentions shape what we do, and hence have causal effects in the physical environment around us.

[13]http://newlearningonline.com/new-learning/chapter-6/donald-on-the-evolution-of-human-consciousness.

7.7.2 Genuine Emergence

Merlin Donald [64, 66] presents a sophisticated conception of a multilayered consciousness drawing much of its power from its cultural matrix. He makes a persuasive case for consciousness as the central player in the drama of mind, as he details the forces, both cultural and neuronal, that power our distinctively human modes of awareness. He proposes that the human mind is a hybrid product, interweaving a super-complex form of matter (the brain) with an invisible symbolic web (culture) to form a distributed cognitive network. This hybrid mind, he argues, is our main evolutionary advantage, for it has allowed humanity as a species to break free of the limitations of the mammalian brain.

This emergence [13] depends on a variety of top-down effects, and the distinction between top-down and bottom-up effects, which is central to understanding the functioning of the brain [99, 147], also plays a central role in experimental psychology [217]. I discuss this briefly in terms of the following:

- The macro-micro connection.
- Irreducibility of the resulting effects.

7.7.2.1 The Macro-Micro Connection

Top-down effects take place in the following situations, among others:

- From society to the brain, e.g., the language we speak.
- From the mind to body, e.g., lifestyle effects on health.
- From higher goals to lower goals, e.g., wanting to help others to training as a doctor.
- From brain to body, e.g., from the goal of playing a guitar to the muscles plucking the strings.
- From psychotherapy to changes in gene expression (see Kande's principles [143] discussed in Sect. 7.3.6.7).

A mind's behaviour is determined by its interaction with other minds and the higher level entities that in fact shape its outcomes, including abstractions such as understanding of mathematics, the value of money, the rules of chess, local social customs, roles, and socially accepted ethical values [10, 18, 166].

Overall the situation is nicely described in an interview with Gazzaniga[14]:

> We are dealing with a layered system, and each layer has its own laws and protocols, just like in physics where Newton's laws apply to one layer of physics and quantum mechanics to another. Think of hardware–software layers. Hardware is useless without software and software is useless without hardware. How are we to capture an understanding of how the two layers interact? For now, no one really captures that reality and certainly no one

[14] "Neuroscience challenges old ideas about free will" Scientific American 15 November 2011 http://www.scientificamerican.com/article/free-will-and-the-brain-michael-gazzaniga-interview/.

> has yet captured how mental states interact with the neurons that produce them. Yet we know the top mental layers and the layers beneath it, which produce it, interact. Patients suffering from depression can be aided by talk therapy (top-down). They can also be aided by pharmacological drugs (bottom-up). When these two therapies are combined the therapy is even better. That is an example of the mind constraining the brain [...] One becomes cognizant there is a system on top of the personal mind/brain layers which is yet another layer—the social world. It interacts massively with our mental processes and vice versa. In many ways we humans, in achieving our robustness, have uploaded many of our critical needs to the social system around us so that the stuff we invent can survive our own fragile and vulnerable lives.

This is based on the underlying neural structures. For example, as explained nicely by Shea [217], in Karl Friston's model of the brain, feedforward signals consist only of prediction errors [96, 97] and it is the top-down signals that represent what is the case. In this way top-down signals directly affect, or even constitute, what we represent in our minds, but without creating a self-reinforcing cycle, because it is not the result of these predictions that is fed forward as input to subsequent processing, but only the difference between prediction and current input. It is for this reason that top-down influences can be seen to be epistemically acceptable [217].

This is related to Tse's criterial causation at the neural level [247], which he claims underlies downward mental causation and free will.

7.7.2.2 Irreducibility

Biology cannot be reduced to physics, because it has an ineliminable teleology component to its explanations [131]: cells and all physiological organs have functions [203], biology at all levels is centred on purpose [127]. High level purposes (such as aiming to catch a bus) cascade down and set up lower level purposes (such as moving a leg), which set up even lower level purposes (contracting a muscle), down to the cellular level (setting ATP generation and transport in motion, and so on) [247]. Subsidiary purposes enable this to happen: the lungs breathe in oxygen, the heart pumps blood, the immune system fights intruders, and so on, in order to keep us alive. And each of these actions cascades down to the cellular level and then to the nuclei and electrons that make up the cell. It is top-down causation that enables the higher levels to have causal powers in their own right, leading to this downward cascade. The mind can have real effects in the physical world [51].

What's wrong with ontological reductionism is that it is wrong. The idea that there are no kinds or classes other than those belonging to the physical sciences is clearly false. Currencies exist, are real, and have causal efficacy, and yet they are not reducible to any physical type or kind, despite the fact that every instance of a currency is made of some physical substance or consists of some physical process or another. They are abstract social constructions [75, p. 70] that are realised in minds, but are not the same as activations in any specific individual's mind because they are communally shaped. Similarly abstract concepts such as plans for an aircraft are causally effective (Sect. 7.5.3).

The Unviability of Reductionism in the Context of the Mind. Social Sciences cannot be reduced to physics, because of the irreducibility of intentionality, as well as the underlying multiple realizability of mental states. Multiple realisability is a key indicator of top-down causation (see [8] and Sect. 3.5). Jerry Fodor discusses the issue very clearly in terms of 'natural kinds' occurring in an explanation [13, pp. 403–407]. Daniel Kauffman[15] explains it as follows. Take any folk-psychological (FP) law, say,

$$\mathrm{FP1} \to \mathrm{FP2}\,. \tag{7.5}$$

The causal relation is described in the notation of an 'if, then' statement. The following would be an example of a commonly employed FP law:

> For any person P, if P desires Q and believes that doing X will achieve Q, then P will do X, *ceteris paribus*.

That will be a sound psychological explanation [114], enabling us to make good predictions, which is what the brain does [129, 150]. In order to reduce this law to a law of biology, and therefore show that the biological law *explains* the relevant phenomenon in the relevant way, one would have to establish bridge laws, in which both FP1 and FP2 are shown to be identical with or materially equivalent to, some biological states. The problem is that FP1 and FP2 are multiply realizable. Thus, such a bridge law will look something like this:

$$\mathrm{FP1} = \mathrm{B1 \text{ or } B2 \text{ or } B3 \text{ or } B4}\ \ldots\,,$$

$$\mathrm{FP2} = \mathrm{B20 \text{ or } B30 \text{ or } B40}\ \ldots\,,$$

resulting in the following reduced statement:

$$\{\mathrm{B1 \text{ or } B2 \text{ or } B3 \text{ or } B4}\ \ldots\} \quad \longrightarrow \quad \{\mathrm{B30 \text{ or } B30 \text{ or } B40}\ \ldots\}\,. \tag{7.6}$$

There are two problems:

- Neither the antecedent nor the consequent describe a biological type.
- The sentence described is not a law of biology.

Thus, one has *not* reduced the folk psychological law to a law of biology and, whatever the folk psychological law *explains*, has not been explained in terms of any law(s) of biology [13, pp. 403–407]. Folk psychological explanations are indeed explanations, and often very powerful ones, much better than anything one could get at a lower level of description. For example, they are what are involved in 'mind reading' [99]. It is not true that in the end the only *real* explanations/predictions are going to come at the level of atoms [189], or genes or cells, for that matter. The psychological level, driven by feelings, is where the choices take place.

---

[15] Writing as 'Aravis Tarkheena' in Massimo Pigliucci's now defunct blog *Scientia Salon*, 29 August 2014.

7.7.3 Crick's Fallacy

Francis Crick's wrote in *Astonishing Hypothesis: The Scientific Search for the Soul* [48]:

> You, your joys and your sorrows, your memories and your ambitions, your sense of personal identity and free will, are in fact no more than the behavior of a vast assembly of nerve cells and their associated molecules.

But nerve cells and molecules are made of electrons plus protons and neutrons, which are themselves made of quarks, so why not say:

> You, your joys and your sorrows, your memories and your ambitions, your sense of personal identity and free will, are in fact no more than the behavior of a vast assembly of quarks and electrons?

And these themselves are possibly vibrations of superstrings. His argument is an unacceptable partial reduction: he reduces the phenomenon of interest to an arbitrarily chosen level, and then stops, despite the fact that is not the bottom level. But a true reductionist could not possibly justify stopping there at an intermediate level. On what grounds does he treat the particle physicists so badly, denying the full reduction to their level of description? The answer is he believes that nerve cells and their associated molecules have real causal power, because that is the level he deals with and understands. I agree. But the implication is that they have causal powers over their constituent atoms, protons, and electrons, in order to make their own causal powers effective. He is assuming the viability of top-down causation to lower levels, a proof that it exists and is significant in the operation of the brain. But if we accept this, we must recognise that assigning real causal powers to an intermediate level such as nerve cells and their associated molecules, acting top-down on the levels below in order to have this power, only makes sense if we assign real causal powers to every other level as well. There is no preferred level of causation [189]. That is the only interpretation that makes sense, and it is my position.

And then there is no reason to deny the reality of causal powers of your joys and your sorrows (the emotional systems), or of your memories (allowing the predictions that underlie our sensory systems) and your ambitions (your set of goals that are the decision framework for what you do). The threat to your sense of personal identity and free will posed by Crick's quote is undermined, because the psychological level is a genuinely emergent level, based on the underlying physics, chemistry, genetics, cell biology, neurology, and physiology, but not dictated by them because they are not the only causal factors in operation.

7.7.4 Top-Down Action and the Free Will Debate

Over the course of history there have been at least six different attacks on free will, as summarised in Table 7.2 (Sect. 1.3.1). They are (in somewhat cartoon form):

Table 7.2 Arguments against free will

	Agent	Mode
Argument 6	God/fate	Top-down
Argument 5	Society	Top-down
Argument 4	Unconscious	Same level
Argument 3	Evolution/genetics	Bottom-up
Argument 2	Neurons	Bottom-up
Argument 1	Physics	Bottom-up

1. **The Physics Challenge**. Atomic level structure determines all. If we just had enough computing power, we could calculate all future behaviour of the mind from complete initial data for the brain.
2. **The Genetic Challenge**. It's all in the genes. We are ruled by behavioural imperatives that they determine (ultimately coming from our evolutionary history as hunter-gatherers).
3. **The Neurophysiology Challenge**. It's all determined bottom-up by our neurons and brain physiology, and consciousness is just an illusion floating on top of unconscious processes.
4. **The Behaviorist Challenge**. The mind is a mechanistic device operating according to predictable macroscopic rules. There is no need for any hypothesis about the effectiveness of mental states.
5. **The Sociology/Cultural Challenge**. What we do is determined by the society in which we live, which operates on a blank slate. Our understandings and social practices are written into us by cultural influences.
6. **The Theological/Fatalist Challenge**. God's will determines all, and/or everything is predetermined by fate.

Each of these has, over time, been stated pretty dogmatically by someone or other as the major determinant of our behaviour, indeed perhaps the only one. Recent attacks on free will include those by Wegner [255] and Harris [126]. Defenses of free will in the light of present day neuroscience are given in [66, 180, 247]. A balanced contemporary account is given in [149]. I will briefly discuss each of the threats listed in Table 7.2 except the first, which is non-scientific [149, pp. 147–162], and then comment on the issue in general and in particular on the necessity of a meaningful degree of free will if both daily life and the scientific enterprise are to proceed as normal (see Sect. 1.7).

Multiple Causes. The first point is that these six arguments against free will cannot all be right because they contradict each other.

> **The Single Causal Fallacy**. Claims that any single one of items 1–5 in Table 7.2 is the sole cause of brain activity is undermined by the fact that each of them can be demonstrated to be causally effective to some degree. Hence, arguments against free will based on the sole powers of any one of them on the brain is contradicted by the causal powers of the other factors in the list.

The fact that each of them has been strongly claimed as undermining free will suggests that each is a partial cause of brain function, but none is the whole cause. This undermines all the usual arguments against free will, which assume one or other of these factors is the only one acting. But each of the others is also significant. Those claiming any one of these is the sole determinant of what happens are ignoring other important causal factors.

Some of the alleged problems in Table 7.2 are based on top-down causation and some are bottom-up. The effects of bottom-up causation are now usually assumed to be the real threat to free will, but in the past some have assumed that the real threat was top-down. However, if all levels are equally real [189], as just discussed, the level of the mind is real even though it is influenced in both a bottom-up and a top-down way (see [180, pp. 238–305] and [247]). In particular, concerns regarding free will at the neuroscience level 3 need to be addressed. A fine refutation of such argumentation is given by Merlin Donald [66].

7.7.4.1 Argument 2: Society

It has been argued in this book that society has a major top-down effect on the human mind, as emphasized for example by Berger and Luckman [18] and Donald [66]. This is sometimes seen as enabled by operant conditioning [114], for example, carefully contrived by the state and media that are run by powerful corporations who mould what we think. The extreme case is put in Huxley's novel *Brave New World* [135], where one is, from conception, conditioned by the World State through a wide variety of means to take one's place happily in a stratified society. Your effective freedom is removed.

However, this is not the only kind of influence on the mind: while culture is crucially important, as emphasized by anthropologists and sociologists, and political manipulation can indeed take place, as emphasized by political scientists, biology is also key to how the human mind operates. The mind is an embodied mind. Ignoring genetics and neuroscience omits important causal factors in brain function. Furthermore, top-down influences from society are much more than operant conditioning. They can give one the intellectual tools to be an informed intelligent being, able to make meaningful choices [66].

7.7.4.2 Argument 3: Unconscious

Same level causation is seen as the key in the modern version of behaviourism, supported by the Libet experiments [121] on the one hand and much evidence of confabulation on the other [224, 255]. Solms emphasizes [223] that Freud talked about unconscious conflicts as causes of behavior, and Skinner talked about environmental contingencies, but either way we are not free to decide.[16] In his book

[16]A nice short summary is given by Seth Schwartz in Psychology Today: http://www.psychologytoday.com/blog/proceed-your-own-risk/201311/do-we-have-free-will.

The Illusion of Conscious Will [255], psychologist Daniel Wegner calls our impression of free will delusional, after-the-fact explanations (a form of confabulation).

Indeed, much is unconscious, as has been known at least since Freud [147, pp. 461–472], and automatisation is a crucial part of how the brain works (see Sect. 7.2.2.5), but that's not all that is going on. Libet's clock experiment is a poor probe of free will, because the subject has made the decision in advance to push the button, which is where the real choice was made. In the laboratory, she merely chooses when to push, which is a poor imitation of what is involved in meaningful decision making [65, 180, 181]. Noble et al. [190] describe the issue as follows:

> As regards intentions of an action, can a full characterization of the physical (neuronal, muscular) processes involved in the act of pointing count as having provided the neurophysiological basis of such an act in such a way that somehow explains away the idea of intention. Such a demonstration would be similar to the way in which the famous experiment by Libet et al. (1983) is often claimed to show that an earlier mechanical neuronal event was the real cause of an intention. The problem here is that an intention is not that kind of thing, if indeed it is a thing at all. The best that the neuroscientist in the story can do is to explain the neuronal and muscular events that occur during the movement of pointing. That may be spectacular as neurophysiology but it is not to explain the act of pointing since being an act precisely requires the possibility of intention. Having an intention is a process and such processes necessarily involve social interactions. Their explanation therefore becomes one that 'jumps out' of the context of neurophysiology to become an interpersonal question. [...] the relevant boundaries must be set to include the social and environmental interactions that enable intentions to arise and be fulfilled. Intentions cannot be interpreted as functions of a brain alone.

In fact, at times, we clearly make rational decisions which are what they seem to be, despite some biases in how we do so (which can be characterised in a rational way, as shown by recent economics literature [140]). They are based on the externalist nature of the brain/mind and in the distributed cognition that characterizes society [130]. The Libet kind of experiment plays no role.

The celebrated role of the unconscious [147, 224] is indeed one of the factors one must take into account, but it is just one of the factors in operation. If that were not the case there would be no point in engaging in conscious efforts at changing behaviour that are the concern of psychoanalysis [143–145], or prescribing drugs like chlorpromazine.

7.7.4.3 Argument 4: Evolution/Genetics

There are bottom-up proposals by evolutionary theorists that genes determined by our evolutionary history are in charge of our choices. They ignore top-down effects of epigenetics [106, 188, 189] and culture [18, 66] which are crucial in shaping what we do. They are certainly a factor but far from decisive on their own (see Sect. 7.3.7). They shape key influences but do not cause unique outcomes.

7.7.4.4 Argument 5: Neurons

The bottom-up view proposed by Crick [48], whereby neurons determine all, ignores top-down effects of epigenetics and culture, as in the previous case, and in any case is incoherent as a reductionist position (Sect. 7.7.3). The neural basis for free will is convincingly elucidated by Tse [247], as briefly discussed below (and see Sect. 7.7.3).

7.7.4.5 Argument 6: Physics

My physics colleagues believe that bottom-up effects from the underlying physics must uniquely determine the operation of the brain, and hence of the mind. The reason this does not work in practice is that we do not yet have enough computing power to carry out the program.

This not only ignores all the top-down effects considered in this chapter, but also the ontological randomness at the bottom quantum physical level, which undermines any claim of unique predictions from the physics side (see [247] and Sect. 8.1). This view is a certainly not taken seriously by neuroscientists (for example, it does not figure in Kandel's books [146, 147]): if it were true, their careers would not make sense.

7.7.4.6 A Viewpoint

In contrast to these various suggestions, the view put here has three pillars:

- Although bottom-up influences occur and enable the brain to function in terms of physics, chemistry, microbiology, genetics, and cell biology, the mind is not determined by these influences in a bottom-up way alone—top-down influences also occur.
- Although top-down influences occur and enable the brain to function in terms of relation to the outside world and society, the mind is not determined by the these effects in a top-down way alone, because bottom-up influences also occur.
- Same-level psychological processes can occur, including rational planning and decision-making, and exert both bottom-up and top-down influences on the mind, body, and world. They enable us to act meaningfully as human beings.

A particular feature here is that one can make decisions *not* to do something (sometimes called free won't), which fits in with the adaptive selection view I propose in this volume: we select what we wish to do from a range of options according to higher level selection criteria such as our sense of ethics and meaning. Brass and Haggard identify the corresponding neural correlate [28]:

> Our results suggest that the human brain network for intentional action includes a control structure for self-initiated inhibition or withholding of intended actions. The mental control of action has an enduring scientific interest, linked to the philosophical concept of 'free will'. Our results identify a candidate brain area that reflects the crucial decision to do or not to do.

Compatibilism. David Hume is widely recognized as providing the most influential statement of the 'compatibilist' position in the free will debate, i.e., the view that freedom and moral responsibility can be reconciled with causal determinism [208].

In his book *Freedom Evolves* [62], Dennett adopts a compatibilist position with an evolutionary twist, i.e., the view that, although in the strict physical sense our actions might be pre-determined, we can still be free in all the ways that matter, because of the abilities we evolved. Free will is simply our ability to perceive, mull over, and act upon choices, even though they are causally determined. Free will is both an 'objective phenomenon' and dependent on our belief in and perception of it, like language, music, money, and other products of society.

The present book by contrast, along with Merlin Donald [66], Peter Tse [247], and Julian Baggini [10] refutes the view that our actions might be predetermined in the strict physical sense. There are other causally effective entities in action than those that are encompassed by physics, including emotions, ideas, theories, social constructions, and intentions. It cannot make sense to suggest an aircraft or a teapot could come into existence unless this were true [76]. Dennett is supposing all causation is bottom-up, but that is not the case.

John Horgan in *Cross-Check*,[17] replying to Harris' book [126], explains this nicely:

> My choices are constrained, by the laws of physics, my genetic inheritance, upbringing and education, the social, cultural, political, and intellectual context of my existence. And as Harris keeps pointing out, I didn't choose to be born into this universe, to my parents, in this nation, at this time. I don't choose to grow old and die. But just because my choices are limited doesn't mean they don't exist. Just because I don't have absolute freedom doesn't mean I have no freedom at all. Saying that free will doesn't exist because it isn't absolutely free is like saying truth doesn't exist because we can't achieve absolute, perfect knowledge. Harris keeps insisting that because all our choices have prior causes, they are not free; they are determined. Of course all our choices are caused. No free-will proponent I know claims otherwise. The question is how are they caused?
>
> Harris seems to think that all causes are ultimately physical, and that to hold otherwise puts you in the company of believers in ghosts, souls, gods, and other supernatural nonsense. But the strange and wonderful thing about all organisms, and especially our species, is that mechanistic physical processes somehow give rise to phenomena that are not reducible to or determined by those physical processes. Human brains, in particular, generate human minds, which while subject to physical laws are influenced by non-physical factors, including ideas produced by other minds. These ideas may cause us to change our minds and make decisions that alter the trajectory of our world. [...] We are physical creatures, but we are not just physical. We have free will because we are creatures of mind, meaning, ideas, not just matter.

A Pseudo-Problem. Daniel Kaufman commenting on two postings by Gregg Caruso on free will in a blog,[18] responds as follows:

[17] http://blogs.scientificamerican.com/cross-check/2012/04/09/will-this-post-make-sam-harris-change-his-mind-about-free-will/.

[18] Scientia Salon, 23 December 2014: 7:15 pm http://scientiasalon.wordpress.com/2014/12/23/free-will-skepticism-and-its-implications-an-argument-for-optimism-part-2/.

> The ascription of agency is not the attribution of some power or force to an actor. Rather, it is a background condition for the intelligibility of not just moral discourse but for the ascription of intentionality and intentional explanation. It is not, therefore, subject to rational justification, but rather, is part of the framework of concepts by which we justify/or condemn action. [...] There are no free will skeptics, only pretend ones. Every time one utters a sentence like 'you should have called instead of standing me up', one has implicitly ascribed agency and agency must be presupposed for the utterance to be intelligible. The author utters scores of sentences like these a day. His skepticism, therefore, is of the variety described and put in its proper place by Hume—the sort that cannot survive even a day's common activity and discourse—the sort that disappears, once we've played a game of backgammon, talked with our friends, and gone for a stroll.

In response to the worry that 'my neurons made me do it', hence we are not responsible for our actions, Murphy and Brown respond as follows [180, pp. 13–14]:

- Organisms are often the causes of their own behaviour.
- Humans are capable of using and understanding the meaning of language.
- Humans act for reasons, not merely on the basis of causes.
- Mature humans are able to act on the basis of moral concepts.

They write [180, pp. 3–4]:

> Our goal here is not to argue that humans are rational and responsible. We could not make sense of what we are doing writing this book if we did not assume common sense views of the role of reason and reasons, that is, of our having good reasons for writing it and of our readers accepting our conclusion on the basis of reason. Rather we hope to show how our neurobiological equipment makes rationality, responsibility, and free will possible.

This chapter has made the case that this is indeed possible because of the occurrence of top-down causation in the brain and in the relation of the brain to society.

7.7.4.7 The Neural Underpinnings

The neural underpinnings of free will are discussed illuminatingly by Tse [247], basing his thesis on the idea of criterial causation, where neurons fire on the basis of the present wiring, but act to change the future wiring. Thus adaptive plasticity is happening all the time as we think. Furthermore, ontological uncertainty at the foundations of physics ensures that neural outcomes cannot be uniquely prescribed by the underlying physics alone. Indeed, randomness plays a key role in the brain, as emphasized by Glimcher [109], because it provides an ensemble of options from which we can choose, and this is the foundation of adaptive selection.

Overall. The point is that the existence of both bottom-up and top-down causation in relation to the mind and brain removes the force of the bottom-up arguments against free will that are now fashionable in some quarters, and opens up the way for more humane visions of the nature of humanity [66].

7.7.5 *Neuroscience and Humanity*

What is at stake is our conception of humanity. How we see ourselves influences culture, society, and politics. There are strong reductionist attacks on commonsense views of humanity, and some vigorous responses that aim at a broader view.[19]

7.7.5.1 Reductionist Attacks

Overclaims for reductive neuroscience are being made all the time, as described by Sally Satel and Scott Lilienfeld in *Brainwashed: The Seductive Appeal of Mindless Neuroscience* [212]. They are based on strong reductionist views of humanity. Semir Zeki, head of the UCL Institute of Neuroesthetics University College London, writes as follows:

> It is only by understanding the neural laws that dictate human activity in all spheres—in law, morality, religion and even economics and politics, no less than in art—that we can ever hope to achieve a more proper understanding of the nature of man.

The vision is that neuroscientists will take over economics as well as ethics and aesthetics, which are seen as having nothing to do with philosophy or the human sciences.[20] This implies that reductive views on the nature of humanity such as that by Crick [48] quoted above (Sect. 7.7.3), and among others the writings of Daniel Dennett [61] and Susan Blackmore [22, p. 237]:

> All human actions, whether conscious or not, come from complex interactions between memes, genes and all their products, in complicated environments. The self is not the initiator of actions, it does not 'have' consciousness, and it does not 'do' the deliberating. There is no truth in the idea of an inner self inside my body that controls the body and is conscious. Since this is false, so is the idea of my conscious self having free will.

7.7.5.2 The Response

Merlin Donald's *A Mind So Rare* is a comprehensive response to these kinds of attacks. Building on evolutionary principles, cognitive science, developmental psychology, and cultural anthropology he makes a convincing case for consciousness as the central core of the mind. More than this, Donald also argues that we cannot divorce 'mind' from its cultural context. Mind is a product both of genetic and environmental forces and also a by-product of human culture.

He responds to the hardline position as follows [66, pp. 29, 36]:

> Hardliners, led by a vanguard of rather voluble philosophers, believe not merely that consciousness is limited, as experimentalists have been saying for years, but that it plays no significant role in human cognition. They believe that we think, speak, and remember entirely

[19]The reader should note that this final section is of a polemical nature. But it has not abandoned science: it is solidly rooted in the underlying science.

[20]See "Neuroaesthetics is killing your soul" by Philip Ball [9] for a strong response.

outside its influence. Moreover, the use of the term 'consciousness' is viewed as pernicious because (note the theological undertones) it leads us into error. They support the downgrading of consciousness to the status of an epiphenomenon, a superficial manifestation of mental activity that plays no role in cognition.

Or again [66, pp. 31, 35]:

Dennett is actually denying the biological reality of the self. Selves, he says, hence self-consciousness, are cultural inventions—the initiation and execution of mental activity is always outside conscious control—consciousness is an illusion and we do not exist in any meaningful sense. But, they apologize at great length, this daunting fact does not matter. Life will go on as always, meaningless algorithm after meaningless algorithm, and we can all return to our lives as if nothing has happened. This is rather like telling you your real parents were not the ones you grew to know and love but Jack the Ripper and Elsa, She-Wolf of the SS. But not to worry.

The practical consequences of this deterministic crusade are terrible indeed. There is no sound biological or ideological basis for selfhood, willpower, freedom, or responsibility. The notion of the conscious life as a vacuum leaves us with an idea of the self that is arbitrary, relative, and much worse, totally empty because it is not really a conscious self, at least not in any important way.

7.7.5.3 The Opposing Views

In essence we have two radically opposed concepts of humanity (Strawson [234] gives a very nice historical overview of their interaction). One view sees us as biologically programmed by DNA, operating largely on an unconscious level with our minds creating an illusion of consciousness and hence an image of self. We are nothing but biological machines, with the emphasis that 'machine' has on purposelessness and mindlessness.

The terrible danger is that if we truly see people in that way, it will inevitably eventually come to shape the way we treat people. If people are *nothing but* machines, why should we look after the sick and the elderly, or Down's syndrome children? However, much it may be denied, this is certainly the logical implication of that worldview.

7.7.5.4 The Holistic View

In contrast, the holistic view developed by Donald [66], Zeman [258], Murphy and Brown [180], Humphreys [133], Kandel [147], Kagan [138], Tse [247], and others, and supported in this book, is in harmony with the great traditions of humanity, ethics, and philosophy [147]. It is that while much brain activity is unconscious, we are nevertheless fully self-conscious beings, capable of rational choice and engaged in cultural communities which extend our cognitive abilities in order to fully express our shared humanity.

Our intentionality is not an illusion: we are intentional beings able to undertake moral action. We can lead a full human life precisely because we are not purposeless machines. We are beings with values and purpose that guide our actions and our lives, and a living spirit that shines through our being.

The issue can be crystallised by considering the following passage from *Flight to Arras* by Antoine de Saint-Exupéry [53], and a photograph by Sid Luckett that expresses the same sentiment in a very different medium. Saint-Exupéry writes [53]:

> I say to myself as I watch the niece, who is very beautiful: in her this bread is transmuted into melancholy grace. Into modesty, into a gentleness without words [...] Sensing my gaze, she raised her eyes towards mine, and seemed to smile [...] A mere breath on the delicate face of the waters, but an affecting vision. I sense the mysterious presence of the soul that is unique to this place. It fills me with peace, and my mind with the words: 'This is the peace of silent realms.' I have seen the shining light that is born of the wheat.

The photo of a woman (Fig. 7.17) conveys the same feeling. Either one sees in that passage and photo something meaningful, or one does not. If one does, those who proclaim that the mind is based simply on meaningless algorithms, implying meaning is an illusion, are dead wrong. This chapter has sought to provide a footing for this opposite view.

Fig. 7.17 Portrait. Credit: Sid Luckett, with permission

References

1. A. Adhikari, T.N. Lerner, J. Finkelstein, S. Pak, J.H. Jennings, T.J. Davidson, E. Ferenczi, L.A. Gunaydin, J.J. Mirzabekov, L. Ye, S.-Y. Kim, A. Lei, K. Deisseroth, Basomedial amygdala mediates top-down control of anxiety and fear. Nature **527**, 179–185 (2015)
2. U. Alon, *An Introduction to Systems Biology: Design Principles of Biological Circuits* (Chapman and Hall /CRC, London, 2007)
3. D.M. Amodio, C.D. Frith, Meeting of minds: the medial frontal cortex and social cognition. Nat. Rev. Neurosci. **7**, 268–277 (2006)
4. M. Arbib, *The Handbook of Brain Theory and Neural Networks* (MIT Press, Cambridge, Mass, 1998)
5. R. Ardrey, *The Territorial Imperative: A Personal Inquiry into the Animal Origins of Property and Nations* (Kodansha, 1966)
6. O. Arias-Carrión, M. Stamelou, E. Murillo-Rodríguez, M. Menéndez-González, E. Pöppel, Dopaminergic reward system: a short integrative review. Int. Arch. Med. **3**, 24 (2010). http://www.biomedcentral.com/1755-7682/3/24/
7. D. Ariely, *Predictably Irrational: The Hidden Forces that Shape Our Decisions* (Harper, 2009)
8. G. Auletta, G. Ellis, L. Jaeger, Top-down causation: from a philosophical problem to a scientific research program. J. R. Soc. Interface **5**, 1159–1172 (2008). arXiv:0710.4235
9. P. Ball, Neuroaesthetics is killing your soul. Nat. News Comment (2014). http://www.nature.com/news/neuroaesthetics-is-killing-your-soul-1.12640
10. J. Baggini, *Freedom Regained: The Possibility of Free Will* (University of Chicago Press, 2015)
11. A.M. Bastos, W.M. Usrey, R.A. Adams, G.R. Mangun, P. Fries, K.J. Friston, Canonical microcircuits for predictive coding. Neuron **76**, 695–711 (2012)
12. M. Beasley, D. Sabatinelli, E. Obasi, Neuroimaging evidence for social rank theory. Front. Hum. Neurosci. (2012). doi:10.3389/fnhum.2012.00123
13. M.A. Bedau, P. Humphreys (eds.), *Emergence: Contemporary Readings in Philosophy and Science* (MIT Press, Cambridge, Mass, 2008)
14. S. Beer, *Brain of the Firm* (Wiley, Chichester, 1981)
15. M.E. Beer, B.W. Connors, M.A. Paradiso, *Neuroscience: Exploring the Brain* (Lippincot, Williams and Wilkins, Philadelphia, 2007)
16. B.K. Bergen, N. Chang, Embodied construction grammar in simulation based language understanding, in *Construction Grammars(s): Cognitive and Cross-Language Dimensions*, ed. by J.O. Östman, M. Fried (John Benjamins, Amsterdam, 2003), pp. 147–190
17. P.L. Berger, *Invitation to Sociology* (Anchor books, 1963)
18. P.L. Berger, T. Luckmann, *The Social Construction of Reality: A Treatise in the Sociology of Knowledge* (Penguin, London, 1991)
19. D. Bickerton, *Language and Human Behaviour* (University of Washington Press, Seattle, 2001)
20. D. Bickerton, *More than Nature Needs: Language, Mind, and Evolution* (Harvard University Press, Harvard, 2014)
21. C.M. Bishop, *Neural Networks for Pattern Recognition* (Oxford University Press, Oxford, 1999)
22. S. Blackmore, *The Meme Machine* (Oxford University Press, 1996)
23. C. Bloch, *Chloe's Story: First Steps to Literacy* (Juta, 1997)
24. D.T. Blumstein et al., Toward an integrative understanding of social behavior: new models and new opportunities. Front. Behav. Neurosci. (2010). doi:10.3389/fnbeh.2010.00034
25. M. Boden, *The Creative Mind: Myths and Mechanisms* (Abacus, 1994)
26. G. Booch, *Object-Oriented Analysis and Design with Applications* (Addison-Wesley, Menlo Park, 2007)
27. C. Booker, *The Seven Basic Plots: Why We Tell Stories* (Bloomsbury, London, 2014)
28. M. Brass, P. Haggard, To do or not to do: the neural signature of self-control. J. Neurosci. **27**, 9141–9145 (2007)

29. N. Bray, Influences from above on memory. Nat. Rev. Neurosci. **16**, 703 (2015)
30. S.L. Bressler, A.R. McIntosh, The role of neural context in large-scale neurocognitive network operations, in *Springer Handbook of Brain Connectivity* (Springer, 200), pp. 403–419
31. S.L. Bressler, W. Tang, C.M. Sylvester, G.L. Shulman, M. Corbetta, Top-down control of human visual cortex by frontal and parietal cortex in anticipatory visual spatial attention. J. Neurosci. **28**, 10056–10061 (2008)
32. D. Bristow, G. Rees, C.D. Frith, Social interaction modifies neural response to gaze shifts. Soc. Cogn. Affect Neurosci. **2**, 52–61 (2007)
33. J. Bronowski, *The Ascent of Man* (Little Brown and Co, 1973)
34. J.T. Buhle, B.L. Stevens, J.J. Friedman, T.D. Wager, Distraction and placebo: two separate routes to pain control. Psychol. Sci. **23**, 246–253 (2012)
35. G. Buzsáki, *Rhythms of the Brain* (Oxford University Press, Oxford, 2006)
36. J.T. Cacioppo, G.G. Berntson, R. Adophs, C.S. Carter, R.J. Davidson, M.K. McClintock, B.S. Mcewan, M.J. Meaney, D.L. Schacter, E.M. Sternberg, S.S. Suomi, S.E. Taylor (eds.), *Foundations in Social Neuroscience* (MIT Press, Cambridge, Mass, 2002)
37. J.T. Cacioppo, J. Decety, Social neuroscience: challenges and opportunities in the study of complex behavior. Ann. N. Y. Acad. Sci. **1224**, 162–173 (2011)
38. S.D. Cardoso, M.C. Teles, R.F. Oliveira, Neurogenomic mechanisms of social plasticity. J. Exp. Biol. **218**, 140–149 (2015)
39. P. Carruthers, J.B. Ritchie, P.M. Churchland, Plato's camera: how the physical brain captures a landscape of abstract universals. Notre Dame Philos. Rev. (2012). http://ndpr.nd.edu/news/32035-plato-s-camera-how-the-physical-brain-captures-a-landscape-of-abstract-universals/
40. J.-P. Changeux, A. Connes, *Conversations on Mind, Matter, and Mathematics* (Princeton University Press, 1995)
41. J.-P. Changeaux, P. Courrege, A. Danchin, A theory of the epigenesis of neuronal networks by selective stabilization of synapses. Proc. Nat. Acad. Sci. USA **70**, 2974–2978 (1973)
42. W. Choi, R.H. Desai, J.M. Henderson, The neural substrates of natural reading: a comparison of normal and non-word text using eyetracking and fMRI. Front. Hum. Neurosci. **8**, Article 1024 (2014)
43. P.M. Churchland, *Plato's Camera: How the Physical Brain Captures a Landscape of Abstract Universals* (MIT Press, Cambridge, Mass, 2013)
44. A. Clark, *Supersizing the Mind: Embodiment, Action, and Cognitive Extension* (Oxford University Press, Oxford, 2008)
45. M.C. Corballis, The evolution of language: from hand to mouth, in *Evolutionary Cognitive Neuroscience*, ed. by S.M. Platek, J.P. Keenan, T.D. Shackelford (MIT Press, Cambridge, Mass, 2007), pp. 403–432
46. L. Cozolino, *The Neuroscience of Human Relationships* (W W Norton, 2006)
47. R.P. Crease, The paradox of trust in science. Phys. World **18** (2004)
48. F. Crick, *Astonishing Hypothesis: The Scientific Search for the Soul* (Scribner, 1995)
49. A. Damasio, *Descarte's Error* (Harper Collins, 2000)
50. A. Damasio, *The Feeling of What Happens: Body, Emotion and the Making of Consciousness* (Vintage, London, 2000)
51. A. Dardis, *Mental Causation: The Mind-Body Problem* (Columbia University Press, New York, 2008)
52. R.J. Davidson, B.S. McEwen, Social influences on neuroplasticity: stress and interventions to promote well-being. Nat. Neurosci. **15**, 689–695 (2012)
53. A. de Saint-Exupéry, *Flight to Arras* (Reynal and Hitchcock, New York, 1942)
54. T. Deacon, *The Symbolic Species: The Co-evolution of Language and the Human Brain* (Penguin Books, London, 1997)
55. T. Deacon, Universal grammar and semiotic constraints, in *Language Evolution*, ed. by M. Christiansen, S. Kirby (Oxford University Press, Oxford, 2003), pp. 111–139
56. J. Decety, W. Ickes, *The Social Neuroscience of Empathy* (MIT Press, Cambridge, Mass, 2009)

57. J. Decety, P.L. Jackson, A social neuroscience perspective on empathy. Curr. Dir. Psychol. Sci. **15**, 54–58 (2006)
58. G. Deco, E.T. Rolls, Neurodynamics of biased competition and cooperation for attention: a model with spiking neurons. J. Neurophysiol. **94**, 295–313 (2005)
59. G. Deco, E.T. Rolls, Attention, short-term memory, and action selection: a unifying theory. Prog. Neurobiol. **76**, 236–256 (2005)
60. S. Dehaene, *Reading in the Brain: The New Science of How We Read* (Penguin, London, 2010)
61. D. Dennett, Quining qualia, in *Mind and Cognition: A Reader*, ed. by W. Lycan (Blackwell, Oxford, 1990)
62. D. Dennett, *Freedom Evolves* (Penguin Books, 2004)
63. D. Deutsch, *The Fabric of Reality* (Penguin Books, 1998)
64. M. Donald, The neurobiology of human consciousness: an evolutionary approach. Neuropsychologia **33**, 1087–1102 (1995)
65. M. Donald, The central role of culture in cognitive evolution: a reflection on the myth of the isolated mind, in *Culture, Thought and Development*, ed. by L. Nucci (Lawrence Erlbaum Associates, 2000), pp. 19–38
66. M. Donald, *A Mind so Rare: The Evolution of Human Consciousness* (W W Norton, New York, 2001)
67. M. Donoso, A.G.E. Collins, E. Koechlin, Foundations of human reasoning in the pre-frontal cortex. Science **344**, 1481–1489 (2015)
68. R.J. Douglas, K.A.C. Martin, Neuronal circuits of the neocortex. Annu. Rev. Neurosci. **27**, 419–451 (2004)
69. J. Driver, R.S.J. Frackowiak, Neurobiological measures of human selective attention. Neuropsychog **39**, 1257–1262 (2001)
70. R.I.M. Dunbar, The social brain: mind, language and society in evolutionary perspective. Ann. Rev. Anthropol. **32**, 163–181 (2003)
71. R. Dunbar, *Human Evolution* (Pelican Books, London, 2014)
72. G.M. Edelman, *Neural Darwinism: The Theory of Group Neuronal Selection* (Oxford University Press, Oxford, 1989)
73. G.M. Edelman, *Brilliant Air, Brilliant Fire: On the Matter of Mind* (Basic Books, New York, 1992)
74. G.M. Edelman, G. Tononi, *Consciousness: How Matter Becomes Imagination* (Penguin Books, London, 2001)
75. D. Elder-Vass, *The Causal Power of Social Structures: Emergence, Structure and Agency* (Cambridge University Press, Cambridge, 2010)
76. G.F.R. Ellis, Physics, complexity, and causality. Nature **435**, 743 (2005)
77. G.F.R. Ellis, Top-down causation and emergence: some comments on mechanisms. J. R. Soc. Interface Focus **2**, 126–140 (2012)
78. G.F.R. Ellis, Multi-level selection in biology. http://lanl.arxiv.org/abs/1303.4001
79. G.F.R. Ellis, D. Noble, T. O'Connor, Top-down causation: an integrating theme across the sciences? Interface Focus **2**, 1–3 (2012)
80. G.F.R. Ellis, W.R. Stoeger, Language infinities. http://www.mth.uct.ac.za/~ellis/Language
81. G.F.R. Ellis, J.A. Toronchuk, Neural development: affective and immune system influences, in *Consciousness and Emotion*, ed. by R.D. Ellis, N. Newton (John Benjamins, 2005), pp. 81–119
82. J.A. Feldman, *From Molecule to Metaphor: A Neural Theory of Language* (MIT Press, Cambridge, Mass, 2008). http://www.icsi.berkeley.edu/NTL/
83. C. Fernando, E. Szathmáry, Natural selection in the brain, in *Towards a Theory of Thinking on Thinking* (Springer, 2010), pp. 291–322
84. C. Fernando, E. Szathmáry, P. Husbands, Selectionist and evolutionary approaches to brain function: a critical appraisal. Front. Comput. Neurosci. **6**, 24 (2012)
85. A.D. Flurkey, E.J. Paulson, K.S. Goodman, *Scientific Realism in Studies of Reading* (Lawrence Erlbaum, 2008)

86. B. Flyvbjerg, T. Landman, S. Schram, *Real Social Science: Applied Phronesis* (Cambridge University Press, Cambridge, 2012)
87. R.W. Folensbee, *The Neuroscience of Pyschological Therapies* (Cambridge University Press, Cambridge, 2007)
88. V. Frankl, *Man's Search for Meaning* (Rider and Co, 2008)
89. S.M. Frankland, J.D. Greene, An architecture for encoding sentence meaning in left-superior temporal cortex. PNAS **112**, 11732–11737 (2015)
90. S.S. Franklin, *The Psychology of Happiness: A Good Human Life* (Cambridge University Press, Cambridge, 2010)
91. J.B. Freeman, N.O. Rule, R.B. Adams, N. Ambady, Culture shapes a mesolimbic response to signals of dominance and subordination that associates with behavior. NeuroImage **47**, 353–359 (2008)
92. G. Frege, The thought: a logical inquiry. Mind **65**, 289–311 (1956)
93. B.S. Frey, Happy people live longer. Science **331**, 542 (2011)
94. P.H. Fries, Words, context and meaning in reading, in *Scientific Realism in Studies of Reading*, ed. by A. Flurkey, K. Goodman, E. Paulson (NJ, Lawrence Erlbaum, Mahwah, 2008), pp. 53–82
95. K. Friston, Hierarchical models in the brain. PLoS Comput. Biol. **4**, e1000211 (2008)
96. K. Friston, The free-energy principle: a unified brain theory. Nat. Rev. Neurosci. **11**, 127–138 (2010)
97. K. Friston, K.E. Stephan, Free-energy and the brain. Synthese **159**, 417–458 (2007)
98. K.J. Friston, G. Tononi, G.N. Reeke, O. Sporns, G.M. Edelman, Value-dependent selection in the brain: simulation in a synthetic neural model. Neuroscience **59**, 229–243 (1994)
99. C. Frith, *Making up the Mind: How the Brain Creates our Mental World* (Blackwell, Malden, 2007)
100. C. Gamble, *Settling the Earth: The Archaeology of Deep Human History* (Cambridge University Press, Cambridge, 2013)
101. D. Garisch, *Eloquent Body* (Modjaji Books, 2012)
102. A. Gazzaley, A.C. Nobre, Top-down modulation: bridging selective attention and working memory. Trends Cogn. Sci. **16**, 129–135 (2012)
103. M. Gazzaniga, *Human: The Science of What Makes Your Brain Unique* (Harper, New York, 2008)
104. S.A. Gerson, N. Mahajan, J.A. Sommerville, L. Matz, A.L. Woodward, Shifting goals: effects of active and observational experience on infants: understanding of higher order goals. Front. Psychol. Dev. Psychol. (2015). doi:10.3389/fpsyg.2015.00310
105. M. Gershon, *The Second Brain: A Groundbreaking New Understanding of Nervous Disorders of the Stomach and Intestine* (Harper Perennial, 1999)
106. S.F. Gilbert, D. Epel, *Ecological Developmental Biology* (Sinhauer, Sunderland, Mass, 2009)
107. C.D. Gilbert, W. Li, Top-down influences on visual processing. Nat. Rev. Neurosci. **14**, 351 (2013)
108. C.D. Gilbert, M. Sigman, Brain states: top-down influences in sensory processing. Neuron **54**, 677–696 (2007)
109. P.W. Glimcher, Indeterminacy in brain and behavior? Annu. Rev. Psychol. **56**, 25–56 (2005)
110. K.S. Goodman, Reading: a psycholinguistic guessing game. J. Read. Spec. **6**, 126–135 (1967)
111. K. Goodman, *What's Whole in Whole Language: 20th*, Anniversary edn. (RDR Books, Muskegon, MI, 2005)
112. K. Goodman, P. Fries, S. Strauss, E. Paulson, *Reading: The Grand Illusion. How and Why Readers Make Sense of Print* (Routledge, 2016)
113. J. Gottschalk, *The Story-Telling Animal: How Stories Make Us Human* (Mariner books, 2012)
114. P. Gray, *Psychology* (Worth, New York, NY, 2011)
115. R.J. Greenspan, *An Introduction to Nervous Systems* (Cold Spring Harbor Laboratory Press, Cold Spring Harbor, 2007)
116. S. Greenspan, S. Shanker, *The First Idea: How Symbols, Language, and Intelligence Evolved from Our Primate Ancestors to Modern Humans* (Da Capo Press, Cambridge, Mass, 2004)

117. P.E. Griffiths, K. Stotz, How the mind grows: a developmental perspective on the biology of cognition. Synthese **122**(1–2), 29–51 (2000)
118. P. Grossman, L. Niemann, S. Schmidt, H. Walach, Mindfulness-based stress reduction and health benefits: a meta-analysis. J. Psychosom. Res. **57**, 35–43 (2004)
119. A.J. Gruber, R.J. McDonald, Context, emotion, and the strategic pursuit of goals: Interactions among multiple brain systems controlling motivated behavior. Front. Behav. Neurosci. **03** (2012)
120. D.A. Hackman, M.J. Farah, Socioeconomic status and the developing brain. Trends Cogn. Sci. **13**, 65–73 (2008)
121. P. Haggard, B. Libet, Conscious intention and brain activity. J. Conscious. Stud. **8**, 47–63 (2001)
122. M.N. Haidle, Building a bridge: an archeologist's perspective on the evolution of causal cognition. Front. Psychol. **5**, Article 1472 (2014)
123. J. Haidt, S. Kesebir, Morality, in *Handbook of Social Psychology*, ed. by S. Fiske, D. Gilbert, G. Lindzey (Wiley, Hoboken, NJ, 2010)
124. E. Harmon-Jones, P. Winkielman, *Social Neuroscience* (Guilford, 2007)
125. K.D. Harris, G.M. Shepherd, The neocortical circuit: themes and variations. Nat. Neurosci. **18**, 170–181 (2015)
126. S. Harris, *Free Will* (Free Press, 2012)
127. L.H. Hartwell, J.J. Hopfield, S. Leibler, A.W. Murray, From molecular to modular cell biology. *Nature* **402**, C47–C52 (1999). Supplement (2 December 1999)
128. M.D. Hauser, N. Chomsky, W.T. Fitch, The faculty of language: what is it, who has it, and how did it evolve? Science **298**, 1569–1579 (2002)
129. J. Hawkins, *On Intelligence* (Holt Paperbacks, New York, NY, 2004)
130. C. Herrmann-Pillath, *Foundations of Evolutionary Economics* (Max Weber Center for Advanced Cultural and Social Studies, University of Erfurt) (2011). http://ssrn.com/abstract=1776652
131. E.P. Hoel, L. Albantakis, G. Tononi, Quantifying causal emergence shows that macro can beat micro. PNAS **110**, 19790–19795. http://www.pnas.org/content/110/49/19790.abstract
132. N. Humphrey, Placebo effect, in *Oxford Companion to the Mind* (Quercus, 2005)
133. N. Humphrey, *Soul Dust: The Magic of Consciousness* (Quercus, 2012)
134. N. Humphrey, J. Skoyles, The evolutionary psychology of healing: a human success story. Curr. Biol. **22**, 8–R695 (2012)
135. A. Huxley, *Brave New World* (Rosetta Books, 2010)
136. H.-A. Jeon, Hierarchical processing in the prefrontal cortex in a variety of cognitive domains. Front. Syst. Neurosci. **8**, Article 223, 1–8 (2014)
137. A. Juarrero, *Dynamics in Action: Intentional Behavior as a Complex System* (Bradford Books, 2002)
138. J. Kagan, *The Human Spark: The Science of Human Development* (Basic Books, 2013)
139. N.L. Kamorova, M.A. Nowak, Language, learning, and evolution, in *Language Evolution*, ed. by M.H. Christensen, S. Kirby (Oxford University Press, Oxford, 2005), pp. 317–337
140. D. Kahneman, *Thinking Fast, Thinking Slow* (Farrar, Straus and Giroux, 2011)
141. D. Kahneman, P. Slvic, A. Tversky, *Judgement under Uncertainty: Heuristics and Biases* (Cambridge University Press, Cambridge, 1982)
142. D. Kahneman, A. Tversky, *Choices, Values and Frames* (Cambridge University Press, Cambridge, 2000)
143. E. Kandel, A new intellectual framework for psychiatry. Am. J. Psych. **155**, 457–469 (1998). reprinted in E. Kandel, *Psychiatry, Psychoanalysis, and the New Biology of Mind* American Psychiatric Publishing, Washington, DC, 2005), pp. 33–58
144. E. Kandel, Biology and the future of psychoanalysis: a new intellectual framework for psychiatry revisited. Am. J. Psychiatric **156**, 505–524 (1999). reprinted in E. Kandel, *Psychiatry, Psychoanalysis, and the New Biology of Mind* (American Psychiatric Publishing, Washington, DC, 2005), pp. 63–106

145. E. Kandel, *Psychiatry, Psychoanalysis, and the New Biology of Mind* (American Psychiatric Publishing, Washington, DC, 2005)
146. E. Kandel, *In Search of Memory: The Emergence of a New Science of Mind* (W W Norton, New York, 2006)
147. E. Kandel, *The Age of Insight: The Quest to Understand the Unconscious in Art, Mind, and Brain, from Vienna 1900 to the Present* (Random House, 2012)
148. E. Kandel, J.H. Schwartz, T.M. Jessell, S.A. Siegelbaum, A.J. Hudspeth, *Principles of Neural Science* (McGraw Hill Professional, 2013)
149. R. Kane, *A Contemporary Introduction to Free Will* (Oxford University Press, 2005)
150. G. Kelly, *A Theory of Personality: The Psychology of Personal Constructs* (Norton, 1963)
151. G. Kempermann, H.G. Kuhn, F.H. Gage, More hippocampal neurons in adult mice living in an enriched environment. Nature **386**, 493–495 (1997)
152. R.R. Kerr, D.B. Grayden, D.A. Thomas, M. Gilson, A.N. Burkitt, Goal-directed control with cortical units that are regulated by both top-down feedback and oscillatory coherence. Front. Neural Circuits (2014). doi:10.3389/fncir.2014.00094
153. C. Keyworth, J. Knopp, K. Roughley, C. Dickens, S. Bold, P. Coventry, A mixed-methods pilot study of the acceptability and effectiveness of a brief meditation and mindfulness intervention for people with diabetes and coronary heart disease. Behav. Med. **40**, 53–64 (2014)
154. R. Kingsley, *Concise Text of Neuroscience* (Lippinscott, Williams and Wilkins, Philadelphia, 2000)
155. J. Kleijnen, A.J.M. de Craen, J. van Everdingen, L. Krol, Placebo effect in double-blind clinical trials: a review of interactions with medications. Lancet **344**, 1347–1349 (1994)
156. C. Koch, *The Quest for Consciousness: A Neurobiological Approach* (Roberts and Company, 2004)
157. S.D. Krashen, T.D. Terrell, *The Natural Approach: Language Acquisition in the Classroom* (Alemany Press, San Francisco, 1983)
158. B.-C. Kuo, M.G. Stokes, A.M. Murray, A.C. Nobre, Attention biases visual activity in visual short-term memory. J. Cogn. Neurosci. **26**, 1377–1389 (2014)
159. G. Lakoff, Mapping the brain's metaphor circuitry: metaphorical thought in everyday reason. Front. Hum. Neurosci. **8**, Article 958 (2014)
160. G. Lakoff, M. Johnson, *Metaphors We Live by* (University of Chicago Press, Chicago, 1980)
161. G. Lakoff, M. Johnson, *Philosophy in the Flesh: The Embodied Mind and Its Challenge to Western Thought* (Basic Books, 1980)
162. G. Layher, F. Schrodt, M.V. Butz, H. Neumann, Adaptive learning in a compartmental model of visual cortex: How feedback enables stable category learning and refinement. Front. Psychol. **05** (2014). fpsyg.2014.01287
163. I. Lee, C.-H. Lee, Contextual behavior and neural circuits. Front. Neural Circuits (2013). doi:10.3389/fncir.2013.00084
164. Y. Li, J. Yosinski, J. Clune, H. Lipson, J. Hopcroft, Convergent learning: do different neural networks learn the same representations? (2015). http://arxiv.org/abs/1511.07543
165. V. LoBue, T. Nishida, C. Chiong, J.S. DeLoache, J. Haidt, When getting something good is bad: even three-year-olds react to inequality. *Social Development* (Blackwell, 2009)
166. J.E. Longres, *Human Behaviour in the Social Environment* (F E Peacock, 1990)
167. W. Lotter, G. Kreiman, D. Cox, Unsupervised learning of visual structure using predictive generative networks (2015). http://arxiv.org/abs/1511.06380
168. J. MacCormack, *9 Algorithms that Changed the Future: The Ingenious Ideas that Drive Today's Computers* (Princeton University Press, Princeton, 2012)
169. E.A. Maguire, K. Woollett, H.J. Spiers, London taxi drivers and bus drivers: a structural MRI and neuropsychological analysis. HIPPOCAMPUS **16**, 1091–1101 (2006)
170. M.M. Mano, C.R. Kime, *Logic and Computer Design Fundamentals* (Pearson/Prentice Hall, 2008)
171. D. Marr, *Vision: A Computational Investigation Into the Human Representation and Processing of Visual Information* (WH Freeman, San Francisco, 1982)

172. D. Marr, T. Poggio, From understanding computation to understanding neural circuitry. MIT Artif. Intell. Lab: Memo **357** (1976)
173. M.P. Mattson, Superior pattern processing is the essence of the evolved human brain. In: Front. Neurosci. *22* (2014). doi:10.3389/fnins.2014.00265
174. H.S. Mayberg, J.A. Silva, S.K. Brannan, J.L. Tekell, R.K. Mahurin, S. McGinnis, P.A. Jerabek, The functional neuroanatomy of the placebo effect. Am. J. Psychiatry **159**, 728–737 (2002)
175. B.S. McEwen, Central effects of stress hormones in health and disease: understanding the protective and damaging effects of stress and stress mediators. Eur. J. Pharmacol. **583**, 174–185 (2008)
176. L.F.F. Meyer, M.J. Braga, Cognition and norms: Toward a developmental account of moral agency in social dilemmas. Front. Psychol. **5**, Article 1528 (2015)
177. G. Miller, Why loneliness can be hazardous for your health. Science **331**, 138–140 (2011)
178. A.H. Modell, *Imagination and the Meaningful Brain* (MIT Press, 2003)
179. E. Morsella, M. Lanska, C.C. Berger, A. Gazzaley, Indirect cognitive control through top-down activation of perceptual symbols. Eur. J. Soc. Psychol. **39**, 1173–1177 (2009)
180. N. Murphy, W. Brown, *Did My Neurons Make Me Do It? Philosophical and Neurobiological Perspectives on Moral Responsibility and Free Will* (Oxford University Press, New York, 2007)
181. N. Murphy, G.F.R. Ellis, T. O'Connor (eds.), *Downward Causation and the Neurobiology of Free Will* (Springer, 2009)
182. D.G. Myers, *Intuition: Its Powers and Perils* (Yale University Press, 2003)
183. K. Nakabayashi, C.H. Liu, Development of holistic vs. featural processing in face recognition. Front. Hum. Neurosci. (2014). doi:10.3389/fnhum.2014.00831
184. H. Nakata, K. Sakamoto, R. Kakigi, Meditation reduces pain-related neural activity in the anterior cingulate cortex, insula, secondary somato sensory cortex, and thalamus. Front. Psychol. **5**: Article 1489 (2014)
185. P. Namburi et al., A circuit mechanism for differentiating positive and negative associations. Nature **520**, 675–678 (2015)
186. J.G. Nicholls, A.R. Martin, B.G. Wallace, P.A. Fuchs, *From Neuron to Brain* (Sinauer, 2000)
187. S. Nishida, T. Shibata, K. Ikeda, Object-based selection modulates top-down attentional shifts. Front. Hum. Neurosci. (2014). doi:10.3389/fnhum.2014.00090
188. D. Noble, *The Music of Life: Biology Beyond Genes* (Oxford University Press, Oxford, 2008)
189. D. Noble, A theory of biological relativity: no privileged level of causation. Focus Interface **2**, 5564 (2012)
190. D. Noble, R. Noble, J. Schwaber, What is it to be conscious?, in *The Claustrum*, ed. by J. Smythies, L. Edelstein, V. Ramachandran (Elsevier, 2014)
191. H. Okon-Singer1, T. Hendler, L. Pessoa, A.J. Shackman, The neurobiology of emotion–cognition interactions: Fundamental questions and strategies for future research. Front. Hum. Neurosci. (2015). doi:10.3389/fnhum.2015.00058
192. J. Panksepp, *Affective Neuroscience: The Foundations of Human and Animal Emotions* (Oxford University Press, London, 1998)
193. J. Panksepp, L. Biven, *The Archaeology of Mind: Neuroevolutionary Origins of Human Emotion* (W W Norton and Company, New York, 2012)
194. R. Penrose, *The Emperor's New Mind: Concerning Computers, Minds and the Laws of Physics* (Oxford University Press, New York, 1989)
195. R. Penrose, *The Large, the Small and the Human Mind* (Cambridge University Press, Canto, 2000)
196. M.A. Peterson, Vision: Top-down effects. *Encyclopedia of Cognitive Science* (Wiley, 2003). doi:10.1002/0470018860.s00633
197. M.A. Peterson, L. Cacciamani, Toward a dynamical view of object perception, in *Shape Perception in Human and Computer Vision*, ed. by S. Dickinson, Z. Pizlo (Springer, 2013), pp. 443–457
198. S.M. Platek, J.P. Keenan, T.D. Shackelford, *Evolutionary Cognitive Neuroscience* (MIT Press, Cambridge, Mass, 2007)

199. T.A. Polk, H.P. Lacey, J.K. Nelson, E. Demiral, L.I. Newman, D. Krauss, A. Raheja, M.J. Farah, The development of abstract letter representations for reading: evidence for the role of context. Cogn. Neuropsychol. **26**, 70–90 (2009)
200. G.K. Pullum, B.C. Scholz, Empirical assessment of stimulus poverty arguments. Linguist. Rev. **19**, 9–50 (2002)
201. D. Purves, *Brains: How They Seem to Work* (FT Press Science, Upper Saddle River, 2010)
202. L.A. Remmelzwaal, J. Tapson, G.F.R. Ellis, Salience-affected neural networks (2010). http://arxiv.org/ftp/arxiv/papers/1001/1001.3246.pdf
203. R. Rhoades, R. Pflanzer, *Hum. Physiol.* (Saunders College Publishing, Fort Worth, 1989)
204. P.J. Richerson, R. Boyd, *Not by Genes Alone: How Culture Transformed Human Evolution* (University of Chicago Press, Chicago, 2005)
205. C.S. Rosati, Moral motivation. *The Stanford Encyclopedia of Philosophy* ed. by E.N. Zalta (Spring, 2014). http://plato.stanford.edu/archives/spr2014/entries/moral-motivation/
206. E. Rosset, It's no accident: our bias for intentional explanations. Cognition **108**, 80–771 (2008)
207. P. Rozin, J. Haidt, C.R. McCauley, Disgust: The body and soul emotion in the 21st century, in *Disgust and Its Disorders*, ed. by D. McKay, O. Olatunji (American Psychological Association, Washington, DC), pp. 9–29 (2008)
208. P. Russell, Hume on free will. *The Stanford Encyclopedia of Philosophy*, ed. by E.N. Zalta (Winter, 2014). http://plato.stanford.edu/archives/win2014/entries/hume-freewill/
209. S. Salcuni, *New frontiers and applications of attachment theory* (Front, Psychol, 2015)
210. R.M. Sapolsky, Social status and health in humans and other animals. Ann. Rev. Anthropol. **33**, 393–418 (2004)
211. R.M. Sapolsky, The influence of social hierarchy on primate health. Science **308**, 648–652 (2005)
212. S. Satel, S.O. Lilienfeld, *Brainwashed: The Seductive Appeal of Mindless Neuroscience* (Basic Books, 2013)
213. G. Schelling, M. Richter, B. Roozendaal, H.-B. Rothenhusler, T. Krauseneck, C. Stoll, G. Nollert, M. Schmidt, H.-P. Kapfhammer, Exposure to high stress in the intensive care unit may have negative effects on health-related quality-of-life outcomes after cardiac surgery. Crit. Care Med. **31**, 1971–1980 (2003)
214. A. Scott, *Stairway to the Mind* (Springer, New York, 1995)
215. J.R. Searle, *Making the Social World: The Structure of Human Civilisation* (Oxford University Press, Oxford, 2011)
216. A. Sen, *Development as Freedom* (Oxford University Press, Oxford, 2001)
217. N. Shea, Distinguishing top-down from bottom-up effects, in *Perception and Its Modalities*, eds. S. Biggs, M. Matthen, D. Stokes (Oxford University Press, Oxford, 2014)
218. G.M. Shepherd, S. Grillner, *Handbook of Brain Microcircuits* (Oxford University Press, Oxford, 2010)
219. M. Sigman, H. Pan, Y. Yang, E. Stern, D. Silbersweig, C.D. Gilbert, Top-down reorganization of activity in the visual pathway after learning a shape identification task. Neuron **46**, 823–835 (2005)
220. H.A. Simon, The architecture of complexity. Proc. Am. Phil. Soc. **106** (1962)
221. F. Smith, *Understanding Reading: A Psycholinguistic Analysis of Reading and Learning to Read* (Routledge, 2004)
222. E. Sohoglu, J.E. Peelle, R.P. Carlyon, M.H. Davis, Predictive top-down integration of prior knowledge during speech perception. J. Neurosci. **32**, 8443–8453 (2012)
223. M. Solms, Affect theory today. in *Psychoanalytic Psychology* (2016) (In press)
224. M. Solms, O. Turnbull, *The Brain and the Inner World* (Other Press, 2003)
225. E.M. Sternberg, Neural-immune interactions in health and disease. J. Clin. Invest. **100**, 2641–2647 (1997)
226. E.M. Sternberg, *The Balance Within: The Science Connecting Health and Emotions* (W H Freeman and Co, 2000)
227. E.M. Sternberg, Neural regulation of innate immunity: a coordinated nonspecific host response to pathogens. Nat. Rev. Immunol. **6**, 318–328 (2006)

228. A. Stevens, J. Price, *Evolutionary Psychiatry: A New Beginning* (Taylor and Francis, New York, 2000)
229. I. Stewart, *Seventeen Equations that Changed the World* (Profile Books, 2012)
230. M.G. Stokes, K. Atherton, E.Z. Patai, A.C. Nobre, Long-term memory prepares neural activity for perception. Proc. Nat. Acad. Sci. **109**, E360–367 (2012)
231. K. Stotz, Extended evolutionary psychology: the importance of transgenerational developmental plasticity. Front. Psychol. (2014). http://journal.frontiersin.org/Journal/10.3389/fpsyg.2014.00908/abstract
232. S.L. Strauss, *The Linguistics, Neurology, and Politics of Phonics: Silent E Speaks Out* (Routledge, 2004)
233. S.L. Strauss, K.S. Goodman, E.J. Paulson, Brain research and reading: how emerging concepts in neuroscience support a meaning constructionist view of the reading process. Educ. Res. Rev. **4**, 21–33 (2009)
234. G. Strawson, Consciousness myth. Times Literary Suppl. (2015). http://www.the-tls.co.uk/tls/public/article1523413.ece
235. G.F. Striedter, *Principles of Brain Evolution* (Sinauer Associates, Sunderland, Mass, 2005)
236. C. Stringer, *The Origin of Our Species* (Penguin, 2012)
237. C. Taylor, *Sources of the Self: The Making of Modern Identity* (Harvard University Press, Cambridge, Mass, 1989)
238. D. Taylor, *Tell Me a Story: The Life-Shaping Power of Our Stories* (Bog Walk Press, 2001)
239. M. Tomasello, *The Cultural Origins of Human Cognition* (Harvard University Press, Cambridge, Mass, 1999)
240. M. Tomasello, The human adaptation for culture. Ann. Rev. Anthropol. **28**, 509–529 (1999)
241. M. Tomasello, *Constructing a Language: A Usage-Based Theory of Language Acquisition* (Harvard University Press, 2009)
242. M. Tomasello, *A Natural History of Human Thinking* (Harvard University Press, Cambridge, Mass, 2014)
243. G. Tononi, G.M. Edelman, Consciousness and complexity. Science **282**, 1846–1851 (1998)
244. J.A. Toronchuk, G.F.R. Ellis, Affective neuronal selection: the nature of the primordial emotion systems. Front. Psychol. **3**, 589 (2012). http://www.ncbi.nlm.nih.gov/pmc/articles/PMC3540967/
245. R.L. Trask, *Key Concepts in Language and Linguistics* (Routledge, Abingdon, 1999)
246. R.L. Trask, *Language and Linguistics: The Key Concepts* (Routledge, Abingdon, 2007)
247. P.U. Tse, *The Neural Basis of Free Will* (MIT Press, 2013)
248. R. van den Bos, J.W. Jolles, J.R. Homberg, Social modulation of decision-making: a cross-species review. Front. Hum. Neurosci. **7**, 301 (2013)
249. D. van der Westhuizen, M. Solms, Social dominance and the affective neuroscience personality scales. Conscious. Cogn. **33C**, 90–111 (2014)
250. J. van Ekert, J. Wegman, G. Janzen, Neurocognitive development of memory for landmarks. Front. Psychol. Dev. Psychol. (2015). doi:10.3389/fpsyg.2015.00224
251. R. Van Gulick, Who's in charge here? And who's doing all the work?, in *Mental Causation*, ed. by J. Heil, A. Mele (Clarendon Press, Oxford, 1995), pp. 233–256
252. S. van Leeuwen, W. Singer, L. Melloni, Meditation increases the depth of information processing and improves the allocation of attention in space. Front. Hum. Neurosci. (2012). doi:10.3389/fnhum.2012.00133
253. T.D. Wager, J.K. Rilling, E.E. Smith, A. Sokolik, K.L. Casey, R.J. Davidson, S.M. Kosslyn, R.M. Rose, J.D. Cohen, Placebo-induced changes in fMRI in the anticipation and experience of pain. Science **303**, 1162–1167 (2004)
254. B.E. Wampold, T. Minami, S.C. Tierney, T.W. Baskin, K.S. Bhati, The placebo is powerful: estimating placebo effects in medicine and psychotherapy from randomized clinical trials. J. Clin. Psychol. **61**, 835–854 (2005)
255. D. Wegner, *The Illusion of Conscious Will* (MIT Press, 2002)
256. D.S. Wilson, Group level evolutionary processes, in *Oxford Handbook of Evolutionary Psychology*, ed. by R. Dunbar, L. Barrett (Oxford University Press, Oxford, 2009)

257. T.P. Zanto, M.T. Rubens, A. Thangavel, A. Gazzaley, Causal role of the prefrontal cortex in top-down modulation of visual processing and working memory. Nat. Neurosci. **14**, 656–661 (2011)
258. A. Zeman, *Consciousness: A Users Guide* (Yale University Press, New Haven, 2002)

Chapter 8
The Broader View

The conclusion arising from the arguments given in this book is that there are other forms of causation than those encompassed by physics and physical chemistry alone, or even in genetics and neuroscience [60, 62]. Higher levels of structure have causal powers, based on strong emergence of higher level structure and function [37, 38, 150], that can shape what happens in the world. A full scientific view of the nature of reality must recognise this fact, or else it will ignore important aspects of causation in the real world, and so will give a causally incomplete view of things. These forms of causation are based on the interaction of bottom-up and top-down effects: if we neglect either, we will be unable to understand genuinely complex systems. There are many implications for society, including issues in health care and education. This chapter will summarise this broader view as follows:

- Section 8.1. The necessity of true emergence.
- Section 8.2. The sources of emergence: chance, necessity, and purpose.
- Section 8.3. Aristotle and types of causation.
- Section 8.4. Aristotle and types of knowledge.
- Section 8.5. A more holistic view.
- Section 8.6. Implications: learning to read and write.
- Section 8.7. Conclusion: Where is truth? What more to do?

This conclusion emphasizes that there is much to be done to complete what is set out here. But what is presented is a coherent whole, and without it, we will not be able to understand complex causation such as occurs in digital computers, life, and the functioning of the brain.

8.1 The Necessity of True Emergence

The existence of complexity on Earth is possible only because of the cosmological context in which we live: the expanding and evolving universe [59, 63, 103] (see Sect. 6.7.2). Because of quantum uncertainty (see Sect. 6.1.1), it simply is not possible

G. Ellis, *How Can Physics Underlie the Mind?*, The Frontiers Collection,
DOI 10.1007/978-3-662-49809-5_8

that emergence takes place in a deterministic way on the basis of the initial data at the start of the universe. Genuine emergence, with its own higher level causal powers, has to take place. This section sets that contextual scene by discussing:

- Cosmological unpredictability (Sect. 8.1.1).
- Evolutionary history (Sect. 8.1.2).
- The necessary conclusion: genuine emergence must occur (Sect. 8.1.3).
- The alternative: the demiurge (Sect. 8.1.4).

8.1.1 Cosmological Unpredictability

To see the improbability of the claim that all structure emerges in a purely bottom-up way, one can contemplate what is required from this viewpoint when placed in its proper cosmic context. To make the issue clear, one can pose the following question: is it possible that the contents of Crick's book [41] is uniquely determined by the initial state of the universe? That would be the case if necessity held sway on the basis of initial data at the start of the universe. The implication is that the configuration of particles and fields that existed in the very early universe [49] just happened to be placed so precisely as to make it *inevitable* that fourteen billion years later, human beings would exist and Crick and Watson would discover DNA, Crick would write his book [41], Townes would conceive of the laser, Witten would develop M-theory, and so on. This is not possible even in principle for two reasons.

8.1.1.1 The Cosmic Context

The expansion history of the universe [49, 103, 173, 174] is represented in Fig. 8.1. Time runs from left to right. An extraordinarily rapid initial accelerating period of expansion ('inflation') gives way to a hot big bang era with matter and radiation strongly interacting, until matter and radiation decouple at the last scattering surface (LSS), 300 000 years after the hot big bang. Cosmic blackbody background radiation (CBR) was released at the LSS and then travelled freely towards us. This decoupling was followed by dark ages until the first stars formed and galaxies came into being through gravitational attraction acting on very small fluctuations on the LSS (Fig. 8.2). Some massive first generation stars, made only of hydrogen and helium, met a fiery end as supernovae, spreading clouds of heavy elements in space. These clouds then allowed second generation stars to form, surrounded by planets. The expanding universe is the environment creating the conditions for life to exist today. The CBR sky we observe (Fig. 8.2) is an image of density fluctuations on the LSS that are the precursors of galaxies that exist today.

Now the question is this: if we were able to detect every micro-fluctuation on the LSS down to the smallest scale, could we then in principle run that data forwards to predict the specific words on the page you are reading now? And going back even earlier, are these words uniquely implied by the state of the universe at the start of

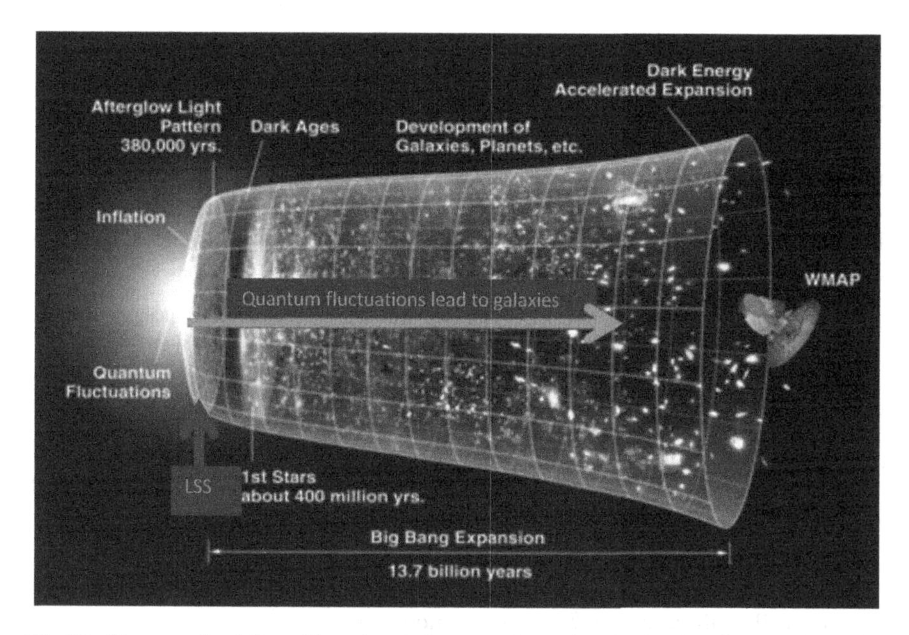

Fig. 8.1 The unpredictability of the universe. Quantum fluctuations lead to the existence of galaxies. Image credit: NASA/WMAP science team

Fig. 8.2 CMB sky image of the LSS measured by the Planck Satellite. *Red* represents higher density and *blue* lower density on the LSS. Image credit: NASA/WMAP science team

inflation? That would be the case if physical determinism was true, and determined the present day state of my brain and hence uniquely caused the details of the print on this page (which is a set of physical markings on paper). In short, is it true that in the real universe, the future is uniquely predicted by the past? In fact, this is not possible for two reasons.

8.1.1.2 Impossibility Due to Inflationary Perturbations

Firstly, according to the standard inflationary model of the very early universe, which is the present standard theory of cosmology [49, 103, 173, 174], we cannot predict the specific large-scale structure existing in the universe today from data at the start of the inflationary expansion epoch, because density inhomogeneities at later times have grown out of random quantum fluctuations in the effective scalar field that is dominant at very early times (Hinshaw [108]):

> Inflation offers an explanation for the clumpiness of matter in the universe: quantum fluctuations in the mysterious substance that powered the [inflationary] expansion would have been inflated to astrophysical scales and therefore served as the seeds of stars and galaxies.

Scalar (density) perturbations are generated from quantum fluctuations of the inflaton field during its evolution. After inflation, a reheating process occurs, and the resulting inhomogeneities cause fluctuations that seed the large-scale structure such as clusters of galaxies:

1. These quantum fluctuations are random in nature. If we know the universe completely at the time T_1 when inflation starts, this does not determine what fluctuations will be there at time T_2 when inflation ends.
2. At a much later time T_3 (after decoupling of matter and radiation), these fluctuations seed large scale structure such as clusters of galaxies.
3. Therefore, the specific galaxies that occur at late times in the real universe are not determined by the state of the universe at time T_1, because of the first point above.

Thus the inhomogeneities that occurred on the LSS were the outcome of quantum fluctuations during inflation, blown up by that enormous expansion to macroscopic scales [49, 148]. They were not determined uniquely by the state of the universe at the start of inflation, because until the relevant quantum fluctuations had become crystallized in classical fluctuations that resulted in density inhomogeneities on the LSS (Fig. 8.2), the outcome was unpredictable even in principle [120, 161], due to the quantum uncertainty described in Sect. 6.1 (the small fluctuations in Fig. 8.2 are in effect similar to the dark markings on the screen in Fig. 6.1).

We may sum up as follows:

> The quantum fluctuations that are amplified to galactic scale by this process are unpredictable in principle. Thus the existence of our specific galaxy, let alone the planet Earth, was not determined by initial data in the very early universe.

For this reason, it is impossible that the initial data in the early universe uniquely specified any particular human action or thought whatever, as neither the existence of the Earth nor even the existence of our galaxy is uniquely implied by that initial data. The state of matter on the LSS is not uniquely determined by the state of the universe at the start of inflation; but it was that state that determined what specific structures formed in the universe at later times. Hence the existence of our specific galaxy, and the Sun and Earth in it, is also not so determined. They are the outcome of unpredictable random events.

As the existence of the Earth is not a specific outcome of that initial data, so what is written in this book cannot possibly be a specific outcome either. Because of irreducible quantum chance [73, 120], necessity cannot explain the details of the present day universe.

8.1.2 Evolutionary History

Secondly, quantum fluctuations can change the genetic inheritance of animals [157] and so influence the course of evolutionary history on Earth. Indeed that is what occurred when cosmic rays—whose emission processes are unpredictable because they are subject to quantum uncertainty—caused genetic damage in the distant past [169]:

> The near universality of specialized mechanisms for DNA repair, including repair of specifically radiation induced damage, from prokaryotes to humans, suggests that the Earth has always been subject to damage/repair events above the rate of intrinsic replication errors [...] radiation may have been the dominant generator of genetic diversity in the terrestrial past.

The emission of a specific cosmic ray at a particular time and place is a quantum event, unpredictable even in principle (Sect. 6.1). Consequently, the specific evolutionary outcomes of life on Earth (the existence of dinosaurs, giraffes, humans) cannot even in principle be uniquely determined by causal evolution from conditions in the early universe, or from detailed data at the start of life on Earth. Quantum uncertainty prevents this, because it significantly affected the occurrence of radiation-induced mutations in this evolutionary history. The specific outcome that actually occurred was determined as it happened, when quantum emission of the relevant photons took place. The prior uncertainty in their trajectories was resolved by the historical occurrence of the emission event, resulting in a specific photon emission time and trajectory that was not determined beforehand, with consequent damage to a specific gene in a particular cell at a particular time and place that cannot be predicted even in principle. But this sequence of events changed evolutionary outcomes.

8.1.3 Conclusion: Genuine Emergence Must Occur

So from where do higher level ideas, theories, and behaviour arise, given that they cannot be uniquely determined by physics data in the early universe? The obvious explanation is that they arise from the autonomous behavior of the human mind acting in an intelligent way, supervenient on but not causally determined by the underlying physics. Top-down action from the mind to the electrons in the brain allows this to happen, with theories such as Einstein's theory of gravitation and Darwin's theory of evolution shaping neural connections in Einstein's and Darwin's minds as they considered the logical options and evidence in their own terms. There thus has to be the causal openness at the lower levels needed to allow this to occur, with abstract theories shaping synaptic strengths [35], for otherwise this demonstrable outcome could not arise. The answer has to be that what appears to be the case is indeed the case:

Thesis 1: Genuine Emergence. Strong emergence [37] *must* occur over time and lead to the macro-levels of order and meaning we see. Processes of evolution and development lead to the emergence of brains and minds with their own causal powers of abstraction and reasoning, which then lead to these theories coming into existence. These outcomes are not uniquely determined by the underlying microphysics acting on the initial state of the universe, because that initial state does not even predict the existence of our galaxy, the Earth, or human brains.

If we had a much more powerful telescope than the one carried by the Planck satellite and could see every micro detail of the LSS, we would not somehow find the general theory of relativity encoded there. What we have there are random Gaussian fluctuations. They do not encode the present day uniquely: rather they encode the possibility of what exists today developing from those conditions. What we have to examine in order to understand this is the processes that lead to genuine emergence of higher levels of causation and meaning, encoded in language and symbolism, which undoubtedly have come into existence in the real universe.

8.1.4 The Alternative: The Demiurge

Suppose the argument presented here were not the case: forget quantum uncertainty and assume everything in the history of the Earth is indeed written into the fluctuations on the LSS, such as those observed by the Planck satellite. We then have to explain how the theory of general relativity, the Mona Lisa, the international banking system, and so on and so forth, could all have been encoded there in the initial data that determined the later universe.

What would be required for this to be the case is a transfer function $T(k, a)$ that could give rise to these results from data on the last scattering surface via the relation

$$\Phi(k, a) = T(k, a)\Phi(k, a_i) \ , \tag{8.1}$$

where a_i corresponds to the LSS [193, p. 31]. This transfer function has to uniquely give the micro-connections in individual brains in order to uniquely lead to these outcomes.

There are two options. The first is that $T(k, a)$ generates these results in a deterministic way from random Gaussian fluctuations on the LSS. But firstly, diffusion processes take place, random collisions occur between particles, and so on, and we have difficulty even in showing that the transfer function will result in the present observed details of the distribution of dark matter halos [187]. The idea that there is some way that the unique English language on this page should be embodied in $T(k, a)$ acting on random Gaussian fluctuations is not remotely plausible, for there is no hint of how this could happen in standard discussions of the transfer function, e.g., [12].

The second option is that in fact after all the fluctuations on the LSS are not random Gaussian fluctuations, as commonly believed [148]: rather they explicitly encode the formulae of the theory of general relativity as written in Einstein's 1915 paper, the words in Darwin's majestic opus on the origin of species, and so on. The issue then is, who or what could have written all this, rather than random Gaussian fluctuations, into the molecules on the LSS precisely so as to get these results? To determine every thought that Maxwell, Einstein, Karl Marx had? Suppose the initial data were ordered so as to produce these highly structured outcomes, then what kind of demiurge could have been responsible?

This appears to be a form of intelligent design theory. The argument presented in this book, namely that genuine emergence took place, leading to intelligent outcomes not uniquely determined by this initial data, is far more plausible.

8.2 The Sources of Emergence

So how did complexity arise? It is possible because adaptive evolutionary and developmental processes [30, 84] lead to the existence of genuine complexity, where higher levels of structure, including brains, come into being and have causal power that leads to physical outcomes, for example, by creating computers that change the world. This emergence of higher level causal powers is possible because of randomness at lower levels that allows for selection of functionality on the basis of higher level selection criteria. And that is a form of top-down causation, adapting emergent life to its environment. There are essentially three ways that emergent properties come into being:

- Self-assembly: emergence in the natural world (Sect. 8.2.1).
- Natural selection: emergence in the biological world (Sect. 8.2.2).
- Design and construction in the man-made world (Sect. 8.2.3).

The first is limited in what it can lead to. Indeed, we may say that:

- Self-assembly is limited in what it can achieve (Sect. 8.2.4).
- Physics principles by themselves cannot generate life (Sect. 8.2.5).

Rather for intelligence and its outcomes to come into being, we need a mixture of causal effects to occur:

- The interconnected causes: chance, necessity, and purpose (Sect. 8.2.6).

8.2.1 *Self-Assembly: Emergence in the Natural World*

Simple components can spontaneously self-assemble to produce interesting higher level entities if the conditions are right. Thus sodium and chlorine can form salt crystals, hydrogen and oxygen can form water, carbon can form a diamond, water molecules can form ice crystals, or a sheet of ice, quartz and feldspar can combine to form granite, and so on. This depends on energy minimisation in chemical bonds. In addition, quite complex patterns can form via equations such as the reaction–diffusion equation underlying the Belousov–Zhabotinsky reaction, and studied by Turing as the chemical basis of morphogenesis [192]. For example Maini et al. [137] study the stability of pattern formation during embryonic development arising in response to a spatial pre-pattern in biochemicals on the basis of Turing's equation

$$\frac{\partial u}{\partial t} = D\nabla^2 u + f(u) , \tag{8.2}$$

where u is a vector of chemical concentrations, D a matrix of constant diffusion coefficients, and $f(u)$ the reaction kinetics (typically nonlinear). In biochemistry, basic organic molecules and even some macromolecules can self-assemble if conditions are right [130, 131]. These effects all depend on non-linear interactions between component parts [11].

However, this outcome always depends on context. Temperature and pressure affect the result and phase changes can occur if one alters these conditions. The patterns resulting from the reaction–diffusion equation depend crucially on boundary conditions and the shape of the boundary. The processes in molecular reactions depend on availability of materials and sources of energy, for example, macromolecule self-assembly depends on the presence of the molecular components to be assembled. Thus self-assembly is a form of algorithmic top-down causation (TD1), depending on boundary and initial conditions.

8.2.2 *Natural Selection: Emergence in the Biological World*

Natural selection is the basic source of emergent properties in biology [30]. As explained in Sect. 4.3, this can only occur because of adaptation to the environment: hence data must flow from the environment into the living entity and alter its structure and behaviour [197]. Thus this is based on top-down causation TD3. It is not just a process of energy minimisation, and so is not the same as a bottom-up process of self-assembly.

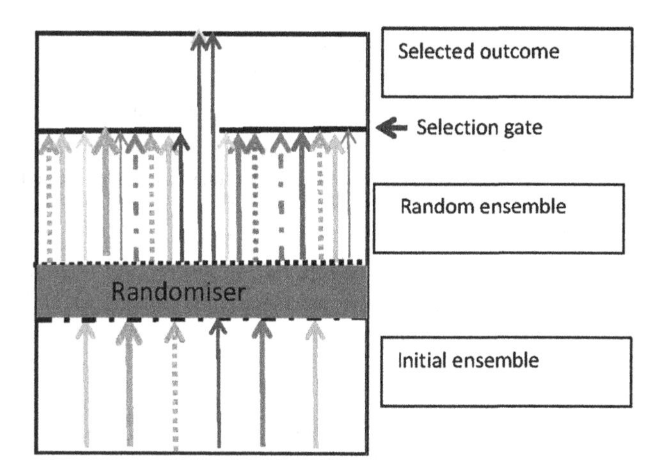

Fig. 8.3 A randomizer provides the random ensemble from which selection then takes place. This random element is what makes the process unpredictable

8.2.2.1 General Adaptive Selection

The generic process of adaptive selection [111] is shown in Fig. 8.3. Whatever the initial ensemble E_i may be, some kind of generator or randomiser $\mathcal{R}$ provides an ensemble of varied entities E_r from which selection takes place according to some selection principle $\mathcal{S}$, giving a projection operation $\Pi(\mathcal{E}, \mathcal{S})$ that leads to an output set of states E_o, where the selector depends on the environment $\mathcal{E}$, which determines what niches are available, as well as the selection principle $\mathcal{S}$. Thus the basic process is one of randomisation:

$$\mathcal{R}(\mathcal{E}) : E_\mathrm{i} \to E_\mathrm{r} , \tag{8.3}$$

where the randomizer $\mathcal{R}$ generically depends on the environment $\mathcal{E}$, followed by projection:

$$\Pi(\mathcal{E}, \mathcal{S}) : E_\mathrm{r} \to E_\mathrm{o} , \tag{8.4}$$

with the entities not selected being discarded, leading to the combined process

$$\Pi(\mathcal{E}, \mathcal{S})_{\mathcal{R}(\mathcal{E})} : E_\mathrm{i} \to E_\mathrm{o} . \tag{8.5}$$

The outcome is not predictable because of the randomisation process $\mathcal{R}$ deriving E_r from E_i (whatever that process is). This is top-down in two ways: $\Pi(\mathcal{E}, \mathcal{S})$ depends on the environment $\mathcal{E}$ as well as the selection principle $\mathcal{S}$. While E_r is a set of physical states, $\mathcal{S}$ is not. It is a set of criteria that are desired as an outcome of the selection process (for example, surviving longer, or being able to form strong groups through communication via language). $\mathcal{E}$ may or may not be physical (it may have social or economic elements that are not themselves physical).

Table 8.1 The contexts of biological emergence

Context	Timescale	Effect
Evolutionary	10^6 years to 10^3 years	Life on Earth emerges from no life
Developmental	10^2 years to weeks	Single cells develop into complex animals
Functional	Minutes to milliseconds	Molecules function to create living beings

However, as well as occurring on evolutionary timescales, biological emergence also occurs on developmental and functional timescales (Sect. 4.3), and it occurs at various levels in the hierarchy of emergence.

8.2.2.2 Timescales of Emergence Processes

The timescales of processes of emergence is indicated in Table 8.1. Emergence selected to adapt to the environment occurs both diachronically (over time) and synchronically (at each moment in time), through three rather different processes occurring on different timescales:

- **Evolution**. Initially, in the early stages of the universe, there was no life, but now it exists: life emerged from a barren initial state.
- **Development**. Each animal starts off as a single cell, and these divide until they may consist of up to the order of 10^{13} cells, each adapted to their function in the body.
- **Function**. Each animal consists of a huge number of particles assembled in such a way that they form a hierarchical system that functions in an ongoing way to satisfy biological purpose.

In each case, adaptation to the environment is taking place, so what is occurring is a top-down adaptive selection process as indicated above. Note that this adaptation takes place in both physiological and behavioural terms. This is a multilevel purpose [139], for example the process of learning is an adaptive process at the psychological or behavioral level, enabled by adaptive plasticity at the neural level.

8.2.2.3 Darwinian Evolution

What is special about the evolutionary process over geological timescales, that has led to the origin of species [30]? It is a diachronic process of adaptive selection as described above, with reproduction and variation repeated so that the outcome of each round of selection is the input to the next round of variation and selection based on survival (Fig. 8.4). Repeated thousands of times over geological timescales, it is this relentless repetition of the top-down process of adaptive selection [29] that gives Darwinian evolution its enormous power.

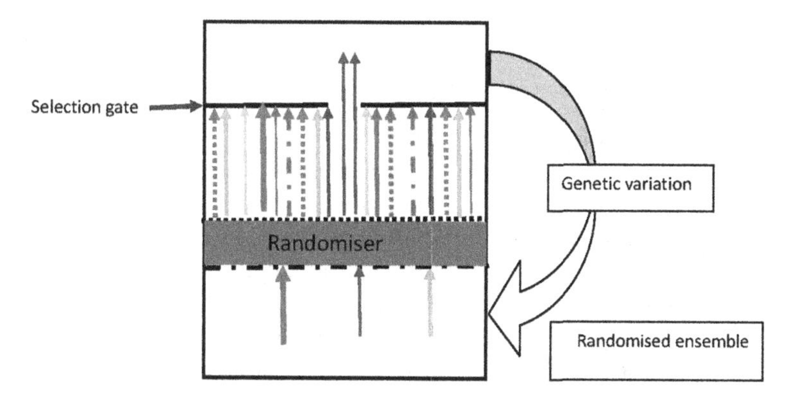

Fig. 8.4 Darwinian selection. Repeated duplication with variation provides the ensemble from which the adaptive choice is made

One should note that genetic drift also takes place: there are neutral variations that do not affect survival prospects. However, there is no doubt that animals end up beautifully adapted to their environments [30], and that is only possible if the projection process (8.4) does indeed take place [81]. There is also convergence on all levels in the way evolution takes place, with the same solutions reached to biological problems along different evolutionary pathways [40, 141]. This is because evolutionary processes are constrained by physiological possibilities [196]. The outcome is still an acceptable adaptation to the environment. After all, if that were not the case, the relevant animal would not be here.

8.2.2.4 Levels of Adaptive Processes

At which levels do adaptive processes work in biology? Considering a simplified version of the life sciences hierarchy [30], all the levels above the level of atomic physics are adaptive (Sect. 4.3.6), see Table 8.2. This applies in particular to human physiology [164].

Thus the biosphere is different if the atmosphere is methane or oxygen, the ecosystem is different in a savannah and in Alaska, a collective of animals is different if they live on land or in the sea, individual animals are adapted to living in herds by having evolved to be social animals, physiological systems in animals such as vision, the cardiovascular system, and so on are adapted to fulfil specific functions in the individual animal, the brain works adaptively, shaping its plasticity [70], and different types of cells are adapted to enable those systems to work, for example neurons are adapted to enable brain function. This in turn is enabled by genes that are adapted to shaping these cells, while specific molecules such as kinesin and dynesin have evolved to enable cell function. However, the atomic level down is fixed and unchanging, i.e., there is no adaptive selection of properties of an individual proton or electron.

Table 8.2 The Levels at which adaptive processes take place

Level	Entity	Example	Adaptive?
Level 10	Biosphere	Oxygen-based	Yes
Level 9	Ecosystem	Savannah	Yes
Level 8	Group	Herd, swarm, school	Yes
Level 7	Individual	Buffalo, lion	Yes
Level 6	Systems/organs	Vision, heart	Yes
Level 5	Cells	Nerve cells, blood cells	Yes
Level 4	Genes	PKD1 gene	Yes
Level 3	Molecules	DNA, haemoglobin, kinesin	Yes
Level 2	Atomic physics	Carbon, oxygen, sodium	No
Level 1	Particle Physics	Electrons, quarks	No

8.2.3 Design and Construction: Emergence in the Man-Made World

Design of physical systems followed by subsequent manufacture or construction is a major source of emergent structure in the world around us. This is 'the science if the artificial' [175]. Some such systems, like automobiles, aircraft, power networks, or digital computer systems, can be very complex. The same design principles for dealing with complexity [21, 175] that hold for them also hold for abstract systems we create such as companies, legal systems, governments, and economic systems [13]. These are all the outcome of abstract plans in the human mind, which are obviously causally effective (Sects. 7.5.3 and 7.5.4). Thus they represent top-down action from the mind to atoms in the world, in order to fulfil some purpose.

Obtaining the desired physical outcomes depends crucially on the variety of ways we create and shape the materials that underlie all our manufactured items. These are carefully tailored to their intended task. Miodownik gives a fascinating discussion of how the desired higher level properties emerge from the lower level physics [146]. His examples are, (1) steel, (2) paper, (3) concrete, (4) chocolate, (5) foam/gel, (6) plastic, (7) glass, (8) graphite/carbon fibres, (9) porcelain, and (10) medical implants. Each material has been carefully designed and manufactured for its specific purpose. These are just some of the many substances that underlie our everyday lives, products of the human mind that have been shaped for their desired function through a process of trial and error, leading to the rise of civilisation as brilliantly described by Bronowski [25]. Design takes place as regards technology such as computers and aircraft (Sect. 7.5.3) and in the processes of art (Fig. 8.5), which depend on an understanding of material as well as a vision of what is to be achieved.

Fig. 8.5 Greek sculpture of the head of a woman from 300 BC (Carole Bloch)

8.2.4 How Far Can Bottom-Up Emergence Succeed?

The big debate in the evolutionary emergence of life is to what extent statistically based self-assembly of molecules can succeed in leading to the emergence of the complex structures that sustain life. The case I make is that, while purely bottom-up effects can produce key ingredients needed for the existence of life, and while they can produce many interesting structures and patterns, what they can achieve is nevertheless strictly limited: it cannot lead to the existence of life. This needs the initiation of the top-down causal transfer of information that is required for adaptive selection to take place [198].

Bottom-up emergence suffices for simple systems where the details of the environment are irrelevant, and statistical physics gives all we need to know, such as the perfect gas law $PV = nRT$, relating coarse-grained variables P, V, n, and T, where P is the pressure, V the volume, n the amount of gas (in moles), R the ideal gas constant, and T the temperature. This macro pressure–density relation is based on statistical physics alone, and the shape of the container is irrelevant.

Self-assembly and self-structuring based on bottom-up action alone can do more interesting things. It can lead to emergence of structures such as crystals and simple biological molecules, dynamical systems with attractors leading to entities such as stars and galaxies, and more complex phenomena such as Bénard cells, patterns associated with the reaction–diffusion equation and sand piles, the dynamics of the Game of Life, properties of slime mould, the existence of ant colonies, and the behaviour of flocks of birds.

8.2.4.1 Limits on Bottom-Up Emergence

However, these do not extend to truly complex systems such as a single living cell, where context matters. Developing very complex systems such as those occurring in biology requires top-down causation, needed in order to build up the necessary biological information [128, 165], which is derived by adaptive selection (Sect. 8.2.2). This information cannot be derived in a bottom-up way, because it necessarily embodies information about the relevant environmental niche for the organism. It would be different in a different environment. This is true in functional, developmental, and evolutionary contexts.

This contrast is illustrated in Figs. 8.6 and 8.7. The complexity of human life at the macro-level is vastly greater than that of systems that assemble bottom-up, for the former embody purpose and meaning, but the latter do not (compare Fig. 8.6 left

Fig. 8.6 Emergence in physics and in life. The macro-level. *Left* Spiral galaxy M81 (Hubble Space Telescope). Credit: NASA/STScI. *Right* Maggie Ellis and Carole Bloch

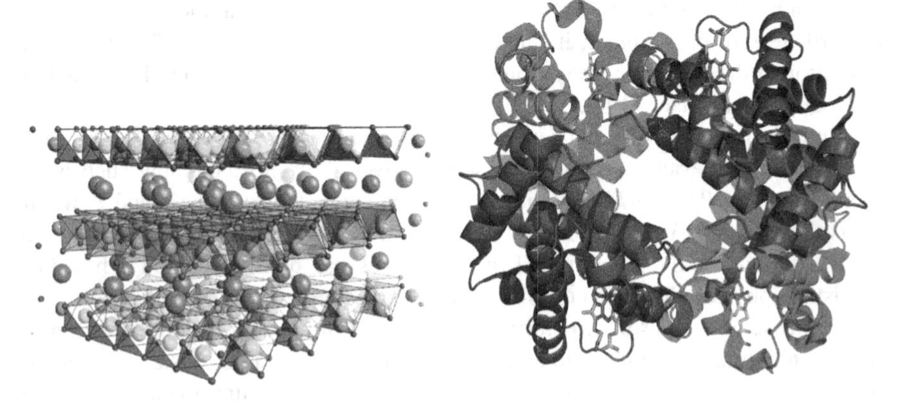

Fig. 8.7 Emergence in physics and in life. The micro-level. *Left* Inorganic crystal. Structure of sodium cobaltate. Credit: Wikimedia Commons (Julien Bobroff). *Right* Organic molecule. Haemoglobin. Credit: Wikimedia Commons (Zephyris)

and right). And this is equally true at the microlevel. The inorganic crystal sodium cobaltate in Fig. 8.7 (left) (blue Na, green Co, red O) becomes a superconductor when water is inserted in the structure. However, it has no function or purpose. By contrast, proteins have functions [158]. Thus the molecular structure of haemoglobin in Fig. 8.7 right has a function, namely to transport oxygen to cells in the body, with the purpose of keeping the body alive. It has been adaptively selected for that purpose [196]. It could not arise in a purely bottom-up way.

As regards evolutionary emergence, an important question is to ask to what extent statistically based self-assembly can lead to the emergence of complexity. It can lead to the existence of simple biomolecules if the right component atoms are there together with a supply of energy, and it can lead to some structures such as vesicles which arise through self-assembly of biomolecules in aqueous solution with hydrophilic heads and hydrophobic tails. However, it seems likely that there is a limit to the level of complexity that can arise through self-assembly:

> **Thesis 2: Contextual Emergence**. While purely bottom-up self-assembly can lead to a certain level of complexity, adaptive organisms need to alter behaviour and internal structures suitably in response to the environment. Hence, they need a flow of information from the environment into the organism in order to enable this adaptation to be take place. Thus higher level biological emergence (existence of eyes responsive to the wavelengths emitted by the Sun, for instance) is crucially based on adaptive top-down effects. It cannot take place in a purely bottom-up way based on physics alone. It is a contextual affair diachronically on evolutionary and developmental timescales, as well as synchronically on functional timescales.

Hence top-down effects (from the environment to the organism) are crucial to biological adaptation. Similar issues arise in development. This is the basic reason that self-assembly by itself, based on attaining configurations that minimise energy, can only go so far: it cannot lead to entities properly adapted to their environment. Bottom-up self-assembly is a necessary part of the process, but it is not sufficient for biological development. The information in DNA is needed to shape what happens, guided by epigenetic processes.

8.2.5 *Not by Physics Alone: The Missing Elements*

The implication is that attempts to explain life in a purely bottom-up way on the basis of physics principles alone, for example, the kinetic theory kind of model by Bellomo et al. [14], cannot succeed, since they omit necessary elements for life.

8.2.5.1 The Necessary Elements

Such proposals lack the following key elements that are necessary for life to exist [30, 104, 164]:

- **Function**. The element of function or purpose that is a central element of biology [104].
- **Adaptive Selection**. The essential role of adaptive selection, which requires top-down transfer of information from the environment to the plant or animal in order that its structure and function can be adapted to the specific context in which it is situated [29].
- **Information**. The key causal role of information in the nature of life [165], as for example in the functioning of RNA and DNA [200].

Concerning the latter, Walker and Davies [198] express it this way, as regards temporal (diachronic) emergence:

> Although it has been notoriously difficult to pin down precisely what it is that makes life so distinctive and remarkable, there is general agreement that its informational aspect is one key property, perhaps the key property. The unique informational narrative of living systems suggests that life may be characterized by context-dependent causal influences, and in particular, that top-down (or downward) causation—where higher-levels influence and constrain the dynamics of lower-levels in organizational hierarchies—may be a major contributor to the hierarchical structure of living systems. Here we propose that the origin of life may correspond to a physical transition associated with a shift in causal structure, where information gains direct, and context-dependent causal efficacy over the matter it is instantiated in. Such a transition may be akin to more traditional physical transitions (e.g., thermodynamic phase transitions), with the crucial distinction that determining which phase (non-life or life) a given system is in requires dynamical information and therefore can only be inferred by identifying causal architecture.

These are what purely bottom-up accounts for the origin of life miss out.

> **Thesis 3: Limits of Physics Explanation**. Physics *per se*, as encapsulated in the laws of dynamics, thermodynamics, electrodynamics, and statistical physics, and in particular energy conservation and structural energy minimisation, cannot by itself lead to biology in a bottom-up way. Biological principles must come into play that are related to biological emergence and function, enabled by the underlying physics but not captured by any physics laws.

8.2.6 The Interconnected Causes: Chance, Necessity, and Purpose

Chance, necessity, and purpose intertwine in the real world around us. Jacques Monod famously claimed that all that matters in biology is chance and necessity [147]. But this misses the key element of purpose or goal-seeking, which is crucial to life [104]. By omitting this, which among other things explained why he wrote his book, Monod's analysis [147] was prevented from relating adequately to deep philosophical issues, even though he claimed to answer them. The same comment applies to more recent books by Susskind, Hawking and Mlodinow, Krauss, and others. This is related to the larger issue of what types of causation shape what happens in the physical and biological world [58].

In this section, I shall consider these issues:

- Necessity.
- Chance.
- Purpose.
- How they fit together.

8.2.6.1 Necessity

Necessity has an inexorable impersonal quality. It is the heart of physics and chemistry. It can be successfully described by mathematical equations such as Newton's equations of motion and Maxwell's equations. The basic process is one of unique prediction from an initial state $u_{\rm i}$ to a final state $u_{\rm f}$:

$$\mathcal{P}(\mathcal{E}, \mathcal{S}) : u_{\rm i} \rightarrow u_{\rm f} , \tag{8.6}$$

where the evolution process $\mathcal{P}$ generically depends on the environment $\mathcal{E}$ (the shape of a drum for example) and/or structural relations $\mathcal{S}$ (the wiring in a computer for example).

Reversible or Not. While this may be a one-to-one, hence reversible, relation, it may also be a many-to-one relationship, which is not reversible (for example, if there is friction as in the case of a real free pendulum, the final state will be static, whatever the initial state). The underlying Hamiltonian dynamics in classical physics is time reversible, but its macroscopic outcomes are not. Unitary Schrödinger or Dirac evolution in quantum physics is time reversible, but this is no longer true when the a measurement (collapse of the wave function) takes place.

Basis of Necessity. The profound basis of necessity is the existence of the laws of physics, characterized by timeless unchanging possibility spaces. These Platonic spaces specify what is and what is not possible: they represent all solutions to the equations of motion. Equation (8.6) characterises all possibilities as $u_{\rm i}$ ranges over all possible values, given the boundary conditions $\mathcal{E}$ and structural relations $\mathcal{S}$, so it characterizes the set of all possible motions. Specific initial conditions $u_{\rm i}$ determine what the actual outcome $u_{\rm f}$ of the dynamics is for a specific system with that particular starting state. The specific starting state determines what happens in a specific case. Thus the laws (8.6) do not by themselves determine the outcomes that will occur in any specific physical case. They only do so when $\mathcal{E}$ (environmental conditions), $\mathcal{S}$ (structural relations), and $u_{\rm i}$ (specific intial conditions) are given.

Outcomes of Necessity. In principle necessity implies that we can uniquely determine outcomes from initial data, obtaining reliable predictions of future states from our knowledge of the present. In practice this may not work because one may be dealing with a chaotic system (such as the weather), one may not have enough computing power to handle the calculations, or one may not be able to access all the necessary data (as for example in climate change or detailed synaptic dynamics of the brain). And of course one may be dealing with chaotic systems, or systems where a catatstrophe occurs so that very small changes in initial data lead to large changes in outcomes.

Reliable Outcomes. However, there are cases where such predictability works out reliably in practice, for example planetary motion in the Solar System. This is true particularly in the case of machines, precisely because they are designed to be reliable (think an on–off electric light switch, the brakes on an automobile, or a clock).

8.2.6.2 Chance

The concept of chance embodies the idea of randomness, implying a lack of purpose or meaning. Things just happen that way, not because it's inevitable, but because it's possible, and maybe probable. It is prevalent in the real universe because of the large number of unrelated causes that influence events, and in particular because of the vast numbers of micro-events that underlie all macroscopic outcomes.

The basic process is one of statistical prediction for an initial state u_{i} to a range of final states u_{f} :

$$\mathcal{P}(\mathcal{E},\mathcal{S}) : u_{\mathrm{i}} \;\rightarrow\; u_{\mathrm{f}}|_{p(u_{\mathrm{f}},\mathcal{E},\mathcal{S})} \;, \tag{8.7}$$

where again the evolution process $\mathcal{P}, \mathcal{S}$ generically depends on the environment $\mathcal{E}$ and structural relations $\mathcal{S}$, and $p(u_{\mathrm{f}}, \mathcal{E}, \mathcal{S})$ is the probability of obtaining the result u_{f} given $\mathcal{E}$ and $\mathcal{S}$. This is always a one-to-many relationship. It is generally not reversible, because it is not a deterministic relationship.

Foundational Quantum Unpredictability. Randomness may be due to irreducible quantum randomness, which occurs in physics at the quantum level (see Chap. 7 for experimental evidence: irreversible unpredictable quantum effects do indeed take place in the real universe). Acting on a given ensemble of initial states, it will produce an ensemble of final states where the statistics of the final state are predictable, but the outcome of each individual event that makes up the ensemble is not. This foundational indeterminism is a key aspect of the nature of physical reality. Its existence is based on deeper levels of being, i.e., the possibility spaces that determine what is and what is not possible in the physical universe. It is the way quantum physics works when specific classical outcomes emerge form the underlying quantum dynamics.

Statistical Interlevel Randomness. But secondly, randomness can also simply be the result of statistical interlevel relations in the hierarchy of complexity: the granular nature of lower level structures and large number of entities involved will result in fluctuations at the higher level. It is a key element in the randomness that is central to Darwinian evolution (when genetic mutations take place). Although this is not foundational randomness, it is genuinely effective randomness in terms of interlevel relations, giving the basis for higher level outcomes to not be predictable on the basis of the initial data. This is the 'chance' that Monod [147] describes in his book.

Outcomes of Randomness. In general randomness in a process means that the outcome is not predictable from initial data. When one has social or engineering systems, randomness is a problem to be handled and as far as possible limited by careful design, so that the desired outcome will be attained despite random events intervening in the dynamics. This is not always successful: in particular, digital computers are notoriously susceptible to the smallest error: a single wrong full-stop can bring an immensely complex program to a crashing halt. However, social systems such as the economic and legal systems and technological artefacts such as modern aircraft are generally more robust: they are designed to handle reasonable classes of random events without disaster occurring. Feedback control in cybernetic or homeostatic systems is specifically designed to tame randomness, but it remains an enemy to be handled with care. It has the potential to derail everything and prevent us from attaining the desired goal.

The Case of Biology. That random processes are a core feature of biological functioning is indicated by many kinds of evidence [33, 56, 140]. There is noise in gene expression [163, 168]. At the micro-level, biological systems do not live in a carefully controlled environment: they face rampant randomness all the time. It turns out that they take advantage of the storm of randomness encountered at the molecular level: there is much evidence that molecular machinery in biology is designed to use that randomness to attain its desired results [110]. This is true also in terms of macro-levels of behaviour, and in particular as regards the way the brain functions [45, 86, 167]. Randomness is harnessed through the process of adaptive selection, which allows higher levels of order and meaning to emerge. It is then a virtue, not a vice. It allows purpose to be an active agent by selecting desired outcomes from a range of possibilities.

8.2.6.3 Purpose

Function or purpose is the core of all life [104], enabled by suitable macro and micro physiological structures [30]. Desired goals are attained via inbuilt homeostatic systems, instinctive nervous systems, or unconscious and conscious operations of the mind [96, 122, 124, 125, 164].

Physiological Functioning. Based on attaining specific goals (a desired temperature, blood pressure, resting action potential, and so on), this enables the stability of living systems under the pressure of all sorts of perturbations. Life depends on the macro and micro homeostatic systems based on feedback control loops that fulfil some purpose or goal [119]. Information is causally effective via feedback control loops, because without the checks of outcome involved in such loops (comparing the actual situation with the desired situation), the attainment of desired goals will be unreliable. Homeostatic systems maintain constant conditions in the internal environment in the human body (Guyton [98, pp. 4–5]):

> Essentially all the organs and tissues of the body perform functions that help to maintain these constant conditions. For instance the lungs provide oxygen as required by the cells, the kidneys maintain constant ion concentration, and the gut provides nutrients.

Homeostasis is ubiquitous in plants and animals [30]: it is the purpose of macroscopic physiological systems [164] and numerous control systems in cell biology [104]. Most of the processes to maintain homeostasis in cells occur in the form of diffusion of materials across membranes, including osmosis, passive transport, and active transport. The human body maintains blood pH within the very narrow range from 7.35 to 7.45. One can live only a few hours with a blood pH below 7.0 or above 7.7, but the body's metabolism produces a variety of acidic waste products that tend to drive pH out of the safe range. Core body temperature is normally 37.2–37.6 °C, and if one goes below 33 °C or above 42 °C death is likely. In both cases physiological systems keep the body in the desired range. That is the purpose of these systems. Other systems regulate blood sugar concentration, water balance, blood pressure, and so on, and brain function is only possible because of voltage-gated Na^+, K^+, and Ca^{2+} ion channels that maintain resting nerve cell potentials. Delayed negative feedback channels govern many rhythms that are crucial to life, such as respiration and heart beat [152, pp. 68–71]. The purpose of the system is to maintain those rhythms.

All these systems came into being through natural selection, because they promoted survival. Once in existence, passed from generation to generation by genes and developed in each body by developmental processes, they have specific functions or purposes that are allowed by and indeed implemented through the underlying physics. But that physics knows nothing of these purposes.

Mental Functioning. Directed at attaining specific goals and utilising the brain's predictive ability [106], this enables the causal power of the mind in the world [58]. This attained great power, firstly when pressures of living together led to development of social institutions that enabled society to function [16, 135], and secondly when cumulative understanding was built up through social interactions that developed a society and science-based technology. Indeed it led to the ascent of man [25]. Purposeful design underlies all the features we expect in life today: automobiles, aircraft, buildings, roads, tapwater, books, clothes, houses, furniture, refrigerators, stoves, washing machines, TV, cell phones, the internet, medicines, and industry.

This is based on learning at the macro-level and underlying brain plasticity at the lower levels. However, ultimately it is shaped by one's high level understanding of meaning and choice of ethical goals, for these shape all lower level goals by delineating what is desirable and what is not acceptable [78]. In mental terms it is based on the way language can be used to describe situations and explore options. What made this possible was that brain systems enabling such functioning came into being through natural selection, because this capacity promoted survival. Once in existence, passed from generation to generation by genes and developed in each body by developmental processes, these structures enable generic pattern recognition, predictions, and generalization processes that underlie effective thought. This is allowed by and indeed implemented through the underlying physics, but that physics knows nothing of these plans and theories.

8.2.6.4 The Combination of Chance, Necessity, and Purpose

All three kinds of causation occur in an intricate interplay in the real universe. The fact that physics is not the only form of causation in the real world has been demonstrated above by numerous examples. Chance, necessity, and purpose all occur in living systems. It is the relation between them that is at issue [64], and this is where information comes in (Hartwell et al. [104]):

> To describe biological functions, we need a vocabulary that contains concepts such as amplification, adaptation, robustness, insulation, error correction and coincidence detection. For example, to decipher how the binding of a few molecules of an attractant to receptors on the surface of a bacterium can make the bacterium move towards the attractant (chemotaxis) will require understanding how cells robustly detect and amplify signals in a noisy environment.

And how does purpose fit in? An element of randomness at the bottom does not mean that all that happens is just pure chance. Rather it is one of the foundations that, together with necessity, opens up the possibilities of purposeful function and meaningful mental life, realised through physical existence. It does not have to have the connotation of meaningless so often ascribed to it. It is the gateway to variety and possibility.

Adaptive Selection and Randomness. Selection criteria shape what happens at lower levels according to higher level purposes when adaptive selection takes place (TD3), as shown in Fig. 8.3 and represented by Eq. (8.4). For adaptive selection to work, it needs some kind of random ensemble to work on. This must be generated by some kind of randomising process. According to Abbot, Davies, and Shalizi [2]:

> We now recognize that noise plays an indispensable role in many creative processes by providing a disturbing or enervating influence that can shunt a physical system randomly through a selection of states. In many systems, living and nonliving, there is an optimal state, defined according to some criterion of fitness (in the biological case that being the most suitable adapted organism). Noise will then enable the system to 'discover' the optimal state and maintain it. This principle of random shuffling toward an optimal state provides the basis for the powerful techniques of genetic algorithms, which have application to a wide range of practical design problems.

At the Lower Levels: Genes and Cells. Noise has many roles in biological function, including generation of errors in DNA replication leading to mutation and evolution, noise-driven divergence of cell fates, noise-induced amplification of signals, and maintenance of the quantitative individuality of cells [162]. Promoter decoding of transcription factor dynamics involves a trade-off between noise and control of gene expression [101]. The control circuits are adapted to take this into account. Elowitz et al. comment as follows [67]:

> Clonal populations of cells exhibit substantial phenotypic variation. Such heterogeneity can be essential for many biological processes and is conjectured to arise from stochasticity, or noise, in gene expression [...] proteins are produced from an activated promoter in short bursts of variable numbers of proteins that occur at random time intervals. As a result, there can be large differences in the time between successive events in regulatory cascades across a cell population. In addition, the random pattern of expression of competitive effectors can produce probabilistic outcomes in switching mechanisms that select between alternative regulatory paths. The result can be a partitioning of the cell population into different phenotypes as the cells follow different paths.

Molecular recognition is a case in point [82], as are molecular machines [34]. Hoffmann describes [110] how molecular machines extract order from chaos ('the molecular storm'):

> Beneath the calm, ordered exterior of a living organism lies microscopic chaos, molecules in liquid continuously crash into each other as part of their thermal motion. Powered by energy, microscopic molecular machines the ratchets of the title work autonomously to create order out of the chaos. Tiny electrical motors turn electrical voltage into motion, tiny factories custom-build other molecular machines, and mechanical machines twist, untwist, separate and package strands of DNA [...] Life emerges from the random motions of atoms filtered through these sophisticated structures of our evolved machinery which harness the disorder of the molecular storm.

The Brain. Through these processes, randomness is crucial in the brain. According to Glimcher, randomness is apparent in neuroscience at the microlevel and behaviour at the macrolevel [86]:

> Recent advances in the psychological, social, and neural sciences, however, have caused a number of scholars to begin to question the assumption that all of behavior can be regarded as fundamentally deterministic in character [...] The theory of games makes it clear that an organism with the ability to produce apparently indeterminate patterns of behavior would have a selective advantage over an animal that lacked this ability [...] at the level of action potential generation, cortical neurons could be described as essentially stochastic [...] the evidence that we have today suggests that membrane voltage can be influenced by quantum level events, like the random movement of individual calcium ions [...] the vertebrate nervous system is sensitive to the actions of single quantum particles. At the lowest levels of perceptual threshold, the quantum dynamics of photons, more than anything else, governs whether or not a human observer sees a light?

Deco, Rolls, and Romo [46] show that, in the dynamics of neural processing, noise breaks deterministic computations and has many advantages. They show how computations can be performed through stochastic dynamical effects, including the role of noise in enabling probabilistic jumping across barriers in the energy landscape describing the flow of the dynamics in attractor networks. *The Noisy Brain: Stochastic Dynamics as a Principle of Brain Function* [167] describes approaches that provide a foundation for this understanding, including integrate-and-fire models of brain and cognitive function that incorporate the stochastic spiking-related dynamics, and mean-field analyses that are consistent in terms of the parameters with these, but allow formal analysis of the networks which include some of the effects of noise on the operation of the system. Unlike digital computers, brain function cannot be understood as a deterministic noiseless system [143, 144].

In Summary. Lower level random processes allow adaptive selection to work, creating purposeful order, on the basis of physical and chemical laws, embodying necessity. Physics provides the possibility space for what happens, but does not determine the outcome. Top-down causation allows higher level causes be what they appear to be: real effective causes that select lower level outcomes. Adaptive selection creates new classes of information and new instances of those classes, e.g., the genetic code and DNA that uses that code. Random fluctuations plus quantum uncertainty provide the freedom at the bottom needed to allow this to happen.

> **Thesis 4: Randomness and Purpose.** It is the existence of random processes at lower levels that enables purposeful actions at higher levels to take place through selection of preferred outcomes according to some higher level selection criterion. This enables processes of adaptation and learning in accordance with the logic of some higher level purpose.

Thus randomness is indeed key to the existence of complex life forms, and is important in molecular and developmental processes [110]. The iron grip of bottom-up determinism is broken by this mechanism, which enables true emergence to occur.

8.3 Types of Causation

The previous section looked at the key element of purpose or goal-seeking, which is crucial to life [104]. This is related to the larger issue of what types of causation shape what happens in the physical and biological world, a topic that has been debated since the time of the Greek philosophers. I consider here the following issues:

- Levels of causation and Aristotle (Sect. 8.3.1).
- Multiple causes and contextual factors (Sect. 8.3.2).
- Causal effects of Platonic (non-emergent) entities (Sect. 8.3.3).

8.3.1 Levels of Causation and Aristotle

Reductionist analysis 'explains' the properties of the machine by analysing its behaviour in terms of the functioning of its component parts (the lower levels of structure). Systems thinking tries to understand the properties of the interconnected complex whole [3, 4, 36, 75], and 'explains' the behaviour or properties of an entity by determining its role or function within the higher levels of structure.

8.3.1.1 The Example of an Aircraft

For example, the question 'Why does an aircraft fly?' can be answered as follows (Fig. 7.13):

- **In Bottom-Up Terms**. It flies because air molecules impinge against the wing with slower moving molecules below creating a higher pressure as against that due to faster moving molecules above, leading to a pressure difference described by Bernoulli's law, which counteracts gravity, etc.
- **In Terms of Same-Level Explanation**. It flies because the pilot is flying it, after a major process of training and testing that developed the necessary skills, and she is doing so because the airline's timetable dictates that there will be a flight today at 16h35 from London to Berlin, as worked out by the airline executives on the basis of need and carrying capacity at this time of year.
- **In Top-Down Terms**. It flies because it is designed to fly (Fig. 7.14). This was done by a team of engineers working in the historical context of the development of metallurgy, combustion, lubrication, aeronautics, machine tools, computer-aided design, etc., all of which was needed to make this possible, and in an economic context of a society with a transportation need and complex industrial organisations able to mobilise all the necessary resources for design and manufacture. A brick does not fly because it was not designed to fly.
- **Ultimate Explanation**. And why was it designed to fly? Because it will make a profit for the manufacturers and the airline company! Without the prospect of that profit, it would not exist.

These are all simultaneously true non-trivial explanations. The plane would not be flying if they were not all true at the same time. The higher-level explanations involving goal choices rely on the existence of the lower level explanations involving physical mechanisms in order that they can succeed, but are clearly of a quite different nature than the lower level ones, and are certainly not reducible to them nor dependent on their specific nature. The bottom-up kind of explanation would not apply to a specific context if the higher-level explanations, the result of human intentions, had not created a situation that made it relevant.

8.3.1.2 Aristotle and Multiple Levels of Causation

This situation was captured by Aristotle through his proposal of four different kinds of causation. According to Falcon [68], they are:

- **The Material Cause**. That out of which, e.g., the bronze of a statue.
- **The Formal Cause**. The form, the account of what-it-is-to-be, e.g., the shape of a statue.
- **The Efficient Cause**. The primary source of the change or rest, e.g., the artisan, the art of bronze-casting the statue, the man who gives advice.
- **The Final Cause**. The end, that for the sake of which a thing is done, e.g., health is the end of walking, losing weight, purging, drugs, and surgical tools.

The last is a teleological explanation, i.e., an explanation that makes a reference to *telos* or purpose. Additionally, circular causation is possible: things can be causes of one another through a relation of reciprocal influence.

Table 8.3 Aristotle's four kinds of cause related to levels of causation

Cause 1	Final	Topmost
Cause 2	Efficient	Next higher
Cause 3	Formal	Same level
Cause 4	Material	Next lower

These four kinds of causes correspond broadly to those identified above in the case of the aircraft flying. Indeed, we can adapt Aristotle's categorisation to the hierarchical context considered here, by seeing the material cause as the lower level (mechanical) cause, the formal cause as the same level (immediate) cause, the efficient cause as the immediate higher (contextual) cause, and the final cause as the ultimate higher level cause, as shown in Table 8.3. The key point is that all four kinds of causation will be at work in all everyday circumstances.

Example: Physics Experiments. Successful completion of a physics experiment, such as observing particle production in a collider like the Large Hadron Collider (LHC), involves all the reinterpreted Aristotelian forms of causation. The material (mechanical) cause is the particle interactions that lead to the production of new particles. The formal (immediate) cause is that the experimenter turns the accelerator and measuring equipment on at a particular time. The efficient (contextual) cause is that the collider was designed and manufactured so that the collisions would take place and outcomes could be observed. The final cause might simply be that the experimenter wanted to determine whether the data supported a physical theory that predicted the existence of a Higgs boson, or it might be because she aspired to attaining a Nobel Prize.

8.3.2 Multiple Causes and Contextual Factors

The bottom line is that one cannot understand what is happening to the elements of the system without taking the system as a whole into account. This for example underlies systems biology [5, 51, 152], but it in particular underlies the understanding of brain function as a whole. The lower levels enable things to happen, but context determines what happens: physics *per se* neither causes the aircraft to come into existence, nor to fly, and neither does biochemimstry, cell biology, evolutionary theory, neuroscience *per se*, or cosmology. It is the operation of the human mind in a social context that does so.

8.3.2.1 The Contextual Factors

This takes place in a multiply layered context, and one cannot sensibly discuss their nature out of that context. There are four kinds of contexts for physical processes, as mentioned in the introduction: the natural world, life, humanity, and artefacts. In the case of the aircraft, the relevant higher level structure is the society in which it is constructed. Practical contextual issues are:

- Is the relevant design capacity available?
- What about machine tools?
- Are the needed materials available?
- What about components such as engines and avionics?
- What about finance?

And of course, the list continues. The societal context is crucial. Given all these factors, there is still a range of design decisions that crucially shape the actual form of the physical aircraft:

- Will it be long haul or short haul aircraft?
- What priority is placed on carbon emissions?
- Will it aim to provide comfort for a few people, or cheap transport for a lot of people?

And once again, the list continues. These decisions, shared by the design team, are abstract choices that will determine the physical outcome of the process. Thus they are also contextual variables: together with the economic and engineering context, they are the environment that leads to the specific nature of the aircraft coming into existence.

8.3.2.2 The Causal Network

The key point about causality in real-world contexts, then, is that simultaneous multiple causality (inter-level, as well as within each level) is always in operation in complex systems. Indeed, there will always be a network of causes in operation, leading to multiple explanations for any effect, including those identified here, but also the overall historical and physical and social environment, without which the identified events would not take place (for example, the laws of physics are as they are, the Earth exists, scientists are able to do experiments, measuring apparatus can be devised reliably, and so on).

> **Thesis 5: Multiple Explanations**. There will always be multiple valid explanations for any physical outcome. Thus one can have top-down system explanations as well as bottom-up and same level explanations, corresponding to the four kinds of Aristotelian explanation, and all being simultaneously applicable.

In terms of the brain, this can be summarised as follows (Hyman in [123, p. 200]):

> The implications for psychiatry of Kandel's work and that of others who have worked on brain plasticity is that life experience and indeed all types of learning, including psychotherapy, influence thinking, emotion, and behaviour by modifying synaptic connections in particular brain circuits. Moreover, as many scientists have shown, these circuits are shaped over a lifetime by multiple complexly interacting factors including genes, illness, injury, experience, context, and chance.

As has been discussed in this book, these factors include theories, ideas, plans, and values.

8.3.3 *Causal Effects of Platonic (Non-Emergent) Entities*

Some aspects of complex systems are emergent from the interaction between the underlying particles and forces, but others are not emergent: rather they arise from the nature of the external environment. A crucial point then is that this environment includes abstract Platonic entities, such as mathematical forms and the laws of logic, which are not reducible to or emergent from any physical entities. They do, however, have causal power. They are transcendent entities in that they are timeless and universal, but not of a physical nature. They are not coarse-grained lower level variables, and exist independently of the mind, even though they are discovered and comprehended by the mind [35]. These have been covered above in Sect. 7.6, so I will just mention them:

- **Mathematics**. The paradigmatic example is mathematics (Sect. 7.6.1): we are confident that all intelligent species in the universe will discover the same set of mathematical objects and relationships, albeit expressed in different notations. And this is plausibly because they are of a Platonic nature [32, 156]. The key point is that these entities are discovered, not created, a hallmark property being that what we find is often surprises that were not expected or desired by their discoverers (for example, the fact that both the square root of 2 and π are irrational).
- **Logic and Algorithms**. Logic is a further example: Boolean algebra will be discovered by any logician anywhere in the universe. Logic is the basis of computational algorithms [126], so they too are of an abstract nature that is discovered, even though we talk of someone inventing an algorithm. The point is that they cannot develop it unless it works. From the viewpoint of the logic involved, it is a discovery of the existence and nature of timeless and eternal possibilities (Sect. 7.6.2).
- **Language Constraints**. Terence Deacon has plausibly argued [44] that any language system must obey necessary semiotic constraints on their grammatical structure. These too are discovered by groups of humans as they develop their language over time, because they are based on the underlying logic of what is possible in effective symbolic representation of the real world. It is these constraints that lead to the language uniformities characterised by Chomsky as deep grammar structure.
- **Physics Theories**. Scientific theories that characterise our understandings of the behaviour of the physical world are also of this nature: we believe, for example, that scientists on any other planet will also discover the existence of the same four fundamental forces we have discovered. Thus they will plausibly understand electromagnetism and gravity much as we do. These physics theories too are discovered rather than being created: we find out the way things are in the real world through the scientific method of experimentation, and represent them as the mathematical laws of physics we teach our students. These theories, albeit expressible in many forms, may be expected to transcend time, place, and culture. These theories have huge practical consequences: they underlie all engineering and technology, for example nuclear power stations, aircraft, power lines, dam safety, bridge design, weapons design, and so on (Sect. 7.5.3).

- **Possibility Spaces**. These various possibilities can be thought of in terms of possibility spaces that characterise in an eternal and unchanging way what is and what is not possible. Thus the phase spaces of physical theories characterise the entire set of possible physical motions, while the space of mathematical theorems characterizes the relations between possible mathematical theorems [24], and so on. There are even timeless and eternal Platonic possibilities delimiting what will ever be possible in biology [196].[1]

8.3.3.1 The Effectiveness of Platonic Entities

These abstract entities cannot be omitted from a complete description of causation, as they have changed society.

> **Thesis 6: The Effectiveness of Platonic Entities**. They are understood by the human mind through a process of logical exploration, enabled by the neural net structure of the cortex, as explained by Churchland [35].

Thus it is not true that all entities that have physical causes are themselves physical. Abstract Platonic possibility spaces are part of the context in which we live. They are causally effective through the nature of the brain and operation of the mind (Sect. 7.6.3), thereby affecting the world. This causal connection is shown in Figs. 7.6 and 7.16.

8.4 Aristotle and Types of Knowledge

A key issue is how causation relates to knowledge. As pointed out by Flyvbjerg et al. [78, pp. 1, 16], Aristotle considered knowledge to have three very different components:

- *Epistemé*, or universal truth. This abstract and universal truth is the domain of science.
- *Techné*, or technical know-how associated with practising a particular craft. This is the domain of engineering and technology.
- *Phronesis*, or practical wisdom on how to address and act on social problems in a particular context. This comes from an intimate familiarity with the contingencies and uncertainties of any particular social practice.

Phronesis is seen as the most important of the intellectual virtues, because it is needed for the management of human affairs, including the management of *epistemé* and *techné*, which cannot manage themselves [78, p. 1]. It is knowledge that is sensitive to its application in specific settings, and hence has prominence in social action and thought.

[1] See http://aeon.co/magazine/philosophy/natures-library-of-platonic-forms for an accessible description.

In his book *Making Social Science Matter* (MSSM) [77], Flyvbjerg argues that, as the social sciences study human interactions that involve human consciousness, volition, power, and reflexivity, attempts to build generalizable, predictive models such as those for the natural world are misplaced and even futile. Rather one must recognize the context dependence of socially relevant forms of knowledge on the one hand, and their dependence on value systems that shape what we do, on the other [78, p. 2]:

> MSSM argued that, given their subject matter, the natural sciences are better at testing hypotheses to demonstrate abstract principles and law-like relationships, while the social sciences are better at producing situated knoweldge abut how to understand and act in contextualised settings, based on deliberation about specific sets of values and interests. Such deliberation about values and interests is central to social, political, and economic development in any society. [...] The natural sciences excel at conducting decontextualised experiments to understand abstract and generalizable law-like relationships, while the social sciences can conduct contextualised studies involving field research that produces intimate knowledge of localised understandings of subjective human relationships, and especially in relationships to the values and interests that drive human relationships.

This is clearly in agreement what is stated above in Sects. 7.4–7.6, and particularly Fig. 7.6, where values are supreme. These same level relationships in the mind proceed on the basis of their own logic, which exists independently of the underlying physics. It is this kind of logic that determines what happens in the real world [78], not the laws of physics. How do we show this is the case? Contemplate the daily news media and analyse what is happening!

8.5 A More Holistic View

This section puts together the holistic view that has been emerging from the rest of the book. I discuss in turn the following issues:

- Recognising emergence and top-down causation (Sect. 8.5.1).
- Other causal influences than physics (Sect. 8.5.2).
- The main thesis (Sect. 8.5.3).
- The counter view: scientific reductionism (Sect. 8.5.4).

8.5.1 Recognising Emergence and Top-Down Causation

Emergence and whole–part or top-down causation are closely related, in that the latter enables the former. There are other causal influences at work than just physics. This is summarized in the main thesis that follows below (Sect. 8.5.3).

8.5.1.1 Emergence

The issue is whether there is real emergence of higher levels, with genuine causal powers in their own right ('strong emergence' [37]), or whether the higher levels are epiphenomena, with no real power of their own: they are dancing to the tune of the lower levels.

This book, through many examples, makes the case that emergence has to be real in both diachronic terms (over time) and synchronic terms (at each moment in time): the higher levels do indeed have genuine causal power in their own right. Support for this view comes both from outcomes, and from the nature of the mechanisms in operation:

- It has to be so because of the cosmic context: it is not possible that initial data in the universe alone led to the words on this page (Sect. 8.1).
- It has to be so because of the causal power of thoughts (Sect. 7.5.3) and social constructions such as money [57] (Sect. 7.5.4).
- It has to be so because of the causal power of digital computer programs over the lower level physical structure in a computer (Sect. 2.7).
- It has to be so because the mechanisms leading to emergence in microbiology and physiology are contextually dependent, with the higher levels controlling what happens in the lower levels through physiological [152] and epigenetic effects [85], which means that those higher levels are indeed causally effective.
- It has to be so if we believe that genes or neurons have real causal effects, because they must then control what happens at the levels below (Sect. 7.7.3).
- It has to be because it is emergent networks of components and interactions that have the causal power that decides what will happen [5], not the individual elements themselves.

Hartwell et al. express the last point in the following way [104]:

> Much of twentieth-century biology has been an attempt to reduce biological phenomena to the behaviour of molecules [...] Despite the enormous success of this approach, a discrete biological function can only rarely be attributed to an individual molecule, in the sense that the main purpose of haemoglobin is to transport gas molecules in the bloodstream. In contrast, most biological functions arise from interactions among many components. For example, in the signal transduction system in yeast that converts the detection of a pheromone into the act of mating, there is no single protein responsible for amplifying the input signal.

The causal power that decides what will happen lies at the level of interactions, not at the level of components. The latter are causally effective in terms of making the higher level purposes come true, but the former shape the sequence of events that take place. The same is true in brain function, which is contextual all the way, as demonstrated conclusively by the examples of perception (Sect. 7.3.3) and language (Sect. 7.3.4).

Consequently the only sensible view is that strong emergence does indeed take place [37, 38, 57, 149]: the higher levels are as real as the lower ones [153].

8.5.1.2 Modularity and Levels

The importance of hierarchy and modularity for emergence, clearly stated by Simon [175], Flood and Carson [75], and Booch [21], is emphasized by molecular biology studies [104]:

> We argue here for the recognition of functional 'modules' as a critical level of biological organization. Modules are composed of many types of molecule. They have discrete functions that arise from interactions among their components (proteins, DNA, RNA and small molecules), but these functions cannot easily be predicted by studying the properties of the isolated components. We believe that general 'design principles', profoundly shaped by the constraints of evolution, govern the structure and function of modules. Finally, the notion of function and functional properties separates biology from other natural sciences and links it to synthetic disciplines such as computer science and engineering.

Those design principles for modules are clearly stated by Booch [21]: abstraction, encapsulation, information hiding, and effective constrained interfaces (Sect. 3.16).

Each module is a real entity that is affected by and affects its environment. It is made up of real subcomponents that also obey the same principles. Together they form the hierarchical levels of the system. Note that this modular structure may be physical, functional, or logical. The same precepts apply in each case. In the case of living systems, cells are the key modular structures, but they are themselves comprised of submodules.

8.5.1.3 Whole–Part Causation

The important realisation then is that as well bottom-up action in this hierarchy, there is whole–part or top-down causation in this hierarchy of structure: the top levels influence what happens at the lower levels, and this is what enables the higher levels to have causal powers in their own right. They do so by setting the context in which the lower level actions function, thereby organising the way lower level functions integrate together to give higher level functions. The higher levels of the hierarchy structure what happens at the lower levels in a coordinated way, enabling self-organisation of complex systems. Boundary effects (linking the system to the environment), as well as structural relations in the system itself, effect top-down causation by changing both context and the nature of the constituent parts. They change the interaction patterns of the parts, and may shape the results of adaptive selection or embody the goals of feedback control systems. These effects are prevalent in the real physical world and in biology, because no real physical or biological system is isolated. As stated in [201]:

> The reductionist perspective is needed, but so is the perspective of the biologist interested in physiology, or whole organisms, or the relation between organisms and their environment, or their evolution. A satisfying biological explanation is one that uses different levels, and so unifies, or makes connections between, some of the research styles. Anti-reductionists do not dispute the revelatory powers of molecular techniques, but argue that the higher levels cannot be ignored. Understanding of these higher levels cannot be reduced to theories that

apply to lower levels. Emergentism states that new properties emerge as you go from level to level, and an attempt to explain these properties in terms of lower levels alone will end in failure.

An important example is human volition: the fact that when I move my arm, it moves because I have 'told it' to do so. In other words, my brain is able to coordinate the action of many millions of electrons and protons in such a way that it makes my arm move as I desire. Every artefact in the room in which you are sitting, as well as the room itself, was created by human volition, so our minds are causally effective in the world around us. Top-down action from the mind to muscle tissue enables the higher levels of the hierarchy to be causally effective.

8.5.2 Other Causal Influences Than Physics

The point then is that physics as it currently stands is causally incomplete. It is not able to describe all the causes and effects shaping what happens in the world. For example, physics cannot explain the curvature of the glass in my spectacles, because it has been shaped on purpose to fit my individual eyes. The vocabulary of physics has no variable corresponding to the intention that has shaped the spectacles. Because of this, physics cannot explain why their glasses have their particular curvature. This is possible because physics at the micro-level has an irreducible random element. This allows higher level selection processes to select lower level outcomes to suit higher level function or purpose (Sect. 8.3).

- **Information**. Information is causally effective, even though it is not a physical but an abstract entity [165]. This is true in biology [128], where the sequence of bases in DNA is an information coding pattern, in engineering where signals coding information are used in control systems, and particularly in our lives and in society where our actions are shaped by the information we receive. Indeed all animals have special sensory systems to receive and analyse information, because this is so important for their survival.
- **Ideas**. Ideas and theories have great causal power, as discussed in Sect. 7.5.3, leading to the whole edifice of technology and the ascent of man [25]. They are informed by our understandings of Platonic realities such as mathematics (Sect. 7.6).
- **Social Constructions**. These, too, are causally effective [16, 57, 135]. A classic example of this is the chess set. Imagine some being coming from Mars and watching chess pieces moving. It is a very puzzling situation. Some pieces can only move diagonally and other pieces can only move parallel to the sides. You imagine the Martian turning the board upside down and looking inside the rook, searching for a mechanism causing this behaviour. But it is an abstraction, a social agreement, that is making the chess piece move that way. Such an agreement, reached by social interaction over many hundreds of years, is not the same as any individual's brain state. It exists in an abstract space of social convention, and yet is causally effective.

- **Money**. Many other social constructions are equally causally effective, perhaps one of the most important being the value of money. This is already enough to undermine any simplistic materialistic views of the world, because these causal abstractions do not have a place in the simple materialist view of how things function.
- **Ethics**. This, too, is causally effective. It constitutes the highest level of goals in the feedback control system underlying our behaviour, because it is the choice of which other goals are acceptable. It relates to the search for meaning [79]. When you have chosen your value system, which depends also on your understanding of meaning (the *telos* or purpose of life), this governs which goals are acceptable to you and which are not. So this abstract entity is causally effective [77]. As an example, if your country believes that a death penalty is acceptable, this will result in the physical realisation of that belief in an electric chair or some causally equivalent physical apparatus. Without the death penalty they will not be there. This lies outside what reductionist physics and chemistry include in their causal schemes.
- **Purpose and Intentions**. Overall, purpose and intentions are causally effective in human life [50], just as function is effective in all biology [104].

8.5.3 The Main Thesis

The conclusion is that these effects are possible, and genuine emergence occurs, because of the conjunction of bottom-up and top-down causation [65]:

> **Main Thesis**. There are other forms of causation than those encompassed by the bottom-up processes described by physics and physical chemistry. This is clearly manifest in physiology and as regards the way the human mind operates. It is particularly clear in those cases where the higher level variables cannot be obtained by coarse-graining of any lower level variables, as is the case both in the causal power of money, and the causal effectiveness of Maxwell's theory of electromagnetism [74] in the world of engineering. It is the combination of both bottom-up and top-down effects that enables this kind of causation to happen.

A full scientific view of the world must recognise this, or else it will ignore important aspects of causation in the real world, and so will give a causally incomplete view of things. Complex causation is based on the interaction of bottom-up and top-down effects: if we neglect either, we will be unable to understand genuinely complex systems

8.5.4 The Counter View: Scientific Reductionism

Recognising these different forms of causation implies that other kinds of causes than physical and chemical interactions are effective in the real world. However, a variety of reductionists claim that all these higher level phenomena are epiphenomena, appealing to one or more of the following reasons:

- They are based solely in the underlying physics, which is causally closed, and so does not allow room for any other influences.
- They are governed by genetics and neurology: we are nothing but molecular machines.

From the viewpoint of this book, the reductionist dynamic is a very important part of what is going on, but it is not all. Strong reductionist claims that it is all that is happening are a denial of some of the causal influences in operation.

Although the scientific perspective *per se* gives a wonderful understanding of mechanisms in operation in the world, it does not encompass all the significant causal mechanisms. In contrast, holism sees the whole relevant interlocking web of causation that is in fact there. That is what is attempted here.[2]

8.5.4.1 Scientific Reductionism

The common physics view is that bottom-up causation is all there is. Micro-forces determine what happens at the lower levels and thereby are the foundation of higher level activity. Electrons attract protons at the bottom level, and this is the basic causal mechanism at work, causing everything else all the way up. This is all there is. In a certain sense that is obviously true. You are able to think because electrons are attracting protons in your neurons. But strong reductionists [8] tell us this is the only kind of causality there is, using the phrase 'nothing but' to emphasize their viewpoint (Webster [201]):

> The major task for the reductionist is to show that nothing important, no essential insight or avenue of research, is lost when some aspect of animal or human behaviour is explained in terms of chemistry: when in short the sociological, psychological or biological is abandoned in favour of the chemical bond. [...] Sometimes criminal or aggressive behaviour is explained in terms of levels of neurotransmitters in the brain [...] the reductionist has to explain not only that serotonin is involved in some way but also that we have no need for sociological theories to understand criminality: that abnormal blood chemistry fully explains abnormal behaviour.

This is a claim that a partial cause is the whole cause. There are other forms of causality in action in the real world. More holistic non-reductionist views of science will take them into account, thus taking emergent properties seriously and freeing us from the straightjacket of strong reductionist world views. What is missing is function, adaptive selection, and information (Sect. 8.2.5).

[2]The reader will recognise the rest of this section as a polemical text, building on the science in the rest of the book.

8.5.4.2 A Partial View

Reductionism comprehends part of the causal nexus and proclaims it to be the whole. It therefore reduces the whole to a part and ignores major factors of importance. This occurs across the spectrum of understanding, resulting in diminished pictures of existence and of human nature. The need is to take the whole into account. It is expressed well by John Dupre as follows [53]:

> Scientific Imperialism is the tendency to push a good scientific idea far beyond the domain in which it was introduced and often far beyond the domain where it can provide much illumination. [...] My own project is to insist that pluralism goes all the way down to the basic metaphysical issues of causality and of what kinds of things there are. This perspective makes the kinds of narrowly focused scientific projects I have been examining look as philosophically misguided as they have proved empirically unrewarding.

Any scientific speciality looks at important aspects of what is going on, but these are only a part of the whole [53]:

> These are important fragments of the picture that we have spent the last few millennia trying to put together. But they are fragments, and trying to make one or even a few such fragments stand for the whole presents us with a deformed image of ourselves. [...] An adequate view of ourselves would include many parts. One of the most traditional objections to such one-sided reductive pictures is that they leave no room for human autonomy or freedom. I have tried to show that the philosophical context in which I consider these reductive views does indeed provide an endorsement of the traditional objection. It would include biological organisation, an account of how societies function, and an account of how aspects of social organisation contribute to the endowment of human individuals with complex capacities that would in principle be beyond the reach of an isolated member of our species.

8.5.4.3 The Problem with Reductionism

Thus the struggle is between strong reductionism, or an integrative wholism. The view here is that the strong reductionist position ignores crucial causal factors:

> **Thesis 7: Scientific Reductionism**. Any claim that physics or chemistry or genetics or molecular biology alone will account for the existence or functioning of the brain, characterised by the catch phrase 'nothing but', is an example of scientific reductionism that proclaims a partial account of causal factors to be the whole truth and ignores many of the causal factors in operation.

Tse comments on how this kind of position arises [191, pp. 4–5]:

> Some philosophers deny that qualia can cause subsequent actions. They say that it is naive to claim that a person goes to the dentist because a tooth hurts. To most postbehaviourist neuroscientists, what seems naive is how epiphenomenalists can deny the facts. The fact is that the toothache causes a person to try to stop the pain [...] Such philosophers impose their first principles (e.g., reductionism, determinism) on reality and argue that pain must be epiphenomenal and acausal (or worse, they explain consciousness away by claiming that it does not exist), because this conclusion follows necessarily and logically from their 'correct' assumptions. Rather than change their first principles, they deny the facts and charge the rest of us with delusion.

Thus the antidote to such reductionism is to embrace the multi-causal nature of things, seeking to understand the web of causation. It involves being ready to see the wider causal patterns, rather than acknowledging only that part one is comfortable with or expert in. In particular it requires relating causation to context, that is, exploring the bi-directional causal effects between emerging levels of structure and function.

8.6 Implications: Learning to Read and Write

The view put forward in this book has substantial social implications. The bottom-up view of causation has pervaded much scientific thought in general, and so for example has been a major factor in medicine and psychiatry. In each of these areas there has in effect been a long-standing tension between bottom-up and top-down views, with major implications for medical practice and effectiveness. The view implied by this book is that it is crucial that top-down effects be taken into account as a major influence, as well as bottom-up (molecular and gene-based) effects. Effectively, this case is made strongly in [129] in the medical context, and in [15] in the psychiatric context.

Closely related issues arise in education: teaching methods are based on and reflect understandings and views about learning. Literacy teaching provides an apt example. When polarized into the 'reading wars' as has been the case in the USA and many other countries, especially in the north, bottom-up and top-down approaches are often both strongly recommended by opposing camps. I deal with the issue here, following [20], as follows:

- The broad context: underlying views of literacy (Sect. 8.6.1).
- The brain, prediction, and reading (Sect. 8.6.2).
- Reading as transacting with texts (Sect. 8.6.3).
- Part to whole: skills-based approach to literacy (Sect. 8.6.4).
- The contextual approach to learning (Sect. 8.6.5).
- Holistic approach to literacy (Sect. 8.6.6).
- Educational implications (Sect. 8.6.7).

This section has been written jointly with my wife Carole Bloch.

8.6.1 The Broad Context: Underlying Views of Literacy

There is a major problem with literacy teaching and learning as mass education aims to create literate citizens across the world (Armbruster et al. [7]):

> In today's schools, too many children struggle with learning to read. As many teachers and parents will attest, reading failure has exacted a tremendous long-term consequence for children's developing self-confidence and motivation to learn, as well as for their later school performance.

The question has been asked how it can be that so many children, who are seemingly effortlessly able to master so much early learning, not least the great feat of learning to listen and speak in a few years, fail to become literate (Cambourne [28]). Key issues are:

- Spoken language and written texts.
- The relation between the spoken and written word.
- The significance of stories.
- The role of emotions in literacy and learning.
- Child's play.
- The sociocultural nature of literacy.
- Learning as social practice.

8.6.1.1 Spoken Language and Written Texts

The same thought or set of ideas ('the cat sat on the mat') can be conveyed through spoken language or written texts. Each occurs in a hierarchically structured way.

Spoken Language. Spoken language takes place in conversations that are components of interactions between individuals, which in turn are part of one's overall life history. Conversations are made up of sentences comprised of phrases made up of words, and these in turn are made up of phonemes. This hierarchical structure is set out in Table 8.4. Sentences are the basic unit of meaning, and while they are made up of the lower level components, they and the conversations in which they are imbedded exert a top-down influence to the phrase and word level, answering questions like: What does 'then' refer to? Who is 'she'? Will the plane fly, or will it smooth wood?

Written Texts. Much the same applies to written texts. The relevant hierarchical nature is in Table 8.5. The basic unit of meaning is the sentence. It attains its contextual meaning from the paragraph, chapter, and book in which it is imbedded. It influences the understanding of words and pronunciation of graphemes in a contextual way.

Table 8.4 The hierarchical nature of spoken language

Level	System	Effect	
7	Life history		
6	Interactions	⇓ Top-down	
5	Conversations	⇓ Top-down	Top-down ⇓
4	Sentences	Unit of meaning	Top-down ⇓
3	Phrases	⇑ Bottom-up	⇐ Top-down ⇓
2	Words	⇑ Bottom-up	⇐ Top-down ⇓
1	Phonemes	⇑ Bottom-up	⇐ ⇐

Table 8.5 The hierarchical nature of written text

Level	System	Effect	
9	Library	⇓ Top-down	
8	Book	⇓ Top-down	
7	Chapter	⇓ Top-down	Top-down ⇓
6	Paragraph	⇓ Top-down	Top-down ⇓
5	Sentences	Meaning	⇐ Top-down ⇓
4	Phrases	⇑ Bottom-up	⇐Top-down ⇓
3	Words	⇑ Bottom-up	⇐ Top-down ⇓
2	Graphemes	⇑ Bottom-up	⇐ ⇐
1	Letters	⇑ Bottom-up	

8.6.1.2 The Relation Between the Spoken and Written Word

One can represent the same concept or idea in a spoken or written way. The two representations are in correspondence with each other in a way that is accurate at the sentence level, but approximate at the word and grapheme/phoneme level.

Underlying the debate about teaching is a claim by Reid Lyons, Shaywitz, and others that there is an essential difference between the spoken and written word: the first is natural and thus learned easily, but the second is not and is thus difficult to learn and has to be taught. Shaywitz states this as follows [170, p. 50]:

> This essential distinction between written and spoken language was best captured by linguist Leonard Bloomfeld: writing is not language, but merely a way of recoding language by visible marks. The written symbols have no meaning in their own, but rather stand as surrogates for speech, or, to be more exact, for the sounds of speech.

But this is wrong: both written and spoken language are alternative representations of thoughts and ideas. When I have a thought in my head, I don't hear it as if spoken, neither do I see it as if written: I think it. I can then represent it in sound or in writing or by pictures or by signing. There are various symbolic ways we can represent the same thoughts [43], although constrained by semiotic requirements [44].

Neither oral nor written language is inbuilt: rather the propensity to learn them both is inbuilt. Developing oral language was a key evolutionary development underlying the development of the social brain [52], and developing writing is just a change in the mode of representation of the underlying thoughts represented by and enabled by spoken language. But that representation is not the same as the thought. The introduction of writing then became central to development of culture, but does not alter the essential relation between the underlying thought and its symbolic representation [43]: it's just a different symbolism. Both are learned in similar ways by interacting with the social world around us [97, 189]. And the profound link between the two is that the primary way over time that humans came to arrange and make sense of their lives is through their ability to imagine, create, and relate to stories, both our own and those of others. First this was oral, particularly round the fireside. Then with writing first and printing second, one could communicate with distant friends, or people one never knew. Understandings and stories could be shared over long distances and long times.

8.6.1.3 The Significance of Stories

Jonothan Gotschall [95] emphasizes that it is stories that make us human [95, pp. xii–xiv]:

> Tens of thousands of years ago when the human mind was young and our numbers were few, we were telling one another stories. And now, tens of thousands of years later, when our species teems across the globe, most of us still hew strongly to myths about the origins of things, and we still thrill to an astonishing multitudes of fictions on pages, on stages, and on screens—murder stories, sex stories, war stories, conspiracy stories, stories true and false. We are, as a species, addicted to stories. Even when the body goes to sleep, the mind stays up all night, telling itself stories.

And this is all related to the search for meaning: our attempt to make sense of it all, and to let us enter other worlds [133, pp. 3–21]. Such story-telling is important in bonding humans together, giving them a sense of communal identity and purpose, and so is important in evolutionary history [52] and in the present day world (albeit the modes of sharing stories have changed greatly with the rise of modern technology).

8.6.1.4 The Role of Emotions in Literacy and Learning

Crucial for considerations about language and literacy learning is the evidence which points to the central role that the emotions play in learning and development. Greenspan and Shanker's observations of babies' development and their ability to create symbols and think lead them to claim [97, p. 210]:

> Emotions are at the very heart of language development [...] A child's first words, her early word combinations, and her first steps towards mastering grammar are not just guided by emotional content, but, indeed, are imbued with it.

Thus language, which comes from 'lived experience', carries the prominence of relationships and shared meaningful experiences. In large part, it is the satisfaction and comfort of the experience of emotional connection that drives children to learn more.

8.6.1.5 Child's Play

Play, a universal feature of early childhood, appears early as a critical aspect of language development. Jean Piaget [159] identified many different forms of play, but it is symbolic (or imaginative) play that has particular relevance for early literacy learning. Research into symbolic play suggests that it underpins and precedes the understanding of written language.

For Vygotsky [195, p. 101], play is important because it is a leading factor in development. He also makes the connection between children's symbolic play and written language as "second order symbolism which gradually becomes direct symbolism" [195, p. 106]. He suggests that the preschool years are the ideal time for a 'natural' and meaningful introduction to learning written language. His view is that [195, p. 111]:

> Symbolic representation in play is essentially a particular form of speech at an earlier stage, one which leads directly to written language.

Bruner [26, pp. 45–63] discusses this key interaction between play and language [26, pp. 45–47]:

> Games are constituted by language and can only exist where language is present [...] they offer the first opportunity to explore how to get things done with words [...] children love to play, and they love to play games.

And when children pretend, their play is story in action [154], and all of their great powers of imagination and prediction come to the fore [93], leading to effective learning, consolidation, and most importantly, the sense-making of the stream of experiences and information they are exposed to.

8.6.1.6 The Sociocultural Nature of Literacy

Both spoken and written language exist in a social context, which shapes how and what we read and write. Our conversations usually take place with family and immediate friends, or in the local community. But they take for granted meaning inherited from the national and international environment. We read texts that may be letters from friends, emails, newspapers, books, and so on. Each contains imbedded and implied meaning from the social environment, as indicated in Table 8.6.

With regards to the way literacy is understood, two overarching views of literacy, broadly defined as reading and writing of text are identified by Brian Street [182]. The first is the 'autonomous model of literacy', where sets of skills are separated out from context and viewed as neutral. The other is the 'ideological model', where reading and writing are viewed as integral to people's social and cultural practices.

Table 8.6 The social context of conversations and written text

Level	System	Effect
6	Global context	⇓ Top-down
5	National context	⇓ Top-down
4	Local community, school	⇓ Top-down
3	Immediate friends	⇓ Top-down
2	Family	⇓ Top-down
1	Individual	Spoken word

Understanding literacy as sociocultural in nature allows it also to be understood as more similar than different to oral language. Language, both oral and written, is primarily about communication and self-expression and occurs in the company of others. No language learning or use would happen in total isolation. It's purpose is to make mutually shared meaning possible.

Although not always appreciated or attended to, these views of literacy have an influence on the way initial reading and writing get taught.

8.6.1.7 Learning as Social Practice

Barbara Rogoff, a social and cultural anthropologist, explains how people learn as participants in communities of practice [166]. Young children learn the culturally valued activities in their communities by being apprenticed to more experienced 'others', and they join in to a whole activity, at the level they are able to, gradually coming to take over more parts, in negotiation with the other.

These points support the view that literacy is social and cultural practice [10, 107, 182, 183]. This implies that to learn to read and write require a process of being immersed in authentic experiences with print, which include, but are not dominated by, learning the technical components (alphabet, letter sound relationships, and so on). The primary impetus for a child is making sense, so language is used for real reasons from it's earliest immature manifestations. Children use a range of cueing systems as they engage with a text: semantic, grapho-phonic, prior knowledge, grammatical (Goodman [91]) so the reader is transacting with the text rather than being a passive recipient: bringing to, as well as taking from it. In this 'ideological' conception, like oral language, literacy has different uses, and people have different reasons for reading and writing, depending on particular histories. Literacy cannot be separated from its uses.

Gradually, through many purposeful and meaningful interactions with people and print over time, their performance moves from immature approximations of reading and writing (pretend reading and emergent writing with invented spelling) to conventionally accepted modes (Ferreiro and Teberosky [71]). The various ways that children are (or are not) introduced and exposed to literacy-related activities at home and in the community thus informs and influences school learning.

So the ability to read and write emerges in a way similar to the way babies learn oral language, by a trial-and-error process with feedback, but always concentrating primarily on the way that language is essentially about conveying meaning [99, 112]. This happens the same way that most learning takes place [80, p. 86]:

> Most learning in childhood happens without a teacher. No one can teach you how to ride a bicycle. You have to learn by doing it yourself. We learn the fundamentals of language before any teaching occurs.

The skills that are learned as a person becomes literate happen in the service of the uses people have for reading and writing. The argument is laid out in detail in [20]. A detailed study of how this occurs for young children's writing is given in Bloch [18].

8.6.2 *The Brain, Prediction, and Reading*

The issue of reading ties into a much larger picture of how the brain functions in a top-down way. The brain is exquisitely constructed to search for meaning [50] and to predict what is likely to happen [106], and this is what shapes the way our senses work. In particular, it applies to the following:

- Vision.
- Listening to music.
- Taking part in a conversation.
- Reading a text.

8.6.2.1 Vision

This happens particularly in vision: it is not true that vision can be understood simply as data coming in from our eyes and being interpreted by the brain. Rather the brain is continually predicting what ought to be there, and filling in what it expects to see on the basis of only some of the data that it actually analyses at any one time. This can conclusively be shown to be the case by analysing visual illusions (see [80], [124, pp. 226–329], [138], and [160]). Low level visual processing establishes the characteristics of a visual scene by locating the position of an object in space and identifying its colour. Intermediate level visual processing assembles simple line segments into contours that define the boundaries of an image and separates the body from its background [124, pp. 270–272]. High level visual processing establishes categories and meaning [124, p. 272]:

> Here the brain integrates visual information with relevant information from a variety of other sources, enabling us to recognise specific objects, faces, and scenes. This top-down processing produces inferences and tests hypotheses against visual experience, leading to conscious visual perception and the interpretation of meaning.

Furthermore [124, p. 284]:

> [...] we live in two worlds at once, and our ongoing visual experience is a dialogue between the two: the outside world that enters through the fovea and is elaborated in a bottom-up way, and the internal world of the brain's perceptual, cognitive, and emotional models that influences information from the fovea in a top-down manner.

Hence, a top-down process of interpretation, based on our expectations and facilitated by specific neuronal connections, modulates and shapes what we actually see. We perceive human faces, hands, and bodies in a Gestalt way as a unified whole as soon as our senses detect them, by a template-matching approach [124, pp. 286–287]. We cannot process individual parts of the face without being influenced by the whole face [124, pp. 295–302]. Top-down processing of information uses memory to find meaning [124, pp. 304–321].

8.6.2.2 Listening to Music

Similar processes happen when listening to music: expectation is a key feature of how we experience music [116, 132], and so music has a holistic nature where the entire context influences how we hear the parts.

8.6.2.3 Taking Part in a Conversation

All conversations are contextually situated and that shapes what we say and how we understand what we hear. This was illustrated in Table 8.6. The brain automatically takes all this context into account as it takes part in a conversation. This includes body language, facial expression and tone of voice. This is why the more distant we are from the communication, from phone to email, etc., the more possibilities there are for misinterpretation of the intended message—as we 'meet' the message with our own meaning.

8.6.2.4 Reading a Text

Crucial to the way we read a text, is that we do not read each phoneme, assembling them all into words, assembling those into phrases, and so on. Rather the eye skips over words, reading whole phrases at a time and filling in the bits that are not actually read. This can be demonstrated by miscue analysis and eye movement research [54, 76, 181].

Our ability to read ambiguous texts derives from the fact that context sets the meaning and even the pronunciation of words: language is driven by word associations rather than by individual words [114]. Consequently, context drives the process of reading: it is not bottom-up, it is a psycholinguistic guessing game [87] This top-down driven process is a fundamental aspect of how the brain works, and is at the core of what reading is about [181]. This has been dealt with in depth in Sect. 7.3.4. The point now is to link this to the social context.

The Social Context of Reading. The context in which this happens is the social context of the reader (Table 8.6) For example, consider the following sentences:

- "She was very sad at the time because of her mother's cancer." You have to know who 'she' was, what time is referred to, the nature of cancer, and what the personal implications of cancer are.

- "The Wright brothers transformed the nature of travel." The reader is supposed to know that the Wright brothers created the first successful aircraft, and so provided the foundations of the development of aircraft that makes intercontinental travel so easy today.
- "Hate speech is liable to lead to another holocaust." The reader is supposed to know that 'the holocaust' refers to a systematic attempt to exterminate an entire nation.

brain automatically takes all this context into account as it reads the text.

Implications. The crucial point is that these pointers to how the brain works strongly suggest that a holistic approach to teaching reading is far more natural to the brain than an initial prioritisation of phonics teaching. It also implies that the more exposure children have to knowledge and information (be it read or told to them), the more they will have in their minds to bring to each text they are presented with as they learn to read. It fits in with the evolutionary adaptations that have made the brain a versatile predictive and meaning-seeking organ [50].

8.6.3 Reading as Transacting with Texts

A holistic way is to look at the structure of language that is recognised through reading, as well as how people read. In this section, I shall look briefly at the following:

- Mappings, thought, and language.
- Textual interaction.
- Corpus studies and priming.

8.6.3.1 Mappings, Thought, and Language

In his book *Mappings in Thought and Language* [69], Fauconnier looks at the mappings between domains that are at the heart of human cognition, and how they relate to meaning construction. He argues that the same mapping operations are at work in elementary semantics, pragmatics, and higher-level reasoning [69, p. 5], with mental spaces being the domains that discourse builds up to provide a cognitive substrate for reasoning and for interfacing with the world [69, p. 34].

As to the learning task [69, p. 189]:

> What children learn is not language structures in the abstract. They acquire entire systems of mappings, blends, and framing, along with their concomitant language manifestations. [...] For a child, to know grammar is not primarily to know which strings are well-formed or ill-formed; it is to know how to apply partial grammatical constructions in context to produce appropriate cognitive configurations.

In this view, the essence of language is the meaning construction system: mappings, frames, and spaces. And this is all contextual [69, pp. 7–8]:

> The interesting cognitive constructions underlying language use have to do with complete situations that include highly structured background knowledge, various kinds of reasoning, online meaning construction, and negotiation of meaning.

8.6.3.2 Textual Interaction

The way that negotiaton takes place is describe by Hoey in his book *Textual Interaction: An Introduction to Written Discourse Analysis* [113]. He defines text as follows [113, p. 11]:

> Text can be defined as the visible evidence of a reasonably self-contained interaction between one or more writers and one or more readers, in which the writer(s) control the interaction and produce most of the language.

Reading interacts with expectations that operate on more than one level [113, p. 23] in what Goodman calls a psycholinguistic guessing game [87]. The hierarchical nature of the text interacts with a hierarchical set of expectations [113, pp. 53–71], which is where much grammatical structure originates. Signals from the writer to the reader give moment by moment guidance [113, p. 27]. As stated by Hoey [113, p. 31]:

> As readers interact with a text they formulate hypotheses about how the text will develop, and these hypotheses help them understand and interpret the text as they continue reading. Learners therefore need to be encouraged to develop the appropriate hypothesis-forming skills and not to treat reading as an exercise in language practice only.

This contextual meaning-based process is what real reading education is about.

8.6.3.3 Corpus Studies

How does this work out at the level of words and sentences? There is now a large volume of data on collocations in text, based on corpora studies (that is, computer studies of large bodies of natural text [115]), that show how our reading at that level is driven by probable patterns of word associations [114].

Collocation. This is a psychological association between words up to four words apart, and is evidenced by their occurrence together in corpora more often than is explicable in terms of random distributions [114, pp. 5, 43–44]. It is a psycholinguistic phenomenon, the evidence for which can be found statistically in computer corpora.

Priming. The cause of the occurrence of collocation (the recurrent co-occurrence of words) is *priming*: every word is primed for collocational use, as explained by Hoey in *Lexical Priming: A New Theory of Words and Language* [114, p. 8]:

> As a word is acquired through encounters with it in speech and writing, it becomes cumulatively loaded with the contexts and co-texts in which it is encountered, and our knowledge of it includes the fact that it co-occurs with certain other words in certain kinds of context. The same applies to sequences built out of these words; these too become loaded with the contexts and co-texts in which they occur.

This hierarchical priming (words can be primed for collocation, semantic association, and colligation [114, p. 43] in a way that allows nesting [114, pp. 58–61] explains a variety of linguistic features [114, pp. 12–14], and is what underlies the contextual ways in which we read text (Sect. 7.3.4).

8.6.4 Part to Whole: Skills-Based Approaches to Literacy

It is widely accepted that, in the initial stages of learning to read and write, children need to be taught phonics as a first priority because this 'cracking the code' is the most important step to literacy learning. This involves an essentially bottom-up view. However, as already mentioned, it has major problems because of its fundamental assumptions that the 'unnatural' written language needs to be decoded into natural oral language for processing by the brain [1].

The view is based on the fact that language is hierarchically structured, so reading is a modular activity, based on the way lower level modules combine to form higher level entities. This idea is supported by various fMRI studies [22, 23, 31]. But then, all senses are modular. However, they are interpreted in a predictive contextual way, rather than in a bottom-up way (Sect. 7.3.3). The same applies to language and reading (Sect. 7.3.4).

Parts to Wholes. Teaching methods based on the bottom-up (part to whole) approach to learning written language tackle detailed technical aspects such as phonics and handwriting as a first step to learning how to read and write, and worries about the functional, communicative roles later, or as a secondary focus, with the assumption that the parts need to come together in some kind of 'building block' way, to ultimately form a meaningful whole.

In this case, if pursued to an extreme, decontextualized exercises are prioritized which ask children to recognize and sound out numerous letter-sound combinations such as ma, me, mi ma, mu, followed by phonetically regular words and even nonsense words such as tok, zat, and fot. When children have managed to learn such restricted texts, they are given opportunities to move on and engage with more meaningful texts. Crucially, in some cases, testing of children also requires children to recognise meaningless words. This omits the core functioning of language and the aim tends easily to become one of just getting the children to pass the tests.

8.6.4.1 The Phonics Approach

The nature of these methods is set out in *The Research Building Blocks for Teaching Children to Read: Put Reading First* (National Institute for literacy) [7]:

> The hallmark of programs of systematic phonics instruction is the direct teaching of a set of letter–sound relationships in a clearly defined sequence. The set includes the major sound–spelling relationships of both consonants and vowels.

There is a huge industry in the beginning stages of children's literacy using only or mainly phonics-based approaches. There is a foundational problem: the top-down predictive way that the mind works is ignored and thus children's great strengths for learning are neglected. Following the National Reading Panel's recommendation of 2008 [176], the sequence which is followed conceptually is: (1) phonemic awareness, (2) phonics, (3) fluency, (4) vocabulary, and (5) text comprehension.

Many in the field argue that teaching sequentially in the above way provides a balance of what is needed, and there are various 'systematic' and 'non-systematic' forms of instruction as described by Armbruster et al. [7]. In a 'balanced' approach, with combinations of the National Reading Panel's five categories, whatever the form of phonics instruction used, many children do of course learn to read—as the saying goes, children learn to read despite the method. But, as discussed previously, this assumes other conditions are in place. There are millions of children who do not learn and do not fulfill their potential when the underlying theory remains a bottom-up one. This is because the emphasis in many classrooms, especially in poor settings where access to high level uses of literacy is limited, tends to be the teaching of skills out of context. The result is not balanced and huge numbers of children miss out on other crucial learning components which will be discussed below.

8.6.4.2 Problems with the Phonics-First Approach

There are a series of problematic issues with teaching methods that insist on an initial emphasis on teaching phonics, and the data that is said to support them. The basic problem is that these methods do not take seriously the way reading works, as described above. The specific issues are as follows:

- The supposed support from its evolutionary base.
- Problems with the relation between phonics and written language.
- Problems with its understanding of reading.
- The supposed support from brain-imaging data.
- The claim that it has been tried and it works.

I shall now examine these in turn.

Evolutionary Bases? As regards the argument for a phoneme-based approach to written language learning as presented from an evolutionary psychology viewpoint, it is summarised by Shawitz as follows [170, p. 50]:

> Reading is not built into our genes, there is no reading module wired into the human brain. In order to read, man has to take advantage of what nature has provided: a biological module for language. For the object of the reader's attention (print) to gain entry to the language module, a truly extraordinary transformation must occur. The reader must somehow convert the print on the page into a linguistic code—the phonetic code, the only code recognized and accepted by the language system. However, unlike the particles of spoken language, the letters of the alphabet have no inherent linguistic connotation.

This is simply wrong. There is no inbuilt biological module for language in our brains (Sect. 7.2.5). Rather there is a basic ability to identify patterns in our environment plus an emotional predisposition that drives us to learn language in order to communicate with our caregivers [97]. It is not true that the human brain is hardwired to process spoken language, but not written language [61]. Both are learnt by the same kinds of pattern association abilities of the brain that underlies all our learning [80]. Having learnt associations of words with ideas, the developmentally acquired language system can associate the same idea ('cat' for example) with any of its representations—spoken, written, pictures, sign language, or mime. The spoken representation is no more privileged than any of these other representations.

Problems with the Relation Between Phonics and Written Language. Additionally, there are deep theoretical problems with the concept of intensive phonics instruction [180]. It tries systematically to simplify the complex and variable nature of the English language which is in fact not strictly alphabetic, but displays significant logographic features [178]. The relation of text to sound is not one-to-one. Thus from an educational viewpoint, this approach has been claimed to have a narrow and limiting character [145, 186]. In multilingual countries, where children have to learn in more than one language, further problems ensue, as curriculum developers and linguists attempt to produce different phonics combinations for different languages, leading to double doses of decontextualised nonsense for young children to digest. This must lead to short term memory overload and anxiety!

Problems with the Understanding of Reading. Above all, prioritising phonics usually implies a misunderstanding of how meaningful reading works, perhaps because many of the workers in the field come from a standpoint based on studies of dyslexia, rather than how fluent readers read. It is based on the bottom-up image of assembling railway cars to form a train [170], instead of the top-down pattern-seeking way it actually works (Sect. 8.6.3). The real problem is that this bit-by-bit immature way of reading is then made the aim of the initial educational project of learning to read. Dehaene makes this explicit in his book on reading [47, p. 200]:

> The child's brain, at this stage, is attempting to match the general shape of the words directly onto meaning, without paying attention to individual letters and their pronunciation—a sham form of reading.

He defines reading the wrong way round, thereby misinterpreting the aim of the learning process! He wants the parts to work rather than the whole, and characterizes as sham reading what is in fact both the aim of fluent readers and the way that young children learn language. What he deplores is precisely what we want children to learn: to read from the whole, not the parts. The testing involved in associated reading programmes is of the same nature as this quote indicates: it penalizes attempts at conventional reading practices.

The Supposed Support from Brain-Imaging Data. There is supposed to be support for this view from brain-imaging data. However, this is problematic, particularly because they are mostly based on nonsense word studies. Brain-imaging data studies (see, for example, Shaywitz et al. [171]) confirm that phonemic processing occurs in particular brain areas, but do not show how reading comprehension occurs, for phonemic processing is only one part of what is involved in reading. In particular, this data does not prove that this process is bottom-up only, as claimed by Shaywitz et al. [171], nor does it show what motivation is crucial for education. The brain-imaging data referred to by Lyons, Shaywitz, and others [136, 171] does not by itself prove that the best approach to written language learning in general is via phonemics.

Obviously, all children have to learn how letters and sounds fit together and develop phonemic awareness, this is not in question. But even then, whole–part processes must remain crucial. According to Ellis [66]:

> This review summarizes a range of theoretical approaches to language acquisition. It argues that language representations emerge from interactions at all levels from brain to society. Simple learning mechanisms, operating in and across the human systems for perception,motor action and cognition as they are exposed to language data as part of a social environment, suffice to drive the emergence of complex language representations. Connectionism provides a set of computational tools for exploring the conditions under which emergent properties arise. I present various simulations of emergence of linguistic regularity for illustration.

Nonsense Word Studies. In particular, brain-imaging studies based on nonsense word strings [47] cannot engage the full range of faculties used in language processing and understanding, and the lack of meaning of such word strings is surely likely to demotivate learners forced to memorise them [89, 90, 203].

The Claim that It Has Been Tried and It Works. What of the claim that phonics-based methods work well? Davis [42] gives a detailed analysis. The introduction to this paper by Michael Hand expresses the outcome as follows:

> "Research has consistently and comprehensively shown", says Michael Gove, "that systematic, phonic instruction by a teacher is the most effective and successful way of teaching children to read" [94]. His confidence in this claim is reflected in the strong emphasis on synthetic phonics in the new National Curriculum for England, due to come into force in September 2014. But is he entitled to his confidence?

Hand notes two problems:

> One problem is that the available empirical research appears to show no such thing. A systematic review of the literature conducted by Carole Torgerson and colleagues [190, p. 10] found that, while there is an association between synthetic phonics and reading accuracy, "the weight of evidence (from RCTs) on reading comprehension was weak, and no significant effect was found for reading comprehension".

The second is the topic of the pamphlet by Andrew Davis:

> Whatever it is that empirical researchers take themselves to be doing when they investigate synthetic phonics, he maintains, they are not investigating a specifiable method of teaching reading. This is for two reasons. First, there are no such things as specifiable methods of teaching. Teaching is a vastly complex human activity involving contextual and reactive practical judgements that are responsive to the myriad contingencies of classroom life. The idea that teachers might proceed by way of prescribed methods rather than practical judgements is, as Davis puts it, simply a fantasy. Second, teaching children to correlate letter combinations with sounds, and to blend sounds into sequences, is not teaching them to read. Reading is a matter of grasping meaning conveyed by text. While sustained attention to letter–sound correspondences can be helpful to some novice readers, we should neither assume that it is helpful to all nor confuse mastery of such correspondences with the ability to read.

In summary, any beginner reader needs certain conditions of learning [28] for the learning to be successful. As with learning to listen and speak, significant 'invisible' literacy learning happens socially and culturally in informal ways. It is likely that any packaged teaching method with an initial focus on technical skills works, in fact, as part of a larger learning project for any child. Children who don't have role models in their lives who can facilitate and mediate regular encounters with the kind of play, stories, and other powerful reasons for reading that inspire and motivate learning in languages they understand are more likely to struggle.

8.6.5 The Contextual Approach to Learning

Taking a holistic approach to learning to read gives rise to a more contextual approach to learning, based on the way the brain functions.

8.6.5.1 Learning and Significance

The key point is set out by Tomasello [188, p. 6]:

> To socially learn the conventional use of a tool or a symbol, children must come to understand why, towards what outside end, the other person is using the tool or symbol. That is to say, they must come to understand the intentional significance of the tool use or symbolic practice—what it is 'for', what 'we', the users of this tool or symbol, do with it.

This applies to all learning of symbolic systems, so it applies in particular both to learning to talk and listen, and to learning to read and write. Halliday famously state that form follows function in language learning [100].

8.6.6 Holistic Approaches to Literacy

According to Krashen and Terrell [127], the natural approach to language acquisition in the classroom is based on the theory that language acquisition occurs only when students receive comprehensible input. The emphasis is on reading and listening comprehension for beginning students, with the intention to motivate the desire to read and behave like a reader.

According to Goodman [88, 89], in the whole language approach, reading is construction of meaning during a transaction between the reader and the text. It is making sense of print. The whole language belief is that learning needs at all times to involve complete meaningful texts. The basis for this is the primary emotional need to understand what is going on. The challenge is to create the conditions that enable children to use their substantial learning abilities to make sense of a complex but meaningful whole.

According to Goodman and Goodman [92], though written language is not the same as oral language, it can be learned in similar ways to the way oral language is learned when it is used in personally meaningful ways to communicate, to understand and be understood. The various aspects of language—talking, listening, reading and writing—are learned together with an emphasis on making meaning [202].

In this emergent literacy approach, a focus on holistic activities, including imaginative play and story reading and telling allows children to develop symbolic representation and rich and complex forms of language. Phonemic and phonological awareness can thus develop as part of and as a consequence of language play. At the same time, children are taught letter sound correspondences, etc., as part of what they need to know to accomplish their intentions. This makes the need for explicit teaching less frequent [112].

8.6.6.1 Does It Work?

What tests are there that this works? Firstly, we can test that reading takes place in the ways indicated above, and this therefore is what we are aspiring to teach. Secondly, we can probe its outcomes as educational practice.

Reading Studies. How we read obviously affects how we should teach learning to read. Apart from our own indubitable experience of how reading works, extensive studies have been made in two ways: miscue analysis and eye movement studies.

- **How Reading Works**. Ths has alreday been explaned in Sect. 7.3.4. Yu read text in holistic gestalt way, and undesrtand it despte grammar nd speeling errars.
- **Miscue Analysis**. This has shown that we read in a top-down meaning-based way [54].
- **Eye Movement Studies**. These have confirmed that we do not read by systematically reading phonemes and words in their printed order on the page [54].

Sounding Words and Reading Words. There is also a whole neuroscience literature on the relation between sounding words and reading words. Undoubtedly one initially learns what printed words mean by hearing them sounded out if one understands the language being used. But then the brain can learn to recognize directly the pattern of connections between print and meaning and hardwire them without conscious attention to the detail, so we do not have to convert written into oral language first before understanding it, but rather make the connection directly. After all, an accomplished reader does not sound the words out. She reads them directly in a top-down manner.

Learning Studies. However, one also needs to study how learning to read and write actually takes place on the basis of the previous items. What about claims that it has been tried and has not worked? It has not been adequately tried on any scale in school contexts with the right conditions set up, so it has not been possible to test it properly. Because of assessment requirements, it would appear that attempts to introduce genuine whole-language, meaning-based syllabi into schools often get subverted into attempts to prepare children to pass batteries of assessment tests. What we need are:

- **Detailed Observation**. See *Gnys at Wrk* [17] and *Chloe's Story* [18] for careful observation of how the actual process of learning to read works when it is integrated with learning to write as an emergent process, based on interaction with and constructing of meaningful texts. This kind of field observation of the real nature of the learning process sets the context for any meaningful educational strategies.
- **Class Trials**. Try holistic reading methods in class and assess them by finding out what children can actually read and understand.
- **Reading Clubs**. Try enthusing children about reading via reading clubs such as the Nal'ibali reading-for-enjoyment campaign,[3] run by the Project for the Study of Alternative Education in South Africa (PRAESA), and monitor the results. This is under way at present.
- **fMRI Studies**. fMRI studies that relate to real reading tasks and avoid meaningless tests. This is not usually done, see, e.g., Dehaene [47]. However, some studies have been completed that support the view put forward here [117, 118].

Hutton et al. presented a paper entitled 'Parent–child reading increases activation of brain networks supporting emergent literacy in 3–5 year-old children: An fMRI study' [117][4] at the Pediatric Academic Societies (PAS) annual meeting in San Diego looking at the relevant issues:

> Emergent literacy depends on integration of visual, association, and language brain networks during sensitive developmental stages. Disparities in home cognitive environment during childhood can have dramatic impact on achievement and health. Parent–child reading has been shown to improve certain emergent literacy skills, though its effect on the brain has not yet been shown.

Their brain-imaging data is shown in Fig. 8.8.

The result is:

> Greater parent–child reading during early childhood is associated with increased activation of brain areas involved with visual imagery and applying meaning to language in preschool

[3] http://nalibali.org.

[4] Described in http://medicalxpress.com/news/2015-04-mri-association-young-children-brain.html.

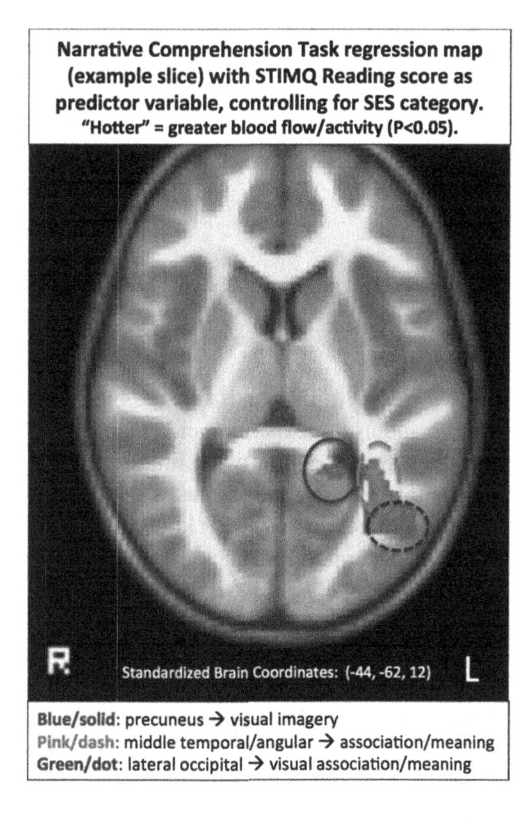

Fig. 8.8 Parent–child reading increases activation of brain networks supporting emergent literacy in 3–5 year-old children: An fMRI study. Credit: From Hutton et al. [117]

> children listening to stories, independent of SES. To our knowledge, this is the first study applying fMRI in this age range in the context of home reading environment to assess brain networks supporting emergent literacy, providing neural biomarkers for future studies of reading development and intervention.

They comment that this is the first paper they are aware of that undertakes the kind of fMRI studies required to test real reading learning tasks, where the interaction of the child is with meaningful texts. More studies with different variables in a range of different contexts are now needed, and they have indeed already written up a fuller study [118].

8.6.7 *Educational Implications*

8.6.7.1 The Top-Down Link

The causative link in reading is top-down from meaning to grammar, syntax, and phonemes, not the other way round (Table 8.7). At the bottom level, it is driven by collocations rather than grammatical rules [114]. This should be reflected in the

Table 8.7 The contextual nature of reading

Level	System	Effect	
8	Book	⇓ Top-down	
7	Chapter	⇓ Top-down	
6	Paragraph	Meaning	Top-down ⇓
5	Sentences	Meaning	Top-down ⇓
4	Phrases	⇑ Bottom-up	Top-down ⇓
3	Words	⇑ Bottom-up	⇐⇐
2	Graphemes	⇑ Bottom-up	⇐⇐
1	Letters	⇑ Bottom-up	⇐⇐

way we teach reading and writing: base it on authentic activities which include creating and using texts first, and teach the technical details later. This view ties in with the way perception works in general [80, 123, 138], as has been adequately demonstrated above, and so makes sense in terms of the way the brain functions. It gives clear direction as to what kinds of reading and writing programs might succeed for children from diverse backgrounds.

8.6.7.2 Learning to Speak and to Write Are Not Different

There is no essential difference between the way children learn to listen and speak, and the way they learn to read and write (see Fig. 8.9). This view should inform us as to how we should interact with children when they are learning to become proficient with language.

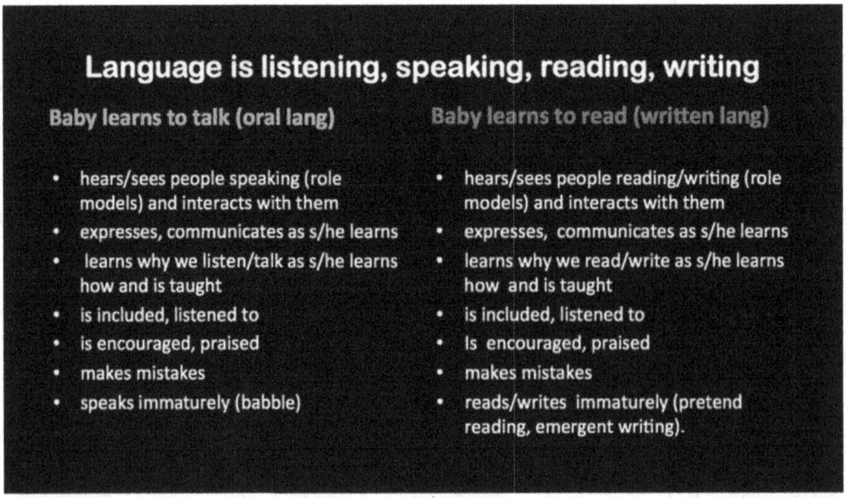

Language is listening, speaking, reading, writing

Baby learns to talk (oral lang)

- hears/sees people speaking (role models) and interacts with them
- expresses, communicates as s/he learns
- learns why we listen/talk as s/he learns how and is taught
- is included, listened to
- is encouraged, praised
- makes mistakes
- speaks immaturely (babble)

Baby learns to read (written lang)

- hears/sees people reading/writing (role models) and interacts with them
- expresses, communicates as s/he learns
- learns why we read/write as s/he learns how and is taught
- is included, listened to
- Is encouraged, praised
- makes mistakes
- reads/writes immaturely (pretend reading, emergent writing).

Fig. 8.9 Language learning: oral and written. Credit: Carole Bloch

8.6.7.3 The Aim of Language Teaching

Overall, the view presented here is that the core skill of language is communication: spelling, pronunciation, syntax are important, but their importance follows, rather than leads, a meaning-making process. The aim of language teaching should be to attain this competency, not to make the centre of language teaching detailed issues of phonics, grammar, and spelling. Yes they need to be got right eventually: and this comes with practice, and particularly with extensive reading and authentic writing.

8.6.7.4 The Debate

The debate between top-down and bottom-up views is a crucial ongoing debate with major practical implications for education. Many children in well-resourced, literate homes grow up where the conditions of learning described by Cambourne [28] are appropriate, and they are immersed in story-reading and play with written language. Such children learn, as if by osmosis, many of the essential 'concepts of print' that are often neglected at the beginning of primary school. Children from home backgrounds which are unable to provide such experiences, such as in many African settings [19], find themselves flailing when their introduction to print concentrates on the technicalities alone, often delivered in a foreign language. Many tend to lose interest in what they see as meaningless activities and this results in a lesser ability to read in a successful way. They need to be given access to the power of reading by experience in meaningful contexts. This will open up their innate abilities to learn to produce and interrogate texts in powerful, imaginative and critical ways.

8.6.7.5 The Political Dimension

However, there is a major political battle underlying what is going on as regards reading programs and phonics: it is not just about educational theories and practice, it is also about politics, power, and money [179]. This is another example of top-down influence in society, from the political level to what happens in classrooms and in the brains and neural connections of the students. Top-down causation is a key factor in the social world.

8.7 Conclusion

This section concludes by considering the following:

- The theses of the book (Sect. 8.7.1).
- To be done (Sect. 8.7.2).
- Where is truth (Sect. 8.7.3)?

8.7.1 The Theses of this Book

The argument of this book is summarised in a main thesis (Sect. 8.5.3):

> **Top-Down Causation and the Brain.** As well as bottom-up causation, top-down causation of all five kinds (TD1–TD5) occurs in the human brain, together enabling same level autonomous causation at all levels of the brain structure and enabling the physical brain to be the vehicle for causal effectiveness of abstract entities. Consequently, higher level mental functions are not epiphenomena: they are causally effective autonomous entities in their own right, even though they are made possible by the underlying physical states.

This is amplified by seven subsidiary theses:

- **Thesis 1**. Because of the cosmic context, strong emergence must occur in order to explain life and the human brain (Sect. 8.1.3).
- **Thesis 2**. Contextual emergence takes place via a combination of bottom-up and top-down effects (Sect. 8.2.4.1).
- **Thesis 3**. There are limits to physics explanations, which do not include biological principles (Sect. 8.2.5).
- **Thesis 4**. Randomness opens space for purpose to select desired outcomes, and so breaks the iron grip of bottom-up determinism (Sect. 8.2.6.4).
- **Thesis 5**. Multiple explanations always hold, so each level can be causally effective (Sect. 8.3.2.2).
- **Thesis 6**. Platonic entities are causally effective via the human mind (Sect. 8.3.3.1).
- **Thesis 7**. Denying these forms of causation is the claim that some of the causal effects in operation are the only causal effects there are (Sect. 8.5.4.2).

8.7.2 To Be Done

This book covers a great many themes, and obviously cannot do so in depth: it does so in a way that aims to highlight main causal threads and indicate relevant variables and mechanisms, but much needs to be done to take forward what is presented here. Specific issues that need further development are the following:

- How does this proposal relate to long-standing philosophical debates [37, 48, 134]?
- How do we distinguish bottom-up and from top-down causation [172]? How do we identify multi-level top-down selection processes [177]?
- What kinds of experimental and observational tests can be made of what is proposed here [9]?
- What is the relation to information [197] and computation [105]?
- What kind of mathematical developments [109, 199] and modelling [151] will best take this forward?

- What kinds of implications does it have in the real world (developing among other things the themes of this chapter)?
- Overall, how can there be a transition from regarding this as just a philosophical proposal to a stage where it is recognised as scientific [9]?

My hope is that reading this book may inspire others to develop all the different aspects that are touched on so briefly here, and to do so in a more adequate way than is done here.

8.7.2.1 Evidence

Further evidence is required to show that what is stated here is correct. The argument of this book is that top-down causation is indeed real.

- **Existence of Top-Down Causation**. The basic point is that one demonstrates existence of top-down causation whenever manipulating a higher level variable can be shown to alter lower level variables.

Manipulation of higher level variables cannot generically determine which specific microstate will result as a consequence of manipulation of some macro variable. By such manipulation we can only access the underlying equivalence class. For example, if we change the temperature of a system, this will change the micro state to any one of the class of micro states that correspond to the new temperature.

- **Deterministic Top-Down Causation (TD1)**. This is commonplace in physics, chemistry, biology, and engineering, and is, for example, evidenced by the reliable functioning of electrical machines and digital computers.
- **Homeostasis or Feedback Control (TD2)**. This is the foundation of physiology in plants and animals [30, 164] and so is confirmed by all the evidence of its causal effectiveness.
- **Adaptive Selection (TD3)**. This is evidenced [83, 111] by the adaptation of animals to their ecological niches [30] and by all the evidence for the plasticity of the brain [124].
- **Adaptive Selection of Goals (TD4)**. This is confirmed by evidence of Pavlovian operant conditioning [96].
- **Adaptive Selection of Selection Criteria (TD5)**. This is abundantly clear in the history of human technology [25] and ideas [185].

The nature of mechanisms in operation and their outcomes also confirms top-down causation, as evidenced in the many examples given above. Some variables are intrinsically higher level variables that cannot emerge by coarse-graining of lower level variables (Sects. 1.3.4 and 1.3.5). One can show that some of these variables affect the structure of lower level entities, as in the following:

- **Biological Macromolecules**. Molecules such as kinesin and dynesin are not emergent variables: they are adapted to their function [110].

- **Digital Computer Programs**. These are not physical entities and they are not coarse-grained variables. There is an intricate relation between higher and lower level programs (written in different languages) and variables [184].
- **Thoughts and Plans**. Occurring in the brain, these are causally effective, as has been extensively discussed above (Sect. 7.5.3).
- **Social Neuroscience**. Here social variables such as roles in society can be shown to affect neuronal connections [6, 27], and hence the flow of electrons in the dendrites and axons of neurons in the brain. This is top-down action from the social level to the level of neurons (Sect. 7.5.4).

Bottom-up emergence cannot by itself lead to existence of the relevant higher level variables in each of these cases (for example, physics *per se* cannot lead to the existence of mathematical theorems, or indeed to theories of physics). This emergence can only take place by developmental and learning processes enabled by top-down causation from abstract possibility spaces to the human mind [35].

8.7.2.2 Experimental Tests

These considerations are in my opinion conclusive, but are based on understanding and explaining what one already knows. What one likes in an experimental science is a prediction of something new that can then be verified by experiment or observation. What new experiments or observations can we propose that will substantiate or disprove the causal efficacy of higher level variables? There seem to be three main streams of possibility here:

- **Convergent Evolution**. When top-down causes drive what happens in evolutionary contexts, one often gets convergent evolution: different evolutionary pathways devise similar means of meeting the same higher level need [39, 142]. A famous case is the development of eyes by various evolutionary paths, driven by the need of animals to see, which clearly improves their survival capacity. One cannot explain such convergent evolution by a bottom-up causation alone: it is driven by a combination of specific higher level needs in conjunction with restrictions on how they can be achieved physiologically [194]. Hence, new evidence of convergent evolution is evidence for top-down causation.
- **Computer Simulations**. Top-down causation can be demonstrated by computer simulations of complex systems where higher level variables are shown to determine the outcome, for example the simulations of heart physiology by Noble [153]. Changing the higher level variables demonstrably changes the lower level dynamics and hence the outcome. This is also the case in studies of structure formation in the expanding universe [148].
- **Equivalence Classes**. As has been emphasized above (Sect. 1.3.4), the concept of equivalence classes of lower level variables is crucial to the nature of and physical implementation of top-down causation [9], and indeed their existence can be taken as convincing evidence that top-down causation is at work. It would be good to have a new prediction of as yet undiscovered equivalence classes that can then

be verified by experiment. This is at least in principle possible in microbiology, where the existence of very interesting cases of equivalence is already established [121], and one can hope to plan experiments that create new kinds of lower level members of an equivalence class satisfying some specific higher level need in cellular biology.

8.7.3 Where Is Truth?

Underlying what is presented in this book is an ancient discussion about what is reality and where is truth. The view here is that all levels are real, because true emergence takes place. That is the true nature of reality.

8.7.3.1 Everyday Life and the Scientific World

Eddington looks in detail at the issue of reality and illusion in his book *The Nature of the Physical World* [55]:

> I have settled down to the task of writing these lectures and have drawn up my chairs to my two tables. Two tables! Yes, there are duplicates of every object about me—two tables, two chairs, two pens. One of them has been familiar to me from earliest years. It is a commonplace object of that environment which I call the world. How shall I describe it? It has extension; it is comparatively permanent; it is colored; above all it is substantial. By substantial I do not merely mean that it does not collapse when I lean upon it; I mean that it is constituted of 'substance' and by that word I am trying to convey to you some conception of its intrinsic nature. It is a thing; not like space, which is a mere negation; nor like time, which is—Heaven knows what! But that will not help you to my meaning because it is the distinctive characteristic of a 'thing' to have this substantiality, and I do not think substantiality can be described better than by saying that it is the kind of nature exemplified by an ordinary table. And so we go round in circles.
>
> Table No. 2 is my scientific table. It is a more recent acquaintance and I do not feel so familiar with it. It does not belong to the world previously mentioned—that world which spontaneously appears around me when I open my eyes, though how much of it is objective and how much subjective I do not here consider. It is part of a world which in more devious ways has forced itself on my attention. My scientific table is mostly emptiness. Sparsely scattered in that emptiness are numerous electric charges rushing about with great speed; but their combined bulk amounts to less than a billionth of the bulk of the table itself. Notwithstanding its strange construction it turns out to be an entirely efficient table. It supports my writing paper as satisfactorily as table No. 1; for when I lay the paper on it the little electric particles with their headlong speed keep on hitting the underside, so that the paper is maintained in shuttlecock fashion at a nearly steady level. If I lean upon this table I shall not go through; or, to be strictly accurate, the chance of my scientific elbow going through my scientific table is so excessively small that it can be neglected in practical life.
>
> I need not tell you that modern physics has by delicate test and remorseless logic assured me that my second scientific table is the only one which is really there—wherever 'there' may be. On the other hand I need not tell you that modern physics will never succeed

> in exorcizing that first table—strange compound of external nature, mental imagery, and inherited prejudice—which lies visible to my eyes and tangible to my grasp.

He develops this profound relation between everyday life and the underlying scientific world view in depth in this book.

8.7.3.2 The Reality of All Levels

Actually, the only view that makes sense is that the table is real, the atoms are real, and the protons and neutrons in the nuclei are real. The evidence is overwhelming that human beings are living embodiment of emergence of higher level effective structures, enabled by a combination of bottom-up and top-down causation ranging from the particle physics level to the level of social interaction. We could not be what we are if this were not the case.

Each level in the hierarchy is real [153], even though it is made of lower level entities that are also real. To do an experiment to show this, bang your hand on the table so hard that you feel pain. You would not feel the pain if the table was not real and the hand not real, and of course the pain too is real. Every level of the hierarchy that we can experiment on or experience is real. Even your thoughts are real (they may or may not give an accurate description of the world around you, but that is a different matter).

8.7.3.3 Which Is More Fundamental?

In his book *The Character of Physical Law* [72, pp. 124–125], Richard Feynman summarises the hierarchy of structure, starting with the fundamental laws of physics and their application to protons, neutrons, and electrons, going on to atoms and heat, and including waves, storms, stars, as well as frogs and concepts like 'man', 'history', 'political expediency', 'evil', 'beauty', and 'hope'. He then says the following [72, pp. 125–126]:

> Which end is nearer to God, if I may use a religious metaphor. Beauty and hope, or the fundamental laws? I think that the right way, of course, is to say that what we have to look at is the whole structural interconnection of the thing; and that all the sciences, and not just the sciences but all the efforts of intellectual kinds, are an endeavour to see the connections of the hierarchies, to connect beauty to history, to connect history to man's psychology, man's psychology to the working of the brain, the brain to the neural impulse, the neural impulse to chemistry, and so forth, up and down, both ways. And today we cannot, and it is no use making believe we can, draw carefully a line all the way from one end of this thing to the other, because we have only just begun to see that there is this relative hierarchy.
>
> And I do not think either end is nearer to God. To stand at either end, and to walk off that end of the pier only, hoping that out in that direction is the complete understanding, is a mistake. And to stand with evil and beauty and hope, or with fundamental laws, hoping that way to get a deep understanding of the whole world, with that aspect alone, is a mistake. It is not sensible for the ones who specialize at one end, and the ones who specialize at the other, to have such disregard for each other ... The great mass of workers in between, connecting one

step to another, are improving all the time our understanding of the world, both from working at the ends and from working in the middle, and in that way we are gradually understanding this tremendous world of interconnecting hierarchies.

8.7.3.4 What Is Truth?

So where does truth lie in this complex context? Here is a view by Isaac Pennington (1653) (see [155]):

> All Truth is shadow except the last, except the utmost; yet every Truth is true in its own kind. It is substance in its own place, though it be but shadow in another place (for it is but a reflection from an intenser substance); and the shadow is a true shadow, as the substance is a true substance.

That expresses it beautifully.

Fig. 8.10 Genuine emergence: life, conversation, and everyday objects 14 billion years after the big bang. Credit: Carole Bloch

The daily world in which we live came about by imaginative investigation of possibilities, discarding those that don't work: the adaptive process that is a central theme of this book, enabled by a modicum of randomness at the macro- and micro-levels, interacting with necessary physical processes. And it is these processes that also allow the emergence of the ordinariness of everyday life (Fig. 8.10): which actually is quite extraordinary. Bottom-up effects are crucial to emergence. Physics underlies all. Nevertheless, the vitality of life, which arises from physics, transcends it.

References

1. H. Abadzi, *Efficient Learning for the Poor: Insights from the Frontier of Cognitive Neuroscience* (The World Bank, Washington, DC, 2006)
2. D. Abbott, P.C.W. Davies, C.R. Shalizi, Order from disorder: the role of noise in creative processes. Fluc. Noise Lett. **02**, 1 (2002)
3. R.L. Ackoff, Systems thinking and thinking systems. Syst. Dyn. Rev. **10**, 175–188 (1994)
4. R.L. Ackoff, F.E. Emery, *On Purposeful Systems: An Interdisciplinary Analysis of Individual and Social Behavior as a System of Purposeful Events* (Aldine, 2005)
5. U. Alon, *An Introduction to Systems Biology: Design Principles of Biological Circuits* (Chapman and Hall, 2006)
6. N. Ambady, The mind in the world: culture and the brain. Assoc. Psychol. Sci. **24**(5–6), 49 (2011)
7. B.B. Armbruster, F. Lehr, J. Osborn, *The Research Building Blocks for Teaching Children to Read: Put Reading First* (National Institute for literacy, 2000). http://lincs.ed.gov/publications/pdf/PRFbooklet.pdf
8. P.W. Atkins, The limitless power of science, in *Nature's Imagination: the Frontiers of Scientific Vision*, ed. by J. Cornwell (Oxford University Press, Oxford), pp. 122–132
9. G. Auletta, G. Ellis, L. Jaeger, Top-down causation: from a philosophical problem to a scientific research program. J. Roy. Soc. Interface **5**, 1159–1172 (2007). arXiv:0710.4235
10. D. Barton, *Literacy: An Introduction to the Ecology of Written Language* (Blackwell, Malden, Massachusetts, 1994)
11. Y. Bar-Yam, *"General Features of Complex Systems" Encyclopedia of Life Support Systems* (EOLSS UNESCO Publishers, Oxford, UK, 2002)
12. C. Baugh, Correlation function and power spectra in cosmology, in *Encyclopedia of Astronomy and Astrophysics*, ed. by P. Murdin (IOP Publishing, 2006)
13. S. Beer, *Brain of the Firm* (Wiley, 1994)
14. N. Bellomo, A. Elaiw, A.M. Althiabi, M.A. Alghamdi, On the interplay between mathematics and biology: Hallmarks toward a new systems biology. Phys. Life Rev. **12**, 44–64 (2015)
15. R.P. Benthal, *Doctoring the Mind: Why Psychiatric Treatments Fail* (Penguin, 2009)
16. P.L. Berger, *Invitation to Sociology* (Anchor Books, 1963)
17. G.L. Bissex, *Gnys at Wrk: A Child Learns to Write and Read* (Harvard University Press, Cambridge, 1985)
18. C. Bloch, *Chloe's Story: First Steps into Literacy* (Juta Academic, 1997)
19. C. Bloch, *Enabling Effective Literacy Learning in Multilingual South African Early Childhood Classrooms* (University of Cape Town, PRAESA, 2005)
20. C. Bloch, *Theory and Strategy of Early Literacy in Contemporary Africa with Special Reference to South Africa* (PRAESA, University of Cape Town, 2006). Occasional Papers No. 25
21. G. Booch, *Object-Oriented Analysis and Design with Applications* (Addison Wesley, New York, NY, 1994)

22. R. Borowsky, J. Cummine, W.J. Owen, C.K. Friesen, F. Shih, G.E. Sarty, fMRI of ventral and dorsal processing streams in basic reading processes: insular sensitivity to phonology. Brain Topogr. **18**, 233–239 (2006)
23. R. Borowsky, C. Esopenko, J. Cummine, G.E. Sarty, Neural representations of visual words and objects: a functional MRI study on the modularity of reading and object processing. Brain Topogr. **20**, 89 (2007)
24. V. Brattka, M. Hendtlass, A.P. Kreuzer, On the uniform computational content of computability theory. http://arxiv.org/abs/1501.00433
25. J. Bronowski, *The Ascent of Man* (Little Brown and Co., Boston, 1973)
26. J. Bruner, *Child's Talk: Learning to Use Language* (Norton, New York, 1983)
27. J.T. Cacioppo, J. Decety, Social neuroscience: challenges and opportunities in the study of complex behavior. Ann. N. Y. Acad. Sci. **1224**, 162–173 (2011)
28. B. Cambourne, Toward an educationally relevant theory of literacy learning: twenty years of inquiry. Read. Teach. **43**, 182–190 (1995). doi:10.1598/RT.49.3.1
29. D.T. Campbell, Downward causation, in *Studies in the Philosophy of Biology: Reduction and Related Problems*, ed. by F.J. Ayala, T. Dobhzansky (University of California Press, Berkeley, 1974)
30. N.A. Campbell, J.B. Reece, *Biology* (Benjamin Cummings, San Francisco, 2005)
31. S.T. Chan, S.W. Tang, K.W. Tang, W.K. Lee, S.S. Lo, K.K. Kwong, Hierarchical coding of characters in the ventral and dorsal visual streams of Chinese language processing. Neuroimage **48**, 423 (2009)
32. J.-P. Changeux, A. Connes, *Conversations on Mind, Matter, and Mathematics* (Princeton University Press, 1995)
33. T. Chouard, Breaking the protein rules: if dogma dictates that proteins need a structure to function, then why do so many of them live in a state of disorder? Nature **471**, 151 (2011). doi:10.1038/471151a
34. D. Chowdhury, Stochastic mechano-chemical kinetics of molecular motors: a multidisciplinary enterprise from a physicist's perspective. Phys. Rep. **529**, 1–197 (2013)
35. P. Churchland, *Plato's Camera: How the Physical Brain Captures a Landscape of Abstract Universals* (MIT Press, 2012)
36. C.W. Churchman, *The Systems Approach* (Delacorte Press, 1968)
37. P. Clayton, *Mind and Emergence: From Quantum to Consciousness* (Oxford University Press, Oxford, 2004)
38. P. Clayton, P.C.W. Davies (eds.), *The Re-emergence of Emergence* (Oxford University Press, Oxford, 2006)
39. S. Conway Morris, *Life's Solution: Inevitable Humans in a Lonely Universe* (Cambridge University Press, Cambridge, 2005)
40. S. Conway Morris (ed.), *The Deep Structure of Biology* (Templeton Foundation Press, 2008)
41. F. Crick, *Astonishing Hypothesis: The Scientific Search for the Soul* (Scribner, 1995)
42. A. Davis, To read or not to read: decoding synthetic phonics, in *IMPACT: Philosophical Perspectives on Education Policy*, vol. 20 (Wiley, 2013). http://onlinelibrary.wiley.com/doi/10.1111/2048-416X.2013.12000.x/epdf
43. T. Deacon, *The Symbolic Species: The Co-evolution of Language and the Human Brain* (Penguin Books, London, 1997)
44. T. Deacon, Universal grammar and semiotic constraints, in *Language Evolution*, ed. by M. Christiansen, S. Kirby (Oxford University Press, Oxford, 2003), pp. 111–139
45. G. Deco, E.T. Rolls, Neurodynamics of biased competition and cooperation for attention: a model with spiking neurons. J. Neurophysiol. (2005). doi:10.1152/jn.01095.2004
46. G. Deco, E.T. Rolls, R. Romo, Stochastic dynamics as a principle of brain function. Prog. Neurobiol. **88**, 1–16 (2009)
47. S. Dehaene, *Reading in the Brain: The New Science of How We Read* (Penguin, London, UK, 2010)
48. H.L. de Jong, Levels of explanation in biological psychology. Philos. Psychol. **15**, 441–462 (2002)

49. S. Dodelson, *Modern Cosmology* (Academic Press, San Diego, 2003)
50. M. Donald, *A Mind so Rare: The Evolution of Human Consciousness* (W. W. Norton, New York, 2001), pp. 29–36
51. W. Dubitzky, O. Wolkenhauer, H. Yokota, K.-H. Cho (eds.), *Encyclopedia of Systems Biology* (Springer, New York, 2013)
52. R. Dunbar, *Human Evolution* (Pelican Books, London, 2014)
53. J. Dupré, *Human Nature and the Limits of Science* (Oxford University Press, New York, 2002)
54. A. Ebe, What eye movement and miscue analysis reveals about the reading process of young bilinguals, in *Scientific Realism in Studies of reading*, ed. by A.D. Flurkey, E.J. Paulson, K.S. Goodman (Taylor and Francis, London, 2008), pp. 131–152
55. A.S. Eddington, *The Nature of the Physical World* (MacMillan, 1928)
56. A. Eldar, M.B. Elowitz, Functional role for noise in genetic circuits. Nature **467**, 167–173 (2010)
57. D. Elder-Vass, *The Causal Power of Social Structures: Emergence, Structure and Agency* (Cambridge University Press, Cambridge, 2010)
58. G.F.R. Ellis, Physics, complexity, and causality. Nature **435**, 743 (2005)
59. G.F.R. Ellis, Issues in the philosophy of cosmology, in *Handbook in Philosophy of Physics* ed. by J. Butterfield, J. Earman (Elsevier, 2006), pp. 1183–1285. http://arxiv.org/pdf/astro-ph/0602280
60. G.F.R. Ellis, On the nature of causation in complex systems. Trans. Roy. Soc. S. Afr. **63**, 69–84 (2008)
61. G.F.R. Ellis, Commentary on 'An Evolutionarily Informed Education Science' by David C Geary. Educ. Psychol. **43**, 206–213 (2008)
62. G.F.R. Ellis, Top-down causation and emergence: some comments on mechanisms. J. Roy. Soc. Interface Focus **2**, 126–140 (2012)
63. G.F.R. Ellis, On the philosophy of cosmology. Stud. Hist. Philos. Sci. Part B. Stud. Hist. Philos. Mod. Phys. **46**, 5–23 (2013)
64. G.F.R. Ellis, Necessity, purpose, and chance: the role of randomness and indeterminism in nature from complex macroscopic systems to relativistic cosmology (2014). http://www.mth.uct.ac.za/~ellis/George_Ellis_Randomness.pdf
65. G.F.R. Ellis, D. Noble, T. O'Connor, Top-down causation: an integrating theme within and across the sciences? Interface Focus **2**(1) (2012). http://rsfs.royalsocietypublishing.org/content/2/1/1.short
66. N.C. Ellis, Emergentism, connectionism and language learning. Lang. Learn. **48**, 631–664 (1998)
67. M.B. Elowitz, A.J. Levine, E.D. Siggia, P.S. Swain, Stochastic gene expression in a single cell. Science **297**, 1183–1186 (2002)
68. A. Falcon, *Aristotle on Causality, The Stanford Encyclopedia of Philosophy* (Spring 2015 Edition). ed. by Edward N. Zalta. http://plato.stanford.edu/archives/spr2015/entries/aristotle-causality
69. G. Fauconnier, *Mappings in Thought and Language* (Cambridge University Press, Cambridge, 1997)
70. C. Fernando, E. Szathmáry, P. Husbands, Selectionist and evolutionary approaches to brain function: a critical appraisal. Front. Comput. Neurosci. **6**, 24 (2012). doi:10.3389/fncom.2012.00024
71. E. Ferreiro, A. Teberosky, *Literacy Before Schooling* (Heinemann Educational Books, London, 1993)
72. R. Feynman, *The Character of Physical Law* (Modern Library, 1994)
73. R.P. Feynman, R.B. Leighton, M. Sands, *The Feynman Lectures on Physics: Quantum Mechanics* (Addison-Wesley, Reading, Mass, 1965)
74. D. Fleisch, *A Student's Guide to Maxwell's Equations* (Cambridge University Press, Cambridge, UK, 2008)
75. R.L. Flood, E. Carson, *Dealing with Complexity: An Introduction to the Theory and Application of Systems Science* (Springer, USA, 1993)

76. A.D. Flurkey, E.J. Paulsen, K.S. Goodman, *Scientific Realism in Studies of Reading* (Laurence Erlbaum, New York, NY, 2008)
77. B. Flyvbjerg, *Making Social Science Matter: Why Social Enquiry Fails and How it Can Succeed Again* (Cambridge University Press, Cambridge, 2001)
78. B. Flyvbjerg, T. Landman, S. Schram, *Real Social Science: Applied Phronesis* (Cambridge University Press, Cambridge, 2012)
79. V.E. Frankl, Man's Search for Meaning (Beacon Press. Boston **1963**, (2006)
80. C. Frith, *Making up the Mind: How the Brain Creates Our Mental World* (Blackwell, Malden, 2007)
81. M. Gellman, *The Quark and the Jaguar* (Abacus, London, 1995)
82. S.H. Gellman (ed.), Molecular recognition. Chem. Rev. **97**(5), 1231–1232 (1997)
83. M. Gellman, *The Quark and the Jaguar* (Abacus, 2002)
84. S. Gilbert, *Developmental Biology* (Sinauer, Sunderland, 2013)
85. S.F. Gilbert, D. Epel, *Ecological Developmental Biology* (Sinauer, Sunderland, Mass, 2009)
86. P.W. Glimcher, Indeterminacy in brain and behavior. Annu. Rev. Psychol. **56**, 25–56 (2005)
87. K.S. Goodman, Reading: a psycholinguistic guessing game, in *Language and Literacy: The Selected Writings of Kenneth Goodman*, vol. 1, ed. by F.V. Gollasch (London, UK, Routledge and Kegan Paul, 1967), pp. 33–44
88. K. Goodman, *What's Whole in Whole Language: 20th*, Anniversary edn. (RDR Books, Muskegon, MI, 2005)
89. K. Goodman, "Afterword: whole language and the pedagogy of the absurd. What's whole in whole language" 20th anniversary edition, in *What's Whole in Whole Language: 20th Anniversary Edition* (RDR Books, Muskegon, MI, 2005)
90. K.S. Goodman, *The Truth About DIBELS: What It Is-What It Does* (Heinemann, Portsmouth, NH, 2006)
91. K. Goodman, *Suffer Little Children to Come to Be DIBELed* (2008). http://www.u.arizona.edu/~kgoodman/sufferlittlechildren.htm
92. K.S. Goodman, E.B. Smith, R. Meredith, Y.T. Goodman, *Language and Thinking in School: A Whole Language Curriculum* (Richard C. Owen Publishers, New York, NY, 1986)
93. A. Gopnik, A. Meltzoff, P. Kuhl, *The Scientist in the Crib: What Early Learning Tells Us about the Mind* (Perennial, New York, 1999)
94. M. Gove, Speech on improving the quality of teaching and leadership, given on 5 Sept 2013 at Policy Exchange, London (2013), http://www.gov.uk/government/speeches/michael-gove-speaks-about-theimportance-of-teaching. Accessed 3 Nov 2013
95. J. Gottschall, *The Storytelling Animal: How Stories Make Us Human* (Boston, Mariner Books, 2012)
96. P. Gray, *Psychology* (Worth, New York, NY, 2011)
97. S.I. Greenspan, S.G. Shanker, *The First Idea: How Symbols, Language, and Intelligence Evolved from our Primate Ancestors to Modern Humans* (Da Capo Press, Cambridge, Mass, 2004)
98. A.C. Guyton, *Basic Human Physiology* (W. B. Saunders, Philadelphia, 1977)
99. N. Hall, *Writing with Reason* (Hodder and Stoughton, London, UK, 1989)
100. M. Halliday, *Learning How to Mean* (Edward Arnold, London, 1975)
101. A.S. Hansen, E.K. O'Shea, Promoter decoding of transcription factor dynamics involves a trade-off between noise and control of gene expression. Mol. Syst. Biol. **9**, 704 (2013)
102. C.M. Harris, D.M. Wolpert, Signal-dependent noise determines motor planning. Nature **394**, 780–784 (1998)
103. E. Harrison, *Cosmology: The Science of the Universe* (Cambridge University Press, Cambridge, 2000)
104. L.H. Hartwell, J.J. Hopfield, S. Leibler, A.W. Murray, From molecular to modular cell biology. Nature **402**, C47–C52 (1999). Supplement (2 December 1999)
105. D. Haslacher, Beyond the computational–representational brain: why affective neuroscience tells us attitudes must be explained on multiple levels. Front. Behav. Neurosci. **8**, 419 (2014)
106. J. Hawkins, *On Intelligence* (Holt Paperbacks, New York, NY, 2004)

107. S.B. Heath, *Ways with Words* (Cambridge University Press, Cambridge, 1983)
108. G. Hinshaw, WMAP data put cosmic inflation to the test. Phys. World **19**, 16–19 (2006)
109. E.P. Hoel, L. Albantakis, G. Tononi, Quantifying causal emergence shows that macro can beat micro. PNAS **110**, 19790–19795 (2013). http://www.pnas.org/content/110/49/19790.abstract
110. P.M. Hoffmann, *Life's Ratchet: How Molecular Machines Extract Order from Chaos* (Basic Books, 2012)
111. J.H. Holland, *Adaptation in Natural and Artificial Systems* (MIT Press, Cambridge, MA, 1992)
112. D. Holdaway, *The Foundations of Literacy* (Ashton Scholastic, Sydney, 1979)
113. M. Hoey, *Textual Interaction: An Introduction to Written Discourse Analysis* (Routledge, 2001)
114. M. Hoey, *Lexical Priming: A New Theory of Words and Language* (Routledge, London, UK, 2005)
115. S. Hunston, *Corpora in Applied Linguistics* (Cambridge University Press, Cambridge, 2002)
116. D. Huron, *Sweet Anticipation: Music and the Psychology of Expectation* (MIT Press, Cambridge, MA, 2007)
117. J.S. Hutton, T. Horowitz-Kraus, T. De Witt, S. Holland, Parent–child reading increases activation of brain networks supporting emergent literacy in 3–5 year-old children: an fMRI study. Presentation at *Pediatric Academic Societies* (PAS) annual meeting in San Diego. Session: General Pediatrics and Preventive Pediatrics—Prevention and Early Intervention, 25 Apr 2015
118. J.S. Hutton, T. Horowitz-Kraus, A.L. Mendelsohn, T. DeWitt, S.K. Holland, Home reading environment and brain activation in preschool children listening to stories. Pediatrics **136**(3), 466–478 (2015). doi:10.1542/peds.2015-0359
119. P.A. Iglesias, B.P. Ingalis, *Control Theory and Systems Biology* (MIT Press, Cambridge, MA, 2010)
120. C.J. Isham, *Lectures on Quantum Theory: Mathematical and Structural Foundations* (Imperial College Press, London, 1995)
121. L. Jaeger, E.R. Calkins, Downward causation by information control in micro-organisms. Interface Focus **2**, 26–41 (2012)
122. J. Kagan, *The Human Spark: The Science of Human Development* (Basic Books, 2013)
123. E. Kandel, *Psychiatry, Psychoanalysis, and the New Biology of Mind* (American Psychiatric Publishing, Washington, DC, 2005)
124. E. Kandel, *The Age of Insight: The Quest to Understand the Unconscious in Art, Mind, and Brain, from Vienna 1900 to the Present* (Random House, 2012)
125. E.R. Kandel, J.H. Schwartz, T.M. Jessell, *Principles of Neural Science* (McGraw Hill, New York, 2000)
126. D.E. Knuth, *Selected Papers on Design of Algorithms* (Center for the Study of Language and Information, Stanford, CA, 2010)
127. S.D. Krashen, T.D. Terrell, *The Natural Approach: Language Acquisition in the Classroom* (Alemany Press, San Francisco, CA, 1983)
128. B.-O. Küppers, *Information and the Origin of Life* (The MIT Press, Cambridge, Mass, 1990)
129. J. Le Fanu, *The Rise and Fall of Modern medicine* (Abacus, 2011)
130. J.-M. Lehn, Perspectives in supramolecular chemistry: from molecular recognition towards molecular information processing and self-organization. Angew. Chem. Int. Ed. Engl. **27**, 89–121 (1988)
131. J.-M. Lehn, *Supramolecular Chemistry* (Wiley-VCH, Weinheim, 1995)
132. D.J. Levitin, *This Is Your Brain on Music: The Science of a Human Obsession* (Plume, London, UK, 2007)
133. C.S. Lewis, *Of Other Worlds: Essays and Stories* (Harcourt, 1966)
134. O. Lombardi, The ontological autonomy of the chemical world: facing the criticisms, in *Philosophy of Chemistry*, ed. by E. Scerri, L. McIntyre, vol. 306. Boston Studies in the Philosophy and History of Science
135. J.E. Longres, *Human Behaviour in the Social Environment* (F.E, Peacock, 1990)

136. G.R. Lyon, J.M. Rumsey (eds.), *Neuroimaging: A Window to the Neurological Foundations of Learning and Behaviour in Children* (Brookes, Baltimore, 1996)
137. P.K. Maini, T.E. Woolley, R.E. Baker, E.A. Gaffney, S.S. Lee, Turing's model for biological pattern formation and the robustness problem. Interface Focus **2**, 487–496 (2012)
138. D. Marr, *Vision: A Computational Investigation into the Human Representation and Processing of Visual Information* (W.H. Freeman, San Francisco, 1982)
139. M. Martinez, A. Moyo, Natural selection and multi-level causation. Philos. Theor. Biol. **3**, 2 (2011). http://quod.lib.umich.edu/p/ptb/6959004.0003.002/--natural-selection-and-multi-level-causation?rgn=main;view=fulltex
140. H.H. McAdams, A. Arkin, Stochastic mechanisms in gene expression. Proc. Nat. Acad. Sci. **94**, 814–819 (1997)
141. G. McGhee, *Convergent Evolution: Limited Forms most Beautiful* (MIT Press, Cambridge, Mass, 2010)
142. G. McGhee, *Convergent Evolution: Limited Forms Most Beautiful* (MIT Press, Cambridge, Mass, 2011)
143. A.R. McIntosh, N. Kovacevic, R.J. Itier, Increased brain signal variability accompanies lower behavioral variability in development. PLOS Comput. Biol. **4**, e1000106 (2008)
144. A.R. McIntosh, N. Kovacevic, S. Lippe, D. Garrett, C. Grady, V. Jirsa, The development of a noisy brain. Arch. Ital. Biol. **148**, 323–337 (2010)
145. R. Meyer, *Phonics Exposed: Understanding and Resisting Systematic Direct Intense Phonics Instruction* (Erlbaum, London, 2002)
146. M. Miodownik, *Stuff Matters: Exploring the Marvelous Materials that Shape our Man-Made World* (Houghton Mifflin Harcourt, Boston, 2014)
147. J. Monod, *Chance and Necessity: An Essay on the Natural Philosophy of Modern Biology* (Penguin, 1997)
148. V.F. Mukhanov, H.A. Feldman, R.H. Brandenberger, Theory of cosmological perturbations. Phys. Rep. **215**, 203–333 (1992)
149. N. Murphy, W. Brown, *Did My Neurons Make me Do it? Philosophical and Neurobiological Perspectives on Moral Responsibility and Free Will* (Oxford University Press, New York, 2007)
150. N. Murphy, G.F.R. Ellis, T. O'Connor (eds.), *Downward Causation and the Neurobiology of Free Will* (Springer, Heidelberg, 2009)
151. J.K. Nicholson, E. Holmes, J.C. Lindon, I.D. Wilson, The challenges of modeling mammalian biocomplexity. Nat. Biotechnol. **22**, 1268–1274 (2004)
152. D. Noble, *The Music of Life: Biology Beyond Genes* (Oxford University Press, Oxford, 2008)
153. D. Noble, A theory of biological relativity: no privileged level of causation. Interface Focus **2**, 55–64 (2012)
154. V.G. Paley, *The Boy Who Would Be a Helicopter: The Uses of Storytelling in the Classroom* (Harvard University Press, Cambridge, MA, 1990)
155. Quaker Faith and Practice, 27.22 (Religious Society of Friends, London)
156. R. Penrose, *The Large, the Small and the Human Mind* (Cambridge University Press, Canton, 2000)
157. I. Percival, Schrödinger's quantum cat. Nature **351**, 357 (1991)
158. G.A. Petsko, D. Ringe, *Protein Structure and Function* (Oxford University Press, Oxford, 2009)
159. J. Piaget, *Play, Dreams and Imitation in Childhood* (Norton, New York, 1962)
160. D. Purves, *Brains: How they Seem to Work* (FT Press Science, Upper Saddle River, NJ, 2010)
161. A. Rae, *Quantum Physics: Illusion or Reality?* (Cambridge University Press, Cambridge, 1994)
162. C.V. Rao, D.M. Wolf, A.P. Arkin, Control, exploitation and tolerance of intracellular noise. Nature **420**, 231–237 (2002)
163. J.M. Raser, E.K. O'Shea, Noise in gene expression: origins, consequences, and control. Science **309**, 2010–2013 (2005)
164. R. Rhoades, R. Pflanzer, *Human Physiology* (Saunders College Publishing, Fort Worth, 1989)

165. J.G. Roederer, *Information and Its Role in Nature* (Springer, Heidelberg, 2005)
166. B. Rogoff, *The Cultural Nature of Human Development* (Oxford University Press, New York, 2003)
167. E.T. Rolls, G. Deco, *The Noisy Brain: Stochastic Dynamics as a Principle of Brain Function* (Oxford University Press, Oxford, 2010)
168. M.S. Samoilov, G. Price, A. Arkin, From fluctuations to phenotypes: the physiology of noise. Sci. STKE **2006**(366), re17 (2006)
169. J. Scalo, J.C. Wheeler, P. Williams, Intermittent jolts of galactic UV radiation: mutagenetic effects, in *Frontiers of Life, 12th Rencontres de Blois* ed. by L.M. Celnikier (2001). arXiv:astroph/0104209
170. S. Shaywitz, *Overcoming Dyslexia: A New and Complete Science-Based Program for Reading Problems at any Level* (Vintage, 2003)
171. S. Shaywitz, B.M. Shaywitz, K.R. Pugh, P. Skudlarski, R.K. Fulbright, R.T. Constable, R.A. Bronen, J.M. Fletcher, A.M. Liberman, D.P. Shankweiler, L. Katz, C. Lacadie, K.E. Marchione, J.C. Gore, The neurobiology of developmental dyslexia as viewed through the lens of functional magnetic resonance imaging technology, in *Neuroimaging: A Window to the Neurological Foundations of Learning and Behaviour in Children*, ed. by G.R. Lyon, J.M. Rumsey (Baltimore, Brookes, 1996), pp. 80–94
172. N. Shea, Distinguishing top-down from bottom-up effects, in *Perception and Its Modalities*, ed. by S. Biggs, M. Matthen, D. Stokes (Oxford University Press, Oxford, 2013)
173. J. Silk, *A Short History of the Universe* (Henry Holt and Company, 1997)
174. J. Silk, *The Big Bang* (Henry Holt and Company, 2000)
175. H.A. Simon, *The Sciences of the Artificial* (MIT Press, Cambridge, Mass, 1992)
176. C. Snow et al., *Developing Early Literacy: Report of the National Early Literacy Panel. A Scientific Synthesis of Early Literacy Development and Implications for Intervention* (National Institute for Literacy, 2008)
177. K. Stotz, Extended evolutionary psychology: the importance of transgenerational developmental plasticity. Front. Psychol **5**(1), 908 (2014)
178. S.L. Strauss, *The Linguistics, Neurology, and Politics of Phonics: Silent E Speaks out* (Lawrence Erlbaum, Mahwah, New Jersey, 2005)
179. S.L. Strauss, The political economy of dyslexia. Mon. Rev. **66**(04) (2014). http://monthlyreview.org/2014/09/01/the-political-economy-of-dyslexia
180. S.L. Strauss, B. Altwerger, The logographic nature of English alphabetics and the fallacy of direct intensive phonics instruction. J. Early Child. Lit. **7**, 299–317 (2007)
181. S.L. Strauss, K.S. Goodman, E.J. Paulson, Brain research and reading: how emerging concepts in neuroscience support a meaning constructionist view of the reading process. Educ. Res. Rev. **4**, 21–33 (2009)
182. B. Street, *Literacy in Theory and Practice* (Cambridge University Press, Cambridge, 1984)
183. B. Street, *Social Literacies: Critical Approaches to Literacy Development, Ethnography and Education* (Longman, London, 1995)
184. A.S. Tanenbaum, *Structured Computer Organisation* (Prentice Hall, Englewood Cliffs, 2006)
185. C. Taylor, *Sources of the Self: The Making of Modern Identity* (Harvard University Press, Cambridge, MA, 1989)
186. D. Taylor, *Beginning to Read and the Spin Doctors of Science: The Political Campaign to Change America's Mind about How Children Learn to Read* (National Council of Teachers of English, Urbana, 1998)
187. J.L. Tinker, B.E. Robertson, A.V. Kravtsov, A. Klypin, M.S. Warren, G. Yepes, S. Gottlober, The large scale bias of dark matter halos: numerical calibration and model tests. Astrophys. J. **724**, 878–886 (2010). arXiv:1001.3162
188. M. Tomasello, *The Cultural Origins of Human Cognition* (Harvard University Press, Boston, 1999)
189. M. Tomasello, *Constructing a Language: A Usage-Based Theory of Language Acquisition* (Harvard University Press, Boston, 2003)

190. C.J. Torgerson, G. Brooks, J. Hall, *A Systematic Review of the Research Literature on the Use of Phonics in the Teaching of Reading and Spelling* (DfES, London, 2006)
191. P.U. Tse, *The Neural Basis of Free Will* (MIT Press, 2013)
192. M.A. Turing, The chemical basis of morphogenesis. Philos. Trans. Roy. Soc. B **237**, 37–72 (1952)
193. J.-P. Uzan, The big bang theory: construction, evolution, and status. L'UNIVERS, Séminaire Poincaré XX **1–69**, (2015)
194. S. Vogel, *Cats' Paws and Catapults: Mechanical Worlds of Nature and People* (W. W. Norton, 2000)
195. L.S. Vygotsky, *Mind in Society* (Harvard University Press, Cambridge, MA, 1978)
196. A. Wagner, *Arrival of the Fittest: Solving Evolution's Greatest Puzzle* (Current, 2014)
197. S.I. Walker, P.C.W. Davies, The algorithmic origins of life. J. Roy. Soc. Interface **10** (2013). http://arxiv.org/abs/1207.4803
198. S.I. Walker, L. Cisneros, P.C.W. Davies, Evolutionary transitions and top-down causation, in Proceedings of Artificial Life XIII (2013), pp. 283–290. http://arxiv.org/abs/1207.4803
199. S.I. Walker, P.C.W. Davies, G.F.R. Ellis (eds.), *Information and Causality: From Matter to Life* (Cambridge University Press, Cambridge and New York, 2016)
200. J.D. Watson, *Molecular Biology of the Gene* (Cold Spring Harbor Laboratory Press, 2007)
201. S. Webster, *Thinking about Biology* (Cambridge University Press, Cambridge, 2003)
202. G. Wells, *The Meaning Makers* (Heinemann, Portsmouth, NH, 1987)
203. J. Willis, The neuroscience of joyful education, *Educational Leadership* (ASCD). Summer 2007, vol. 64, *Engaging the Whole Child* (2007). ASCD website, http://www.ascd.org/portal/site/ascd

Author Index

G. Ellis, *How Can Physics Underlie the Mind?*, The Frontiers Collection,
DOI 10.1007/978-3-662-49809-5

Index

G. Ellis, *How Can Physics Underlie the Mind?*, The Frontiers Collection,
DOI 10.1007/978-3-662-49809-5

C

D

Titles in This Series

Quantum Mechanics and Gravity
By Mendel Sachs

Quantum-Classical Correspondence
Dynamical Quantization and the Classical Limit
By Dr. A.O. Bolivar

Knowledge and the World: Challenges Beyond the Science Wars
Ed. by M. Carrier, J. Roggenhofer, G. Küppers and P. Blanchard

Quantum-Classical Analogies
By Daniela Dragoman and Mircea Dragoman

Life—As a Matter of Fat
The Emerging Science of Lipidomics
By Ole G. Mouritsen

Quo Vadis Quantum Mechanics?
Ed. by Avshalom C. Elitzur, Shahar Dolev and Nancy Kolenda

Information and Its Role in Nature
By Juan G. Roederer

Extreme Events in Nature and Society
Ed. by Sergio Albeverio, Volker Jentsch and Holger Kantz

The Thermodynamic Machinery of Life
By Michal Kurzynski

Weak Links
The Universal Key to the Stability of Networks and Complex Systems
By Csermely Peter

The Emerging Physics of Consciousness
Ed. by Jack A. Tuszynski

G. Ellis, *How Can Physics Underlie the Mind?*, The Frontiers Collection,
DOI 10.1007/978-3-662-49809-5

Quantum Mechanics at the Crossroads
New Perspectives from History, Philosophy and Physics
Ed. by James Evans and Alan S. Thorndike

How Should Humanity Steer the Future
Ed. by Anthony Aguirre, Brendan Foster and Zeeya Merali

Mind, Matter and the Implicate Order
By Paavo T.I. Pylkkanen

Particle Metaphysics
A Critical Account of Subatomic Reality
By Brigitte Falkenburg

The Physical Basis of the Direction of Time
By H. Dieter Zeh

Asymmetry: The Foundation of Information
By Scott J. Muller

Decoherence and the Quantum-To-Classical Transition
By Maximilian A. Schlosshauer

The Nonlinear Universe
Chaos, Emergence, Life
By Alwyn C. Scott

Quantum Superposition
Counterintuitive Consequences of Coherence, Entanglement, and Interference
By Mark P. Silverman

Symmetry Rules
How Science and Nature are Founded on Symmetry
By Joseph Rosen

Mind, Matter and Quantum Mechanics
By Henry P. Stapp

Entanglement, Information, and the Interpretation of Quantum Mechanics
By Gregg Jaeger

Relativity and the Nature of Spacetime
By Vesselin Petkov

The Biological Evolution of Religious Mind and Behavior
Ed. by Eckart Voland and Wulf Schiefenhövel

Homo Novus—A Human without Illusions
Ed. by Ulrich J. Frey, Charlotte Störmer and Kai P. Willfiihr

Brain-Computer Interfaces
Revolutionizing Human-Computer Interaction
Ed. by Bernhard Graimann, Brendan Allison and Gert Pfurtscheller

Extreme States of Matter
On Earth and in the Cosmos
By Vladimir E. Fortov

Searching for Extraterrestrial Intelligence
SETI Past, Present, and Future
Ed. by H. Paul Shuch

Essential Building Blocks of Human Nature
Ed. by Ulrich J. Frey, Charlotte Störmer and Kai P. Willführ

Mindful Universe
Quantum Mechanics and the Participating Observer
By Henry P. Stapp

Principles of Evolution
From the Planck Epoch to Complex Multicellular Life
Ed. by Hildegard Meyer-Ortmanns and Stefan Thurner

The Second Law of Economics
Energy, Entropy, and the Origins of Wealth
By Reiner Köummel

States of Consciousness
Experimental Insights into Meditation, Waking, Sleep and Dreams
Ed. by Dean Cvetkovic and Irena Cosic

Elegance and Enigma
The Quantum InterviewsThe Quantum Interviews
Ed. by Maximilian Schlosshauer

Humans on Earth
From Origins to Possible Futures
By Filipe Duarte Santos

Evolution 2.0
Implications of Darwinism in Philosophy and the Social and Natural Sciences
Ed. by Martin Brinkworth and Friedel Weinert

Probability in Physics
Ed. by Yemima Ben-Menahem and Meir Hemmo

Chips 2020
A Guide to the Future of Nanoelectronics
Ed. by Bernd Hoefflinger

From the Web to the Grid and Beyond
Computing Paradigms Driven by High-Energy Physics
Ed. by Rene Brun, Federico Carminati and Giuliana Galli Carminati

The Language Phenomenon
Human Communication from Milliseconds to Millennia
Ed. by P.-M. Binder and K. Smith

The Dual Nature of Life
By Gennadiy Zhegunov

Natural Fabrications
By William Seager

Ultimate Horizons
By Helmut Satz

Physics, Nature and Society
By Joaquín Marro

Extraterrestrial Altruism
Ed. by Douglas A. Vakoch

The Beginning and the End
By Clément Vidal

A Brief History of String Theory
By Dean Rickles

Singularity Hypotheses
Ed. by Amnon H. Eden, James H. Moor, Johnny H. Søraker and Eric Steinhart

Why More Is Different
Philosophical Issues in Condensed Matter Physics and Complex Systems
Ed. by Brigitte Falkenburg and Margaret Morrison

Questioning the Foundations of Physics
Ed. by Anthony Aguirre, Brendan Foster and Zeeya Merali

It From Bit or Bit From It?
Ed. by Anthony Aguirre, Brendan Foster and Zeeya Merali

Trick or Truth?
Ed. by Anthony Aguirre, Brendan Foster and Zeeya Merali

The Challenge of Chance
Ed. By Klaas Landsman

Quantum (Un)Speakables II
Half a Century of Bell's Theorem
Ed. by Reinhold Bertlmann, Anton Zeilinger

Energy, Complexity and Wealth Maximization
Ed. by Robert Ayres

Printed in the United States
By Bookmasters